Resource Recovery from Industrial Wastewater through Microbial Electrochemical Technologies

Resource Recovery from Industrial Wastewater through Microbial Electrochemical Technologies

Edited by

Sovik Das and Maulin P. Shah

IWA PUBLISHING

Published by **IWA Publishing**
Unit 104–105, Export Building
1 Clove Crescent
London E14 2BA, UK
Telephone: +44 (0)20 7654 5500
Fax: +44 (0)20 7654 5555
Email: publications@iwap.co.uk
Web: www.iwaponline.com

First published 2024

British Library Cataloguing in Publication Data
A CIP catalogue record for this book is available from the British Library

ISBN: 9781789063806 (paperback)
ISBN: 9781789063813 (eBook)
ISBN: 9781789063820 (ePub)

Doi: 10.2166/9781789063813

This eBook was made Open Access in September 2024.

Contents

doi: 10.2166/9781789063813_0001

Chapter 1

Introduction to different types of microbial electrochemical technologies

*Md. Nazir, Rajdeep Shaw and Rajib Bandopadhyay**

Microbiology Section, UGC Centre for Advanced Study, Department of Botany, The University of Burdwan, Golapbag, Bardhaman, West Bengal 713104, India
**Corresponding author: rajibindia@gmail.com*

ABSTRACT

The global energy crisis due to the diminishing of fossil fuels with an increasing demand for power has a major impact on the environment and the health of humans. On the contrary, waste production has been increasing for many decades. Among wastewaters, industrial wastewater is a challenging threat in today's era. To generate bioenergy as an alternative energy source, wastewater treatment and nutrient recovery by microbial electrochemical technologies (METs) are very efficient. Different METs are being developed, some of them being very efficient; for example, microbial electrolysis cells produce hydrogen or methane from organic matter when electric current is applied. Microbial electrochemical snorkel is a microbial fuel cell which has a short circuit. *Geobacter* sp. produces 556 mW/m^2 power from food wastes. An alga *Shewanella* sp. produces 50 mW/m^2 from dairy wastewater. Therefore, resource recovery technology using desirable microorganisms is very convenient.

Keywords: microbial electrochemical technology, wastewater treatment, application, valuable gas production

1.1 INTRODUCTION

1.1.1 Bacterial electrochemical technologies

Fuel-cell technology is being developed using modern skills although faced with challenges. Therefore, challenges must be overcome for an increased utilization of fuel, such as reduction of the cost of catalysts, increase in hydrogen production and distribution of fuel, and so on (Sommer *et al.*, 2012). Conventional fossil fuels fulfil a major portion of energy demand. Among the problems, there are two major energy resources that are related to non-sustainable energy resources, that is, environmental pollution and depletion, so the current focus of research is to locate alternative energy sources (Jafary *et al.*, 2015). Today, fundamental and practical research has begun to focus on electrochemical energy systems or in other words microbial electrochemical technology (MET) which works with biological mediation. There have been reports on MET applications in a variety of fields which include waste treatment, bioenergy, sequestration of carbon dioxide (CO_2) recovery of energy resource, bioelectronics, purification, and so on. Its main benefit is its adaptability to valorize waste in any form such as liquid, solid, and gaseous, and then produce a range of products such as biofuel, electricity, and biogas in a sustainable manner. The systems are essentially multidisciplinary hybrid systems that integrate knowledge from many other fields including microbiology, electrochemistry, environmental engineering, biochemistry material science, and many more. On the basis of their uses, METs can be divided into microbial desalination cells (MDCs) to separate ions; microbial fuel cells (MFCs) for the production of bioelectricity; microbial electrolysis cells (MECs) to produce methane and hydrogen gas; bioelectrochemical treatment (BET) which is used to treat high potency wastewaters and complex recalcitrant; microbial electrosynthesis snorkels (MESs) for the production of platform chemicals; and electro fermentation for high production of a bio-based product. Each application provides notable benefits from economic, environmental, and technological points of view (Mohan *et al.*, 2019). Simultaneous energy generation and treatment of wastewater were facilitated from organic sources of waste by capturing renewable energy. Electricity has been seen as a promising alternative to fossilized fuels for refilling the energy demand of the world which is growing gradually. The process of conversion of electricity from organic materials by microorganisms is carried out using MFCs. Thus, MFCs are regarded as a potential new technology that can cleanse wastewater while also generating power (Al-Mamun *et al.*, 2018; Sangeetha & Muthukumar, 2011).

1.1.2 Developmental background of MFCs

Galvani reported that biologically catalysed reactions were the source of electrical effects. In 1789 it was observed that frog muscles twitched with zinc–copper couples, which might be the latest discovery related to the 'physiological process accompanied by chemical changes associated with electrical changes'. Over a protracted period, M.C. Potter discovered reduction of electrodes which is microbially induced in 1911, that made the easy way for the fabrication of biological fuel cells, afterwards known as 'microbial fuel cells'. Potter conducted research using a unique kind of galvanic cell electrode which contains platinum to ascertain the electrically generated driving force by microorganisms through fermentative activity (Mohan *et al.*, 2019). Cohen (1931) used a similar theory to show that adding an appropriate material, benzoquinone or potassium ferricyanide, to a half-bacterial electrical cell maintains the medium's oxidation–reduction order and improves the overall electrical intensity and capacity. Thirty years later, the production of electricity utilizing various microorganisms was shown by Sisler (1961, 1962), Davis and Yarbrough (1962), and Davis (1963). These authors contended that the culture's enzymatic activity and the reduction–oxidation potential were connected, and this system was called an MFC due to its interaction with the medium. In the early 1990s, an MFC-related mechanism was generally referred to as bioelectrochemical technologies/systems (BET) (Hernandez & Osma, 2020).

1.2 DIFFERENT TYPES OF METs

1.2.1 Microbial fuel cells

The main principle of MFCs is the transformation of chemical energy from inorganic or organic matter to produce electricity using microorganisms as a catalyst. MFCs are available in a variety of forms, but the fundamental designs that are often employed in labs include single-chamber, double-chamber, stacked, and up-flow MFCs. Moreover, different designs have been used for extensive study. Proton-exchange membranes (PEMs) and the materials of electrodes used in MFCs influence a variety of applications, which includes biosensor production, bioelectricity, biologically produced hydrogen, wastewater analysis, and so on (Kumar *et al.*, 2017). Figure 1.1 shows the general structure of an MFC. Microbial respiration through an anodic reaction releases electrons and protons by converting the substrate electron mediator (deviation of the cell membrane) in the anode providing a direct contact with carriers that are electrochemically active materials. Using an external integrated circuit, electrons and protons then continue to the cathode. Each transferred electron produces a corresponding proton to the cathode chamber which migrates through a PEM to maintain neutral

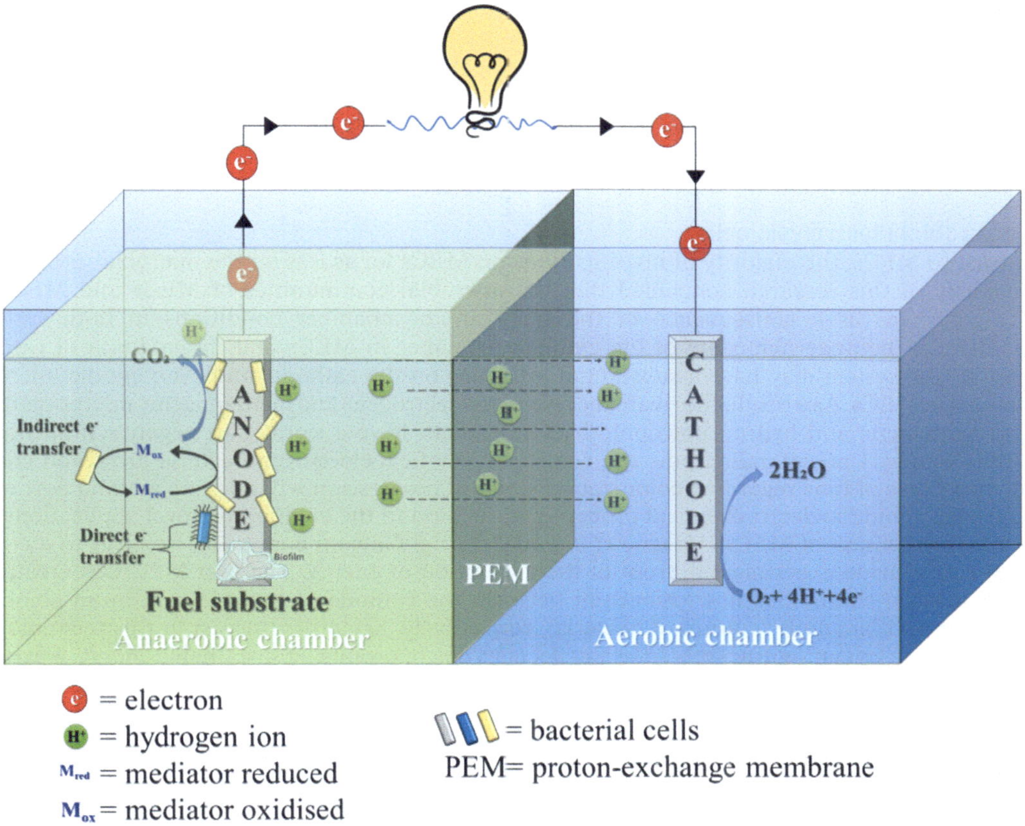

Figure 1.1 Diagrammatic structure of an MFC with an anode and cathode chamber separated by a PEM.

charge. Helpful by-products are produced by the process of oxidation at the anode compartment. Electrons are transported to the cathode by a conducting wire cathode electrode. In addition, during the reduction process, oxygen molecules and protons combine to form the molecules of water. A membrane between the anode and cathode compartments facilitates the movement of protons to the cathode chamber. Electrons are unable to transfer through the membrane but protons pass through the membrane for the successful reaction at the cathode in the presence of oxygen and electrons supplied from the external circuit.

It is interesting to note that power is produced by the movement of electrons through an external circuit from the cathode to anode. The PEM internally divides the anode and cathode chambers. To prevent oxygen from diffusing to the anode from the cathode but allowing the proton migration to the cathode from the anode, the PEM serves as a barrier. Additionally, it permits the movement of protons but blocks the passage of electrons, substrate, or oxygen from the cathodic chamber (Obileke *et al.*, 2021). Using mixed cultures of electroactive bacteria, this microbial electrochemical method mineralizes the organic waste found in wastewater while simultaneously producing electricity. It may help the wastewater treatment process use less energy while recovering biologically gained energy found in various kinds of wastewater. MFC technique is still unable to be used in practical applications because of (1) less power and current production, which reduces its ability to remove pollutants,

(2) its high price, and (3) the less stability of the cathode during the long-term performance. The key to realizing its uses in the real world is to increase stability of the cathode at a low price and high-power generation. Most experts agree that the MFCs' primary performance-limiting component is the cathode. The fabrication of expensive, precious metal-free oxygen reduction reaction (ORR) catalysts is of key interest of various investigations over the recent decade. Several ORR catalysts have been investigated including activated carbon, heteroatom-doped carbon, metal oxides, and catalysts made of Co, Fe, Mn, and Ni (Chen *et al.*, 2018).

1.2.2 Microbial electrolysis cells

In contrast to MFCs, the majority of investigations on MECs focus less on the microbial community. The majority of this research concluded that the microbial communities of MECs and MFCs are identical, perhaps as a consequence of the same anodic chamber conditions in both systems. Comparatively, the anaerobic state of the cathodic chamber in MECs and the addition of external potential to the system may have a substantial influence on the cathodic and even anodic microbial populations in MECs. As a result, the rivalry between exoelectrogens and methanogenic microorganisms (such as acetolactic and hydrogenotrophic bacteria) might have a significant negative influence on MEC functioning. Understanding how the communities of exoelectrogen bacteria grow and change during the course of time regarding colonization series of progression will open up a brand new world of knowledge about bioelectrochemical processes in relation to the creation of novel applications and the commercialization of MECs (Hasany *et al.*, 2016). Luigi Galvani connected frog legs to a copper conductor to provide experimental proof of the generation of energy. The first MFC was created by Potter by using a continuous flow of current between the cathode and anode kept in an abiotic or sterile media and bioanode (culture of bacteria), respectively. Scientific interest in this technological advancement has grown significantly in recent decades due to promising sustainable energy generation and wastewater treatment. Potter observed a decrease in redox potential during bacterial growth that caused the current generation. The current MFCs were invented by Kim *et al.* (2009) who used some bacteria which were active by electrochemical stimulation that can use electrodes to accept electrons.

In 2004, they produced electricity by electrochemical cells based on MFCs for the first time. In a conventional two-chambered MFC, live microorganisms degraded a substrate (such as acetate) bioelectrochemically by releasing electrons in the anode and CO_2 and protons in the anode chamber due to a half-oxidation reaction in a two-chambered MFC (equation (1.1)). To generate bioelectrical current, electrons must travel to the cathode by an external circuit electrode. Protons migrate past the PEM to react with electrons in the cathode chamber to generate water by the half-reduction reaction, completing the circuit (equation (1.2)) in the presence of oxygen (electron acceptor). In a modified MFC when extra voltage is applied to the H^+ ions then it is reduced to hydrogen (H_2) by a half-oxidation reaction when oxygen is not present (equation (1.3)) (Jafary *et al.*, 2015). Figure 1.2 presents a general MEC structure.

$$C_2H_3O_2 + 2H_2O \rightarrow 2CO_2 + 8e^- + 7H^+ \quad (1.1)$$

$$O_2 + 4H^+ + 4e^- \rightarrow 2H_2O \quad (1.2)$$

$$H^+ + 2e^- \rightarrow H_2 \quad (1.3)$$

1.2.3 Microbial electrochemical snorkels

A number of organizations have lately adopted various MES versions and succeeded in achieving several uses, such as nitrate reduction and hydrocarbon elimination in wastewater treatment, even in biogeochemical process cycles. MET is one of the emerging fields with the possibility of rapid development in the near future. These evaluations are designed to draw attention to this innovation for an increase in MES performance. An MES is an MFC that is short-circuited. Furthermore, MFCs are distinguished by a bell-shaped curve containing power and current. A short circuit is considered when

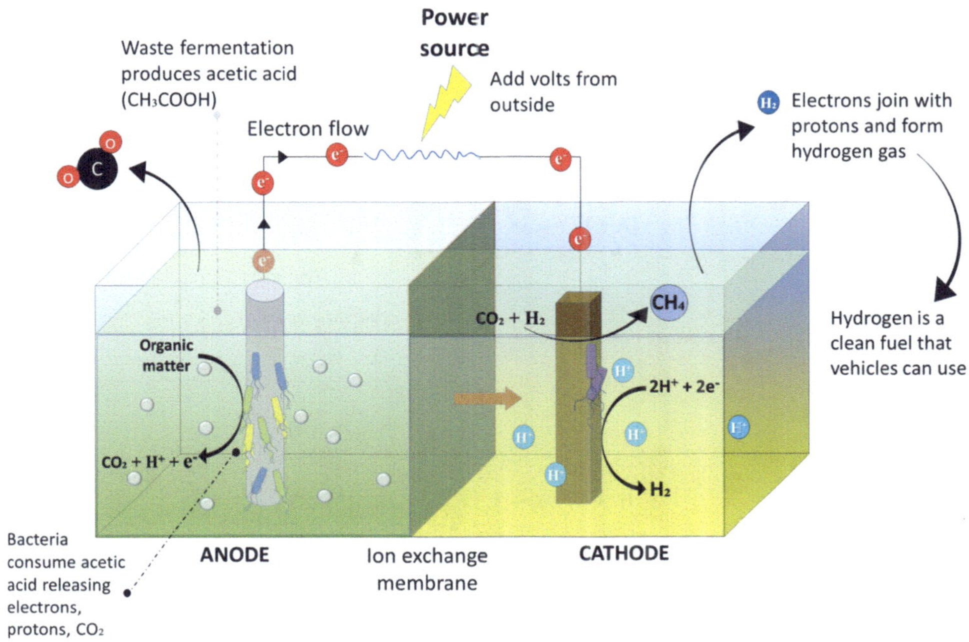

Figure 1.2 MEC diagrammatic structure; CO_2 is produced in the anode chamber by degradation of organic matter.

the zero voltage is present between the cathode and anode; however, it operates at the highest current conceivable. Hence, the fundamental benefit of MESs is to maintain the highest current that an MFC is capable of producing between the cathode and anode, as shown in Figure 1.3. The response rates are increased as high as the system will allow. With increasing electrochemical reaction rates for generating electricity as the goal, this mode of operation is optimal. This is true when reducing organic materials in the environment is the main goal of wastewater treatment, soil or sediment decontamination, and wastewater treatment for removing pollutants from effluents, recovering metals, and so on. In reality, current output of power density of MFCs is so low that the best option should be functioning in an MES set-up. It may be sufficient to use a rod that is conductive and crosses two chemically distinct areas, such as a carbon rod that is moderately planted in an anaerobic environment and a top portion lying bare in water. An anaerobic biofilm that is electroactive grows on the surface and releases electrons to the substance. The aerobic portion of the body receives electrons, which are then converted to oxygen. A system's low cost and minimal maintenance are guaranteed by its simple design. An MES has been described as a process that resembles cable bacteria (filamentous bacteria which are enclosed in cables). Another reason to be upbeat about MESs' ability to adapt to large-scale applications is the fact that they function similarly to natural processes. Applications over a fairly broad spectrum are envisioned. Experiments have previously shown that MESs can improve wastewater treatment, remediate sediments containing hydrocarbon, and remove nitrate or sulphate. Moreover, there is a hypothesis that installing MESs in wetlands or landfills can redirect the flow of electrons away from the anaerobic zone's methane production, providing a method for reducing methane emissions.

The patterns of microbial respiration in soils and sediments may be altered by benthic MESs. To reduce the formation of sulphide and mercury methylation, as well as to enhance the growth of crops such as rice, a rise in redox potential connected to the transfer of electrons by MESs might be a solution (Hoareau *et al.*, 2019).

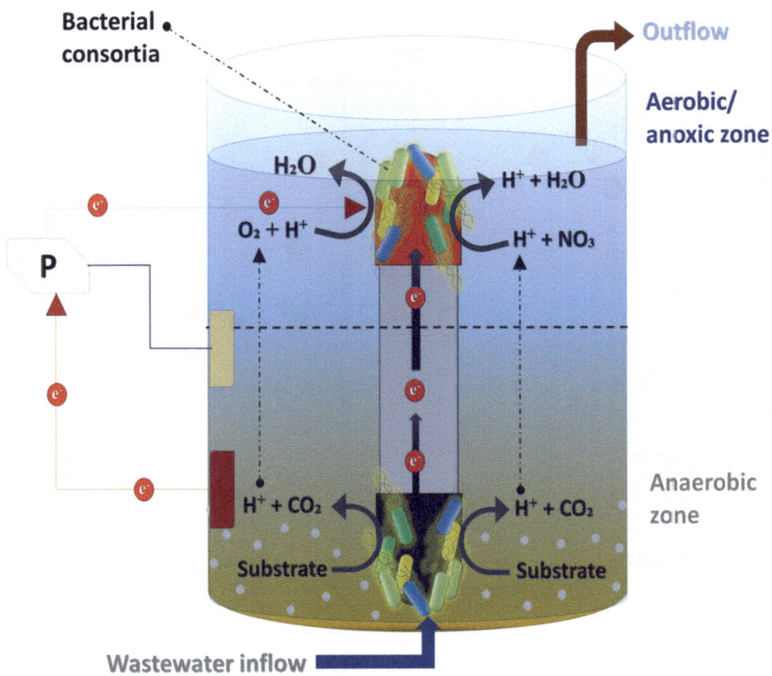

Figure 1.3 Diagrammatic structure of microbial electrochemical snorkels.

1.2.4 Microbial desalination cells

MDCs were developed for wastewater treatment, desalination, and the generation of electrical energy, initially by Cao in 2009. Ion-exchange membranes (IEMs) are used in a three-chambered desalination device. Several types of MDC systems have been developed and implemented in recent years. An MDC system was first recognized to have several drawbacks, including increased internal resistance and an imbalance of pH. These limitations and others have been thoroughly examined and resolved by Jingyu *et al.* (2020) and Luo *et al.* (2012 and 2017), respectively. In an anode chamber where exoelectrogens are present, the MDC's performance primarily relies on these exoelectrogens, which is an essential component of these systems as a biological technology. Exoelectrogens are microorganisms, primarily being bacteria, which produce energy by electrical mechanisms through the oxidation of organic matter and transport of electrons to an electron acceptor situated outside the cell. Therefore, the name 'exo'. Where can one find a wealth of exoelectrogens in the environment? Typically, primary anaerobic silt and anaerobic sludge from home or commercial wastewater treatment facilities; farm soil, municipal effluents, and industrial wastewater contain exoelectrogens that may be separated from the appropriate sources and employed in MDCs as pure culture or mixed culture. Wastes collected from municipalities, homes, and treatment of wastewater from industrial facilities have the ability to generate energy that is produced electrically from chemical or organic sources, including glucose-like simple sugar being used as a source of carbon in MDCs or commercial dyes. As bacteria can reproduce themselves, there is no need for restocking catalysts in the breakdown of organic materials. One of the key factors affecting the effectiveness of MDC systems is exoelectrogens; hence, their ideal survival and development are crucial for both. Many parameters, including salt content, temperature, pH, and medium, define these bacteria. Using appropriate exoelectrogens in an MDC system may improve its effectiveness, particularly in the generation of electrical energy. Known exoelectrogens may be divided into a number of functional categories according to the different forms of anaerobic respiration. Among these exoelectrogens are microorganisms that reduce nitrate (denitrifying); dissimilatory metal-reducing bacteria, such as *Geobacter* sp., *Shewanella* sp., *Geopsychrobacter* sp., and *Geothrix* sp.; as well as denitrifying bacteria such as *Pseudomonas* sp. and *Ochrobactrum* sp.; and microorganisms that reduce sulphate, such as *Desulfuromonas* sp. and *Desulfobulbus* sp. Furthermore, fermentative bacteria such as *Clostridium* sp. and *Escherichia coli* may produce oxygen via anaerobic respiration pathways. Shifts in electrode potentials have been theorized to affect the catabolic and respiratory pathways of the exoelectrogens. Zhu detailed that an optimal electrode potential is required for exoelectrogens to produce electricity more effectively, as shown in Figure 1.4. Hence, the electrode potential encourages the development of an electrically active biofilm in the microbial community and it varies greatly in terms of energy. Exoelectrogens employed in different bioelectrochemical systems (BESs), including MFCs, have been well-characterized, learning about the metabolic processes of exoelectrogens, how certain electrons work, and how an intermediary molecule called pyruvate is produced. After this pyruvate has been broken down to acetyl CoA which enters Kreb's cycle and uses the electron transfer chain to release extra electrons, all of which have an effect on how well an MDC functions overall and are crucial for scaling and developing the technology.

In-depth research has been carried out on the treatment of wastewater purification and electricity generation processes, including the design and mechanism of MDCs for more than 10 years. Kim and Logan (2013) examined the kinds of exoelectrogens and their performances with minimal emphasis on the state of the art in MDC design and performance as well as the safety concerns associated with the usage of MDCs. In a related review research, Huang *et al.* (2017) examined the link between certain elements and by which mechanism they affect the effectiveness and performance of MDCs. Moreover, Saeed *et al.* (2015) conducted a thorough study of MDC technology, including the fundamentals of the standard MDC system as well as the numerous variants of MDCs now in use. Although MDC

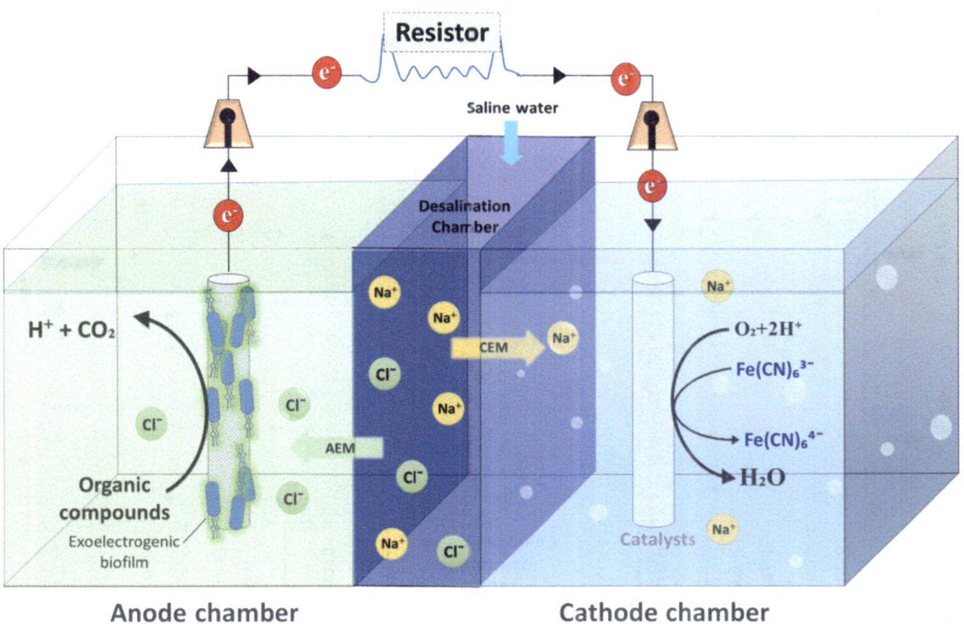

Figure 1.4 Diagrammatic structure of a microbial desalination cell; the microbe utilizes the organic compounds in the anode chamber.

technology has attracted a lot of academic interest over the years, the biological force behind it is still not fully understood. Although few studies discuss their dominant species and how they influence the processes, most studies just reveal about the origin of their MDC culture. There is a significant gap in our knowledge of the nature, mechanism, effects on the system, and characterization of these exoelectrogens employed in MDCs. This chapter examines the traits of the study and looks at driving forces such as exoelectrogens in MDCs and the effects that purification, pH, the output of power, and substrate have on their development and the effectiveness of MDC systems. It compares the effectiveness of cultures that were mixed with some dominant species to pure cultures which are used in MDCs and examines key areas for further research into exoelectrogens in MDCs and the related domains (Guang *et al.*, 2020).

1.2.5 Alkaline membrane fuel cells

Although ORR kinetics is enhanced under alkaline circumstances, alkaline membrane fuel cells (AMFCs) have recognized as a viable alternative to existing technology. Co-precipitation was used to create the first row of transition metal ferrites, MFe_2O_4 (M = Co, Mn, and Zn), which served as ORR electrocatalysts. For both pre- and post-calcined samples, ORR electrocatalysis was investigated. A revolving disc electrode operating at 1600 rpm was used to obtain ORR polarization curves in O_2-saturated KOH. $MnFe_2O_4$ had the greatest onset potential (+0.88 V versus reversible hydrogen electrode (RHE)). According to X-ray diffraction, annealing of the materials reduced the ORR's onset potential and stopped the growth of undesirable crystal phases. Succinic acid was added to the $MnFe_2O_4$ precipitation process, which improved the onset potential, as shown in Figure 1.5. The increased surface area of smaller crystal particles is thought to be the cause of this phenomenon.

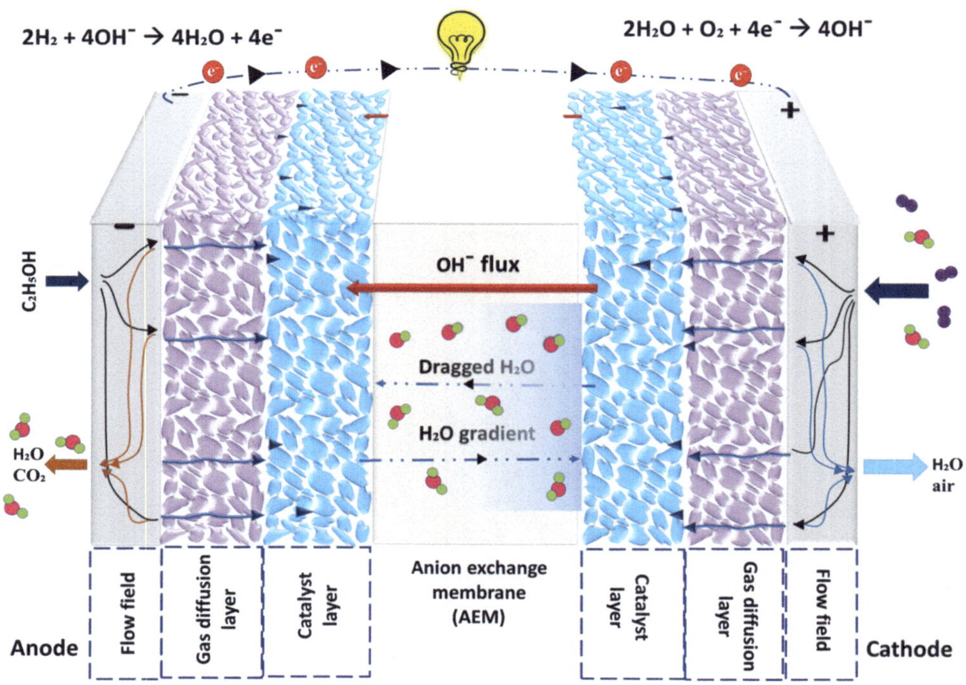

Figure 1.5 AMFC; carbon compound enters the anode flow field gas diffusion layer and catalytic layer.

Future research aims to comprehend the in-situ reactions occurring under ORR circumstances. AMFCs are sometimes referred to as anion-exchange membrane fuel cells that have gained much attention. When compared to proton-exchange membrane fuel cells, AMFC technology has a number of advantages. These include improved oxygen reduction catalysis, which permits the use of Pt-free and non-noble metal catalysts, a wider variety of cell and stack materials, an expanded range of fuels besides hydrogen, and a large range of economical polymers because fluorinated fresh materials are no longer required. A cathode, an anode, and an IEM are the three fundamental elements of an AMFC. Although water and oxygen react at the cathode to form hydroxide ions that will pass through the anion-exchange membrane, water and oxygen react at the anode to oxidize hydrogen and convert it into water. ORR has so far been one of the obstacles to the commercialization and affordability of fuel cells. Nevertheless, non-noble catalysts such as transition metal oxides may be used in AMFCs because of the alkaline pH of the cell environment. Currently, electrocatalysts made of perovskites and spinels are receiving a lot of interest. In contrast to perovskites, spinels are simpler to make at low temperatures and have a number of desired properties, including low electrical resistance, ubiquitous availability, good activity, thermodynamic stability, and environmental friendliness (Del-Pilar and Cabrera, 2019; Pilar *et al.*, 2019).

1.2.6 Plant MFCs

BESs, such as MFCs and electrogenic microorganisms, may transfer electrons from metabolic activities within a cell to acceptors in the environment. Microorganisms (also known as heterotrophic MFCs) may receive electrons from organic substrates, whereas some MFCs can obtain electrons through photosynthetic processes, that is, photosynthetic microbial fuel cells (pMFCs). By electrocatalytic or biocatalytic reductions, the released electrons will flow by an external circuit and are mixed with an electron acceptor at the cathode. These BESs may be 'mediated', in which the redox mediators are movable electron shuttles, or 'mediator-less', in which the electrons are transmitted continuously from the bacterium to the electrode. This process, known as a direct electron transfer, involves sending electrons from the cathode to the anode through the modified structure which appears like pilus ('nanowires') or electron transport enzymes. Both MFCs and pMFCs are available in a variety of designs where the electrode is in the same or different chambers, presence or absence of a PEM, and where biologically related activities may take place at either or both of the fuel cell's electrodes. Cyanobacteria, which have a more primitive basic cellular structure, as well as algae which are eukaryotic and more complex, may both exhibit light-dependent electrogenic activity in pMFCs. Electrons are directly provided by cyanobacterial electron transport networks which operate photosynthetically. As a consequence, pMFCs experience an electrogenic effect that is reliant on light and cause current to be produced instantly and directly in reaction to light. Just lately, the mystery surrounding the mechanism by which the electrons are transferred to acceptors on the outside of microbes began to be revealed. The electrogenic species of cyanobacteria have cellular appendages which are hair-like structures called pili, probably in charge of delivering electricity.

They are also called 'bacterial nanowires', and have great conductivity. Most recent studies have revealed their conductivity which is more or less similar to metals. They also include redox-active peptides and proteins, as well as maybe other moieties with electron shuttling capacity. Although at least three explanations or combinations of explanations are feasible, the exact reasons why bacteria transfer electrons externally are yet unclear. The first and most plausible scenario is the release of low power for prevention of intracellular reactive oxygen species formation, as shown in Figure 1.6. However, electron transport chains, together with those involved in light-independent cellular respiration, lose some electrons which are about 2% when oxidative stress results. This light-driven oxidative stress is very important in the case of photosynthetic microorganisms. In research utilizing chemical photosynthesis inhibitors regarding electrogenic light-dependent action in pMFC cyanobacteria, protection against oxidative stress has been shown. The second explanation is that the metals have higher bioavailability when their concentration in the environment is reduced, which is a

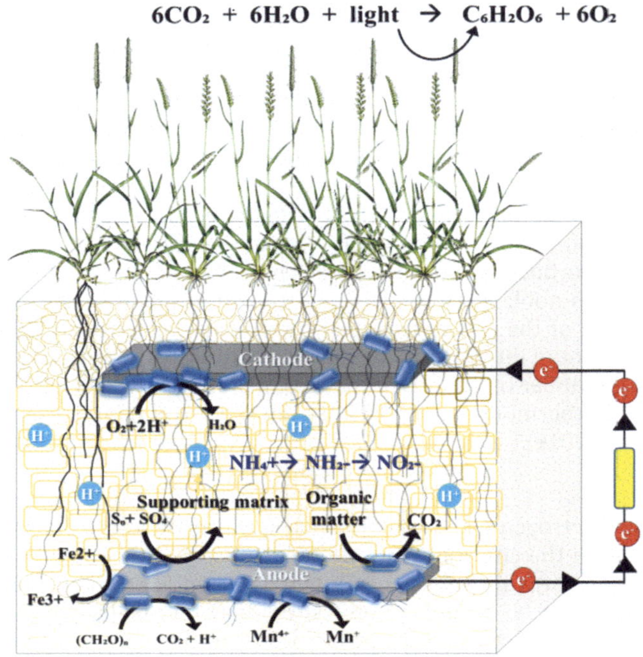

Figure 1.6 Diagrammatic representation of plant microbial fuel cells; in the anode plate microbes convert carbon compound or organic matter to CO_2.

phenomenon of external active electron transport of the microorganism. The last theory reports that the release of electrons is a component of communication among microbe channels, although this is presently only a theory because there is little to no experimental evidence to support it. According to this, several electrogenic species naturally inhabit the intricate biological niches found within biofilms. Together with chemical communication, electrical communication between the microbes also appears conceivable. To test individual algae and cyanobacterial cultures for their capacity to exhibit a photosynthetic electrogenic effect, Luimstra *et al.* (2014) fabricated an easy and economically viable pMFC and showed its benefits. As has been conducted with heterotrophic MFCs, the device was also helpful straightforwardly for providing pressure which was selective in a pattern of a 'redox poise' on pMFC's electrode to extract electrons from photosynthetically active bacteria. As the power produced by MFCs is substantially less than that of chemically operated fuel cells and pMFCs, they could only be effective for producing an electrical current in special cases. Among more possible uses is biologically synthesized electricity, which modifies the redox state within the microbe-containing cell to encourage the production of target compounds using the MFC, or for bioremediation of waste-containing metals such as chromium and selenite as well as aromatic hydrocarbons. It will be easier to identify prospective applications if one is aware of the fundamental elements of the light-dependent electrogenic phenomena. Recent research has shown that the light-dependent electrogenic effect, which occurs when the off:on conversion phase of artificial light occurs, was sustained for several weeks in both microalgae and cyanobacteria. This inspired us to look into the possibility of using the phenomena as an environmental biosensor. Biosensors are analytical tools that employ a biological component as a physicochemical detector to detect an analyte. The sensing component is biologically sourced and may take the form of an enzyme, antibody, cell, receptor of the cell, tissue, or nucleic acid. The biologically generated substance or mimicry of a biological sample combines with the being studied analytical molecule by attaching to it or in some other way recognizing it. Currently, the detecting element in most biosensors is generally particular and preset, notably in the most popular types employing antibodies or enzymes. In the current study, we examine if photosynthetic metabolism may be utilized to integrate and perceive environmental information. More specifically, we examine whether it is possible to detect the effects of harmful substances by disrupting the consistent light-dependent electrogenic impact. When electrons are transferred from the photosynthetically active microbe to the anode of a pMFC, called an electrogenic positive response, it is reflected in recording traces of declining movement as the pMFC is customarily linked to the recording apparatus, and highly negatively charged electrons are followed by this reduction in potential (Labro *et al.*, 2017).

1.2.7 Constructed wetland-microbial fuel cells

In the case of conventional constructed wetland-microbial fuel cells (CW-MFCs), the two electrodes are isolated by a PEM, fabric materials, and ground. One of the electrodes serves as an anode electrode (anodic chamber), and another is a cathode electrode (cathodic chamber). The biochemical processes that generate electrons and protons take place in the anodic chamber in which electrically active bacteria grow over an anode electrode. Bacterial growth occurs through rhizodeposition via separating materials, the protons are transferred to the cathodic chamber by an externally situated resistor, and the electric circuit is closed as electrons are transferred to the anode. In the cathodic chamber, in which oxygen is reduced, both electrons and protons are combined to create water vapour, which is then released into the environment. Granulated support methods that permit the plant's root development and the action of various flow types are used to stabilize aquatic plants. Combining an electrochemical reactor with a wetland involves a variety of diverse subjects, including bacterial study, and electrochemical processes. Hence, combining these aspects into a solitary biological system with abiotic and biotic elements to generate biological energy is the most effective method to comprehend the interactions between different sectors. A CW-MFC may be thought of as a particular kind of biosystem that is split into biocontrol and bioprocess structures. A plant is a part of the biocontrol framework, where sunlight is provided for the generation of electricity by photosynthetic activities (exudates) as shown in Figure 1.7. A community of microbes, which uses root exudates to generate energy by the metabolism of microbes, is referred to as the bioprocess structure. Biological and physiochemical processes may be classified as the variables that influence the CW-MFC system's performance. Biological variables include plants, algae, and bacteria, whereas physicochemical factors include conductivity, pH, humidity, and temperature.

For CW-MFCs to operate as efficiently as possible, it is essential to optimize operational variables and elements of the environment such as hydraulic retention time (HRT), redox parameters, loading of organic substances, temperature, and humidity. The experiments to generate bioelectricity via the plants have employed CW-MFC designs with double chamber and solitary chamber, marshy wetlands, and rice fields. The benefits of combining a CW with an MFC include the ability to produce bioelectricity and treat wastewater simultaneously, produce energy devoid of any external natural substrate, release reduced methane, extract biologically produced electricity from naturally formed water bodies, and act as reactors for the integration of the environment. Several CW-MFC designs, including sequencing batch cells, down-flow, vertical up-flow, horizontal subsurface flow, and mixed flow have been reported in the literature. Various configurations may be categorized as double- or single-chamber systems. A CW-MFC system is the most popular, with the cathode on the surface of plant or in its rhizosphere and the anode buried under the support of the plant. These unique designs provide the greatest oxygen availability in the cathode area by minimizing oxygen dissolved at the anode. An up-flow CW-MFC arrangement offers an adequate redox profile and is appropriate for a CW functioning. The drawback of using the flow regime is that the electrode distance is significant, which increases the ohmic resistance in a CW-MFC. Comparing CW-MFC systems to traditional MFCs, the former produces a greater internal resistance. By using CW-MFCs (horizontal subsurface flow) with a

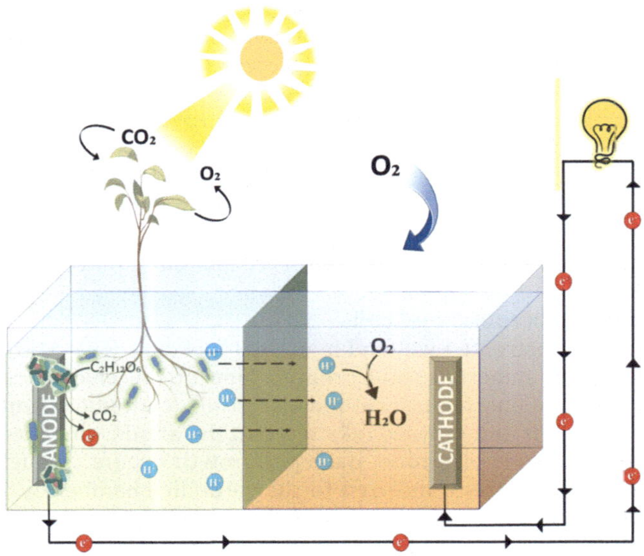

Figure 1.7 Diagrammatic structure of CW-MFCs; the microbes in the anode chamber metabolized the carbon matter extracted from the roots of the plant.

graphite plate and *Phragmites australis* plant, Villaseor *et al.* (2013) observed an internal resistance of 120 Ω. The electrode was separated by a calcium bentonite membrane. A traditional MFC with multielectrodes has an internal resistance, according to Ahn and Logan. To minimize the distance between the anode and cathode, Doherty *et al.*, (2015) employed a fibreglass substance as an electrode separator in a CW-MFC that functioned with simultaneous up-flow and down-flow. They made use of the *P. australis* plant and granular graphite electrodes. According to the scientists, the power density was enhanced by up to 70% with this CW-MFC combination, achieving 10.51 mW/m^2. The usage of a separator such as fibreglass might result in long-term blockage issues because plant roots can penetrate this material. They also noted that the anodic biofilm's potential to thoroughly oxidize the organic materials resulted in impaired bioelectricity generation under large organic loads. Because the range of electron movement lengthens as CW-MFC size increases, the internal resistance also increases. A CW-MFC system is often built rectangular with a tubular design. Gravel and dirt are used as the system's support materials in CW-MFCs. A bed of gravel is an inert substance that may have various element sizes and can let various kinds of CW-MFCs pass through. The horizontal subsurface flow takes advantage of the gravel support. Three CW-MFCs were built by Xu *et al.*, (2016) using various kinds of cathodes and anodes, including carbon stroked as the cathode and carbon which is activated and granular as the anode. A *Canna indica* plant was employed in the three CW-MFC experiments. There were four layers that made up the initial CW-MFC. Gravel was layered initially (lowest layer), then initiated granules of carbon followed by a layer of intermediate gravel, and finally carbon felt was layered on top. A layer of gravel was used to construct the second system at the CW base. In these MFCs, the cathode was positioned at the air–water interface after the anode had been positioned around the plant's roots and covered by a gravel layer. The supporting components of CW-MFC 1 and 2 were combined in the third CW-MFC, which also included an intermediary layer between the anode and the cathode within the gravel. Multielectrode designs for CW-MFC systems are uncommon. Granular graphite was employed as the anode and carbon fabric multielectrode as the cathode in a CW-MFC by Xu *et al.* (2017) The CW-MFC received synthetic fuel, which produced a 26.16 mW/

m^2 density of power. The inoculum, substrate, kind of membrane, outside and inside resistance, the ionic forte of the solution, electrode resources, and the distance among electrodes are some of the characteristics that need to be improved in order to boost its efficacy. While building a CW-MFC, it is important to take the location, complexity, and magnitude of the electrodes into account. According to Huan *et al.* (2014) and Takanezawa *et al.* (2010) power production from CW-MFCs using anodes that were submerged 5 cm was three times greater than that of using anodes that were submerged only 2 cm into CW-MFCs. This leads to the conclusion that in order to create anoxic situations and utilize the carbon-based rhizodeposition chemicals released through the root zone to produce bioelectricity from a CW-MFC, the determination of the proper anodic zone is crucial. Greater chemical oxygen demand (COD) reductions from wastewater have been recorded for tubular designs. Energy efficiency can be increased in some of these designs; the combined skill is mostly becoming accessible to change the conformation where better effectiveness may be attained (Guadarrama-Pérez *et al.*, 2019).

Several types of patents regarding METs are tabulated in Table 1.1 after consulting the literature from Espacene patent research website: https://worldwide.espacenet.com/patent/search?q=microbial%20 fuel%20cell.

1.3 EFFICIENCY AND APPLICATION TO TREAT DIFFERENT TYPES OF WASTE

The most researched MET application is the generation of electric power via MFCs. Electron acceptor bacteria oxidize organic molecules. An MFC with an annular single chamber and a curved microbial bacteriological electrochemical cell is the result of efforts to create novel forms and combinations of conductor materials to improve the performance of an MFC. One such particular MFC apparatus is constructed with a cylindrical Plexiglas chamber, a coiled anode conductor made of a stainless-steel web that has been varnished with graphite, and a concentrically cylindrical cathode made of either carbon cloth or stainless-steel mesh that has been treated with Pt (0.5 mg/cm^2) to improve the material's electro-conductive performance (Kim *et al.*, 2009). The anodic surface area is increased while the distance between electrodes is decreased in this sort of spiral design. Given the relatively inexpensive cost of the applied constituents, it has the ability to be scaled up. These qualities lead to exceptional results in terms of treatment, wastewater attainment of up to 91% of COD elimination, and a maximum power compactness of 20 W/m^3 (anode-occupied size). Compared to conventional MFC configurations, the so-called tubular MFCs are another novel idea. A vertical flow reactor is a variation of this system that consists of two punctured polypropylene cylinders serving as a twofold explosive, a layer of carbon-cloth cathode inserted that is exposed to the air, a layer of hydrogel functioning as a transitional layer, and an interior ion conversation membrane. Opylene tubes function as a double shell, with an internal ion exchange membrane, an embedded layer of carbon cloth cathode that is exposed to air, and an intermediate layer made of hydrogel. This reactor's interior contains a concentric monolithic activated carbon anode. An external circuit with a resistor (1000 Ω at startup; 150 Ω at normal operating circumstances) connects the anode and cathode of this reactor, which has a concentric monolithic-activated carbon anode fitted within. In the literature, it was able to achieve COD elimination charges between 51% and 82% with a concurrent energy output of up to 1.75 Wh/g COD by running two tubular MFC units supplied with synthetic wastewater and at varied organic loading rates. These results provide hope for expanding this reactor in complementing modular systems for polishing sewages from treatment anaerobic digester services (Ramírez-Vargas *et al.*, 2018). An MFC is often used to bioremediate organic contaminants. Electroactive microorganisms work by utilizing the anode as an electron acceptor and oxidizing substrates such as organic acids or hydrocarbons. In the presence of a specific catalyst, electrons transfer from the anode to the cathode where electron acceptors with a higher potential are reduced which results in electricity generation. Several oxidized and reduced species need to be removed during bioremediation, which presents chances for METs. Some important electricity-driven products are highlighted for pollution removal in the section that follows. A mixed population was used to exhibit full denitrification to nitrogen gas

Table 1.1 Patents regarding METs.

Patent No.	Components	Implementation	Aspirant	Inventors	Patent Received in the Year
CN113178589A Method for cathode preparation	Cobalt salt	High-production power, operation constancy, low cost, and useful to large-scale production	Fuel Cell Industry and Wuhan Hydrogen Energy Tech Research Institute Co. Ltd.	Liu Qiang; Wu Tong; Lei Gang; Li Hongxiang; Huang Hao	2021
CN112374605A Microbial fuel-cell series test device	Photosynthetic algae, bioreactor-led light sources	Photosynthetic biological systems	University of Sichuan	Pu Wei; Zhouguangwu; Zhouzhicheng; Zhouqinghua; Xu Fei; Qingrenwei	2021
CN203922925U Recycling treatment of organic wastewater	Duckweed culture, photobioreactors, organic wastewater, biofuel methane	Important meaning on energy conservation, emission reduction, and environmental control	University of Inner Mongolia Science and Technology	Jiang Haiming; Pan Jian Gang; Li Xia	2014
CN105463029A Hydrogen production of single chamber	Single chamber, hydrogen production	Fluctuating and intermittent electric energy is directly converted into hydrogen energy	University of Sun Yat-sen	Liu Guangli; Zeng Jia; Luohaiping; Luyaobin; Zhangrenduo; Yang Kunpeng	2016
CN218561635U Hydrogen production synchronous organic sewage	Solar power, photovoltaic power generation	The cathode rinse assemblage is for rinsing the cathode with a catholyte	Powerchina Huadong Engineering Corp. Ltd.	Sun Yi; Ji Shiyu; Chenxiaolin	2023
US2011311887A1 Microbial desalination cells	Anode, a cathode, a saline solution chamber, organic compounds	Microorganisms are maintained and the electricity production capacity and the desalination efficiency	He Zhen [USA]; UWM Research Foundation Inc. [USA]	He Zhen [USA]	2011

(Continued)

Table 1.1 Patents regarding METs (*Continued*).

Patent No.	Components	Implementation	Aspirant	Inventors	Patent Received in the Year
CN104617322A Microbial capacitive desalination fuel-cell technology	Anode chamber, desalination chamber, and the cathode, carbon-cloth electrodes	Microorganisms are maintained and the electricity production capacity and the desalination efficiency	University of Hunan; Huangkuan	Kuan Huang; Jianyu; Hao L Tang	2015
CN104617322A Microbial capacitive desalination fuel-cell technology	Microbial power generation and capacitive deionization, pH value of the anode chamber	Communication module to allow the real-time collection of data from field deployments	University of Hunan; Huangkuan	Kuan Huang; Jianyu; Hao L Tang	2015
US11226306B2 Microbial sensor system for monitoring and imaging of an environment	Microbial sensors, pesticides, herbicides	Current collector while stiffening the frame so as to reduce thermal expansion	Burge Scott [USA]; Hoffman David A [USA]; Burge Environmental Inc. [USA]	Burge Scott [USA]; Hoffman David A [USA]	2022
CA2644201A1 Electrode structure for stacked alkaline fuel cells	Inlet and outlet manifolds, electrolyte flow	Colour changes of a pH-indicating material that is held within the fuel-cell container	Mku Cyprus Ltd [CY]	Nor Jiri [CA]	2007
WO2009113523A1 Alkaline fuel cells	Alkaline fuel cell, a cathode collector, an anode collector, pH of the alkali-containing liquid fuel	Receives the energy from the direct current intermediate circuit and controls a driving motor	Neccorp [JP]; Sekino Shoji [JP]; Kobayashi Kenji [JP]; Nakamura Shin [JP]	Sekino Shoji [JP]; Kobayashi Kenji [JP]; Nakamura Shin [JP]	2009

(*Continued*)

Table 1.1 Patents regarding METs (*Continued*).

Patent No.	Components	Implementation	Aspirant	Inventors	Patent Received in the Year
DE102009039684A1 Electrical generator and high-temperature proton conversation sheath fuel cells	Proton conversation sheath, electrical generator, current intermediate circuit	Hydrogen conductive ability of proton-exchange film	Voith Patent GmbH [DE]	Tengler Heinz [DE]	2011
CN1805197A Activation apparatus for proton-exchange membrane fuel cell	Proton-exchange film, open ion channel	The plurality of parts and a plant are planted in the planting portion	Shanghai Palton Fuel Cell Systems [CN]	Liu Zhentaiyi [CN]	2006
JP2022033628A Plant microbial fuel cell	Anode electrode, pMFC, carbon electrode	Water vapour is captured from gas streams from a gas-emitting facility, into electricity and methane	University of Yamaguchi	Md Azizul Moqsud	2022
US2011300411A1 Photoelectron methanogenic microbial fuel cell	Methanogenic microbial culture, electron-conductive cathode, photoelectron methanogenic microbial fuel-cell apparatus	Relieve part of the pollution and produce hydrogen and methane	Materiwayne Paul [CA]; Carbonitum Energy Corp. [CA]	Materiwayne Paul [CA]	2011

driven by a cathode, although, nitrate is reduced to nitrite by *Geobacter metallireducens* as reported previously. Since then, several studies have been conducted on this procedure (Gregory & Lovley, 2005). However, MECs had greater denitrification rates as a result of the external energy input. Acetate is in the electron donor form in the bioanode which then removes nitrate from an MFC. The occurrence of intermediates is driven by the cathode voltage, which has been shown in several investigations, to influence electron availability. In fact, the stability of the biocathode and the existence of unfavourable intermediates (such as nitrite and nitrous oxide) are all influenced by the cathode latent. The amount of denitrification and the amount of energy used are also influenced by operational factors such as nitrate load, HRT, and progression design. In-situ MFCs utilized the least energy (0.34 kWh/kg NO_3-N) when yielding groundwater with contaminated nitrate through BES, according to a recent study, but ex-situ MFCs required larger energy (1.6 kWh/kg NO_3-N) because the groundwater needed to be driven out. Due to the external power supply, MECs reported much higher energy expenditures (19 kWh/kg NO_3-N in-situ and 10 kWh/kg NO_3-N ex-situ, respectively) (Wang *et al.*, 2020). With the aforementioned examples, MFCs have been produced in a wide range of sizes, from capacities of micro-litres to tens and hundreds of litres. Because of its relative simple design, the absence of a need of a membrane, and several developments aimed at bringing down material costs, MFCs are expected to be the most practical METs in the near future. In certain applications, such as employing sediment MFCs (sMFCs) to power devices in seawater, the power output of an MFC is a key design element. sMFCs also known as benthic MFC where organic matter containing sediment use as a source of electricity generation using bacterial metabolism. For use in lab-on-a-chip applications or tiny wearable devices, micro-sized MFCs (μMFCs) are also being established. Power generation is needed in other MFC techniques, such as wastewater action, but the removal of the requirement to aerate wastewater and a decrease in sludge creation is of more value. Although the power generated by systems based on oxygen reduction has increased over the years, from 1 mW/m^2 of the predicted anode area to as much as 6.9 mW/m^2 anode of the zone by utilizing a relatively large cathode, the power produced by MFCs will be lower than expected. However, employing high attention to fuel (such as acetate) and well-buffered, highly conductive electrolytes has consistently led to the best power densities under almost perfect circumstances. Although power densities as high as 17–19 W/m^2 have been predicted by taking into account microbial kinetics or decreasing reactor's internal resistance, it is unlikely that these power densities would be attained in actual wastewater treatment systems. We arranged the power density range of MFCs into following categories: sMFCs or small power source MFCs used for remote power supply, different substrate and laboratory media-based MFCs, and complex media-based actual wastewater treating MFCs or MFC(WW). These determine the upper limit for performance of the air cathode. To deliver a more accurate scenario of their application in a controlled manner, the μMFC data also involved systems with a ferricyanide catholyte. This is because the applications planned for these devices might theoretically utilize final electron acceptors rather than oxygen (O_2). On the contrary, it seems that MFCs with specified substrates (the majority with acetate) generated more power at smaller sizes as opposed to larger ones. Although there was a connection between the two domains based on other criteria which impact power output, such as reactor design (electrode-specific surface area) and operating conditions, the power densities produced with specified substrates significantly surpassed those reported for wastewater. As they were intended to have capacities of <2 mL, μMFCs were virtually well isolated as compared to other systems. The diameters of sMFCs were larger than those of the other MFCs, and generally, it seemed that power densities increased with size. When power densities of a given area were compared using volumetric power density instead, a somewhat different scenario became apparent. Here, we can observe that, in terms of volumetric density, MFCs employing specified substrates had definitely surpassed MFC(WW) findings, and that sMFCs had the lowermost volumetric power densities. Again, there are no discernible changes in the volumetric power generation of MFCs used for wastewater treatment. Although comparing these statistics based on energy density would be helpful, MFC studies have not provided enough information on energy recovery and efficiency. Analyses of the available data revealed that the energy recoveries of tiny MFCs (<100 mL) with high power densities are not significantly different from those of larger systems. There may be some underlying causes for variations in power generation, as evidenced by the discrepancies in these patterns depending on volume with volumetric power concentration. One issue is undoubtedly the fuel: sMFCs essentially utilize highly diluted bases of organic substance within the sediment, and their ability to generate power is probably limited by the pace at which bacteria yield the soluble substrates that they need for producing current. It is possible to reduce the restrictions based on fuel accessibility for power production through the anode in MFCs employing single substrates which is extensively used in μMFC technology. There did not seem to be a pattern in the volumetric power density of power production for MFCs handling wastewater. Nonetheless, there is evidence that the cathode definite surface zone is the key design element in all of these systems for volumetric power, as we shall demonstrate below. We especially looked into how cathode layouts affected to analyse how a specific zone of the cathode area performed (Logan *et al.*, 2015). The main focus of researchers is to gain energy from waste by using wastewater as an energy source that is renewable. An MFC is one such kind of recovery system that uses biocatalyst-like bacteria in different redox electrochemical procedures to obtain electricity from energy-containing organic waste. This biofuel cell can handle a variety of effluents, including monomer sugars (such as acetate and glucose), amalgamated industrial wastewater, and lignocellulosic run off; however, a potential waste supply, namely human faeces, has not been extensively used in MFCs. Few publications are available on the conduct of human excretions and urine in MFCs, in addition to the treatment of complicated industrial effluent. Cow muck, cow pee, boar wastewater, urine of elephants, and imitation of man-made waste have all been successfully treated in MFCs in the past. Human waste is similar to animal waste in that it contains significant levels of nutrients and carbon. Nevertheless, pharmaceutical residue elements and infective load are two important extra anxieties that are within the human waste, which make management of human waste difficult (Pandey *et al.*, 2016). Human waste is rich in organic and nutritive particles, including ammonium-nitrogen initiated in urine, which has a high energy value. Human urine has 2% of its mass in urea. In contrast to protons in water, each urea molecule has four hydrogen atoms that are chemically attached to it in a moveable configuration. In electrochemical processes, infringement of such biochemical pledge generates protons and electrons more efficiently and with less energy input. Similarly to this, the high concentration of carbonaceous substances in human faces may cause microbial oxidation to liberate electrons. To generate direct current, these electrons are captured by an exterior electrical (load manner) circuit in MFCs. Even the sludge that has collected in septic tanks may be resolved successfully and used as a fodder material for biocatalytic corrosion in MFCs. Unused human excretions must thus be seen as a valuable resource that can be reused as part of the circular economy's transition from 'better sanitation' to 'sustainable sanitation' by using this bioelectrochemical technology. Regarding the actual use of MFC technology, it may be utilized in either onsite sanitation facilities or sewage treatment plants. Ge and He evaluated 96 MFC modules that have a 200 L volume for the principal effluent received from the neighbourhood common effluent treatment facility. The system proved difficult to operate because of its low power output, variable wastewater properties, and complex design. These restrictions can be reduced by using MFCs to treat waste produced by toilet systems and utilizing higher catalytic activity containing cathodes for ORRs by using an improved system to manage the power and properly engineered system design to reduce the substrate complexity and distribution of flow. Therefore, adding MFCs to an existing or new infected cistern system is a workable approach for better hygiene services. Using MFCs for human waste treatment may contribute to improving sanitary facilities, especially for rural regions, while also producing energy, given a simple substrate and practically steady flow. Currently, MFCs are used as experimental magnitude with both genuine municipal wastewater and substrate based on synthetic acetate (85 L MFC) (Jadhav *et al.*, 2020). For the last 20 years, METs have been the focus of much research, with the majority of the applications envisioned being the management of several types of wastewater streams. The two main constraints that have, to yet, plagued both MFCs and MECs are (1) the low ionic conductivity of most wastewaters (usually about 1–2 mS/cm) and (2) the poor buffering bulk of most wastewaters (generally about 5–10 mM). A biofilm is soon rendered inactive by anode acidification because an anodic process generates protons; however, the cathode has the reverse issue with a substantial pH rise (Rozendal *et al.*, 2008). In laboratory trials, the problem of poor bulking has been overcome by adding buffering agents to the meal. Nevertheless, practical wastewater treatment systems cannot use this, necessitating the use of alternate technologies, which may have additional downsides. Contrarily, low ionic conductivity restricts the current and voltage that can be used in MFCs and frequently requires all-out anode–cathode spacing of only limited millimetres in order to ensure successful functioning. Higher voltages may be applied to MECs to solve the issue; however, this would considerably increase the energy intake. Unlike most other wastewater streams, source-separated pee is unique. By using an ionic conductivity of 20 mS/cm and urea with a concentration of 20 g/L, the above-mentioned problem can be solved because by hydrolysis urea acts as a buffer that neutralizes acid by oxidation. These figures suggest that urine is the best substrate for METs, which opens the door to a wide range of applications for this new technology (Ledezma *et al.*, 2015). Due to the significant amount of wastewater produced and its intricate composition, textile manufacturing is often regarded as one of the greatest polluting industrial divisions. More than a million tonnes of dye are formed every year, and 60–70% of the dyes that are used widely are azo compounds, which have several azo linkages (–N=N–) with different by-products (Yurtsever *et al.*, 2016). Azo dyes and their breakdown products have been shown to be poisonous, carcinogenic, and mutagenic, which limit the development of aquatic life. The widespread release of azo dyes into the environment has an influence on photosynthetic functions as well as aesthetics. Moreover, sulphide often exists in dye effluents due to the inclusion of sodium sulphide for the conversion of other sulphur compounds, such as sulphate, sulphite, and thiosulphate, which are commonly used in fabric-dyeing progressions, or for the reduction of azo compounds. The staining of sulphur dyes (N90%), which are known for having sulphide structures, has been demonstrated to require significant quantities of sodium sulphide, producing 15–20% sulphide-containing effluents. A serious danger to human health is sulphur (aqueous sulphide and gas sulphide), which is poisonous, efflorescent, and corrosive at high concentrations. Hence, prior to ultimate discharge into natural water bodies, treatment of azo dye effluents containing sulphide is essential. Many practical methods for treating textile wastewater effluents, including adsorption, coagulation, flocculation, and advanced oxidation, have been investigated in recent years in an effort to reduce or remove harmful elements. Traditional physicochemical methods for removing sulphur dioxide include chemical dosing oxidation, moist air oxidation, and high-temperature oxidation. More ecologically friendly and sustainable solutions are required because these procedures often demand significant amounts of energy. The metabolism of microorganisms that change dangerous substances into less damaging and ecologically acceptable by-products requires less chemical dosing when using biological approaches, which is an interesting finding. An MFC has received a prodigious deal of interest because of its capacity to turn organic and inorganic contaminants into bioelectricity. MFCs are a promising technology because they completely use the metabolic functions of microbes to produce power efficiently while also removing pollutants. According to Zhao *et al.* (2008) sulphate is transformed into sulphide, which is then further transformed into elemental sulphur and soluble polysulphide, both of which may be eliminated from the bacterial solution. In a double-chambered MFC, simultaneous anaerobic sulphide and nitrate removal was effectively accomplished, according to Cai *et al.* (2016) The impulsive elimination of aqueous sulphide and azo dyes in MFCs has, however, hardly been studied. Early research showed that a mix of chemical and biological mechanisms was used to degrade azo dyes. There are many chemical processes to degrade azo dyes but biological processes must be carried out by enzymatic reactions directly or by co-factor-mediated enzymatic reactions which are generated by bacteria biologically such as sulphate reduction bacteria (SRB) for breaking of azo bonds. Sulphate, sulphide, polysulphide, and other sulphur compounds are often found in dyeing effluents. Earlier studies have shown that sulphide (H_2S, HS^-, and S^{2-}) and sulphate (SO_4^{2-}) have an impact on the decolorization of azo dyes. It has been shown that the degradation of different azo dyes depends on both chemical and biogenic sulphides. In addition, owing to the enhancing action of redox intermediaries from cell lysis harvests, biogenic sulphides serve a significant role in the degradation of azo dyes under anaerobic SRB-enriched situations. Yet, at various sulphate concentrations, sulphate has an ambiguous impact on the biodegradation of azo dyes. Moreover, the literature on the biological, chemical, and electrochemical components of degradation of azo dye and elimination of sulphide is insufficient. To decolorize azo dyes in single-compartment air-breathing cathode MFCs that concurrently generate bioelectricity and damage azo dyes, this research set out to understand the mechanism of sulphides (biochemical sulphide and biogenic sulphide). On the removal of sulphide and azo dyes, the belongings of sulphide concentrations (120, 180, 240, 300, and 360 mg/L) and dye attention (100, 200, 300, 400, and 500 mg/L) were assessed. On the basis of data from ultraviolet–visible spectrophotometry, high-performance liquid chromatography, and liquid chromatography–mass spectrometry, the intermediate products of dye decolorization were investigated, and a potential Congo red degradation route was suggested in MFCs. To identify the major bacteria responsible for the impulsive removal of sulphide and dyes in this MFC, the microbial population was also analysed (Dai *et al.*, 2020).

1.4 PRODUCTION OF VALUABLE GASES

By-product recovery from reactions operated bioelectrochemically minimizes the need of extra energy for manufacturing of products. By using external voltage in MFCs to increase production such as hydrogen peroxide, methane, caustic soda, and hydrogen gas in the cathode at a lower energy cost, BES-based technology is significantly economical regarding manufacturing and market value. When carbon dioxide is captured and stored by photoautotrophic microorganisms in a microbial solar cell, they typically synthesize organic compounds utilizing CO_2 as a carbon source. Additionally, the cathodic chamber might promote photosynthesis to produce algal biomass as a by-product and provide oxygen for cathodic reduction. After undergoing a variety of pre-treatments such as ultrasonic waves and microwaves with heat several chemicals either acidic or alkaline residues were separated after extraction of algal oil. The recovered algal biomass may efficiently be used as a substrate in photo bioelectrochemical cells. Algal biomass recovery has been shown to be possible when utilizing carbon compounds which the microbial cell captures. Microbial cells capture carbon. In these cells, CO_2 may be sequestered by sprinkling generated anodic CO_2 gas into a catholyte that contains *Chlorella vulgaris* algae, generating a biocathode. Additionally, the marine algal species *Chaetoceros* sp. (Shaw *et al.*, 2023) possesses algal biomass that may be utilized as a substrate in MFCs for anodic oxidation and perhaps hinders the development of the methanogens in the mixed inoculum. The main goal of MFC operation is to produce as much electricity as possible, although the majority of the energy is wasted in the production of methane and hydrogen gas. Utilizing a pure culture of *Methanobacterium palustre*, the effectiveness of the microbial biocathode was assessed for the repossession of methane via reduction of carbon dioxide in the cathodic chamber of the MEC. It showed a greater rate of methane production and hydrogenophilic methanogenic culture with over 80% efficiency in electron capture. Methane was generated by both the biological fermentation of organic materials and through abiotically formed hydrogen gas (also known as hydrogenophilic methanogenesis). According to reports, hydrogen gas generated during the procedure of hydrogenotrophic methanogenesis is used in the majority of cases when methane is produced in MECs. Hydrogenotrophic methanogens utilize the CO_2 produced during acetoclastic methanogenesis to generate methane (CH_4) gas. An MFC–MEC system was used to collect gases including H_2, CO_2, and carbon monoxide (CO) from wastewater, and these gases may now be used in industrial settings. Environmental electrochemistry has been the subject of study and development for around 10 years, and during that time, METs have undergone significant modifications and improvements that have significantly increased their usefulness. In addition to the major products (hydrogen, power, etc.), BES demonstrates the capacity to recover additional significant heavy metals, nutrients, engineering chemicals, and gaseous energies.

BES has excellent opportunities for resource recovery, but it still faces certain biotechnological and financial obstacles that must be overcome before it can be widely used and be thought of as a feasible method different from the current wastewater handling techniques. If BES expertise is used to convalesce the power from animate carbon and phototrophic types of machinery are used to recover nutrients and other by-products, the negative energy equilibrium of current wastewater treatment schemes may be reversed. To improve the MFC performance by increasing power output, BES-based life-cycle assessment was conducted by Foley *et al.* (2010). This technology is used to recuperate the energy from animate carbon. The usefulness of microbial electrochemical machinery has significantly increased. Over the course of nearly 10 years of research and development in the area of ecological electrochemistry, the performance of the machinery has vastly improved. Along with the primary outputs (hydrogen, power, etc.), BES demonstrates the capacity to recover other useful outputs such as heavy metals, nutrients, manufacturing chemicals, and gaseous oils. As was mentioned in this appraisal chapter, BES offers excellent potential for resource recovery; however, before it can be widely applied and thought of as a viable alternative to the existing wastewater treatment techniques, some biotechnological barriers and economic challenges need to be overcome. If BES expertise is used to recuperate the power from organic carbon and phototrophic strategies are used for the recovery of nutrients and other by-products, the current wastewater treatment systems' negative energy balance may be turned around (Jadhav *et al.*, 2017; Shaw *et al.*, 2024). Methanogens exhibit a dual behaviour in their metabolic pathways; extracellular electron transfer-mediated electrogenesis or methanogen-mediated methanogenesis occurs in MFCs where anode or Fe(III)-like electron acceptor is present. Research on methane-based MFCs by Rother and Metcalf shows when methane is oxidized it produces energy, known as methane reversal. In the comparison of standard hydrogen electrode under natural conditions, CO_2 forms by the oxidation of methane at $E_0' = 0.245$ V redox potential which is theoretical under unrestricted conditions. In conjunction with sulphate-reducing microorganisms (SRMs), certain archaea in the deep-sea biocatalyse process occur spontaneously. The SRM receives the electrons produced during the oxidation of CH_4 by direct interspecies electron transfer. However, certain methods and adjustments must be made so electrons are harvested for use in applied applications. The treatment of wastewater using an anaerobic digester, methane-fed MFC, and biophotovoltaic named AMMB is suggested here as a hybrid system. The purification of biogas in an AMMB hybrid system is very difficult because it also includes ammonia, hydrogen sulphide, nitrogen, and hydrogen. Small anoxic degraders produced may be transformed to bioelectricity utilizing the AMMB method by employing a diffusion membrane for CH_4 that permits the discriminatory passageway of CH_4. The main source of marine methane production is methane emitted from hot springs on a seabed. Thus, these methane-oxidizing consortia may be employed in floatable MFCs or sMFCs, which are used to run low-power devices in deep ocean, to further tap into this massive methane resource (Nath *et al.*, 2021).

1.5 HYBRID METs

It is often more costly to use carbon material or carbon tabloid for MFCs; thus, an alternate, less expensive carbon web material was considered as a cheaper option for the anode material in an MFC. To guarantee adequate MFC function, the carbon mesh must be pretreated. First, the carbon mesh must be heated in a muffle furnace for 30 min at 450°C. The aforementioned anode which is heat treated may produce 922 mW/m^2 (46 W/m^3) power density and which is about 3% more power than the acetone-cleaned mesh anode (893 mW/m^2; 45 W/m^2). This energy mass obtained by heating is just 7% lower than the power density obtained by treating carbon fabric with ammonia gas at high temperatures (988 mW/m^2; 49 W/m^3). Power rose to 1015 mW/m^2 (51 W/m^3) after the management of NH_3 gas of the carbon mesh. As a result, the reduction in atomic O/C ratio on cleaned or heated outsides indicated the elimination of impurities that interfered with charge transfer. Due to nitrogen-related functional groups that aided electron transport, the ammonia gas treatment also resulted in an increase in the atomic N/C proportion. These findings demonstrate that the anode in an MFC may be fabricated using heat-treated carbon mesh materials, which lower the costs. Oxidation of $K_4Fe(CN)_6$ increased the electrochemical activities of treated carbon mesh anodes, producing power densities that are less costly than carbon cloth. The O/C ratio for the various anode treatments, including heating, washing, and ammonia gas management, was reduced. The fact that cleaning raised the electrochemically energetic surface area to 20 cm^2 is also crucial. The active area is expanded to 54–58 cm^2 for further heat treatments or ammonia handling, with no change in power density that might be attributed to variations in the active zone. The carbon mesh's cheap cost and effective performance are especially promising because they enable the near arrangement of the anodes and cathodes. High volumetric energy masses are accomplished by using a cloth separator and closely spaced carbon-cloth electrodes. Although the carbon mesh might replace the carbon cloth as the anode, using it as a cathode might not be practical. The performance of MFC is enhanced by using a phosphate buffer to boost the solution's conductivity. This is accomplished by treating a carbon-cloth anode with ammonia gas, which significantly increases the electrode's surface charge. There were four diffusion layers and a platinum (Pt) catalyst in the carbon-cloth cathode (0.5 mg/cm^2). Platinum has relatively high catalytic activity, selectivity for the ORR, and resistivity to chemical and electrochemical degradation. The anticipated surface area of both electrodes was 7 cm^2. Ammonia gas was handled by a thermogravimetric analyzer on carbon cloth. The electrode spacing used to build single-compartment air-cathode MFCs was 2 cm. Domestic wastewater and phosphate-buffered nutrient wastes were used to inoculate MFCs. When the voltage fell below 20 mV, the feed solution had to be changed. A multimeter with a data-collecting device was used to measure the voltage of the cell across an exterior resistor. Maximum power output was discovered to have been attained after 150 h of operation with an untreated carbon-cloth anode. However, a relatively shorter amount of time (60 h) is needed when using a carbon-cloth anode that has been treated with ammonia (Choudhury *et al.*, 2017).

1.6 CONCLUSION

Energy generation by MET is a cutting-edge method that has seized the consideration of scientists all over the globe. METs are ideal for power production in distant areas because of their exceptional capacity to generate energy from waste without the usage of outside or extra energy, except for MECs. This technique can directly use microbial metabolism, making it suitable for usage in any environment. Considering better marketing of MET, high power output for usage in large devices needs to develop. The electrolyte and electrode modification along with technical development are necessary for consistent power generation. The majority of the literature included topics such as mechanisms, reactor designs, MET-affecting variables, and optimization of operating conditions. More research needs to be conducted on potential microbes or good electrodes for better efficiency of MET. Additional investigation into the mechanisms governing the allocation of electrons from exoelectrogens to electrodes in other exoelectrogens is also intended, as only *Geobacter* spp. and *Shewanella* spp. have been thoroughly investigated. The commercial use of METs may be increased by careful optimization of its key components. We think that as science develops, MET will eventually be able to economically address all of the world's issues via the development of innovative electrodes and microbial optimization.

ACKNOWLEDGEMENTS

Authors extremely acknowledge the UGC Centre for Advanced Studies, Department of Botany, the University of Burdwan and DST-FIST Phase-II for providing research facilities. MN acknowledges Swami Vivekananda Merit Cum Means Scholarship for providing financial support; RS is grateful to the Council of Scientific & Industrial Research (CSIR-JRF) for financial assistance.

REFERENCES

Al-Mamun A., Ahmad W., Baawain M. S., Khadem M. and Dhar B. R. (2018). A review of microbial desalination cell technology: configurations, optimization and applications. *Journal of Cleaner Production*, **183**, 458–480, https://doi.org/10.1016/j.jclepro.2018.02.054

Cai, J., Zheng, P. and Mahmood, Q. (2016). Effect of cathode electron acceptors on simultaneous anaerobic sulfide and nitrate removal in microbial fuel cell. *Water Science and Technology*, **73**(4), 947–954, https://doi.org/10.2166/wst.2015.570

Chen S., Patil S. A. and Schröder U. (2018). A high-performance rotating graphite fiber brush air-cathode for microbial fuel cells. *Applied Energy*, **211**, 1089–1094, https://doi.org/10.1016/j.apenergy.2017.12.013

Choudhury P., Prasad Uday U. S., Bandyopadhyay T. K., Ray R. N. and Bhunia B. (2017). Performance improvement of microbial fuel cell (MFC) using suitable electrode and bioengineered organisms: a review. *Bioengineered*, **8**(5), 471–487, https://doi.org/10.1080/21655979.2016.1267883

Cohen, B. (1931). The bacterial culture as an electrical half-cell. *J. Bacteriol*, **21**(1), 18–19.

Dai Q., Zhang S., Liu H., Huang J. and Li L. (2020). Sulfide-mediated azo dye degradation and microbial community analysis in a single-chamber air cathode microbial fuel cell. *Bioelectrochemistry (Amsterdam, Netherlands)*, **131**, 107349, https://doi.org/10.1016/j.bioelechem.2019.107349

Davis, J. B. and Yarbrough Jr., H. F. (1962). Preliminary experiments on a microbial fuel cell. *Science*, **137**(3530), 615–616, https://doi.org/10.1126/science.137.3530.615

Davis, J. B. (1963). Generation of electricity by microbial action. *Advances in applied microbiology*, **5**, 51–64, https://doi.org/10.1016/S0065-2164(08)70006-6

Del-Pilar J. and Cabrera C. R. (2019). Non-precious metal-based electrocatalytic cathode materials for alkaline membrane fuel cells (AMFC). *ECS Transactions*, **92**(8), 661, https://doi.org/10.1149/09208.0661ecst

Doherty, L., Zhao, Y., Zhao, X. and Wang, W. (2015). Nutrient and organics removal from swine slurry with simultaneous electricity generation in an alum sludge-based constructed wetland incorporating microbial fuel cell technology. *Chemical Engineering Journal*, **266**, 74–81, https://doi.org/10.1016/j.cej.2014.12.063

Foley, J., De Haas, D., Hartley, K. and Lant, P. (2010). Comprehensive life cycle inventories of alternative wastewater treatment systems. *Water research*, **44**(5), 1654–1666, https://doi.org/10.1016/j.watres.2009.11.031

Gregory K. B. and Lovley D. R. (2005). Remediation and recovery of uranium from contaminated subsurface environments with electrodes. *Environmental Science & Technology*, **39**(22), 8943–8947, https://doi.org/10.1021/es050457e

Guadarrama-Pérez O., Gutiérrez-Macías T., García-Sánchez L., Guadarrama-Pérez V. H. and Estrada-Arriaga E. B. (2019). Recent advances in constructed wetland-microbial fuel cells for simultaneous bioelectricity production and wastewater treatment: a review. *International Journal of Energy Research*, **43**(10), 5106–5127, https://doi.org/10.1002/er.4496

Guang L., Koomson D. A., Jingyu H., Ewusi-Mensah D. and Miwornunyuie N. (2020). Performance of exoelectrogenic bacteria used in microbial desalination cell technology. *International Journal of Environmental Research and Public Health*, **17**(3), 1121, https://doi.org/10.3390/ijerph17031121

Hasany M., Mardanpour M. M. and Yaghmaei S. (2016). Biocatalysts in microbial electrolysis cells: a review. *International Journal of Hydrogen Energy*, **41**(3), 1477–1493, https://doi.org/10.1016/j.ijhydene.2015.10.097

Hernandez C. A. and Osma J. F. (2020). Microbial electrochemical systems: deriving future trends from historical perspectives and characterization strategies. *Frontiers in Environmental Science*, **8**, 44, https://doi.org/10.3389/fenvs.2020.00044

Hoareau M., Erable B. and Bergel A. (2019). Microbial electrochemical snorkels (MESs): a budding technology for multiple applications. A mini review. *Electrochemistry Communications*, **104**, 106473, https://doi.org/10.1016/j.elecom.2019.05.022

Huan, D. E. N. G., Yi-Cheng, W. U., Zhang, F., Huang, Z. C., Zheng, C. H. E. N., Hui-Juan, X. U. and Feng, Z. H. A. O. (2014). Factors affecting the performance of single-chamber soil microbial fuel cells for power generation. *Pedosphere*, **24**(3), 330–338, https://doi.org/10.1016/S1002-0160(14)60019-9

Huang, W., Chen, J., Hu, Y., Chen, J., Sun, J. and Zhang, L. (2017). Enhanced simultaneous decolorization of azo dye and electricity generation in microbial fuel cell (MFC) with redox mediator modified anode. *International Journal of Hydrogen Energy*, **42**(4), 2349–2359, https://doi.org/10.1016/j.ijhydene.2016.09.216

Jadhav D. A., Ray S. G. and Ghangrekar M. M. (2017). Third generation in bio-electrochemical system research – a systematic review on mechanisms for recovery of valuable by-products from wastewater. *Renewable and Sustainable Energy Reviews*, **76**, 1022–1031, https://doi.org/10.1016/j.rser.2017.03.096

Jadhav D. A., Das I., Ghangrekar M. M. and Pant D. (2020). Moving towards practical applications of microbial fuel cells for sanitation and resource recovery. *Journal of Water Process Engineering*, **38**, 101566, https://doi.org/10.1016/j.jwpe.2020.101566

Jafary T., Daud W. R. W., Ghasemi M., Kim B. H., Jahim J. M., Ismail M. and Lim S. S. (2015). Biocathode in microbial electrolysis cell; present status and future prospects. *Renewable and Sustainable Energy Reviews*, **47**, 23–33, https://doi.org/10.1016/j.rser.2015.03.003

Jingyu H., Miwornunyuie N., Ewusi-Mensah D. and Koomson D. A. (2020). Assessing the factors influencing the performance of constructed wetland–microbial fuel cell integration. *Water Science and Technology*, **81**(4), 631–643, https://doi.org/10.2166/wst.2020.135

Kim, Y. and Logan, B. E. (2013). Microbial desalination cells for energy production and desalination. *Desalination*, **308**, 122–130, https://doi.org/10.1016/j.desal.2012.07.022

Kim J. R., Premier G. C., Hawkes F. R., Dinsdale R. M. and Guwy A. J. (2009). Development of a tubular microbial fuel cell (MFC) employing a membrane electrode assembly cathode. *Journal of Power Sources*, **187**(2), 393–399, https://doi.org/10.1016/j.jpowsour.2008.11.020

Kumar R., Singh L. and Zularisam A. W. (2017). Microbial fuel cells: types and applications. In: Waste Biomass Management – A Holistic Approach, L. Singh, and V. C. Kalia (eds.), Springer, pp. 367–384, https://doi.org/10.1007/978-3-319-49595-8, eBook ISBN: 978-3-319-49595-8.

Labro J., Craig T., Wood S. A. and Packer M. A. (2017). Demonstration of the use of a photosynthetic microbial fuel cell as an environmental biosensor. *International Journal of Nanotechnology*, **14**(1–6), 213–225, https://doi.org/10.1504/IJNT.2017.082467

Ledezma P., Kuntke P., Buisman C. J., Keller J. and Freguia S. (2015). Source-separated urine opens golden opportunities for microbial electrochemical technologies. *Trends in Biotechnology*, **33**(4), 214–220, https://doi.org/10.1016/j.tibtech.2015.01.007

Logan B. E., Wallack M. J., Kim K. Y., He W., Feng Y. and Saikaly P. E. (2015). Assessment of microbial fuel cell configurations and power densities. *Environmental Science & Technology Letters*, **2**(8), 206–214, https://doi.org/10.1021/acs.estlett.5b00180

Luimstra, V. M., Kennedy, S. J., Güttler, J., Wood, S. A., Williams, D. E. and Packer, M. A. (2014). A cost-effective microbial fuel cell to detect and select for photosynthetic electrogenic activity in algae and cyanobacteria. *Journal of applied phycology*, **26**, 15–23, https://doi.org/10.1007/s10811-013-0051-2

Luo, H., Xu, P., Roane, T. M., Jenkins, P. E. and Ren, Z. (2012). Microbial desalination cells for improved performance in wastewater treatment, electricity production, and desalination. *Bioresource Technology*, **105**, 60–66, https://doi.org/10.1016/j.biortech.2011.11.098

Luo S., Jain A., Aguilera A. and He Z. (2017). Effective control of biohythane composition through operational strategies in an innovative microbial electrolysis cell. *Applied Energy*, **206**, 879–886, https://doi.org/10.1016/j.apenergy.2017.08.241

Mohan S. V., Sravan J. S., Butti S. K., Krishna K. V., Modestra J. A., Velvizhi G., … and Pandey A. (2019). Microbial electrochemical technology: emerging and sustainable platform. In: Microbial Electrochemical Technology, S. Venkata Mohan, S. Varjani, and A. Pandey (eds.), Elsevier, pp. 3–18, ISBN:978-0-444-64052-9.

Nath D., Chakraborty I. and Ghangrekar M. M. (2021). Integrating microbial electrochemical technologies for methane-to-bioelectricity and water-splitting to impart self-sustainability to wastewater treatment plants. *Bioresource Technology Reports*, **13**, 100644, https://doi.org/10.1016/j.biteb.2021.100644

Obileke K., Onyeaka H., Meyer E. L. and Nwokolo N. (2021). Microbial fuel cells, a renewable energy technology for bio-electricity generation: a mini-review. *Electrochemistry Communications*, **125**, 107003, https://doi.org/10.1016/j.elecom.2021.107003

Pandey P., Shinde V. N., Deopurkar R. L., Kale S. P., Patil S. A. and Pant D. (2016). Recent advances in the use of different substrates in microbial fuel cells toward wastewater treatment and simultaneous energy recovery. *Applied Energy*, **168**, 706–723, https://doi.org/10.1016/j.apenergy.2016.01.056

Pilar, M. D. F., Marques, M. and Correia, A. (2019). New and growing firms' entrepreneurs' perceptions and their discriminant power in EDL countries. *Global Business and Economics Review*, **21**(3–4), 474–499, https://doi.org/10.1504/GBER.2019.099397

Ramírez-Vargas C. A., Prado A., Arias C. A., Carvalho P. N., Esteve-Núñez A. and Brix H. (2018). Microbial electrochemical technologies for wastewater treatment: principles and evolution from microbial fuel cells to bioelectrochemical-based constructed wetlands. *Water*, **10**(9), 1128, https://doi.org/10.3390/w10091128

Rozendal R. A., Hamelers H. V., Rabaey K., Keller J. and Buisman C. J. (2008). Towards practical implementation of bioelectrochemical wastewater treatment. *Trends in Biotechnology*, **26**(8), 450–459, https://doi.org/10.1016/j.tibtech.2008.04.008

Saeed, M. A., Medina, C. H., Andrews, G. E., Phylaktou, H. N., Slatter, D. and Gibbs, B. M. (2015). Agricultural waste pulverised biomass: MEC and flame speeds. *Journal of Loss Prevention in the Process Industries*, **36**, 308–317, https://doi.org/10.1016/j.jlp.2014.12.007

Sangeetha T. and Muthukumar M. (2011). Catholyte performance as an influencing factor on electricity production in a dual-chambered microbial fuel cell employing food processing wastewater. *Energy Sources, Part A: Recovery, Utilization, and Environmental Effects*, **33**(16), 1514–1522, https://doi.org/10.1080/15567030903397966

Shaw, R., Khan, D., Ghorui, M., Nazir, M. D., Chaudhuri, P. and Bandopadhyay, R. (2024). Risk and Remediation of Microplastic Pollution in Marine Life. In: Remediation of Plastic and Microplastic Waste, R. Shaw, D. Khan, M. Ghorui, M. D. Nazir, P. Chaudhuri and R. Bandopadhyay (eds) pp. 160–185, CRC Press/Taylor & Francis, Boca Raton, https://doi.org/10.1201/9781003449133-10

Shaw, R., Halder, U., Chaudhuri, P. and Bandopadhyay, R. (2023). Antarctica Microbial Communities: Ecological and Industrial Importance. In: Current Status of Marine Water Microbiology, pp. 125–150, Singapore: Springer Nature Singapore, https://doi.org/10.1007/978-981-99-5022-5_6

Sisler, F. D. (1962). Electrical energy from microbiological processes. *Journal of the Washington Academy of Sciences*, **52**(8), 181–187, https://www.jstor.org/stable/24535009

Sisler, H. H., Ahuja, H. S. and Smith, N. L. (1961). Further Studies in the Synthesis of Quaternary Hydrazinium and Aminophosphonium Salts. *The Journal of Organic Chemistry*, **26**(6), 1819–1821, https://doi.org/10.1021/jo01065a030

Sommer E. M., Martins L. S., Vargas J. V. C., Gardolinski J. E. F. C., Ordonez J. C. and Marino C. E. B. (2012). Alkaline membrane fuel cell (AMFC) modeling and experimental validation. *Journal of Power Sources*, **213**, 16–30, https://doi.org/10.1016/j.jpowsour.2012.03.089

Takanezawa, K., Nishio, K., Kato, S., Hashimoto, K. and Watanabe, K. (2010). Factors affecting electric output from rice-paddy microbial fuel cells. *Bioscience, biotechnology, and biochemistry*, **74**(6), 1271–1273, https://doi.org/10.1271/bbb.90852

Villasenor, J., Capilla, P., Rodrigo, M. A., Canizares, P. and Fernández, F. J. (2013). Operation of a horizontal subsurface flow constructed wetland–microbial fuel cell treating wastewater under different organic loading rates. *Water research*, **47**(17), 6731–6738, https://doi.org/10.1016/j.watres.2013.09.005

Wang X., Aulenta F., Puig S., Esteve-Núñez A., He Y., Mu Y. and Rabaey K. (2020). Microbial electrochemistry for bioremediation. *Environmental Science and Ecotechnology*, **1**, 100013, https://doi.org/10.1016/j.ese.2020.100013

Xu, D., Xiao, E. R., Xu, P., Zhou, Y., Zhou, Q. H., Xu, D. and Wu, Z. B. (2016). Effects of influent organic loading rates and electrode locations on the electrogenesis capacity of constructed wetland-microbial fuel cell systems. *Environmental Progress & Sustainable Energy*, **36**(2), 435–441, https://doi.org/10.1002/ep.12481

Xu, L., Zhao, Y., Wang, X. and Yu, W. (2017). Applying multiple bio-cathodes in constructed wetland-microbial fuel cell for promoting energy production and bioelectrical derived nitrification-denitrification process. *Chemical Engineering Journal*, **344**, 105–113, https://doi.org/10.1016/j.cej.2018.03.065

Yurtsever A., Çınar Ö. and Sahinkaya E. (2016). Treatment of textile wastewater using sequential sulfate-reducing anaerobic and sulfide-oxidizing aerobic membrane bioreactors. *Journal of Membrane Science*, **511**, 228–237, https://doi.org/10.1016/j.memsci.2016.03.044

Zhao, F., Rahunen, N., Varcoe, J. R., Chandra, A., Avignone-Rossa, C., Thumser, A. E. and Slade, R. C. (2008). Activated carbon cloth as anode for sulfate removal in a microbial fuel cell. *Environmental science & technology*, **42**(13), 4971–4976, https://doi.org/10.1021/es8003766

doi: 10.2166/9781789063813_0027

Chapter 2

Fundamentals of the electron transfer mechanism and factors influencing the performance of METs

Somil Thakur[1], Sai Chaithanya[2], Swati Sirsant[1] and Rajni Devi[3]*

[1]Department of Civil Engineering, Institute of Technology, Nirma University, Ahmedabad, Gujrat 382481, India
[2]Department of Environment and Water Resources Engineering, School of Civil Engineering (SCE), Vellore Institute of Technology, Katpadi, Vellore, Tamil Nadu 632014, India
[3]Department of Botany and Environmental Science, Sri Guru Granth Sahib World University, Fatehpur Sahib, Punjab 140407, India
**Corresponding author: somil.thakur@nirmauni.ac.in*

ABSTRACT

Microbial electrochemical systems have the capability to produce either bioelectricity from organic matter or industry-relevant products such as acetate, glycerol, hydrogen, and so on. These systems utilize electroactive bacteria (*Geobacter*, *Shewanella*, etc.) to transfer electrons onto the surface of electrodes. The transfer of these electrons from an anode to cathode is then facilitated by the redox potential of cells, which can be attained artificially or naturally depending on the type of microbial electrochemical technology (MET). Although the principles of the electron transfer mechanism (ETM) in a variety of METs are similar, pathways can be generally classified as direct electron transfer and indirect electron transfer. Each pathway leads to a variation in electron–electrode interactions and thus the efficiency of systems varies accordingly. This chapter elucidates the various pathways of electron transfer from bacteria to electrode surface as well as the factors that affect the ETM. It will help researchers to understand the fundamentals of electron transfer in METs and discusses possible ways to improve this phenomenon.

Keywords: electron transfer, microbial electrochemical systems, electroactive bacteria, bioelectricity, resource recovery.

2.1 INTRODUCTION

The recent era of energy transition and renewable energy sources has recently piqued the interest of an assortment of researchers who want to investigate the potential of microbial electrochemical technology (MET). Microorganisms referred to as electroactive microorganisms carry out a variety of oxidation–reduction reactions in regards to intracellular and extracellular electron transfers (EET) during their metabolic processes (Roy *et al.*, 2022). These microorganisms can transport electrons to extracellular insoluble electron acceptors via direct or indirect mechanisms depending on mediators. These bacteria are of immense significance to microbial electrochemical systems (MESs) because their EET mechanisms can be explored for the remediation of organic and inorganic pollutants (Kato, 2016). In an anode chamber, organic matter is oxidized by exoelectrogens to produce protons and electrons. Although electrons are caught on the anode and flow towards the cathode via an external circuit, protons move inside through an ion-exchange membrane towards the cathode. An ion-exchange membrane divides the anode and cathode chambers, which make up the system's basic electrochemical set-up (Thakur & Das, 2021a). Different MES reactor designs have been tested to date depending on the requirement for substrate degradation, kind of application, and integration with another reactor (Herrera-Melián *et al.*, 2020; Thakur & Das, 2021a, 2021b).

MESs have various operational capabilities and functions, including microbial fuel cells (MFCs), microbial electrolysis cells (MECs), constructed wetland-microbial fuel cells (CW-MFCs), and microbial desalination cells (Figure 2.1). The transfer of electrons from the anode to cathode, however, is the most crucial element in order to achieve any of the results from these systems. Any of these systems' performance, whether it be directly or indirectly, can be predicted predominantly due to the transfer of electrons. Depending on how bacteria or electroactive biofilms are used, the two primary forms of MESs are MFCs and MECs.

According to Ucar *et al.* (2017), MFCs use microbes to oxidize organic materials in the anode compartment and produce protons, electrons, and carbon dioxide (CO_2). To produce clean hydrogen (H_2) fuel, methane (CH_4) gas, and electrosynthesis of reduced carbon compounds, MECs need an additional voltage supply. The nature of the targeted waste and the course of microbial action determine whether chemicals should be added to the electrode chamber. The selection of electroactive microbes, electrodes, the accessibility of other microbial consortia, and so on are the primary determinants of biological modifications. This chapter will concentrate on the various processes, such as electron transfer mechanism (ETM), various components, and operational factors that affect the performance of MESs for wastewater treatment.

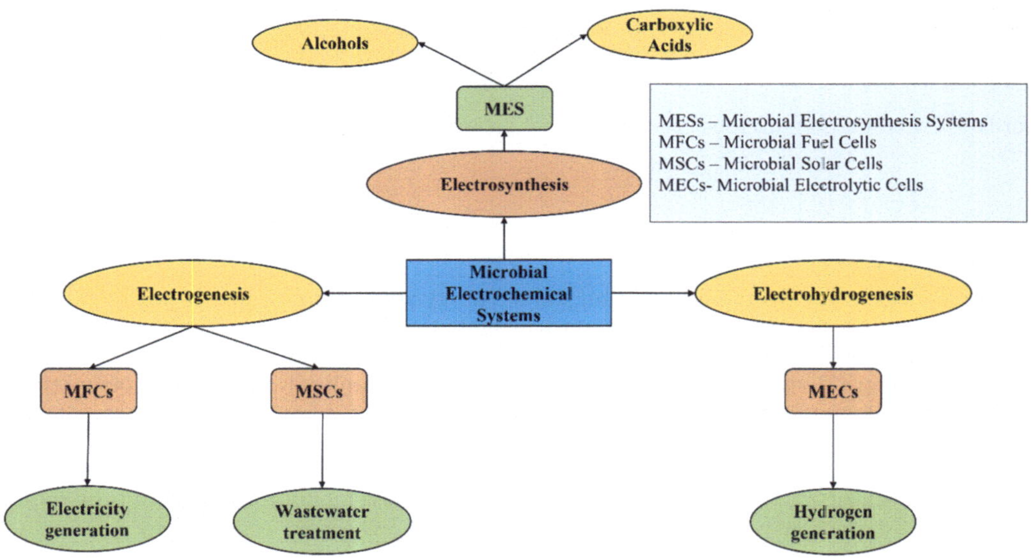

Figure 2.1 Types of MESs.

2.2 COMPONENTS AND BACTERIA INVOLVED

Electrodes play a crucial role in these systems by facilitating electron transfer between the microorganisms and the external circuit. The selection of electrodes depends on the specific application, system design, and operating conditions (Kalathil *et al.*, 2018).

2.2.1 Types of electrodes

Various electrodes commonly used in MESs are:

- Carbon-based materials: carbon-based electrodes, such as graphite, carbon cloth, or carbon nanotubes (CNTs), are often used as cost-effective alternatives to noble metals.
- CNTs: CNT-based anodes offer high conductivity and a large surface area, promoting efficient electron transfer (Ferrier & Honeychurch, 2021).
- Conductive polymers: polymers such as polyaniline or polypyrrole can be used as anode materials, offering specific functionalities and improved performance (Nezakati *et al.*, 2018).
- Platinum (Pt) or other noble metals: these metals are highly efficient catalysts for oxygen reduction and are commonly used in laboratory-scale systems.
- Oxygen diffusion cathodes: these cathodes consist of a gas-diffusion layer coated with catalysts such as carbon black and typically operate with oxygen as the electron acceptor.
- Reference electrodes: a reference electrode is used to measure and control the electrochemical potential in MESs. The most commonly used reference electrode is the Ag/AgCl electrode.
- Counter electrodes: in some MES configurations, a counter electrode is employed to complete the electrical circuit. A counter electrode is usually made of an inert material such as platinum or graphite, ensuring the supply of electrons needed for the reduction reactions at the cathode.

2.2.2 Bacteria involved

Various types of bacteria with different characteristics are involved in the process of electron transfer. These bacteria dwell under different environmental conditions and have different characteristics (Figure 2.2). For example, exoelectrogens are bacteria capable of transferring electrons to solid electrodes. They play a crucial role in electricity generation in MFCs and other METs (Logan *et al.*,

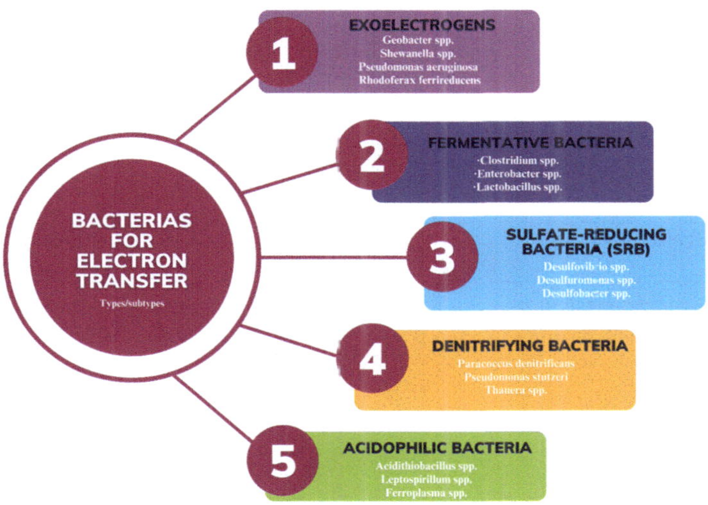

Figure 2.2 Bacteria involved in electron transfer.

2019). Similarly, fermentative bacteria are involved in the fermentation of organic matter and produce electron donors such as acetate and hydrogen, which can be utilized by exoelectrogens (Bhagchandanii *et al.*, 2020). They contribute to the overall microbial community in METs.

Although sulfate-reducing bacteria are found in anaerobic environments and commonly use sulphate as electron acceptors, denitrifying bacteria participate in the denitrification process, converting nitrate or nitrite into nitrogen gas, and can act as electron donors in METs (Sharma *et al.*, 2015; Wang *et al.*, 2016b). They are important in nitrogen removal applications. Some METs operate under acidic conditions, such as bioelectrochemical systems used for metal recovery (Zhang *et al.*, 2021). Acidophilic bacteria are adapted to low-pH environments and play a role in metal solubilization and oxidation. The specific bacterial community composition and diversity depend on the specific application, environmental conditions, and electrode configuration.

2.3 ELECTROMICROBES, ETM, AND THEIR ROLE IN METs

Electromicrobes are a type of microorganisms that have the ability to transfer electrons to or from solid surfaces, including metals (Hassan *et al.*, 2021). There are many classes of microorganisms that have been studied by researchers including α, β, γ, δ, ε-*Proteobacteria*, *Acidobacteria*, *Actinobacteria*, *Cyanobacteria*, and so on (Table 2.1). However, microbial nanowires are specifically prominent in some electromicrobes such as *Shewanella* and *Geobacter* which are strains of *Proteobacteria* (Creasey *et al.*, 2018). The whole process of electron transfer is termed as the extracellular electron transfer (EET), where microbes can either donate or accept electrons from an electrode or another microbe. These microbes can oxidize organic matter and transfer the resulting electrons to an electrode thereby generating electricity and hence are being studied for their potential utilization in MFCs. Similarly, they can be employed in MESs to drive the electrolysis of water.

These electromicrobes' biological and chemical reactions are powered by electron transfer, referring to the transport of electrons from one molecule to another. Traditionally, electron transfer reactions were thought to proceed via indirect methods involving intermediary molecules. However, researchers have found a fascinating phenomenon known as direct electron transfer (DET), which eliminates the need for these intermediates and allows electrons to travel directly between donor and acceptor molecules (Ma & Ludwig, 2019). There are different types of ETMs.

2.3.1 Based on mechanism

2.3.1.1 Direct electron transfer

In a typical electron transfer process, intermediary molecules known as redox mediators enable electron flow between source and acceptor molecules. DET, however, avoids these mediators and

Table 2.1 Use of exoelectrogens for electron transfer in bioelectrochemical systems.

Class	Strain	Electron Transfer Mechanism	Substrate	Reference
Acidobacteria	*Geothrix fermentans*	IET	Acetate	Mehta-Kolte and Bond (2012)
Actinobacteria	*Propionibacterium freudenreichii*	IET	Glycerol	Reiche *et al.* (2016)
Cyanobacteria	*Synechocystis* sp.	DET	CO_2	Madiraju *et al.* (2012)
α-*Proteobacteria*	*Acidiphilium cryptum*	IET	Glucose	Cao *et al.* (2019)
β-*Proteobacteria*	*Rhodoferax ferrireducens*	DET	Glucose	Liu *et al.* (2007)
γ-*Proteobacteria*	*Shewanella putrefaciens*	IET	Glucose	Rewatkar and Goel (2022)
δ-*Proteobacteria*	*Geobacter sulfurreducens*	IET	Acetate	Nevin *et al.* (2008)

allows for direct electron transmission between the interacting species (Ma & Ludwig, 2019). This is majorly performed by *Geobacter* and *Shewanella* species of microorganisms with the help of special structures called as conductive pili. These pili act as a conduit for electron transfer between the cells and electron acceptors. This leads to reduction in losses in the form of energy as well as quick reaction time (Garbini *et al.*, 2023). DET has paved the way for the development of biosensors and other bioelectrochemical devices. These biosensors can be used to identify the presence of various substances such as glucose, cholesterol, and so on (Huang *et al.*, 2023).

2.3.1.2 Indirect/mediated electron transfer and its application

Indirect/mediated electron transfer (IET) relies on intermediate molecules, often referred to as electron shuttles or carriers, which efficiently transport electrons between distant molecules (Glasser *et al.*, 2017). These electron shuttles typically possess redox-active groups, allowing them to accept or donate electrons as required. They act as intermediates in the transfer process, shuttling electrons from an electron donor to an acceptor molecule (Figure 2.3). It plays a crucial role in various biological processes such as cellular respiration and so on. The understanding of IET can revolutionize the fields of energy production and bioengineering. Researchers can improve the performance of MFCs, which employ microorganisms to produce energy through the oxidation of organic materials, by comprehending and modifying electron transport routes. Furthermore, the development of innovative enzymatic systems and biosensors based on IET processes can lead to breakthroughs in a variety of industries, including medical and environmental monitoring (Bollella & Katz, 2020).

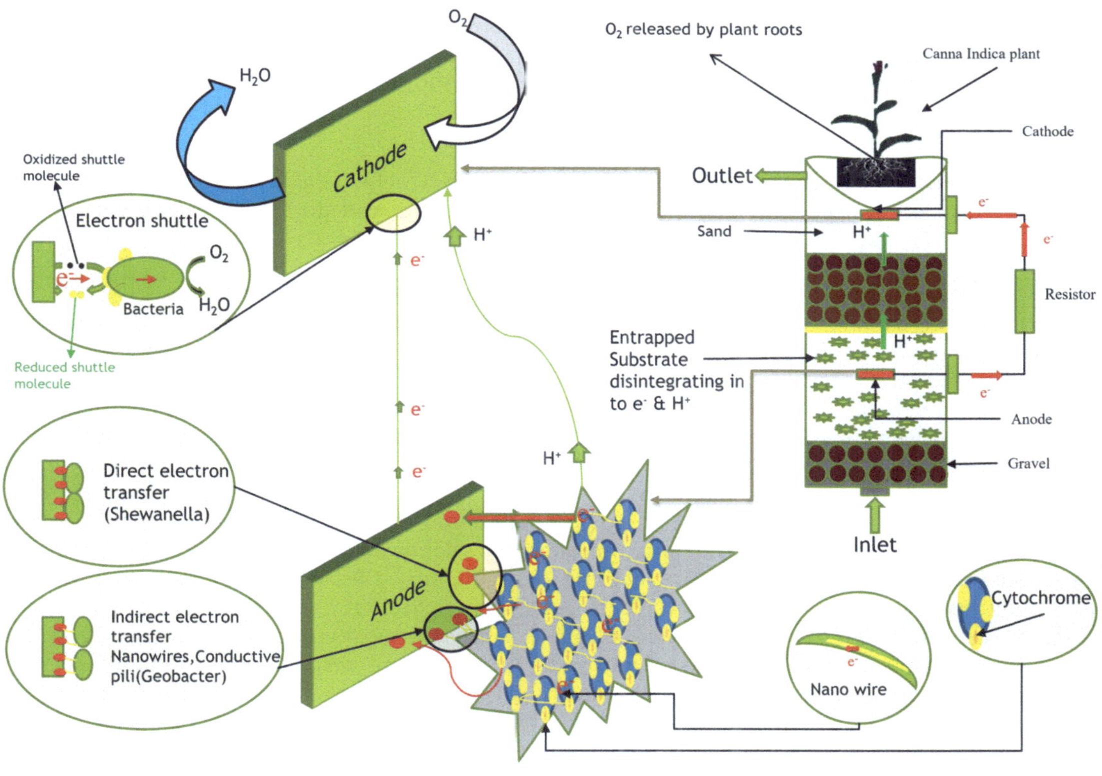

Figure 2.3 Electron transfer mechanism in CW-MFCs (*source*: Thakur *et al.*, 2021).

The study of IET pathways has significant implications beyond the realm of fundamental biology. In the field of renewable energy, researchers are exploring the possibility of mimicking natural electron transfer processes to create efficient artificial photosynthetic systems (Marzolf *et al.*, 2020).

2.3.2 Based on electrode environment

2.3.2.1 Anodic electron transfer

During anodic electron transfer, oxidation occurs at the anode which produces electrical energy through a spontaneous redox reaction (Schröder, 2007). The species being oxidized loses electrons and releases them into the external circuit or the electrolyte solution:

Reaction at anode: $CH_3COOH + 2H_2O \rightleftharpoons 2CO_2 + 8H^+ + 8e^-$

When an external power source drives the redox reaction, the anode is connected to the power source's positive terminal. The positive terminal feeds electrons to the anode, letting the oxidation process to continue. Electrons are released during the oxidation process and pass via an external circuit to provide an electrical energy source (Song *et al.*, 2019). Anodic electron transfer is an essential part of many electrochemical processes, including batteries, fuel cells, electrolysis, corrosion, and various electrochemical reactions.

2.3.2.2 Cathodic electron transfer

This process is facilitated by reduction that occurs at the cathode, which draws and takes in electrons produced either within the system or drawn from an external power source (Kawaichi *et al.*, 2018). Cathodic electron transfer is commonly used in electrochemical processes such as electrolysis, electroplating, fuel cells, and batteries. For example, in a simple electrolysis cell with water, the cathodic electron transfer occurs at the cathode, leading to the reduction of water molecules:

Reaction at cathode: $2H_2O + 2e^- \rightarrow H_2 + 2OH^-$

Overall, cathodic electron transfer is a crucial aspect of electrochemical systems and plays a critical role in several scientific applications involving energy conversion and storage (Choi & Sang, 2016).

2.4 GENETIC ENGINEERING OF EXOELECTROGENS

The performance of bioelectrochemical systems can be enhanced by certain genetic engineering and modification techniques which can improve the capabilities of exoelectrogens for various applications. Exoelectrogens have specific genes responsible for manufacturing electron transfer proteins such as cytochromes and conductive pili (Holmes *et al.*, 2006). Genetic engineering can be used in exoelectrogens to enhance the expression of these particular genes or extra copies of the same genes to improve efficiency of electron transfer. It can be used in exoelectrogens to optimize metabolic pathways to enhance electron generation and transfer (Alper & Avalos, 2018). This can include manipulating key enzymes involved in energy production and electron transport such as those from citric acid cycle or electron transport chain. Moreover, by selectively down-modulating or eliminating expression of certain genes genetic engineering helps identify the specific genes responsible for undesirable traits or inefficiencies in exoelectrogens. This is known as gene knockout/knockdown and leads to development of more efficient and robust strains (Wang *et al.*, 2016a). Similarly, synthetic biology techniques could enhance exoelectrogens to sense and respond to individual cues for perceived threats as well as actual effector states (Glaven, 2019). A method called genetic diversity screening would be used to introduce genetic diversity into populations of exoelectrogens. By creating libraries of genetically diverse exoelectrogen strains, researchers can then screen and identify strains with enhanced electrogenic properties or other desirable traits. Lastly, horizontal gene transfer can facilitate the transfer of exogenous genetic material into exoelectrogens. This can be achieved by introducing plasmids or using gene-editing technologies such as CRISPR-Cas9 to insert or modify specific genes related to exoelectrogenic activity (Hsu *et al.*, 2014).

2.5 FACTORS AFFECTING PERFORMANCE OF METs

2.5.1 Effect of substrate on performance of METs

The efficiency and performance of METs can be significantly influenced by the type of substrate. To optimize METs for individual applications, significant consideration must be given to the intended outcomes, substrate features, and microbial interactions (Sonawane *et al.*, 2022). The choice of substrate determines the type and yield of the product such as electricity, hydrogen, methane, acetate, and so on (Das *et al.*, 2020). These products are formed due to the variations in microbial metabolism and pathways. For instance, using carbon dioxide as a substrate in microbial electrosynthesis can lead to the production of valuable chemicals such as acetate or methane (Bajracharya *et al.*, 2017).

METs utilize microorganisms to catalyse electrochemical reactions, such as the oxidation of organic compounds and the reduction of inorganic compounds, to produce electricity or valuable products. The substrate serves as a fuel source for the microorganisms and directly influences their metabolic activities, which, in turn, affect the overall performance of the system (Thakur & Das, 2021b). The type and composition of the substrate directly influence the biochemical activity of the microorganisms involved. Different substrates can provide varying energy yields and electron transfer pathways, affecting the overall efficiency. The availability and accessibility of the substrate influence microbial growth rates, metabolic activity, and substrate utilization efficiency. Easily degradable substrates are preferred in METs as they can be readily utilized by the microorganisms, resulting in faster and more efficient electron transfer (Thakur *et al.*, 2021). Complex substrates may require additional pre-treatment or microbial consortia to break them down into simpler compounds before they can be utilized effectively. Substrates with higher energy content or higher electron donor capacity can result in increased current production. For example, substrates such as acetate, glucose, or volatile fatty acids are commonly used in MFCs due to their high energy yields and efficient electron transfer (Yu *et al.*, 2012).

2.5.2 Effect of temperature and pH

Temperature can have a great impact on METs and it significantly alters microbial activity, such as favouring the acidification of contaminated groundwater by lowering pH; changing community composition; altering the electrochemical kinetics in communities that use metal electrodes; influencing mass transport; affecting microbial sensitivity; and dramatically affecting the operation of systems (Gadkari *et al.*, 2020). Understanding why conditions govern system performance is essential to apply proper control techniques on temperatures. METs utilize the metabolic activities of microorganisms to catalyse electrochemical reactions. These reactions can include electricity generation in MFCs or chemical production in MECs. Temperature directly influences the metabolic activity of microorganisms and higher temperatures generally increase the rate of microbial reactions, including substrate utilization, electron transfer, and biomass growth (Price & Sowers, 2004). This can result in enhanced electrochemical performance and higher power or chemical production rates in METs. Temperature also affects the composition and diversity of microbial communities in METs. Temperature also affects the kinetics of electrochemical reactions occurring at the electrode surfaces in METs. Higher temperatures generally increase the reaction rates, resulting in faster electron transfer and improved electrochemical performance (Edwards *et al.*, 2018). However, excessively high temperatures can also cause issues such as electrode fouling or thermal degradation of components.

On the contrary, pH is considered as one of the most important parameters in terms of defining the path of ETMs. It influences the survival and activity of microorganisms in the system as pH outside the optimum range can inhibit the metabolic activity, leading to reduced performance of the system. According to Raghavulu *et al.* (2009), acidophilic pH outperformed neutral and alkaline operations in terms of power output. The pH of the environment significantly affects the performance and efficiency of MESs. pH influences the activity, growth, and survival of microorganisms, as well as the electrochemical reactions occurring at the anode and cathode of the system. Different microbial species have distinct pH ranges at which they can thrive. pH values outside the optimal range can inhibit microbial growth and metabolic activity, leading to reduced performance of the system. Therefore, it is crucial to consider the pH requirements of the specific microorganisms used in the system as it affects the electrochemical reactions occurring at the anode and cathode (Jin & Kirk, 2018). For instance, in MFCs, the anode undergoes oxidation whereas the cathode undergoes reduction. The speed of these reactions can be governed by pH. Furthermore, pH influences the conductivity and ionization of species in the electrolyte, affecting the overall system performance. Buffering agents can prevent abrupt pH change with their capability to intake excess hydroxyl and proton ions (Qiang *et al.*, 2011). Therefore, the choice of a suitable buffering system can be extremely beneficial for the maintenance of appropriate pH range.

2.5.3 Effect of electrode material

The selection of an appropriate electrode material is crucial for optimizing the performance, efficiency, and long-term stability of METs. Electrode materials with high electrical conductivity, such as carbon-based materials (e.g. graphite, carbon cloth, CNTs), facilitate efficient electron transfer between the microorganisms and the electrode. This promotes better overall performance of METs (Yu *et al.*, 2021). The surface properties of the electrode material can influence the attachment and growth of microbial biofilms. Different materials exhibit varying degrees of hydrophobicity or hydrophilicity, surface charge, and roughness, which can affect the initial attachment and subsequent biofilm development. Biofilm formation is important for the sustained activity and stability of METs. Electrodes with larger surface areas provide more sites for microbial attachment and biofilm formation. This enhances the microbial activity and increases the overall power output of METs. Materials with a high surface area-to-volume ratios, such as porous carbon materials or nanostructured electrodes, are commonly used for this purpose (Yu *et al.*, 2021). Electrode materials should be chemically stable, mechanically robust, and resistant to corrosion under the operating conditions of METs (Yu *et al.*, 2021). Long-term stability is crucial for sustainable and practical applications of METs. Some commonly used stable materials include graphite, stainless steel, and titanium. Certain electrode materials possess intrinsic catalytic properties that can enhance the electrochemical reactions occurring in METs (Mier *et al.*, 2021). For example, platinum, gold, and other noble metals exhibit excellent catalytic activity for oxygen reduction or hydrogen evolution reactions, which are essential in MFCs and MECs, respectively. However, the high cost and limited availability of these materials restrict their widespread use.

2.5.4 Effect of applied voltage

The effects of applied voltage in METs can be complex and interrelated with other factors such as electrode materials, microbial community composition, and operating conditions. Therefore, careful optimization and control of the applied voltage are required to achieve the desired performance in METs. An increase in the applied voltage generally results in an increase in the generated power. However, there is an optimal voltage range for each MET where the power output is maximized. According to Lim *et al.* (2020), the minimum cell voltage of 0.3 V was adequate to stimulate biofilm formation on the surfaces of both electrodes. The most significant operational voltage was established between 0.9 and 1.8 V in order to maintain the potential of a biocathode low enough for reduction processes while also protecting the ability of a bioanode to undergo oxidation reactions.

Applied voltage provides the driving force for electron transfer between the microorganisms and the electrode. It helps in facilitating the movement of electrons from the microbial metabolism to the electrode surface, where they can be collected and used for electricity generation or other electrochemical processes (Kracke *et al.*, 2015). Higher voltages can enhance the microbial activity by providing a more favourable electrochemical environment for the microorganisms. It can also influence the microbial community structure, favouring certain species or metabolic pathways over others and can even lead to cell rupture (Wang *et al.*, 2017). The applied voltage can affect the availability of electron acceptors or donors in the system. If the voltage is too high, it can lead to limitation of electron acceptor at the cathode, reducing the overall performance of the system. It also influences the kinetics of electrochemical reactions occurring at the electrode surfaces, rates of electron transfer, electrode potential, and reaction rates of various redox reactions involved in METs.

2.6 FUTURE SCOPE

This chapter discussed about the ETM and factors affecting the performance of MESs. Direct and mediated ETMs offer unique opportunities for harnessing the metabolic capabilities of microorganisms and converting their activities into usable electrical energy. Although, there are still a lot of unresolved obstacles to overcome before implementation of these technologies on a large scale, researchers are making significant advances to solve the problems. The functionality of MESs has significantly increased over the past 10 years of research and development, and their performance has improved exponentially. By carefully considering microbial community selection, substrate availability, electrode design, environmental conditions, and electrochemical parameters, researchers and engineers can enhance the efficiency and applicability of METs for sustainable energy production and wastewater treatment, thus advancing the technology to a greener and more sustainable future.

REFERENCES

Alper H. S. and Avalos J. L. (2018). Metabolic pathway engineering. *Synthetic and Systems Biotechnology*, **3**, 1–2, https://doi.org/10.1016/j.synbio.2018.01.002

Bajracharya S., Vanbroekhoven K., Buisman C. J. N., Strik D. P. B. T. B. and Pant D. (2017). Bioelectrochemical conversion of CO_2 to chemicals: CO_2 as a next generation feedstock for electricity-driven bioproduction in batch and continuous modes. *Faraday Discussions*, **202**, 433–449, https://doi.org/10.1039/C7FD00050B

Bhagchandanii D. D., Babu R. P., Sonawane J. M., Khanna N., Pandit S., Jadhav D. A., Khilari S. and Prasad R. (2020). A comprehensive understanding of electro-fermentation. *Fermentation*, **6**, 92, https://doi.org/10.3390/fermentation6030092

Bollella P. and Katz E. (2020). Enzyme-based biosensors: tackling electron transfer issues. *Sensors*, **20**, 3517, https://doi.org/10.3390/s20123517

Cao Y., Mu H., Liu W., Zhang R., Guo J., Xian M. and Liu H. (2019). Electricigens in the anode of microbial fuel cells: pure cultures versus mixed communities. *Microbial Cell Factories*, **18**, 39, https://doi.org/10.1186/s12934-019-1087-z

Choi O. and Sang B.-I. (2016). Extracellular electron transfer from cathode to microbes: application for biofuel production. *Biotechnology for Biofuels*, **9**, 11, https://doi.org/10.1186/s13068-016-0426-0

Creasey R. C. G., Mostert A. B., Nguyen T. A. H., Virdis B., Freguia S. and Laycock B. (2018). Microbial nanowires – electron transport and the role of synthetic analogues. *Acta Biomaterialia*, **69**, 1–30, https://doi.org/10.1016/j.actbio.2018.01.007

Das S., Diels L., Pant D., Patil S. A. and Ghangrekar M. M. (2020). Review – Microbial electrosynthesis: a way towards the production of electro-commodities through carbon sequestration with microbes as biocatalysts. *Journal of the Electrochemical Society*, **167**, 155510, https://doi.org/10.1149/1945-7111/abb836

Edwards M. A., Robinson D. A., Ren H., Cheyne C. G., Tan C. S. and White H. S. (2018). Nanoscale electrochemical kinetics & dynamics: the challenges and opportunities of single-entity measurements. *Faraday Discussions*, **210**, 9–28, https://doi.org/10.1039/C8FD00134K

Ferrier D. C. and Honeychurch K. C. (2021). Carbon nanotube (CNT)-based biosensors. *Biosensors*, **11**, 486, https://doi.org/10.3390/bios11120486

Gadkari S., Fontmorin J.-M., Yu E. and Sadhukhan J. (2020). Influence of temperature and other system parameters on microbial fuel cell performance: numerical and experimental investigation. *Chemical Engineering Journal*, **388**, 124176, https://doi.org/10.1016/j.cej.2020.124176

Garbini G. L., Barra Caracciolo A. and Grenni P. (2023). Electroactive bacteria in natural ecosystems and their applications in microbial fuel cells for bioremediation: a review. *Microorganisms*, **11**, 1255, https://doi.org/10.3390/microorganisms11051255

Glasser N. R., Saunders S. H. and Newman D. K. (2017). The colorful world of extracellular electron shuttles. *Annual Review of Microbiology*, **71**, 731–751, https://doi.org/10.1146/annurev-micro-090816-093913

Glaven S. M. (2019). Bioelectrochemical systems and synthetic biology: more power, more products. *Microbial Biotechnology*, **12**, 819–823, https://doi.org/10.1111/1751-7915.13456

Hassan R. Y. A., Febbraio F. and Andreescu S. (2021). Microbial electrochemical systems: principles, construction and biosensing applications. *Sensors*, **21**, 1279, https://doi.org/10.3390/s21041279

Herrera-Melián J. A., Mendoza-Aguiar M., Alonso-Guedes R., García-Jiménez P., Carrasco-Acosta M. and Ranieri E. (2020). Multistage horizontal subsurface flow vs. hybrid constructed wetlands for the treatment of raw urban wastewater. *Sustainability*, **12**, 5102, https://doi.org/10.3390/su12125102

Holmes D. E., Chaudhuri S. K., Nevin K. P., Mehta T., Methe B. A., Liu A., Ward J. E., Woodard T. L., Webster J. and Lovley D. R. (2006). Microarray and genetic analysis of electron transfer to electrodes in *Geobacter sulfurreducens*. *Environmental Microbiology*, **8**, 1805–1815, https://doi.org/10.1111/j.1462-2920.2006.01065.x

Hsu P. D., Lander E. S. and Zhang F. (2014). Development and applications of CRISPR-Cas9 for genome engineering. *Cell*, **157**, 1262–1278, https://doi.org/10.1016/j.cell.2014.05.010

Huang W., Bin Zulkifli M. Y., Chai M., Lin R., Wang J., Chen Y., Chen V. and Hou J. (2023). Recent advances in enzymatic biofuel cells enabled by innovative materials and techniques. *Exploration*, **3**, 20220145, https://doi.org/10.1002/EXP.20220145

Jin Q. and Kirk M. F. (2018). pH as a primary control in environmental microbiology: 1. Thermodynamic perspective. *Frontiers in Environmental Science*, **6**, 21, https://doi.org/10.3389/fenvs.2018.00021

Kalathil S., Patil S. A. and Pant D. (2018). Microbial fuel cells: electrode materials. In: Encyclopedia of Interfacial Chemistry: Surface Science and Electrochemistry, K. Wandelt (ed.), Institute of Physical and Theoretical Chemistry, University of Bonn, Germany, Heinemann-Elsevier, pp. 309–318, ISBN: 978-0-12-809894-3, https://doi.org/10.1016/B978-0-12-409547-2.13459-6

Kato S. (2016). Microbial extracellular electron transfer and its relevance to iron corrosion. *Microbial Biotechnology*, **9**, 141–148, https://doi.org/10.1111/1751-7915.12340

Kawaichi S., Yamada T., Umezawa A., McGlynn S. E., Suzuki T., Dohmae N., Yoshida T., Sako Y., Matsushita N., Hashimoto K. and Nakamura R. (2018). Anodic and cathodic extracellular electron transfer by the filamentous bacterium *Ardenticatena maritima* 110S. *Frontiers in Microbiology*, **9**, 68, https://doi.org/10.3389/fmicb.2018.00068

Kracke F., Vassilev I. and Krömer J. O. (2015). Microbial electron transport and energy conservation – the foundation for optimizing bioelectrochemical systems. *Frontiers in Microbiology*, **6**, 575, https://doi.org/10.3389/fmicb.2015.00575

Lim S. S., Fontmorin J.-M., Izadi P., Wan Daud W. R., Scott K. and Yu E. H. (2020). Impact of applied cell voltage on the performance of a microbial electrolysis cell fully catalysed by microorganisms. *International Journal of Hydrogen Energy*, **45**, 2557–2568, https://doi.org/10.1016/j.ijhydene.2019.11.142

Liu Z. D., Du Z. W., Lian J., Zhu X. Y., Li S. H. and Li H. R. (2007). Improving energy accumulation of microbial fuel cells by metabolism regulation using *Rhodoferax ferrireducens* as biocatalyst. *Letters in Applied Microbiology*, **44**, 393–398, https://doi.org/10.1111/j.1472-765X.2006.02088.x

Logan B. E., Rossi R., Ragab A. and Saikaly P. E. (2019). Electroactive microorganisms in bioelectrochemical systems. *Nature Reviews Microbiology*, **17**, 307–319, https://doi.org/10.1038/s41579-019-0173-x

Ma S. and Ludwig R. (2019). Direct electron transfer of enzymes facilitated by cytochromes. *ChemElectroChem*, **6**, 958–975, https://doi.org/10.1002/celc.201801256

Madiraju K. S., Lyew D., Kok R. and Raghavan V. (2012). Carbon neutral electricity production by *Synechocystis* sp. PCC6803 in a microbial fuel cell. *Bioresource Technology*, **110**, 214–218, https://doi.org/10.1016/j.biortech.2012.01.065

Marzolf D. R., McKenzie A. M., O'Malley M. C., Ponomarenko N. S., Swaim C. M., Brittain T. J., Simmons N. L., Pokkuluri P. R., Mulfort K. L., Tiede D. M. and Kokhan O. (2020). Mimicking natural photosynthesis: designing ultrafast photosensitized electron transfer into multiheme cytochrome protein nanowires. *Nanomaterials*, **10**, 2143, https://doi.org/10.3390/nano10112143

Mehta-Kolte M. G. and Bond D. R. (2012). *Geothrix fermentans* secretes two different redox-active compounds to utilize electron acceptors across a wide range of redox potentials. *Applied and Environmental Microbiology*, **78**, 6987–6995, https://doi.org/10.1128/AEM.01460-12

Mier A. A., Olvera-Vargas H., Mejía-López M., Longoria A., Verea L., Sebastian P. J. and Arias D. M. (2021). A review of recent advances in electrode materials for emerging bioelectrochemical systems: from biofilm-bearing anodes to specialized cathodes. *Chemosphere*, **283**, 131138, https://doi.org/10.1016/j.chemosphere.2021.131138

Nevin K. P., Richter H., Covalla S. F., Johnson J. P., Woodard T. L., Orloff A. L., Jia H., Zhang M. and Lovley D. R. (2008). Power output and columbic efficiencies from biofilms of *Geobacter sulfurreducens* comparable to mixed community microbial fuel cells. *Environmental Microbiology*, **10**, 2505–2514, https://doi.org/10.1111/j.1462-2920.2008.01675.x

Nezakati T., Seifalian A., Tan A. and Seifalian A. M. (2018). Conductive polymers: opportunities and challenges in biomedical applications. *Chemical Reviews*, **118**, 6766–6843, https://doi.org/10.1021/acs.chemrev.6b00275

Price P. B. and Sowers T. (2004). Temperature dependence of metabolic rates for microbial growth, maintenance, and survival. *Proceedings of the National Academy of Sciences of the United States of America*, **101**, 4631–4636, https://doi.org/10.1073/pnas.0400522101

Qiang L., Yuan L. and Ding Q. (2011). Influence of buffer solutions on the performance of microbial fuel cell electricity generation. *Huan Jing Ke Xue*, **32**, 1524–1528.

Raghavulu S. V., Mohan S. V., Goud R. K. and Sarma P. N. (2009). Effect of anodic pH microenvironment on microbial fuel cell (MFC) performance in concurrence with aerated and ferricyanide catholytes. *Electrochemistry Communications*, **11**, 371–375, https://doi.org/10.1016/j.elecom.2008.11.038

Reiche A., Sivell J. and Kirkwood K. M. (2016). Electricity generation by *Propionibacterium freudenreichii* in a mediatorless microbial fuel cell. *Biotechnology Letters*, **38**, 51–55, https://doi.org/10.1007/s10529-015-1944-8

Rewatkar P. and Goel S. (2022). *Shewanella putrefaciens* powered microfluidic microbial fuel cell with printed circuit board electrodes and soft-lithographic microchannel. *Chemosphere*, **286**, 131855, https://doi.org/10.1016/j.chemosphere.2021.131855

Roy A. S., Sharma A., Thapa B. S., Pandit S., Lahiri D., Nag M., Sarkar T., Pati S., Ray R. R., Shariati M. A., Wilairatana P. and Mubarak M. S. (2022). Microbiomics for enhancing electron transfer in an electrochemical system. *Frontiers in Microbiology*, **13**, 868220, https://doi.org/10.3389/fmicb.2022.868220

Schröder U. (2007). Anodic electron transfer mechanisms in microbial fuel cells and their energy efficiency. *Physical Chemistry Chemical Physics*, **9**, 2619–2629, https://doi.org/10.1039/b703627m

Sharma M., Sarma P. M., Pant D. and Dominguez-Benetton X. (2015). Optimization of electrochemical parameters for sulfate-reducing bacteria (SRB) based biocathode. *RSC Advances*, **5**, 39601–39611, https://doi.org/10.1039/C5RA04120A

Sonawane J. M., Mahadevan R., Pandey A. and Greener J. (2022). Recent progress in microbial fuel cells using substrates from diverse sources. *Heliyon*, **8**, e12353, https://doi.org/10.1016/j.heliyon.2022.e12353

Song H.-L., Zhu Y. and Li J. (2019). Electron transfer mechanisms, characteristics and applications of biological cathode microbial fuel cells – a mini review. *Arabian Journal of Chemistry*, **12**, 2236–2243, https://doi.org/10.1016/j.arabjc.2015.01.008

Thakur S. and Das B. (2021a). Performance evaluation of microbial fuel cell with sewage wastewater and RO concentrate using composite anode made of *Luffa aegyptiaca*. *Environmental Progress & Sustainable Energy*, **40**, e13504, https://doi.org/10.1002/ep.13504

Thakur S. and Das B. (2021b). Bio-electrochemical evaluation of two-stage constructed wetland microbial fuel cells with high strength raw domestic wastewater and simultaneous energy recovery. *Water and Environment Journal*, **35**, 1239–1248, https://doi.org/10.1111/wej.12714

Thakur S., Ghosh D. and Das B. (2021). Comparative performance analysis of constructed wetland-microbial fuel cells operated under batch and continuous mode for treating wastewater with RO concentrate. *Journal of Environmental Engineering*, **147**, 04021049, https://doi.org/10.1061/(asce)ee.1943-7870.0001925

Ucar D., Zhang Y. and Angelidaki I. (2017). An overview of electron acceptors in microbial fuel cells. *Frontiers in Microbiology*, **8**, 643, https://doi.org/10.3389/fmicb.2017.00643

Wang H., Lin J. and Shen Z. X. (2016a). Polyaniline (PANi) based electrode materials for energy storage and conversion. *Journal of Science: Advanced Materials and Devices*, **1**, 225–255, https://doi.org/10.1016/j.jsamd.2016.08.001

Wang J., Song X., Wang Y., Abayneh B., Li Y., Yan D. and Bai J. (2016b). Nitrate removal and bioenergy production in constructed wetland coupled with microbial fuel cell: establishment of electrochemically active bacteria community on anode. *Bioresource Technology*, **221**, 358–365, https://doi.org/10.1016/j.biortech.2016.09.054

Wang K., Sheng Y., Cao H., Yan K. and Zhang Y. (2017). Impact of applied current on sulfate-rich wastewater treatment and microbial biodiversity in the cathode chamber of microbial electrolysis cell (MEC) reactor. *Chemical Engineering Journal*, **307**, 150–158, https://doi.org/10.1016/j.cej.2016.07.106

Yu J., Park Y., Cho H., Chun J., Seon J., Cho S. and Lee T. (2012). Variations of electron flux and microbial community in air-cathode microbial fuel cells fed with different substrates. *Water Science and Technology*, **66**, 748–753, https://doi.org/10.2166/wst.2012.240

Yu B., Feng L., He Y., Yang L. and Xun Y. (2021). Effects of anode materials on the performance and anode microbial community of soil microbial fuel cell. *Journal of Hazardous Materials*, **401**, 123394, https://doi.org/10.1016/j.jhazmat.2020.123394

Zhang X., Wang H., Xia T. and Wang X. (2021). Characterization of a new electrochemically active bacterium phylogenetically related to *Alicyclobacillus hesperidum* and its electrochemical performance in microbial fuel cell. *Biosensors and Bioelectronics*, **175**, 112865, https://doi.org/10.1016/j.bios.2020.112865

doi: 10.2166/9781789063813_0039

Chapter 3

Characteristics and variability of the industrial wastewater and their suitability to be used as substrates in METs

*Vivekanandan Sangeetha[1], Reiva Sibi[1], Sabarija A. Mohandas[1] and Praveena Gangadharan[1,2]**

[1]*Department of Civil Engineering, Indian Institute of Technology Palakkad, Palakkad, Kerala, India 678623*
[2]*Environmental Sciences and Sustainable Engineering Centre, Indian Institute of Technology Palakkad, Palakkad, Kerala, India 678623*
**Corresponding author: praveenag@iitpkd.ac.in*

ABSTRACT

Rapid industrialization and its associated industrial processes demand enormous quantity of water. Consequently, wastewater is produced with undesirable compounds and is detrimental to the environment. Industrial wastewater is complex, as it comprises various types of pollutants such as toxic heavy metals, phenolic organic compounds, surfactants, and persisting organic compounds (pharmaceuticals, pesticides, endocrine disruptors), hydrocarbons, and so on. Conventional treatment methods namely adsorption and coagulation involve enormous chemical requirements; ozonation, membrane filtration, and ion exchange require huge capital investments and are non-energy efficient; and other biological treatments require skilled supervision and startup period. Moreover, conventional methods fail to utilize the high energy content of industrial wastewaters and their resource potential remains unused. Microbial electrochemical techniques (METs) are synergistic, that leverages the microbial metabolism for power production with simultaneous substrate degradation. Further, a wide variety of industrial wastewaters from distilleries, breweries, dairy processing units, paper and pulping mills, pharmaceutical industries, tanneries, and electroplating units differ in their complexity, strength, and characteristics. For instance, lignin and its derivatives in the pulping wastewater facilitate electron transfer resulting in high power densities and impede methanogens in the anode chamber in METs; whereas distillery wastewater with high chemical oxygen demand content is conducive to be used as a substrate in METs. The stable composition and high organic content of industrial wastewaters facilitate high current densities and are therefore suitable for METs. Therefore, this chapter aims to explore the suitability of various industrial wastewaters as substrate for METs.

Keywords: industrial wastewater characteristics, microbial electrochemical techniques, substrate suitability, resource recovery, power production

3.1 INTRODUCTION

Population growth, urbanization, and industrialization nexus is significant pertaining to resources utilization, waste generation, and the consequent environmental pollution. Industries process and/or manufacture useful products such as paper, milk products, fabrics, gasoline products, fuel, pharmaceuticals, and plastics from raw materials. During these physical, chemical, and biological processes, enormous quantities of wastewater is produced comprising biodegradable organics, inorganic compounds, heavy metals, toxic and refractory compounds, nutrients, and compounds of recalcitrant nature (Selvasembian *et al.*, 2022). However, wastewater from different industries such as breweries, distilleries, electroplating, tannery, paper and pulp, petroleum refinery, pharmaceutical, textile and dyeing, and so on differs in their physical and chemical characteristics, strength, and complexity.

For instance, different types of dyes (direct, indirect, acidic, basic, azo) are present only in textile and dyeing wastewater (TDW) (Yaseen & Scholz, 2019). Similarly, recalcitrant hydrocarbons such as xylene, benzene, phenols, and so on are predominant in petroleum refinery wastewater (PRW) (Abu-Reesh *et al.*, 2022) and heavy metals such as chromium (Cr), lead (Pb), arsenic (As), mercury (Hg), copper (Cu), nickel (Ni), and so on are present in electroplating wastewater (EW) (Rajoria *et al.*, 2022). Whereas, the paper and pulp industry wastewater (PPW) comprise cellulose, chitin, and lignin (complex organic polymer) (Elakkiya & Niju, 2021) and on the contrary, pharmaceutical wastewater (PW) contains pharmaceutically active compounds such as antibiotics, anti-inflammatory drugs of toxic and recalcitrant nature, and are considered as emerging pollutants (Thapa *et al.*, 2022). Tables 3.1 and 3.2 present the physical and chemical properties of various industrial wastewaters.

Several conventional aerobic and anaerobic methods are employed for treating industrial wastewaters. For example, PPW undergoes aerobic and anaerobic treatment in an aerobic reactor and upflow anaerobic sludge blanket (UASB) reactor, respectively. However, huge energy requirement (50% of total energy input), secondary sludge treatment, and low methane (CH_4) to electricity conversion in UASB necessitates the need for alternate technologies (Elakkiya & Niju, 2021). Likewise, dairy wastewater (DAW) is treated by physical, chemical, and biological (aerobic) processes with a typical energy requirement of around $8 \pm 1\%$ of the dairy industry energy consumption (Marassi *et al.*, 2020a). The compositional complexity of the industrial wastewaters requires several stages of physical

Table 3.1 Physio-chemical characteristics of various industrial wastewaters (BW, DW, DIW, EW, PPW).

Parameters	BW	DW	DIW	EW	PPW
pH	6.6–6.8[a]	7.53–10.87[d,e]	3.4–4.1[f]	6.1[i]	11.68[j]
Conductivity, mS/cm	–	2.7 ± 0.3[d]	–	14.8[i]	5.32[k]
TDS, mg/L	–	–	2160[g]	–	15 840[j]
Total SS, mg/L	36–450[a]	–	1100[g]	128[i]	24 336[j]
Total organic carbon, mg/L	970[b]	–	–	–	74 661[j]
Total COD, mg/L	2106–2250[a]	5980 ± 203[d]	80 000–10 000[f]	2500[i]	1 00 000–1 80 000[j]
Total BOD, mg/L	1285–1540[a]	3110 ± 144[d]	30 000–45 000[f]	–	30 000–40 000[j]
Total nitrogen, mg/L	35[c]	54.8 ± 7.4[d]	–	–	–
Ammonia nitrogen, mg/L	52–60[a]	–	–	–	–
Nitrates, mg/L	–	21.9 ± 7.5[d]	–	–	–
Sulphates, mg/L	–	5.3 ± 1.0[d]	1300–3700[h]	11.26[i]	10 202[j]
Sulphides, mg/L	–	–	–	2100[i]	–
Chlorides, mg/L	–	–	4100[g]	–	800[j]
Total phosphorus, mg/L	33[c]	16.2 ± 1.3[d]	–	–	–
Phosphate, mg/L	50[b]	–	–	–	–
Oil and grease, mg/L	–	–	–	4–8[i]	–
Total residual chlorine, mg/L	–	–	–	15.58–550[i]	–

[a]Sangeetha *et al.* (2020); [b]Wang *et al.* (2008); [c]Çetinkaya *et al.* (2015); [d]Marassi *et al.* (2020a); [e]Bejjanki *et al.* (2021); [f]Ghosh Ray and Ghangrekar (2015); [g]Mohanakrishna *et al.* (2010); [h]Ha *et al.* (2012); [i]Mirzaienia *et al.* (2017); [j]Shankar *et al.* (2016); [k]Chaurasia *et al.* (2021).

Table 3.2 Physio-chemical characteristics of various industrial wastewaters (PRW, PW, TW, TDW).

Parameters	PRW	PW	TW	TDW
pH	7–9[a]	5–10.7[c]	4–8[f]	6–10[i]
Conductivity, mS/cm	–	–	–	1000[i]
TDS, mg/L	80[b]	1935[d]	4450[f]	1500–6000[i]
Total SS, mg/L	500[a]	530[d]	345[f]	100–5000[i]
Total COD, mg/L	900–2150[b]	1000–10 000[c]	3455[f]	150–12 000[i]
Total BOD, mg/L	400[a]	500–2500[c]	244[f]	–
Total Kjeldahl nitrogen, mg/L	–	–	–	70–80[i]
Nitrates, mg/L	–	1–7150[c]	–	–
Sulphates, mg/L	28[b]	138–9000[c]	2848[g]	500–700[i]
Sulphides, mg/L	25.6[a]	–	–	5–20[i]
Chlorides, mg/L	–	10[e]	431[h]	200–6000[i]
Sodium, mg/L	–	–	–	400–7000[i]
Oil & grease, mg/L	300[a]	–	–	10–30[i]
Phenol, mg/L	20[a]	–	–	–

[a]Abu-Reesh *et al.* (2022); [b]Mohanakrishna *et al.* (2018a); [c]Bagchi and Behera (2020); [d]Nayak and Ghosh (2019); [e]Ismail and Ibrahim (2017); ; [f]Miran and Mumtaz (2022); [g]Elabed *et al.* (2019); [h]Sawasdee and Pisutpaisal (2016); [i]Yaseen and Scholz (2019).

or chemical or biological treatment that are energy intensive. Unlike domestic sewage, industrial wastewater has high COD (chemical oxygen demand)/BOD (biological oxygen demand) ratio (>1.5) (Shah & Ruparelia, 2022), thus demanding advanced treatment technologies involving substrate oxidation processes. Moreover, the recent paradigm shift towards valorization of waste into usable materials, fuel, and energy impulses the need for alternate, sustainable, and integrated technologies (Srikanth *et al.*, 2016). It is reported that about 1 kg of carbohydrate (1.06 kg of COD) can generate 13×10^6 Coulombs or 4.41 kW of energy (Mohanakrishna *et al.*, 2018a).

Recently, bio electrochemical techniques or microbial electrochemical techniques (METs) are adopted for simultaneous wastewater treatment and recovery of valuable products and energy (Pandey *et al.*, 2016). METs include microbial fuel cells (MFC) and its variants such as photosynthetic MFCs (P-MFC), sediment MFC (S-MFC), constructed wetland MFC (CW-MFC); microbial desalination cell (MDC); microbial electro synthesis (MES); and microbial electrolysis cell (MEC). Figure 3.1 presents the increase in research publications on adopting METs for simultaneous treatment and power generation from industrial wastewaters. It could be inferred that MFCs are more pronounced pertaining to industrial wastewater treatment, indicating the suitability to leverage industrial wastewater for clean energy production. The MFCs comprise anode and cathode chamber and are separated by ion-exchange membrane or separators such as glass wool. The anode is inoculated with a single microbial community or consortium of microbes and the substrate of degradation interest. The anode chamber is maintained anaerobic to facilitate substrate oxidation by electrogenic microbes. As oxidative products, electrons, and protons are produced which are transferred to the cathode chamber through external circuit and ion-exchange membrane, respectively. In double-chambered MFCs, sometimes, both the anode and cathode chambers can contain same substrates. For example, the azo dyes in TDW can act as both electron donor in the anode chamber and electron acceptor in the cathode chamber. MDCs are yet another MET, consisting of three chambers – anode, cathode, and desalination chamber, wherein anode and desalination chamber are separated by anion exchange membrane and cathode and desalination chamber are separated by cation-exchange membrane. The anode and cathode reactions are the same as that of MFCs, while in the middle chamber, desalination occurs through anion- and cation-exchange membranes induced by voltage difference, ionic strength,

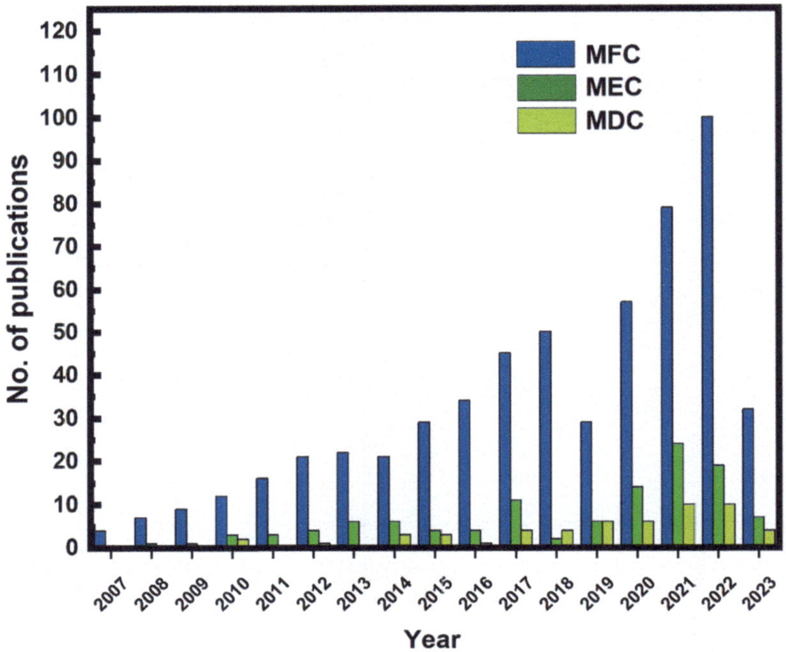

Figure 3.1 Increase in the number of research articles published on METs and industrial wastewaters. (Source: Scopus, data collected on 28 May 2023.)

and ion migration between the anode and cathode (Bejjanki *et al.*, 2021). MECs are analogous to MFCs, but require external voltage for substrate oxidation at anode and reduction at cathode to produce hydrogen (Kadier *et al.*, 2016).

This chapter explores the suitability of various industrial wastewaters as a substrate for METs. The compositional variability of different industrial wastewaters is well presented along with case studies. Also, the existing challenges and future prospects of adopting METs as sustainable resource recovery and treatment technologies are elucidated.

3.2 INDUSTRIAL WASTEWATER AS SUBSTRATE FOR VARIOUS METs

3.2.1 Brewery wastewater

Brewing, also known as beer production, is the process of making beer from cereals at a brewery. Water is used extensively during the various stages of brewing such as malting, mashing, milling, lautering, boiling, fermenting, conditioning, and filtering (Sangeetha *et al.*, 2020). Despite significant technological advancements in the past, the breweries still generate about 3–10 L of effluent per litre of beer produced. A large portion of brewery wastewater (BW) constitute organic components such as sugar, starch, and protein and consequently contributing to high BOD and COD. Biological methods namely aerobic sequencing batch reactor, sludge blanket reactor, and so on, were proven to be effective for BW treatment (Feng *et al.*, 2008). The increased energy requirement for the above-mentioned processes have consistently been identified as a significant limitation and has led research studies to investigate the possibility of METs in treating BW. The high concentration of non-toxic carbohydrates and low ammonium–nitrogen content makes BW suitable to be used as a substrate in various METs. The food-derived organic materials in BW enables the microbes to attach on the anode, easily pull electrons out from substrates and convert them into demineralized products in an anaerobic environment (Dannys *et al.*, 2016).

The effectiveness of treating varied strength BW in single chamber air cathode MFCs was examined by using carbon cloth electrodes and BW diluted with phosphate buffer solution (PBS) as the anolyte. The COD removal efficiency increased from 54 to 98% with the increase in BW concentration from 84 to 1600 mg/L (Feng *et al.*, 2008). On the contrary, when the BW was diluted with PBS, a significant impact on the power output and coulombic efficiency (CE) of MFC were observed. A maximum power density (PD) of 188 mW/m^2 was obtained with undiluted BW, whereas 528 and 483 mW/m^2 were obtained with BW diluted with 200 and 50 mM PBS, respectively. Thus, ionic strength was proven to be more significant for the production of electricity than wastewater strength. The study further pointed out that the kinetics of the MFC's air cathode was also improved by adding PBS to the catholyte. As a result, greater power is obtained due to the maintenance of favourable conditions for the oxygen reduction reaction (ORR). In another study by Yu and researchers the performance of BW fed MFC was compared with that of pure substrate (glucose, acetate, butyrate, and propionate) fed MFCs (Yu *et al.*, 2015). The study suggested that real wastewater fed MFCs produced less power and CE compared with that of pure substrates. MFCs fed with pure fermentable substrates like glucose and acetate produced PD of 1519 mW/m^2 (CE – 62% and R_{int} – 54 Ω) and 1256 mW/m^2 (CE – 71% and R_{int} – 47 Ω), respectively, while BW fed MFC produced only 251 mW/m^2 (CE – 31% and R_{int} – 258 Ω) (Yu *et al.*, 2015). The decreased performance of BW could be attributed to low conductivity, non exo-electrogens invasion, and availability of other electron acceptors.

A 90-litre stackable pilot MFC comprising five independently stackable modules with rolling sheets of activated carbon cathodes and carbon fibre brush anodes was studied for BW treatment (Dong *et al.*, 2015). The modules were placed inside a reactor vessel made of Plexi glass. The pilot MFC was tested using diluted BW and raw BW for COD and suspended solids (SS) removal. An increased hydraulic retention time (HRT) was provided in the case of raw BW. The COD and SS removal efficiencies were 84.7% and 81.7% with diluted BW, and 87.6% and 86.3% with raw BW (Dong *et al.*, 2015). The study found that while treating raw BW (with increased HRT and loading rate), the system achieved relatively higher removal efficiencies. The hydrolysis of carbohydrates in BW into soluble substrates played a major role in SS reduction in the stacked MFC. The stacked system produced sufficient energy to power the pumping system and also net electrical energies of 0.021 and 0.034 kWh/m^3 were harvested using diluted and raw BW, respectively (Dong *et al.*, 2015). Çetinkaya and researchers treated BW in a dual-chambered MFC with tin-coated copper mesh electrodes (Çetinkaya *et al.*, 2015). The study suggested that by adjusting the HRT, the type and population of bacteria can be selected. A reduced HRT increases the substrate strength thereby increasing the overall rate of substrate consumption by bacteria and improved power production. A higher HRT causes membrane fouling by inorganic salt deposition, microbes, and extracellular polymers which will negatively influence the power output from MFCs. The effects of the organic loading rate on the MFCs were investigated by monitoring the COD concentration at various HRTs and the maximum performance was obtained at an HRT of 0.5 d. At an HRT of 0.5 d, 82% COD removal and voltage generation of 0.31 V was obtained. The obtained maximum PD was 8.001 μW/cm^2 and it decreased to 1.069 μW/cm^2 upon increasing the HRT from 0.5 to 1 d (Çetinkaya *et al.*, 2015).

The rich organic content and food-derived nature of BW can also be utilized for the recovery of valuable products such as hydrogen (H_2) and methane (CH_4) using MECs. Lu and research group demonstrated that the choice of a cathode material is the major bottleneck in producing H_2 from BW (Lu *et al.*, 2017). A Ni foam modified with double-layered hydroxide catalyst was used as the cathode (NiFe LDH/Ni) and carbon brush as anode for H_2 generation from a single chamber MEC employing BW. NiFe LDH/Ni proved to enhance system performance and greater stability over time than conventional Pt/Ni cathode. NiFe LDH/Ni achieved comparable H_2 generation rate (2.01 m^3-H_2/m^3/d) than Pt/Ni cathode (2.12 m^3-H_2/m^3/d). A higher H_2 recovery of 76–80% was obtained with

NiFe LDH/Ni while Pt/Ni foam could yield only 55–66% (Lu *et al.*, 2017). A similar experiment, a tubular MEC with granular graphite anode and various mesh cathodes (Ni, Cu, and stainless steel) for CH_4 recovery (Sangeetha *et al.*, 2016) was conducted. Ni mesh outperformed others and yielded 85% COD and 83% (total organic carbon) TOC removal and produced 143 mL/g COD of CH_4 with 8.6 mA of current generation.

3.2.2 Dairy wastewater

DAW is generated from processing of raw milk into milk, cheese, curd, whey protein, butter, and milk powder. For every litre of milk processed, 2–10 L of wastewater is generated (Marassi *et al.*, 2020b). The DAW is high strength wastewater (COD – 2000 to 10 000 mg/L; BOD – 1500 to 4000 mg/L) constituting fermentable organic compounds such as casein, carbohydrates, lipids and proteins and high concentrations of inorganic ions including potassium (K), sodium (Na), calcium (Ca), magnesium (Mg), iron (Fe), phosphate (PO_4^{3-}), nitrate (NO_3^-), nitrite (NO_2^-), ammonium (NH_4^+), and sulphate (SO_4^{2-}) (Marassi *et al.*, 2020a; Thulasinathan *et al.*, 2022).

Simultaneous electricity production and treatment efficiency of MFC for high strength DAW using an acclimatized electrogenic consortium was studied (Marassi *et al.*, 2019). In this study, an electrogenic consortium of *Clostridium butyricum* and *Shewanella oneidensis* was inoculated in the anode chamber. The bacterial strain *C. butyricum* could degrade long-chain fatty acids and proteins into metabolites which further can be dissimilated by *S. oneidensis* (metal reducing strain). During the start-up, the anode chamber was fed with synthetic solution (electrogenic consortium) and 10% DAW, during which an open-circuit voltage (OCV) of 530 mV and PD of 1.31 W/m³ was achieved. In phase 1 (acclimatization phase), 25% DAW and 75% nutrient solution was fed, wherein, the OCV and PD decreased to 472 mV and 1.14 W/m³. However, the system recovered with OCV and PD reaching up to 510 mV and 1.45 W/m³ on the 43rd day, respectively. During phase 2 (45th day onwards), raw DAW was fed, during which, the OCV and PD dropped to 370 mV and 0.84 W/m³, respectively. Nevertheless, the MFC recovered with an upsurge in OCV and PD to 460 mV and 1.25 W/m³ (66th day), respectively and remained stable till 86th day of operation (Marassi *et al.*, 2019). The recovery of the MFC from reduced electrochemical activity was claimed due to the acclimation with mixed substrate constituting fermentable (DAW) and non-fermentable substances (nutrient solution). The polarization curves revealed that ohmic losses were predominant than activation and concentration losses at low and high current density (CD), respectively. The minimized activation losses were due to suitable acclimatization and start-up, and reduced concentration losses was ascribed to the internal recirculation. The removal efficiencies of TBOD (total BOD), TCOD (total COD), TKN (total Kjeldahl nitrogen), and Na were obtained as 90%, 62%, 72%, and 71%, respectively. Also, a maximum PD of 1.45 W/m³ in the acclimation phase and 1.32 W/m³ in the treatment phase was attained (Marassi *et al.*, 2019).

Choudhury and researchers demonstrated the treatment and bioelectricity production from DAW using a single-chambered MFC (SCMFC) inoculated with a single pure culture of *Pseudomonas aeruginosa* (Choudhury *et al.*, 2021). The SCMFC was operated in batch mode for 15 days, after an acclimatization period of 7 days with DAW. A COD removal efficiency of 94.40%, CE of 46.59%, CD of 161 mA/m², and PD of 34.82 mW/m² was achieved. The substrate removal was low and so as the COD removal, initially during lag phase due to acclimatization of *P. aeruginosa*. However, in exponential growth phase, COD removal and the specific growth rate of *P. aeruginosa* was congruent with each other.

In yet another study, a COD removal of 92.2%, BOD removal of 88.02%, total dissolved solids (TDS) removal efficiency of 76.3%, and electricity generation of 644 mV was achieved in a dual-chambered MFC (Sanjay & Udayashankara, 2019). Also, the long-chain fatty acids resulting from lipid hydrolysis impedes the growth of methanogen facilitating power generation by exoelectrogenic bacteria (Elakkiya & Matheswaran, 2013).

Apart from power production, DAW was also used for simultaneous electricity production and desalination in an MDC (Bejjanki *et al.*, 2021). A three-chambered MDC comprising anode, desalination and cathode chambers separated by anion- and cation-exchange membranes was used with plain graphite plate as anode and cathode. The photosynthetic microorganism (*Oscillatoria* sp.) was used as a biocatalyst in the cathode owing to advantages such as in-situ oxygen production, increased desalination efficiency and decreased start-up time. Both the anode and cathode chamber were filled with DAW and desalination chamber was filled with seawater or synthetic saline solution (10–30 g/l). An OCV of 652 ± 10 mV, COD removal efficiency of 80.2 ± 0.5%, and desalination efficiency of 65.8 ± 0.5% was attained demonstrating the suitability of DAW for desalination (Bejjanki *et al.*, 2021). However, transfer of chlorine ions (from seawater) impeded the exoelectrogenic microbial activity, which further lowered the current generation.

3.2.3 Distillery wastewater

Distillery wastewater (DIW) also known as spent wash is the effluent generated from alcohol industries where about 8–15 L of effluent is generated per litre of alcohol produced. The characteristics of DIW depends on the various substrates used for alcohol fermentation such as sugar cane molasses, and cereals such as rice, barley, wheat, maize, and so on. DIW is characterized by high organic content (COD; 80 000–1 00 000 mg/L, BOD; 30 000–40 000 mg/L), SO_4^{2-}; 1300–3700 mg/L, high temperatures of 70–80°C and persistently dark brown in colour (Ha *et al.*, 2012). The presence of melanoidins gives DIW a lasting colour and their high molecular weight limits their removal by traditional methods. On the contrary, MFC is capable of removing colour from DIW which can be largely attributed to the electrochemical oxidation catalysed by microbes.

Bioelectricity generation using DIW as substrate was demonstrated using an MFC assembly with non-catalysed graphitic electrodes. The MFC produced a stable power output up to 360 h of operation using DIW and yielded a maximum volumetric PD of 1.74 W/m^3. The high organic content of DIW facilitated sustained power production for a long duration with 72% substrate degradation and 20.13% SS removal. Apart from substrate degradation, the high salt content of DIW will enable the electrochemical system to generate active chlorine-based oxidants, which can decolorize highly coloured dye and 31.67% colour removal was also obtained in the study (Mohanakrishna *et al.*, 2010). Therefore, using DIW as a substrate in METs will help achieve multimodal treatment of DIW in addition to substrate degradation.

The high temperature of DIW has prompted researchers to experiment on thermophilic MFCs for its treatment. Thermophilic MFCs provide the additional benefits of reduced activation resistance, ohmic potential loss and mass-transfer limitation, and also improved electron production rate. Ha and research group used a simple dual-chambered setup with Nafion CEM and graphite electrodes as thermophilic MFC (Ha *et al.*, 2012). The system achieved high CE up to 89%, SO_4^{2-} reduction (60%) along with oxidizing organic substrates (Ha *et al.*, 2012). The findings imply that in addition to oxidizing complex organic substrates and reducing sulphate, thermophilic MFCs can generate electricity with high efficiency (CD – 2.3 A/m^2, PD – 1 W/m^2) while using less energy to cool the DWW. The effectiveness of using low-pH (<4) wastewater produced from distilleries as anolyte in MFC was investigated (Kim *et al.*, 2014). The experiment demonstrated that even at low pH, MFC could produce 12.9 W/m^3 of PD ($R_{ext} = 5\ k\Omega$) and also good organic removal was achieved by various bacteria including *Proteobacteria*, *Firmicutes*, and *Bacteroidetes* (Kim *et al.*, 2014). The low pH DIW boosted the proton transfer rates in the anode compartment and decreased cathode proton limitations and hence exhibited superior performance.

A compact MEC was designed within an anaerobic digester to investigate the H_2 generation potential and COD removal efficiency of DIW (Samsudeen *et al.*, 2020). The results showed that a cumulative H_2 generation of 14.7 ± 1.2 mL was achieved by DIW as compared to 30.2 ± 0.5 mL by synthetic wastewater. DIW obtained CE and COD removal efficiency of 13.16 ± 0.2 and 72.5 ± 0.5%, respectively, against 18.35 ± 0.15 and 77.5 ± 0.5% of synthetic wastewater (Samsudeen *et al.*, 2020). These results indicate the potential of DIW to be used as substrate in various METs.

3.2.4 Electroplating wastewater

Electroplating industry involves a series of processes such as alkaline cleaning, plating, acid pickling, and rinsing which generates around 1.5 MLD of wastewater. The EW contains highly toxic cyanide, heavy metals like As, cobalt (Co), Cu, Cr, Hg, Fe, Ni, zinc (Zn), Pb, cadmium (Cd), organic solvents, oil and grease, BOD, COD, and turbidity. Thus, it comprises nearly 29% toxic and hazardous waste, with toxic metal ion concentrations higher than permissible levels (Rajoria *et al.*, 2022). Exposure to EW has many health effects namely thyroid dysfunction, kidney failure, rheumatic arthritis, circulatory system, and neural system-related issues, and lung cancer. Conventional methods such as chemical precipitation, chemical oxidation or reduction, ion exchange, and membrane filtration were commonly used to treat EW. However, most of these methods require high doses of chemicals and are energy intensive (Kim *et al.*, 2017).

EW containing hexavalent chromium (Cr^{6+}) is an excellent substrate for MFC, as Cr^{6+} is used as the electron acceptor in place of oxygen, hexacyanoferrate or permanganate in the cathode chamber of MFC, where it accepts electrons and protons to generate energy. In acidic condition, Cr^{6+} accepts six electrons and is reduced to Cr^{3+} (refer equations 3.1 and 3.2).

$$Cr_2O_7^{2-} + 8H^+ + 6e^- \rightarrow Cr_2O_3 + 4H_2O \; (E_0 = 1.33\ V) \tag{3.1}$$

$$Cr_2O_3 + 6H^+ \rightarrow 2Cr^{3+} + 3H_2O \tag{3.2}$$

The chromium oxide (Cr_2O_3) gets deposited on cathode and thus Cr^{6+} is removed from wastewater (Li *et al.*, 2008). As per equation (3.1), the reduction potential for Cr^{6+} (1.33 V against SHE) is higher than oxygen (1.23 V) and hexacynoferrate (0.36 V) making it a more favourable electron acceptor. This possibility of using pollutant in the cathode chamber of MFC in addition to the anode chamber increases the environmental benefits from MFC.

Li and research group treated real EW containing Cr^{6+} of initial concentration 204 ppm, in a dual-chambered MFC and achieved 99.5% Cr^{6+} and 66.2% total Cr removal (Li *et al.*, 2008). The system generated a maximum PD of 1600 mW/m^2 with a CE of 12%. However, the required low pH for the reduction of Cr affected the stable performance of MFC. The proton imbalance between the ion-exchange membrane caused reverse proton transport inhibiting growth of anodic bacterial community and reduced the bio electrochemical reactions. A bipolar membrane was used to reduce the pH imbalance which enhanced a stable bio-electricity generation and Cr^{6+} removal from real EW (Kim *et al.*, 2017). The ability to generate high PD, high Cr^{6+} tolerance, and less toxic waste sludge production in MFC makes it a promising technology for treating EW when compared to other biological treatment techniques (Li *et al.*, 2008).

Another heavy metal, Cu in EW, can be removed by hydroxide precipitation using sodium hydroxide and calcium oxide with negligible energy. The hydroxyl ion release in the catholyte of METs to balance charges in the system can be used as alkalinity donors for facilitating precipitation of heavy metals (Dong *et al.*, 2017). Mirzaienia *et al.* used the middle chamber of MDC to remove Ni (68.1%) and Pb (70.04%) from EW (Mirzaienia *et al.*, 2017). Nanofiltration concentrate from electroplating industry also requires efficient treatment before discharging because of its high salinity and heavy metal concentration. Microbial electrolysis desalination and chemical production cell (MEDCC) combined with Fenton process was used to remove heavy metals from nanofiltration concentrate along with COD reduction (79%), acid–alkali recovery, and low energy consumption (Lan *et al.*, 2019). The desalination chamber of MEDCC removed 94% of Ni^{2+}, 82% of Zn^{2+}, and 91% of Ca^{2+} in 20 h (Lan *et al.*, 2019).

3.2.5 Paper and pulp wastewater

The paper and pulp industry has a noteworthy impact on the socio-economic development of a country. Paper industries utilize agro-based materials as raw materials along with high water consumption and is reported that a normal bagasse-based paper mill typically uses 32 kilo litres per day (KLD) of water per tonne of paper produced, and generates 56 000 m^3 of wastewater (Elakkiya & Niju, 2021). The paper and pulp wastewater (PPW) contains soluble organics and particle materials like cellulose that conventional wastewater treatment systems are ineffective at degrading. Additionally, PPW constitutes high COD (1 00 000 mg/L), BOD (30 000–40 000 mg/L) and colour (Shankar *et al.*, 2016). Apart from paper mills, paper recycling and recovery units also produce considerable amounts of wastewater with similar characteristics.

Elakkiya and Niju treated PPW in a dual-chambered MFC employing graphite plate electrodes (Elakkiya & Niju, 2021). The PPW used in the study was categorized into two depending on the source from which it is collected. Category 1 was the PPW collected from sugarcane bagasse storage and washing units of paper mill (BSW), while category 2 contains PPW from pulping and bleaching units (PBW). Sucrose, a primary constituent of BSW, is a disaccharide of glucose and fructose and require numerous hydrolytic and phosphorolysis enzymes for their catabolism. Fermenters present in the anode surface of MFC are essential for converting sucrose into the glycolytic intermediates' fructose-6-P and glucose-6-P and hence treating PPW in MFC would be ideal. With wastewater treatment efficiency of 85% and CE of 6%, BSW could achieve high PD and CD of 53 mW/m^2 and 173 mA/m^2 at external resistance of 470 Ω, respectively. PBW being a more complex wastewater, exhibited reduced performance when treated in MFC. PBW could yield only 4 mW/m^2 and 16 mA/m^2 of PD and CD, respectively at 10 000 Ω external resistance. A combination of BSW and PW in the ratio of 9:1 (v/v) resulted in greater PD and CD of 73 mW/m^2 and 202 mA/m^2, respectively, at 470 Ω. 18% CE and 82% COD removal was also obtained when the mixture was treated (Elakkiya & Niju, 2021).

Paper recycling plant wastewater was treated in a SCMFC using diluted and undiluted wastewater (Huang & Logan, 2008). Treatment of raw paper recycling wastewater (lower conductivity – 0.8 mS/cm) produced 144 mW/m^2 PD and removed 16%, 29%, and 52% of cellulose, TCOD, and soluble COD (SCOD), respectively, in 350 h batch cycle. On the contrary, diluted wastewater with 50 and 100 mM of PBS resulted in improved conductivity of wastewater. A PD of 501 and 672 mW/m^2 was obtained, respectively with 50 and 100 mM of PBS in 500 h batch cycle. The addition of PBS almost completely removed (96%) cellulose, 76% TCOD, and 73% SCOD (Huang & Logan, 2008). Undiluted wastewater (high substrate concentration) requires more time to fully degrade the substrate as with increased time, more oxygen could diffuse into the system, causing aerobic removal of the substrate with a lowered CE, and this resulted in decreasing the overall treatment efficiency.

A column MFC (Figure 3.2) technique was adopted to treat PPW using the combination of electro coagulation (EC) and MFC technique (Shankar *et al.*, 2016). PPW with a lower biodegradability index (BI) (BI = COD/BOD) was pre-treated in an electrocoagulation unit to improve the BI before adding to column MFC. The study demonstrated on the one hand that using the combined treatment unit PPW with an initial COD of Σ5600 mg/L could be brought to discharge standards, along with 0.02 mA current and 56 mV voltage generation. On the other, the column MFC was efficient only when the PPW COD ≤750 mg/L. A CW-MFC was employed for treating PPW (Narayan *et al.*, 2018). The study demonstrated 89 and 97% removal of COD and BOD, respectively, along with 29 mA of current and 26 mV voltage production in 5 days HRT (Shankar *et al.*, 2016).

Chaurasia and researchers utilized PPW for simultaneous H_2 generation and wastewater treatment using MEC with the view of boosting industrial performance through the creation of value-added products (Chaurasia *et al.*, 2021). Effective functioning of MEC necessitates cost-effective cathodes that can treat industrial wastewater and recover biohydrogen. The study used different cathodes, namely, Ni, Ni–Co and Ni–Co–phosphorous (P) co-deposits on the surface of steel and Cu. The manufactured cathodes effectively treated industrial wastewater under ambient settings with better energy recovery. For a 500 mL wastewater in a 7-day batch cycle, the manufactured Ni–Co–P produced H_2 at a higher rate of 0.16 m^3 per day with stainless steel and 0.14 m^3 per day with Cu and obtained approximately 33–42% of wastewater treatment (Chaurasia *et al.*, 2021).

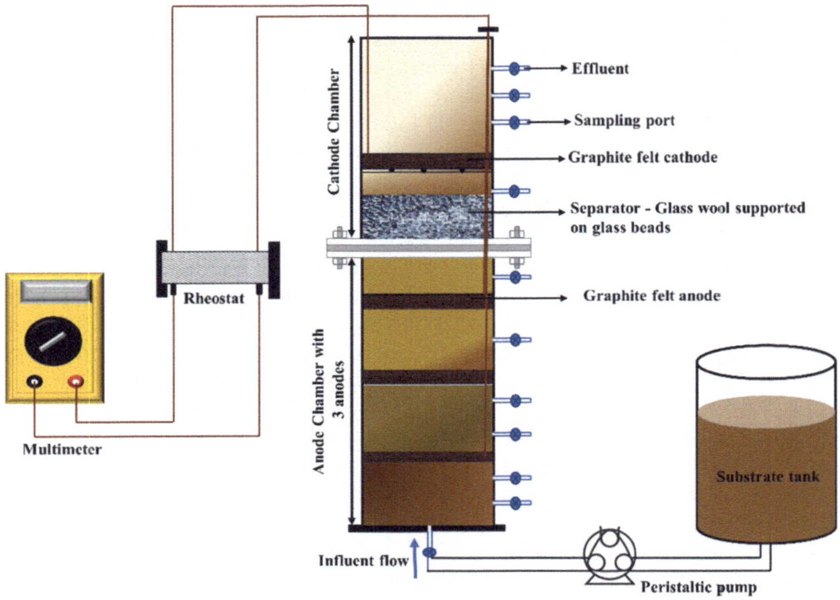

Figure 3.2 Schematic representation of column MFC for PPW. (Redrawn from Shankar *et al.*, 2016.)

3.2.6 Petroleum refinery wastewater

The PRW majorly comprises polycyclic aromatic hydrocarbons (phenols, benzene), cyclic aromatic hydrocarbons (toluene, ethylbenzene, and xylene), and diesel range organics (DRO) (Mohanakrishna *et al.*, 2018b). Petroleum refinery processes produce wastewater around 0.4–1.6 times the volume of crude oil processed (Abu-Reesh *et al.*, 2022) and the low BOD/COD ratio (<0.2) and high TDS of PRW is unfavourable for biological treatment (Mohanakrishna *et al.*, 2018a).

The degradation of hydrocarbons, phenols, and sulphides with power generation from PRW using a SCMFC was studied (Srikanth *et al.*, 2016). Carbon cloth with carbon coating and carbon cloth with platinum coating was used as anode and cathode (air cathode), respectively, and the proton-exchange membrane separated the electrodes. The anode compartment was inoculated with pre-enriched electrogenic mixed culture (Fe reducing bacteria, sulphur reducing bacteria, sulphur oxidizing bacteria, and acid-producing bacteria) and PRW, and was maintained under anaerobic condition. The MFC was operated in batch mode until the stabilization of power generation and COD removal efficiency. Later, continuous mode of operation was adopted with an HRT of 8 h. During the batch process, the voltage and PD decreased from 200 ± 2 mV and 20 ± 1 mW/m^2 to 174 ± 2 mV and 16 ± 2 mW/m^2, respectively, which was due to a drop in pH to 4. A pH drop and consequent reduction in voltage and PD occurred for four cycles in batch after which the voltage stabilized, which indicates the stabilized microbial activity. In continuous mode, although initially the power output reduced (8 ± 0.54 mW/m^2) with 8 h (organic loading rate: 2.5 kg COD/m^3-day), gradually it increased to 219 ± 4 mW/m^2 with increase in HRT (16 h) and organic loading rate of 1.25 kg COD/m^3-day. This could be due to prolonged contact time between the biocatalyst and PRW. The COD, phenol, oil, and grease, and sulphide removal efficiencies were reported to be $84 \pm 1\%$, $80 \pm 1.8\%$, $95 \pm 0.6\%$, and $79.5 \pm 1.2\%$, respectively in continuous mode (HRT – 16 h). Sulphide acts as a redox shuttle mediator between biocatalyst and insoluble electron acceptors. Sulphide oxidation to S is the principal phenomenon behind power generation.

In another investigation, PRW was used as a substrate in MFC, external voltage was applied to enhance the substrate utilization rate (Mohanakrishna *et al.*, 2018b). A 500 mV supplemental voltage was applied which resulted in a maximum PD of 132 mW/m^2, highest substrate removal efficiency (48%), and maximum DROs efficiency of 89%. Also, a 100% removal efficiency of hydrocarbons such as n-decane and n-octacosane was achieved.

3.2.7 Pharmaceutical wastewater

The PW is very complex with varied composition comprising organic materials, metals, non-metallic elements, acids, alkalis, and drug components (antibiotics, antiepileptics, cosmetic compounds). The PW is characterized to have high TDS, COD, BOD, and SS and its treatment is challenging due to their recalcitrant nature, high toxicity, and less biodegradability (Ismail & Habeeb, 2017). Therefore, processes or technologies that can degrade complex drug molecules and organic matter are required as an alternative to conventional biological treatment techniques. Compared to conventional anaerobic reactors, the METs are found to enhance the catalytic activity of microorganisms by 33%, by providing an electrogenic environment (Bagchi & Behera, 2020). The METs combination of oxidation–reduction reactions allows the degradation of wide range of compounds thus making it more suitable for treating organic compounds in the PW. Furthermore, the persistence of antibiotic-resistant gene and hormone biometabolites remains a challenge in conventional techniques for treating PW. The MET, being an anaerobic treatment technique, reduces the possibility of antibiotic-resistant genes and the low residual sludge produced recedes hormone biometabolite carriers to some extent (Xu *et al.*, 2022).

The mechanism of pharmaceutical waste removal may include anode oxidation, anode reduction, adsorption of contaminants, and cathode reduction (Xu *et al.*, 2022). As a part of anode oxidation contaminants degrade and release electrons, whereas some receive electrons and get reduced at the anode. Some contaminants get adsorbed to the electrodes or to the biofilm and some get reduced at the cathode. The mechanism of removal will depend on the characteristic of that particular component in PW. The aromatic compounds (such as benzene, nitrobenzene, phenol, and their derivatives) are degraded by the catabolic flexibility of the biofilm in the anode of MFC and the MFCs are found to be effective in degrading antibiotics such as penicillin, tetracyclin, metronidazole, chloramphenicol, and so on (Thapa *et al.*, 2022). In addition, compounds like penicillin in the substrate results in better PD in MFC. Wen and researchers showed that, the combination of glucose (1 g/L) and penicillin (50 mg/L) in an air-cathode single-chambered MFC generated a PD of 101.2 W/m^3, whereas, a PD of only 14.7 and 2.1 W/m^3 was achieved with glucose (1 g/L) and penicillin (50 mg/L), respectively (Wen *et al.*, 2011). This can be attributed to penicillin's ability to improve the permeability of cell membranes in bacteria, which results in better transport of electron to the anode from bacterial cell membranes (Wen *et al.*, 2011). Thus, the internal resistance of MFC was reduced and thereby increased the PD and in addition, 98% of penicillin was also degraded in this process (Wen *et al.*, 2011). Similar results were also found when glucose-ceftriaxone Na combination was used in MFC. The PD of MFC increased from 19 to 113 W/m^3 when 50 mg/L of ceftriaxone Na was added to 1000 mg/L of glucose (Thapa *et al.*, 2022). Thus, antibiotic effluents have the ability to generate power and can be a valuable resource for power generation in MFC. These studies reveal the suitability of using PW as substrates in MFC.

Various studies have shown that the PW degrading efficiency of MET can be improved by changing operating parameters such as anolyte composition, anode catalyst, acclimatization of biocatalyst, physicochemical pretreatment, electric stimulation or by changing electrode material (Thapa *et al.*, 2022; Xu *et al.*, 2022). An antibiotic, chloramphenicol has chlorine and nitro groups that are toxic in nature and was treated by using bio catalysed cathode. The bio catalysed cathodic reactions improved the chlorine reduction and converted nitro to amino groups and the energy required in this process was minimum. Further electrical stimulations were found to affect the chloramphenicol-resistant bacteria and could be used to maintain an optimum bacterial population. Acclimatization of bacteria to the pollutants to make them resistant to the toxicants is an effective way to improve the performance of the system. Such observations were found in the treatment of sulphamethoxazole in MFC, where an aromatic degrading bacteria, *Thauera* was found to dominate the anode (Bagchi & Behera, 2020).

Rashid and research group developed a unique paraboloid-shaped graphite-based MFC to treat PW which effectively removed 80.5% COD and generated PD of 2.01 W/m^3 (Rashid *et al.*, 2021). The truncated paraboloid shape increased the surface area per unit volume and was found favourable for PW treatment and electricity generation. Nayak and Ghosh used the combination of photo bioreactor (PBR) and P-MFC to remove pollutants and generate bio-energy (bio-electricity and biodiesel) from PW (Nayak & Ghosh, 2019). PBR was used to pretreat the PW to reduce the organic and inorganic contents, which thereby increased the efficiency of P-MFC by reducing membrane fouling. PBR uses microalgae which has the ability to grow in the presence of nutrients, to reduce the complexity of PW. The pretreated PW was then used in P-MFC and generated PD of 838.68 mW/m^3 with a maximum TCOD removal of 97.24% from PW (Nayak & Ghosh, 2019).

3.2.8 Tannery wastewater

Tannery industries process animal hides into commercial products and generate wastewater containing hairs, proteins, alkalis, acids, chromium salts, sulphides, chlorides, solvents, tannins, and dyes. The tannery industry generates 20 000–80 000 L of turbid and foul-smelling wastewater on processing one ton of skin (Ghorab *et al.*, 2022). Tannery wastewater (TW) is characterized by alkaline pH, pungent odour, dark brown colour, high COD, BOD, TDS, and high saline content (Zhao *et al.*, 2022). Many methods such as sedimentation, chemical precipitation, adsorption, froth flotation, filtration, and so on were used to treat the TW. However, these technologies are limited by low toxin degradation efficiency, hazardous by-products production, and energy requirement. The TW contain high organic content and the endogenous bacterial community can be acclimatized to heavy metal loading and sulphur content (Elabed *et al.*, 2019). Hence, TW is found to be a suitable substrate for METs as it comprises high organic content and act as electron donors by feeding the microbes.

The TW was used in an air-cathode SCMFC to generate maximum CD and PD of 120 mA/m^2 and 7 mW/m^2 with simultaneous removal of TKN (50%) and SCOD (88%) (Sawasdee & Pisutpaisal, 2016). Elabed and researchers used pre-treated TW in an MFC as substrate and achieved CD of 11.2 A/m^2 along with COD (90%), BOD$_5$ (84%), and SO_4^{2-} (96%) removal (Elabed *et al.*, 2019). The study revealed that pre-treatment of TW resulted in more efficient microbial anode with better current generation. Even though TW is rich in organic elements and microbial consortia, the isolation and acclimatization of tolerant microbes will be limiting the scaling up of this technique (Elabed *et al.*, 2019). A dual-chambered MFC attained a contaminant removal efficiency of around 85% with a PD of 7371 mW/cm^3 using TW as substrate (Chauhan *et al.*, 2022). Furthermore, studies have compared the treatment efficiencies of tannery waste acclimatized cultures and non-acclimatized microbial cultures in MFC. Acclimatized cultures resulted in approximately 1.5 times more COD removal and higher PD (223 ± 11 mW/m^2) than the non-acclimatized cultures, as they simulated better biodegradation of organic chemicals found in TW (Miran & Mumtaz, 2022). A CW-MFC was used to treat TW, wherein, 53.78% of Cr^{6+} was removed by electrochemical cathode reduction and 21.50% by plant uptake (Liu *et al.*, 2022).

3.2.9 Textile and dyeing wastewater

The TDW arises from a series of processes involving sizing, de-sizing, sourcing, bleaching, mercerizing, dyeing, printing, and finishing (Yaseen & Scholz, 2019). Of all the constituents (metals, salts) of TDW, dyes (natural and synthetic) are of interest pertaining to their need for degradation due to their recalcitrant and toxic nature. Among all other dyes variants (basic, acid, vat, disperse, and direct), azo dyes are the most popular class of dyes and comprise 50% of all dyes used (Saba *et al.*, 2021). Azo dyes comprise one or more –N=N– groups, termed as chromophore, responsible for colour production by absorption of light (Gupta *et al.*, 2020). The degradation of azo dyes is achieved through reduction of –N=N– groups to aromatic amines and subsequent oxidation into completely demineralized products. The operation of MFCs facilitates this sequence of azo dye degradation mechanism. In the anode chamber, azo dye molecules undergo anaerobic reduction to produce naphthalene, benzene, and quinone through electrons generated from the oxidation of co substrates (carbon sources). Co-substrate oxidation also provides supplemental energy to the dye-degrading micro-organisms. The electrons and protons produced as oxidative products are transferred to the cathode chamber through external circuit and proton-exchange membrane, respectively. At the cathode, the electrons are accepted by the electron acceptor, azo dye itself, thereby reducing to aromatic amines and sulphanilic acids (decolourized metabolites). Also, the phenomenon of resonance (delocalization of electrons) enhances the decolourization of the dye as the electron withdrawal from the double bond, makes the azo bond electrophilic (Gupta *et al.*, 2020; Saba *et al.*, 2021). Furthermore, the decolourized metabolites can act as electro shuttle mediators and enhance reductive decolourization of dyes (Chen *et al.*, 2016). Hence, TDW as a substrate in MFC is electrochemically favourable which involves electron transfer for dye decolourization.

A real TDW was studied for its degradation and electricity production in a granular activated carbon MFC (GAC-MFC) (double chambered) (Kalathil *et al.*, 2011). Granular activated carbon with graphite rods as current collectors were used as biocathodes. A glass wool separated anode and cathode chambers and TDW was fed into both anode and cathode chambers. The TDW in GAC-MFC generated 1.7 W/m^3 with an OCV of 0.45 V. Colour and COD removal of 73% and 71% was achieved at the anode and 77% and 76% at the cathode. However, a high internal resistance of 800 Ω and low PD was observed. Further, the GAC-MFC was scaled up (working volume = 2.5 L) and modified into a granular activated carbon – SCMFC (GAC-SCMFC). In GAC-SCMFC, the anode chamber (lower portion) was fed with TDW continuously and the treated TDW is discharged into cathode chamber (top portion) for further treatment. A PD of 8 W/m^3 was achieved with 71% COD removal and 75% decolourization. The study also suggested that subsequent anaerobic and aerobic treatment is required for effective decolourization and toxicity removal of TDW.

Rathour *et al.* studied the treatment of dye wastewater (DW) using a CW-MFC (Figure 3.3) (Rathour *et al.*, 2019). The CW-MFC was filled from the bottom with gravel, pond mud and was planted with

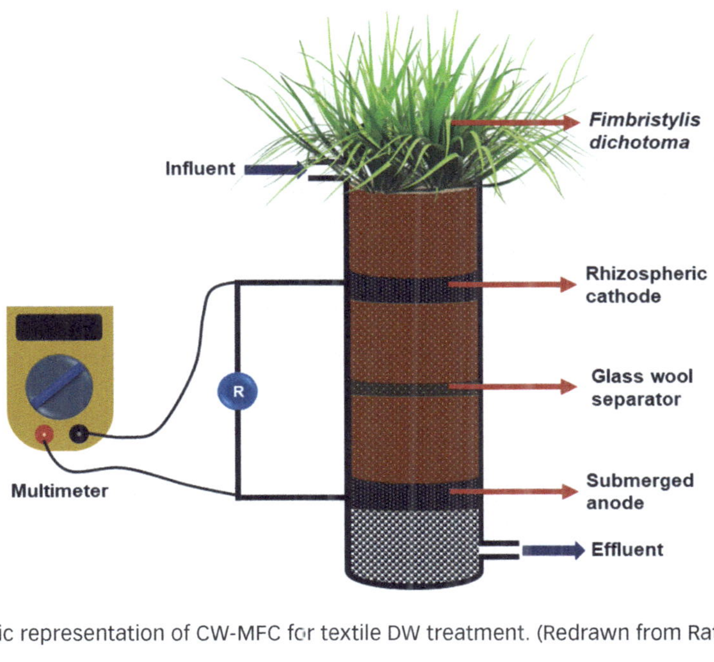

Figure 3.3 Schematic representation of CW-MFC for textile DW treatment. (Redrawn from Rathour *et al.*, 2019.)

Fimbristylis dichotoma. The anode and cathode were stainless-steel grade 316 plates; cathode was placed near the root zone and the anode was placed at the bottom, and were separated by glass wool. In the CW, cathode zone is aerobic at the air–water interface and the anode zone is anaerobic thus exhibiting a stratified redox gradient conducive for operation as MFC. The DW was fed from the top and degradation efficiencies were determined for the effluent (HRT = 9 d). A COD removal efficiency of 70 ± 2% and colour removal efficiency of 82.2 ± 1.7% was achieved. Also reported that, the genus *Desulfobulbus* dominated the anode microbial community which degraded the azo dye and mediated an efficient and direct electron transfer to the anode.

3.3 CHALLENGES AND FUTURE PERSPECTIVES OF METs FOR INDUSTRIAL WASTEWATER TREATMENT

Real wastewater always contains varied type of organic materials as well as the complex nature of the same imposes a significant challenge on the design and operation of METs. The summary of various real industrial wastewaters as substrates in METs is presented in Table 3.3. Despite of the fact that COD removal linearly increased with the increase in the strength of wastewater, ionic strength has turned out to be more crucial for the generation of electricity than wastewater strength (Feng *et al.*, 2008). Hence, diluting wastewaters with solutions like PBS, addition of sodium chloride, and so on, were done to simultaneously achieve efficient substrate degradation and power production from METs. Addition of PBS is not cost effective and also PO_4^{3-} addition would not be appropriate for wastewater discharged at an industrial site (Huang & Logan, 2008). Another major challenge limiting the MET performance is the non-exoelectrogens such as methanogens invasion and availability of other electron acceptors in real industrial wastewaters. Therefore, inoculum pre-treatment turns out to be a mandatory prerequisite in the removal of methanogens from seed culture. In addition to pre-treatment, adding a suppressor to the substrate during long-term operation is necessary to prevent recurrence (Elakkiya & Niju, 2021). Industrial wastewaters have a high organic content; hence, a longer HRT is always necessary to accomplish adequate substrate degradation. Higher HRT results in membrane fouling from microbial growth, extracellular polymers, and inorganic salt deposition, all of which reduce MFC power output. Higher HRT will increase oxygen diffusion into the cell in single-chambered air cathode METs, which will interfere with exoelectrogenic activity in the anode chamber (Çetinkaya *et al.*, 2015). Moreover, high internal resistance, low power output, expensive electrode materials, fabrication of METs hinders the scaling up and implementation in field scale (Kalathil *et al.*, 2012). Also, electrode and membrane fouling while using high strength wastewater (such as DW) and salt deposits on the connectors during long-term operation could increase solution resistance and local electrical resistance, respectively (Marassi *et al.*, 2019). pH fluctuation during the initial phases (adaptation or acclimatization) is inevitable, and could be avoided by using buffer or alkaline solutions (Srikanth *et al.*, 2016). Industrial wastewater with diverse and complex composition as substrate in METs influence substrate degradability and anodic microbial ecosystem (Velasquez-Orta *et al.*, 2011). Also, the oxygen reduction kinetics at the cathode is lethargic and requires expensive platinum catalyst. This could be avoided by using photosynthetic microorganisms as biocatalyst (Bejjanki *et al.*, 2021).

3.4 CONCLUSION

Laboratory scale studies have demonstrated the suitability of industrial wastewater as substrate in METs, irrespective of their strength and complexity. However, more research is required pertaining to enhancement of power production upon scaling up of the METs. Also, the development of biochemical mechanistic pathways of different industrial wastewaters is essential for effective treatment. Moreover, the development of low-cost electrode materials and well devised METs configuration to increase power production is inevitable to reduce capital, operation, and maintenance cost.

Table 3.3 Summary of industrial wastewaters as substrate in various METs.

Wastewater	Target Pollutants	METs Used/ Applicable	Electrode Materials	Substrate as Anolyte/ Catholyte	Removal Efficiency	PD	Remarks	References
BW	COD (carbohydrates)	MFC	Carbonaceous electrodes	Anolyte	87.6%	528 mW/m^2	Power production is congruent with wastewater conductivity	Feng *et al.* (2008)
	(COD) proteins	MEC	Carbonaceous anode, NiFe LDH/Ni cathode	Anolyte	(i) COD – 94% (ii) Proteins – 100%	NR	2.01 m^3-H_2/m^3/d (H_2 generation rate)	Lu *et al.* (2017)
DW	Casein, proteins, lipids, carbohydrates, inorganic ions	MFC	Anode – Graphite Cathode – carbon cloth	Anolyte	(i) TBOD – 90% (ii) TCOD – 62% (iii) TKN – 72% (iv) Sodium – 71%	57.87–67.13 mW/m^2	(i) Acclimatization with mixed substrate (fermentable and non-fermentable) enhanced MFC performance (ii) Ohmic losses were predominant (iii) Activation loss overcome by suitable acclimatization and start-up (iv) Concentration losses was ascribed to the internal recirculation (v) Lipid hydrolysis impede methanogen growth by releasing long-chain fatty acids	Marassi *et al.* (2019)
		MFC	Anode – Carbon cloth with carbon coating Cathode – Carbon cloth with Pt/C coatings	Anolyte	(i) COD – 94.40% (ii) CE – 46.59%	PD – 34.82 mW/m^2		Choudhury *et al.* (2021)

(Continued)

Table 3.3 Summary of industrial wastewaters as substrate in various METs (*Continued*).

Wastewater	Target Pollutants	METs Used/ Applicable	Electrode Materials	Substrate as Anolyte/ Catholyte	Removal Efficiency	PD	Remarks	References
		MDC	Anode and cathode – plain graphite plate		(i) COD – 80.2 ± 0.5% (ii) Desalination – 65.8 ± 0.5%	NR	–	Bejjanki *et al.* (2021)
DIW	COD, TDS, sulphate, colour	MFC	Carbonaceous electrodes	Anolyte	(i) COD – 75% (ii) TDS – 23.96% (iii) Sulphate – 60% (iv) Colour – 31%	124 mW/m³	High temperature and low pH of DIW needs special control during treatment	Mohanakrishna *et al.* (2010), Ha *et al.* (2012)
	COD	MEC	Carbonaceous anode and Pt mesh cathode	Anolyte	COD – 72.5%	NR	0.52 ± 0.03 mmol/l d (H_2 generation rate)	Samsudeen *et al.* (2020)
EW	Cr^{6+}	MFC	Anode – carbon felt Cathode – graphite paper	Catholyte	99.5%	1600 mW/m²	High oxidation potential of Cr^{6+} makes it a good catholyte	Li *et al.* (2008)
	Cr^{6+}	MFC	Carbon electrode	Catholyte	53.1%	150.5 mW/m²	50% diluted chromium waste water was used	Kim *et al.* (2017)
	Ni^{3+} and Pb^{4+}	MDC	Carbon graphite electrodes	Middle chamber solution	Ni – 68.81% Pb – 70.04%	NR	Optimum condition- Do- 4.6 mg/L (Ni) 4.52 mg/L (Pb), 26°C, 120 min HRT	Mirzaienia *et al.* (2017)
	Ni^{2+}, Zn^{2+}, Ca^{2+}	MEDCC	Anode – graphite brush Cathode – stainless-steel mesh with activated carbon and polytetrafluoroethylene	Middle chamber solution	Ni – 94% Zn – 82% Ca – 91%	NR	Treatment of nanofiltration concentrate from electroplating industry	Lan *et al.* (2019)
Paper and pulp wastewater	COD (carbohydrates)	MFC	Carbonaceous electrodes	Anolyte	(i) 52–85%	53–672 mw/m²	Greater HRT required as COD values are very high	Elakkiya and Niju (2021); Shankar *et al.* (2016)
	COD (carbohydrates)	MEC	Carbonaceous anode, Cu/steel coated with Ni cathode	Anolyte	–	NR	0.16 m^3-H_2/m^3/d (H_2 generation rate)	Chaurasia *et al.* (2021)

(*Continued*)

Table 3.3 Summary of industrial wastewaters as substrate in various METs (*Continued*).

Wastewater	Target Pollutants	METs Used/ Applicable	Electrode Materials	Substrate as Anolyte/ Catholyte	Removal Efficiency	PD	Remarks	References
PRW	(i) Polycyclic aromatic hydrocarbons (phenols, benzene) (ii) cyclic aromatic hydrocarbons (toluene ethylbenzene, and xylene)	MFC	Anode – Carbon cloth with carbon coating Cathode – Carbon cloth with Pt/C coatings	Anolyte	(i) COD – 84 ± 1% (ii) Phenol – 80 ± 1.8% (iii) Oil and grease – 95 ± 0.6% (iv) Sulphide – 79.5 ± 1.2%	219 ± 4 mW/m^2	(i) Sulphide acts as a redox shuttle mediator (ii) Sulphide oxidation to elemental sulphur is the principal phenomenon behind power generation.	Srikanth *et al.* (2016)
	(iii) diesel range organics	MFC	Anode – carbon fibre brush Cathode – Carbon cloth with Pt/C coatings	Anolyte	(i) Substrate removal efficiency – 48% (ii) DROs efficiency of 89% (iii) n-Decane and n-Octacosane – 100%	132 mW/m^2	–	Mohanakrishna *et al.* (2018b)
PW	Organic and inorganic contents, nitrates, phosphates	PMFC + PBR (pretreatment)	Graphite rods	Anolyte	COD – 97.24% Nitrate – 84% Phosphate – 81%	838.6 mW/m^2	Maximum values obtained when anolyte is 75% pharmaceutical waste and 25% sewage water	Nayak and Ghosh (2019)
	COD and TDS	MFC (truncated paraboloid shaped)	Paraboloid-shaped graphite	Anolyte	COD – 80.55% TDS – 35.23%	2.01 W/m^3	Increased surface area to unit volume was favourable	Rashid *et al.* (2021)
TW	TKN, NH_4-N and sCOD	MFC (single chambered)	Air cathode	Anolyte	TKN – 50% NH_4-N – 35% sCOD – 88%	7 mW/m^2	Effective method to remove nitrogen pollution	Sawasdee and Pisutpaisal (2016)
	COD, BOD_5, sulphate and chromium	MFC + electrochemical pretreatment	Anode – carbon felt Electrochemical reactor – stainless steel	Anolyte	COD – 90% BOD_5 – 84% Sulphate – 96% Chromium – 100%	NR	TW was pretreated using an electrochemical reactor (with stainless steel)	Elabed *et al.* (2019)
	COD, BOD and TDS	MFC	Graphite sheets	Anolyte	COD – 88% BOD – 74% TDS – 94%	7371 mW/cm^3	–	Chauhan *et al.* (2022)

(*Continued*)

Table 3.3 Summary of industrial wastewaters as substrate in various METs (*Continued*).

Wastewater	Target Pollutants	METs Used/ Applicable	Electrode Materials	Substrate as Anolyte/ Catholyte	Removal Efficiency	PD	Remarks	References
	COD	MFC	Anode – graphite felt Cathode – platinum-coated carbon cloth	Anolyte	76%	223 mW/m^2	Acclimatized cultures gave better efficiency	Miran and Mumtaz (2022)
	Cr^{6+}	CW-MFC	Carbon felt sandwiched with stainless-steel mesh	Catholyte	98.8%	37.8 mW/m^2	Plant uptake removed 21.50% of Cr^{6+}	Liu *et al.* (2022)
TDW	Azo dyes	MFC	Anode – carbon brush Cathode – graphite rods	Anolyte and catholyte	(i) Colour – 73% and COD – 71% (anode) (ii) Colour – 77% and COD – 76% (cathode)	1.7 W/m^3	Subsequent anaerobic and aerobic treatment for effective decolourization and toxicity removal of TDW	Kalathil *et al.* (2011)
		CW-MFC	Anode and cathode – Stainless-steel grade 316 plates	Anolyte and catholyte	(i) COD – 70 ± 2% (ii) Colour – 82.2 ± 1.7%	(i) PD – 198.8 mW/m^2	Genus *Desulfobulbus* dominated the anode microbial community; degraded the azo dye and mediated an efficient and direct electron transfer to the anode	Rathour *et al.* (2019)

Nevertheless, METs are potential candidates for alternate technologies sought by the industries for efficient treatment, resource recovery and energy production. The overall sustainability and viability of the developed systems can be assessed with the aid of tools like techno-economic analysis, life cycle assessment, and exergy economic analysis. Nevertheless, METs are potential candidates for alternate technologies sought by the industries for efficient treatment, resource recovery, and energy production.

REFERENCES

Abu-Reesh I. M., Kunju A. and Sevda S. (2022). Performance of microbial fuel cells in treating petroleum refinery wastewater. *Journal of Water Process Engineering*, **49**, 103029, https://doi.org/10.1016/j.jwpe.2022.103029

Bagchi S. and Behera M. (2020). Pharmaceutical wastewater treatment in microbial fuel cell. In: Integrated Microbial Fuel Cells for Wastewater Treatment, R. Abbassi, A. K. Yadav, F. Khan and V. Garaniya (eds.), Butterworth-Heinemann, United Kingdom, pp. 135–155, https://doi.org/10.1016/B978-0-12-817493-7.00006-0

Bejjanki D., Muthukumar K., Radhakrishnan T. K., Alagarsamy A., Pugazhendhi A. and Naina Mohamed S. (2021). Simultaneous bioelectricity generation and water desalination using *Oscillatoria* sp. as biocatalyst in photosynthetic microbial desalination cell. *Science of the Total Environment*, **754**, 142215, https://doi.org/10.1016/j.scitotenv.2020.142215

Çetinkaya A. Y., Köroğlu E. O., Demir N. M., Baysoy D. Y., Özkaya B. and Çakmakçi M. (2015). Electricity production by a microbial fuel cell fueled by brewery wastewater and the factors in its membrane deterioration. *Cuihua Xuebao/Chinese Journal of Catalysis*, **36**(7), 1068–1076, https://doi.org/10.1016/S1872-2067(15)60833-6

Chauhan S., Sharma V., Varjani S., Sindhu R. and Chaturvedi Bhargava P. (2022). Mitigation of tannery effluent with simultaneous generation of bioenergy using dual chambered microbial fuel cell. *Bioresource Technology*, **351**, 127084, https://doi.org/10.1016/j.biortech.2022.127084

Chaurasia A. K., Shankar R. and Mondal P. (2021). Effects of nickle, nickle-cobalt and nickle-cobalt-phosphorus nanocatalysts for enhancing biohydrogen production in microbial electrolysis cells using paper industry wastewater. *Journal of Environmental Management*, **298**, 113542, https://doi.org/10.1016/j.jenvman.2021.113542

Chen B. Y., Ma C. M., Han K., Yueh P. L., Qin L. J. and Hsueh C. C. (2016). Influence of textile dye and decolorized metabolites on microbial fuel cell-assisted bioremediation. *Bioresource Technology*, **200**, 1033–1038, https://doi.org/10.1016/j.biortech.2015.10.011

Choudhury P., Ray R. N., Bandyopadhyay T. K., Tiwari O. N. and Bhunia B. (2021). Kinetics and performance evaluation of microbial fuel cell supplied with dairy wastewater with simultaneous power generation. *International Journal of Hydrogen Energy*, **46**(31), 16815–16822, https://doi.org/10.1016/j.ijhydene.2020.08.024

Dannys E., Green T., Wettlaufer A., Madhurnathakam C. M. R. and Elkamel A. (2016). Wastewater treatment with microbial fuel cells: a design and feasibility study for scale-up in microbreweries. *Journal of Bioprocessing & Biotechniques*, **6**(1), 1–6, https://doi.org/10.4172/2155-9821.1000267

Dong Y., Qu Y., He W., Du Y., Liu J., Han X. and Feng Y. (2015). A 90-liter stackable baffled microbial fuel cell for brewery wastewater treatment based on energy self-sufficient mode. *Bioresource Technology*, **195**, 66–72, https://doi.org/10.1016/j.biortech.2015.06.026

Dong Y., Liu J., Sui M., Qu Y., Ambuchi J. J., Wang H. and Feng Y. (2017). A combined microbial desalination cell and electrodialysis system for copper-containing wastewater treatment and high-salinity-water desalination. *Journal of Hazardous Materials*, **321**, 307–315, https://doi.org/10.1016/j.jhazmat.2016.08.034

Elabed A., El Abed S., Ibnsouda S. and Erable B. (2019). Sustainable approach for tannery wastewater treatment: bioelectricity generation in bioelectrochemical systems. *Arabian Journal for Science and Engineering*, **44**(12), 10057–10066, https://doi.org/10.1007/s13369-019-04050-y

Elakkiya E. and Matheswaran M. (2013). Comparison of anodic metabolisms in bioelectricity production during treatment of dairy wastewater in microbial fuel cell. *Bioresource Technology*, **136**, 407–412, https://doi.org/10.1016/j.biortech.2013.02.113

Elakkiya E. and Niju S. (2021). Bioelectrochemical treatment of real-field bagasse-based paper mill wastewater in dual-chambered microbial fuel cell. *3 Biotech*, **11**(2), 1–10, https://doi.org/10.1007/s13205-020-02606-6

Feng Y., Wang X., Logan B. E. and Lee H. (2008). Brewery wastewater treatment using air-cathode microbial fuel cells. *Applied Microbiology and Biotechnology*, **78**(5), 873–880, https://doi.org/10.1007/s00253-008-1360-2

Ghorab R. E. A., Pugazhendi A., Jamal M. T., Jeyakumar R. B., Godon J. J. and Mathew D. K. (2022). Tannery wastewater treatment coupled with bioenergy production in upflow microbial fuel cell under saline condition. *Environmental Research*, **212**, 113304, https://doi.org/10.1016/j.envres.2022.113304

Ghosh Ray S. and Ghangrekar M. M. (2015). Enhancing organic matter removal, biopolymer recovery and electricity generation from distillery wastewater by combining fungal fermentation and microbial fuel cell. *Bioresource Technology*, **176**, 8–14, https://doi.org/10.1016/j.biortech.2014.10.158

Gupta S., Mittal Y., Tamta P., Srivastava P. and Yadav A. K. (2020). Textile wastewater treatment using microbial fuel cell and coupled technology: a green approach for detoxification and bioelectricity generation. In: Integrated Microbial Fuel Cells for Wastewater Treatment. R. Abbassi, A. K. Yadav, F. Khan and V. Garaniya (eds.), Butterworth-Heinemann, United Kingdom, pp. 73–92, https://doi.org/10.1016/B978-0-12-817493-7.00004-7

Ha P. T., Lee T. K., Rittmann B. E., Park J. and Chang I. S. (2012). Treatment of alcohol distillery wastewater using a bacteroidetes-dominant thermophilic microbial fuel cell. *Environmental Science and Technology*, **46**(5), 3022–3030, https://doi.org/10.1021/es203861v

Huang L. and Logan B. E. (2008). Electricity generation and treatment of paper recycling wastewater using a microbial fuel cell. *Applied Microbiology and Biotechnology*, **80**(2), 349–355, https://doi.org/10.1007/s00253-008-1546-7

Ismail Z. Z. and Habeeb A. A. (2017). Experimental and modeling study of simultaneous power generation and pharmaceutical wastewater treatment in microbial fuel cell based on mobilized biofilm bearers. *Renewable Energy*, **101**, 1256–1265, https://doi.org/10.1016/j.renene.2016.10.008

Kadier A., Simayi Y., Abdeshahian P., Azman N. F., Chandrasekhar K. and Kalil M. S. (2016). A comprehensive review of microbial electrolysis cells (MEC) reactor designs and configurations for sustainable hydrogen gas production. *Alexandria Engineering Journal*, **55**(1), 427–443, https://doi.org/10.1016/j.aej.2015.10.008

Kalathil S., Lee J. and Cho M. H. (2011). Granular activated carbon based microbial fuel cell for simultaneous decolorization of real dye wastewater and electricity generation. *New Biotechnology*, **29**(1), 32–37, https://doi.org/10.1016/j.nbt.2011.04.014

Kalathil S., Lee J. and Cho M. H. (2012). Efficient decolorization of real dye wastewater and bioelectricity generation using a novel single chamber biocathode-microbial fuel cell. *Bioresource Technology*, **119**, 22–27, https://doi.org/10.1016/j.biortech.2012.05.059

Kim H., Kim B., Kim J., Lee T. and Yu J. (2014). Electricity generation and microbial community in microbial fuel cell using low-pH distillery wastewater at different external resistances. *Journal of Biotechnology*, **186**, 175–180, https://doi.org/10.1016/j.jbiotec.2014.06.012

Kim C., Lee C. R., Song Y. E., Heo J., Choi S. M., Lim D. H., Cho J., Park C., Jang M. and Kim J. R. (2017). Hexavalent chromium as a cathodic electron acceptor in a bipolar membrane microbial fuel cell with the simultaneous treatment of electroplating wastewater. *Chemical Engineering Journal*, **328**, 703–707, https://doi.org/10.1016/j.cej.2017.07.077

Lan J., Ren Y., Lu Y., Liu G., Luo H. and Zhang R. (2019). Combined microbial desalination and chemical-production cell with Fenton process for treatment of electroplating wastewater nanofiltration concentrate. *Chemical Engineering Journal*, **359**, 1139–1149, https://doi.org/10.1016/j.cej.2018.11.067

Li Z., Zhang X. and Lei L. (2008). Electricity production during the treatment of real electroplating wastewater containing Cr^{6+} using microbial fuel cell. *Process Biochemistry*, **43**(12), 1352–1358, https://doi.org/10.1016/j.procbio.2008.08.005

Liu S., Lu F., Qiu D. and Feng X. (2022). Wetland plants selection and electrode optimization for constructed wetland-microbial fuel cell treatment of Cr(VI)-containing wastewater. *Journal of Water Process Engineering*, **49**, 103040, https://doi.org/10.1016/j.jwpe.2022.103040

Lu M., Chen S., Babanova S., Phadke S., Salvacion M., Mirhosseini A., Chan S., Carpenter K., Cortese R. and Bretschger O. (2017). Long-term performance of a 20-L continuous flow microbial fuel cell for treatment of brewery wastewater. *Journal of Power Sources*, **356**, 274–287, https://doi.org/10.1016/j.jpowsour.2017.03.132

Marassi R. J., Hermanny R. S., Silva G. C., Silva F. T. and Paiva T. C. B. (2019). Electricity production and treatment of high-strength dairy wastewater in a microbial fuel cell using acclimated electrogenic consortium. *International Journal of Environmental Science and Technology*, **16**(11), 7339–7348, https://doi.org/10.1007/s13762-019-02391-7

Marassi R. J., Queiroz L. G., Silva D. C. V. R., dos Santos F. S., Silva G. C. and de Paiva T. C. B. (2020a). Long-term performance and acute toxicity assessment of scaled-up air–cathode microbial fuel cell fed by dairy wastewater. *Bioprocess and Biosystems Engineering*, **43**(9), 1561–1571, https://doi.org/10.1007/s00449-020-02348-y

Marassi R. J., Queiroz L. G., Silva D. C. V. R., da Silva F. T., Silva G. C. and de Paiva T. C. B. (2020b). Performance and toxicity assessment of an up-flow tubular microbial fuel cell during long-term operation with high-strength dairy wastewater. *Journal of Cleaner Production*, **259**, 120882, https://doi.org/10.1016/j.jclepro.2020.120882
Miran F. and Mumtaz M. W. (2022). Complex waste acclimatized mixed culture enhances bioelectricity generation and tannery wastewater treatment in microbial fuel cells. *Biomass Conversion and Biorefinery*, **13**, 12467–12473, https://doi.org/10.1007/s13399-021-02095-6
Mirzaienia F., Asadipour A., Jafari A. J. and Malakootian M. (2017). Removal efficiency of nickel and lead from industrial wastewater using microbial desalination cell. *Applied Water Science*, **7**(7), 3617–3624, https://doi.org/10.1007/s13201-016-0505-1
Mohanakrishna G., Venkata Mohan S. and Sarma P. N. (2010). Bio-electrochemical treatment of distillery wastewater in microbial fuel cell facilitating decolorization and desalination along with power generation. *Journal of Hazardous Materials*, **177**(1–3), 487–494, https://doi.org/10.1016/j.jhazmat.2009.12.059
Mohanakrishna G., Abu-Reesh I. M. and Al-Raoush R. I. (2018a). Biological anodic oxidation and cathodic reduction reactions for improved bioelectrochemical treatment of petroleum refinery wastewater. *Journal of Cleaner Production*, **190**, 44–52, https://doi.org/10.1016/j.jclepro.2018.04.141
Mohanakrishna G., Abu-Reesh I. M., Kondaveeti S., Al-Raoush R. I. and He Z. (2018b). Enhanced treatment of petroleum refinery wastewater by short-term applied voltage in single chamber microbial fuel cell. *Bioresource Technology*, **253**, 16–21, https://doi.org/10.1016/j.biortech.2018.01.005
Narayan M. D., Kumar Rabha A. D., Kumar Srivastava R., Narayan M., Solanki P., Kumar Rabha A. and Kumar Srivastava R. (2018). Treatment of pulp and paper industry effluent and electricity generation by constructed wetland coupled with microbial fuel cell (CW-MFC). *Journal of Pharmacognosy and Phytochemistry*, **7**(6), 493–498.
Nayak J. K. and Ghosh U. K. (2019). Post treatment of microalgae treated pharmaceutical wastewater in photosynthetic microbial fuel cell (PMFC) and biodiesel production. *Biomass and Bioenergy*, **131**, 105415, https://doi.org/10.1016/j.biombioe.2019.105415
Pandey P., Shinde V. N., Deopurkar R. L., Kale S. P., Patil S. A. and Pant D. (2016). Recent advances in the use of different substrates in microbial fuel cells toward wastewater treatment and simultaneous energy recovery. *Journal of applied energy*, **168**, 706–723, https://doi.org/10.1016/j.apenergy.2016.01.056
Rajoria S., Vashishtha M. and Sangal V. K. (2022). Treatment of electroplating industry wastewater: a review on the various techniques. *Environmental Science and Pollution Research*, **29**(48), 72196–72246, https://doi.org/10.1007/s11356-022-18643-y
Rashid T., Sher F., Hazafa A., Hashmi R. Q., Zafar A., Rasheed T. and Hussain S. (2021). Design and feasibility study of novel paraboloid graphite based microbial fuel cell for bioelectrogenesis and pharmaceutical wastewater treatment. *Journal of Environmental Chemical Engineering*, **9**(1), 104502, https://doi.org/10.1016/j.jece.2020.104502
Rathour R., Patel D., Shaikh S. and Desai C. (2019) Eco-electrogenic treatment of dyestuff wastewater using constructed wetland-microbial fuel cell system with an evaluation of electrode-enriched microbial community structures. *Bioresource Technology*, **285**, 121349, https://doi.org/10.1016/j.biortech.2019.121349
Saba B., Kjellerup B. V. and Christy A. D. (2021). Eco-friendly bio-electro-degradation of textile dyes wastewater. *Bioresource Technology Reports*, **15**, https://doi.org/10.1016/j.biteb.2021.100734
Samsudeen N., Spurgeon J., Matheswaran M. and Satyavolu J. (2020). Simultaneous biohydrogen production with distillery wastewater treatment using modified microbial electrolysis cell. *International Journal of Hydrogen Energy*, **45**(36), 18266–18274, https://doi.org/10.1016/j.ijhydene.2019.06.134
Sangeetha T., Guo Z., Liu W., Cui M., Yang C., Wang L. and Wang A. (2016). Cathode material as an influencing factor on beer wastewater treatment and methane production in a novel integrated upflow microbial electrolysis cell (Upflow-MEC). *International Journal of Hydrogen Energy*, **41**(4), 2189–2196, https://doi.org/10.1016/j.ijhydene.2015.11.111
Sangeetha T., Rajneesh C. P. and Yan W. M. (2020). Integration of microbial electrolysis cells with anaerobic digestion to treat beer industry wastewater. In: Integrated Microbial Fuel Cells for Wastewater Treatment, R. Abbassi, A. K. Yadav, F. Khan and V. Garaniya (eds.), Butterworth-Heinemann, United Kingdom, pp. 313–346, https://doi.org/10.1016/B978-0-12-817493-7 00015-1
Sanjay S. and Udayashankara T. H. (2019). Dairy wastewater treatment with bio-electricity generation using dual chambered membrane-less microbial fuel cell. *Materials Today: Proceedings*, **35**, 308–311, https://doi.org/10.1016/j.matpr.2020.01.533
Sawasdee V. and Pisutpaisal N. (2016). Simultaneous pollution treatment and electricity generation of tannery wastewater in air-cathode single chamber MFC. *International Journal of Hydrogen Energy*, **41**(35), 15632–15637, https://doi.org/10.1016/j.ijhydene.2016.04.179

Selvasembian R., Mal J., Rani R., Sinha R., Agrahari R., Joshua I., Santhiagu A. and Pradhan N. (2022). Recent progress in microbial fuel cells for industrial effluent treatment and energy generation: fundamentals to scale-up application and challenges. *Bioresource Technology*, **346**, https://doi.org/10.1016/j.biortech.2021.126462

Shah H. and Ruparelia J. (2022). Comparative studies for the treatment of industrial effluents employing advanced processes: towards enhancement of environmental performance. *Discover Water*, **2**(1), 4, https://doi.org/10.1007/s43832-022-00012-y

Shankar R., Varma A. K., Mondal P. and Chand S. (2016). Simultaneous treatment and energy production from PIW using electro coagulation & microbial fuel cell. *Journal of Environmental Chemical Engineering*, **4**(4), 4612–4618, https://doi.org/10.1016/j.jece.2016.10.021

Srikanth S., Kumar M., Singh D., Singh M. P. and Das B. P. (2016). Electro-biocatalytic treatment of petroleum refinery wastewater using microbial fuel cell (MFC) in continuous mode operation. *Bioresource Technology*, **221**, 70–77, https://doi.org/10.1016/j.biortech.2016.09.034

Thapa B. S., Pandit S., Patwardhan S. B., Tripathi S., Mathuriya A. S., Gupta P. K., Lal R. B. and Tusher T. R. (2022). Application of microbial fuel cell (MFC) for pharmaceutical wastewater treatment: an overview and future perspectives. *Sustainability*, **14**(14), https://doi.org/10.3390/su14148379

Thulasinathan B., Jayabalan T., Arumugam N., Rasu Kulanthaisamy M., Kim W., Kumar P., Govarthanan M. and Alagarsamy A. (2022). Wastewater substrates in microbial fuel cell systems for carbon-neutral bioelectricity generation: an overview. *Fuel*, **317**, 123369, https://doi.org/10.1016/j.fuel.2022.123369

Velasquez-Orta S. B., Head I. M., Curtis T. P. and Scott K. (2011). Factors affecting current production in microbial fuel cells using different industrial wastewaters. *Bioresource Technology*, **102**(8), 5105–5112, https://doi.org/10.1016/j.biortech.2011.01.059

Wang X., Feng Y. J. and Lee H. (2008). Electricity production from beer brewery wastewater using single chamber microbial fuel cell. *Water Science & Technology*, **57**(7), 1117–1121, https://doi.org/10.2166/wst.2008.064

Wen Q., Kong F., Zheng H., Cao D., Ren Y. and Yin J. (2011). Electricity generation from synthetic penicillin wastewater in an air-cathode single chamber microbial fuel cell. *Chemical Engineering Journal*, **168**(2), 572–576, https://doi.org/10.1016/j.cej.2011.01.025

Xu W., Zou R., Jin B., Zhang G., Su Y. and Zhang Y. (2022). The ins and outs of pharmaceutical wastewater treatment by microbial electrochemical technologies. *Sustainable Horizons*, **1**, 100003, https://doi.org/10.1016/j.horiz.2021.100003

Yaseen D. A. and Scholz M. (2019). Textile dye wastewater characteristics and constituents of synthetic effluents: a critical review. *International Journal of Environmental Science and Technology*, **16**(2), 1193–1226, https://doi.org/10.1007/s13762-018-2130-z

Yu J., Park Y., Kim B. and Lee T. (2015). Power densities and microbial communities of brewery wastewater-fed microbial fuel cells according to the initial substrates. *Bioprocess and Biosystems Engineering*, **38**(1), 85–92, https://doi.org/10.1007/s00449-014-1246-x

Zhao J., Wu Q., Tang Y., Zhou J. and Guo H. (2022). Tannery wastewater treatment: conventional and promising processes, an updated 20-year review. *Journal of Leather Science and Engineering*, **4**(1), 10, https://doi.org/10.1186/s42825-022-00082-7

doi: 10.2166/9781789063813_0061

Chapter 4

Light-assisted microbial electrochemical technologies for bioelectricity generation and product recovery

*Praveen Rajpurohit and Manaswini Behera**

School of Infrastructure, Indian Institute of Technology Bhubaneswar, Bhubaneswar, Odisha, India
**Corresponding author: manaswini@iitbbs.ac.in*

ABSTRACT

With the increase in demand for the improvement of microbial electrochemical technologies (METs) for bioelectricity generation and product recovery, light-assisted METs have developed as an option. The use of light helps in the electrohydrogenesis process at the cathode. Various variants of light-assisted METs employ photosynthetic bacteria/algae, anode and photocathode assembly, and so on. Microbial fuel cells (MFCs) using photosynthetic bacteria, bioelectrodes, and hybrids of photoelectrocatalytic cells (PECs) and MFCs show superior performance compared to individual MFCs. The dye-sensitized solar cell coupling of MFCs helps enhance electrohydrogenesis and H_2 production.

This book chapter deals with all types of light-assisted METs. The effect of the configuration, electrode material, electrolyte, and physical and chemical factors on the performance of light-assisted METs is discussed. The miniaturizing and stacking of reactors in solar-assisted METs is a current approach showing superior performance. The value-added products formed at the cathodic compartment, carbon-based or H_2 gas, are discussed and reported literature compared with the enhanced recovery of existing METs.

Keywords: microbial electrochemical technologies, microbial fuel cells, light-assisted METs, resource recovery, photocathode, electrohydrogenesis.

4.1 INTRODUCTION

Pollutants from wastewater must be removed before they reach natural water bodies such as rivers, lakes, estuaries, and oceans to maintain proper balance of the environment (Rout *et al.*, 2021). Harmful substances contained in wastewater can harm wildlife as well as aquatic life. The most considerable amount of water can be saved and reused by correctly treating wastewater. The main goal of wastewater treatment plants is to safeguard residents and ecosystems from dangerous compounds found in wastewater. Using physical, chemical, and biological techniques, or conventional wastewater treatment, contaminants are removed from wastewater before it is discharged into the environment (Singh *et al.*, 2021). The following are a few issues with traditional wastewater treatment:

- An increase in the requirement for chemicals such as disinfectants and coagulants.
- Extremely high maintenance expenses result from the need for specialized labour, energy use, and equipment replacement.

- Sludge disposal issues because biomass generated during biological treatment needs to be correctly processed and disposed of; more space and energy are needed since treatment plants demand a lot of land and electricity.
- The impact on the environment, as treated wastewater, may still contain pollutants that have the potential to degrade water quality and endanger aquatic life.
- Degradability concerns, as current practices, may not sufficiently eliminate some emerging and refractory contaminants.

A microbial fuel cell (MFC) is a device that uses microorganisms to transform chemical energy into electrical energy (Ghafari *et al.*, 2008). The redox processes of bacteria that oxidize organic matter to produce carbon dioxide, electrons, and protons serve as the foundation for the MFC's operation. The protons are carried across a membrane that divides the two chambers as an external circuit transfers the electrons from the anode (where oxidation occurs) to the cathode (where reduction occurs). The MFC can produce electricity using industrial wastes or wastewater treatment (Selvasembian *et al.*, 2022). In terms of wastewater treatment and by-product recovery, MFCs are promising. Bioelectrochemical reactions are performed using them to oxidize biodegradable organic materials and produce power that can be used in various ways. To address both the problem of wastewater treatment and the problem of energy production simultaneously, MFCs can be of tremendous assistance. In addition to using MFCs to treat wastewater, these devices are also used to produce biohydrogen, remove contaminants, remove heavy metals, and more (Mahmoodi Nasrabadi & Moghimi, 2023). They can oxidize simple carbonates to carbon dioxide (CO_2) while enabling biological reduction to move electrons. Similar to batteries, this electron transport requires fuel for metabolic conversion, found in substrates on the anode side. MFCs offer numerous advantages, including fewer hazardous products, efficacy, cleanliness, and recyclability. In addition, they are environmentally benign and considerably reduce sludge production. The production of sludge is significantly decreased while also producing energy during wastewater treatment using a novel MFC (Doherty *et al.*, 2015).

4.1.1 Light-assisted MFC

To create electricity or hydrogen from organic materials, a light-assisted MFC combines light energy with microbes. The advantages of solar cells and MFCs are combined in this new technique. Bio-photoelectrochemical cells, hybrid systems, and photosynthetic MFCs are a few examples of the available kinds of light-assisted MFCs. They could be used for environmental restoration, wastewater treatment, and renewable energy creation. It permits self-sustaining, continuous hydrogen gas generation that only uses solar/light and sewage as inputs (Maddalwar *et al.*, 2021). In these systems, solar energy makes bioelectricity or hydrogen production possible. The literature indicates that there are numerous varieties of solar-assisted MFCs (SAMFCs) and some examples of light-assisted MFCs are explained subsequently.

Photosynthetic microbial fuel cells (PMFCs) use photosynthetic microorganisms such as algae, cyanobacteria, or plants as the substrate source at the anode chamber. The photosynthetic microorganisms can directly absorb and convert solar energy while simultaneously fixing CO_2 as carbohydrates through photosynthetic metabolism (Maddalwar *et al.*, 2021). Electrogenic microorganisms can then oxidize the carbohydrates to produce bioelectricity.

These devices use photosynthetic bacteria or algae at the anode or the cathode to generate oxygen for microbial electrogenesis or organic substrates (Kannan & Donnellan, 2021), depending on how the photosynthetic bacteria function, they can be divided into oxygenic and anoxygenic types.

Bio-photoelectrochemical cells (BPECs) use a single device's photocathode and a microbially catalysed anode. The photocathode is a semiconductor material that can absorb visible light and generate electron–hole pairs. The holes at the photocathode can either reduce water to produce hydrogen or accept electrons from another source to enhance hydrogen production (Shlosberg *et al.*, 2022). The anode is a MFC that can degrade organic substrates and release electrons and protons.

The anode and photoanode electrons flow through an external circuit to the photocathode, which reduces water to produce hydrogen.

These gadgets harness solar energy and improve electron transport between microorganisms and electrodes by using semiconductor photoelectrodes at the anode or the cathode. Depending on where the photoelectrodes are, they can be divided into bio photocathode and photoanode (Xiao *et al.*, 2012).

SAMFCs use a MFC and a PEC in a single device. The MFC anode chamber serves as the PEC photoanode, while the MFC cathode chamber serves as the PEC photocathode (Wang *et al.*, 2014). The MFC anode contains photosynthetic microorganisms that can convert solar energy to chemical energy through photosynthesis. The MFC cathode contains a semiconductor material that can absorb visible light and generate electron–hole pairs (Chae *et al.*, 2009). The electrons from the MFC anode and the PEC photoanode flow through an external circuit to the PEC photocathode, where they are used to reduce water to produce hydrogen.

Hybrid systems: These devices combine MFC with solar cells or PECs to supply external bias for hydrogen synthesis or enhance bioelectricity. DSSC-powered MEC, MFC-PV hybrid, and MFC-PEC hybrid types can be distinguished based on the solar or PEC types (Singh *et al.*, 2021).

Both normal MFCs and SAMFCs use electrogenic bacteria to transform organic material into electricity or hydrogen. Regarding their layout, functionality, and advantages, they do differ little; however, there are several vital variations, including:

- In contrast to conventional MFCs, which depend on the potential difference between the anode and the cathode, SAMFCs employ sun's energy to improve electron transfer or provide an external bias for electrochemical reactions (Tharali *et al.*, 2016).
- While normal MFCs employ organic material from wastewater or soil as the fuel source, SAMFCs can use photosynthetic bacteria or algae to provide organic substrates or oxygen for microbial electrogenesis.
- Due to the complementary nature of solar energy and microbial activity, SAMFCs may produce hydrogen at a higher rate and power density than conventional MFCs.
- Normal MFCs frequently encounter these difficulties, which have a negative impact on their performance. SAMFCs, however, can decrease these issues.
- General MFCs may need to replace substrates or electron acceptors periodically, whereas SAMFCs can function continuously and sustainably with little maintenance (Corbella *et al.*, 2015).

4.2 TYPES OF LIGHT-ASSISTED MFCS

4.2.1 Bio-photoelectrochemical cell

A device known as a BPEC generates electricity or chemical compounds by utilizing light and biological components. There are two primary types of BPECs: one that generates electrical energy similarly to a dye-sensitized photovoltaic cell and one that uses light to directly initiate a chemical reaction, such as splitting water into hydrogen and oxygen (Shlosberg *et al.*, 2022). It is also known as an artificial photosynthesis cell or a PEC. A photosensitizer, semiconductor, or aqueous metal is often dissolved in an electrolytic solution that contains biological molecules or cells to create a BPEC (Yi *et al.*, 2022). BPECs come in a variety of forms, including:

(1) A self-contained BPEC in succulent plants that produce hydrogen gas from water oxidation using iron and platinum electrodes, the plant's water content, and its cuticle as the electrolyte and container.
(2) A biological photovoltaic system that generates energy by transferring electrons from a cathode to an anode using photosynthetic cyanobacteria or algae as the cathode.
(3) A two-in-one BPEC that uses photo-enzymes to catalyse both oxidation and reduction reactions, such as the reduction of carbon dioxide and the oxidation of water, and transfers electrons between them (Sridhar *et al.*, 2021). BPECs are a promising method of utilizing solar energy and transforming it into proper forms, absorbing carbon dioxide and creating oxygen (Zhao *et al.*, 2022).

4.2.2 Photosynthetic MFC

An example of a biological electrochemical system that uses photosynthetic microorganisms to produce power from light is the PMFC. These microorganisms include algae and cyanobacteria (Jiang *et al.*, 2012). An anode chamber where bacteria oxidize organic materials and a cathode chamber where photosynthetic microorganisms decrease oxygen are the two chambers that make up a PMFC. Oxygen and electrons are produced by the photosynthetic microorganisms using light, and these products are then delivered to the cathode via an external circuit (Vinayak *et al.*, 2021). As soon as the oxygen crosses a membrane from the anode chamber, it interacts with protons to create water. A voltage and a current are produced when the anode and the cathode have different redox potentials (Zhang *et al.*, 2022).

"Photo-microbial fuel cell" and "plant microbial fuel cell" can also describe a PMFC. It is a technology that has the potential to clean wastewater, produce renewable energy, and sequester carbon dioxide. PMFCs have several benefits, as mentioned below.

They have access to a plentiful and cost-free energy source known as sunshine. They can use seawater or wastewater as an electrolyte, lowering water use costs and environmental impact (Fu *et al.*, 2009). The photosynthetic microbes they use can be used to make lucrative by-products like biomass, biofuels, or chemicals. The need for energy and maintenance is decreased because they can function at ambient pressure and temperature. The obstacles facing PMFCs, in contrast to solar or conventional fuel cells, are that they have poor power density and efficiency.

Changes in light quantity and quality throughout the year impact them (Apollon *et al.*, 2021). They demand meticulous pH, temperature, nutrient delivery, and microbial community composition management. Other bacteria in the system that could absorb oxygen or organic matter pose a threat to them (Bazdar *et al.*, 2018).

4.2.3 Dye-sensitized solar cell coupled microbial fuel cell (DSSC-MFC)

A DSSC-MFC is a type of SAMFC that uses a dye-sensitized solar cell (DSSC) to provide additional voltage or current to the MFC circuit. A DSSC is a type of solar cell that uses a dye as a light absorber and a semiconductor as an electron transporter (Ajayi *et al.*, 2009). A DSSC-MFC can produce bioelectricity or hydrogen from wastewater and light.

The working principle of a DSSC-MFC is as follows.

The anode chamber of the MFC contains electrogenic bacteria that oxidize organic matter in wastewater and release electrons and protons. The electrons flow from the anode to the DSSC through an external circuit, where they are collected by the counter electrode of the DSSC (Ajayi *et al.*, 2009). The counter electrode of the DSSC is coated with a catalyst, such as platinum, that reduces iodide ions to iodine molecules in an electrolyte solution. The iodine molecules diffuse to the photoelectrode of the DSSC, where they are oxidized back to iodide ions by the electrons generated by the dye molecules under light irradiation (Singh *et al.*, 2021). The dye molecules are attached to a semiconductor layer, such as TiO_2 or CdS, that transfers the electrons to a transparent conductive oxide layer, such as FTO or ITO, and then to the external circuit. The protons produced at the anode chamber of the MFC pass through a proton-exchange membrane (PEM) to the cathode chamber, where they combine with electrons and oxygen to form water, as shown in Figure 4.1. Alternatively, if the voltage provided by the DSSC is high enough, a hydrogen evolution reaction (HER) can occur in the cathode chamber instead of an oxygen reduction reaction (ORR) (Phanwilai *et al.*, 2020).

The overall reactions in a DSSC-MFC are

Anode: organic matter + bacteria $\rightarrow CO_2 + H^+ + e^-$
Dye: light + dye $\rightarrow$ dye* $\rightarrow$ dye + + e^-
Photoelectrode: dye + + $I^- \rightarrow$ dye + I_2
Counter electrode: $I_2 + 2e^- \rightarrow 2I^-$
Cathode (ORR): $O_2 + 4H^+ + 4e^- \rightarrow 2H_2O$
Or cathode (HER): $2H^+ + 2e^- \rightarrow H_2$

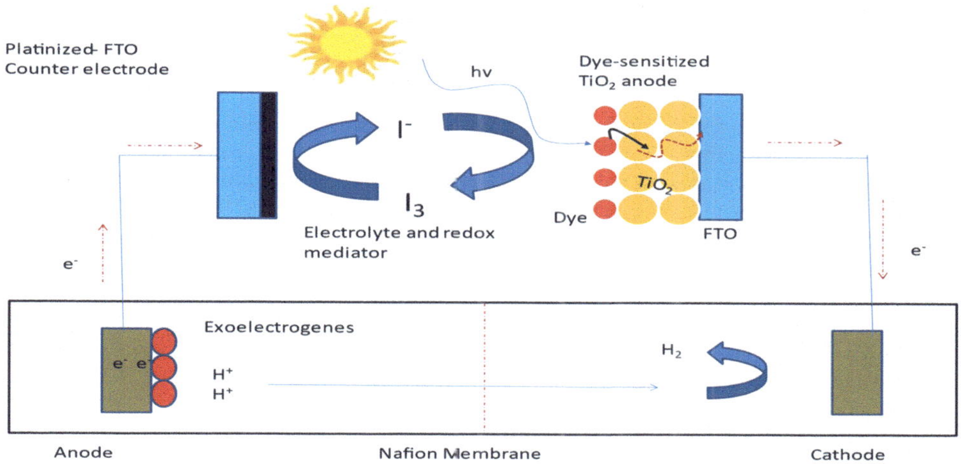

Figure 4.1 A DSSC-MFC, the working principle, and required setup (Chen *et al.*, 2019).

The benefits of a DSSC-MFC include using solar energy as an extra voltage or current source to increase the power output and hydrogen production of a MFC. Using a cheap and effective DSSC as a voltage booster removes the overpotential barrier for hydrogen production in a microbial electrolysis cell (Phanwilai *et al.*, 2020). It can clean wastewater while also recovering energy by using it as a source of organic matter for the production of bioelectricity or hydrogen.

The drawbacks of a DSSC-MFC include the need for a challenging integration of two different devices with various designs and materials. Its performance relies on light availability and intensity, which can change with time and place (Bazdar *et al.*, 2018). It might run into problems with the DSSC's stability, toughness, scalability, and cost-effectiveness.

The first DSSC–MFC system, which utilized a DSSC as an external power source for a single-chamber MFC, was described by Chae *et al.* (2009). With acetate as the substrate, they produced hydrogen at a rate of 0.12 L/L-day and a maximum power density of 0.5 W/m^3. A DSSC–MFC system was created by Zhang *et al.* (2014) that utilized a DSSC as an external power source for a dual-chamber MFC. With glucose as the substrate, they produced hydrogen at a rate of 0.32 L/L-day and a maximum power density of 1.5 W/m^3. A scalable cell fabrication procedure that incorporates inkjet and printing techniques for DSSC-MFC components was proposed by Gong *et al.*, after reviewing recent advancements and limitations in DSSC-MFC research and development. N719 dye, iodide/triiodide electrolyte, and carbon cloth electrodes produced the most significant results, according to Talebi *et al.*, who investigated the impact of various dyes, electrolytes, and electrodes on the performance of DSSC–MFC systems.

4.2.4 PEC coupled MFC

A hybrid SAMFC that couples PECs with MFCs enhances electron transport from bacteria to anode or cathode to oxygen. A photoanode and a photocathode, made of semiconductors like TiO_2 or CdS and covered in catalysts like platinum or cobalt oxide, make up a PEC. From wastewater and light, a PEC coupled MFC can generate bioelectricity or hydrogen (Fischer, 2018). A PEC connected to a MFC operates on the following principles.

Electrogenic bacteria that decompose organic materials in wastewater and release electrons and protons are present in the anode chamber of the MFC. Through an external circuit, electrons move from the anode of the PEC to the photoanode, where the semiconductor layer of the photoanode collects them (Xiao *et al.*, 2012). Light is absorbed and produced by the semiconductor layer of the photoanode, such as TiO_2 or CdS. The transparent conductive oxide layer, such as FTO or ITO, receives the electrons before they are transmitted to the external circuit (Antolini, 2019). Once on the photoanode's surface, the holes oxidize water or other electrolyte species to release oxygen and protons. PEM transports the created protons from the MFCs anode chamber and the PECs photoanode the cathode chamber, where they combine with oxygen and electrons to make water (He *et al.*, 2009). Alternatively, if the PECs voltage output is high enough, a hydrogen evolution reaction (HER) rather than an oxygen reduction process can occur in the cathode chamber (ORR). The semiconductor layer of the photocathode collects the electrons as they pass from the external circuit to the photocathode of the PEC (Liu *et al.*, 2013). Light is absorbed and produced by the semiconductor layer of the photocathode, such as TiO_2 or CdS. The holes are transported to the external circuit from the transparent conductive oxide layer, such as FTO or ITO. To make water or hydrogen, the electrons decrease oxygen or other electrolyte species on the photocathode's surface (Tong *et al.*, 2022).

The overall reactions in a PEC coupled MFC are

Anode: organic matter + bacteria $\rightarrow CO_2 + H^+ + e^-$
Photoanode: $2H_2O + 4H^+ \rightarrow O_2 + 4H^+$
Cathode (ORR): $O_2 + 4H^+ + 4e^- \rightarrow 2H_2O$
Or cathode (HER): $2H^+ + 2e^- \rightarrow H_2$
Photocathode: $2H^+ + 2e^- \rightarrow H_2$
Or photocathode: $O_2 + 4H^+ + 4e^- \rightarrow 2H_2O$

A PEC paired with a MFC has the following benefits: by employing solar energy as an extra voltage or current source, it can increase the power output and hydrogen production of a MFC (Jiang *et al.*, 2012). Using an inexpensive and effective PEC as a voltage booster, the overpotential barrier for hydrogen production in a microbial electrolysis cell can be removed. Wastewater treatment and energy recovery can be accomplished simultaneously by using wastewater as a source of organic matter to produce bioelectricity or hydrogen (Zhang *et al.*, 2019).

A PEC coupled MFC has the following drawbacks: it involves a complex integration of two unique devices with various materials and configurations. Its performance relies on the presence and quality of light, which may change with time and place (Wu *et al.*, 2014). It can encounter difficulties with the PECs and their components' stability, toughness, scalability, and cost-effectiveness. A TiO_2/CdS photoanode and a carbon cloth cathode were employed in the PEC–MFC system that Zhang *et al.* (2019) created. With acetate as the substrate, they reached a maximum power density of 0.68 A/m^2 and a hydrogen generation rate of 0.32 L/L per day. Fischer *et al.* (2011) addressed the elements impacting performance, such as light intensity, dye type, electrolyte type, electrode material, and so on, and reviewed recent advancements and obstacles in PEC–MFC research and development. A PEC–MFC system with a TiO_2 photoanode and a platinum cathode was described by Chae *et al.* (2009). With phenol as the substrate, they reached a maximum power density of 0.5 W/m^3 and a hydrogen generation rate of 0.12 L/L per day.

4.3 COMPARISON BETWEEN DIFFERENT LIGHT-ASSISTED MFCs

SAMFCs are considered the best type of light-assisted MFCs due to their high energy conversion efficiency and low cost. Solar energy is the most abundant and renewable energy source on Earth, and it can be harnessed to generate electricity in SAMFCs. In these systems, solar cells are used to convert sunlight into electrical energy, which is then used to power the MFC. SAMFCs have been shown to have higher power densities and shorter start-up times compared to other types of light-assisted MFCs.

Photo-assisted MFCs, on the other hand, use photosynthetic microorganisms such as algae or cyanobacteria to generate electricity. These microorganisms can convert light energy into chemical energy through photosynthesis, which is then used to produce electricity in the MFC. However,

Table 4.1 Comparison of performance of different light-assisted microbial fuel cells.

Type of Light Assistance	Light Intensity (lx)	Electrode Material	Micro-organism Used	COD/ N/P Ratio	Power Density	COD Removal Efficiency	Hydrogen Production	Reference
Photosynthetic bacteria	N/A	Carbon nanotube	*Rhodopseudomonas* sp.	100:4	0.14 mW/m^2	N/A	N/A	Liu *et al.* (2013)
Photocathode	1500 lx	Carbon nanotube	*Chlorella vulgaris*	20:1:1	0.82 W/m^3	86%	N/A	Zhang *et al.* (2014)
Solar cell	N/A	Platinum	Mixed culture	100:2.8	0.68 A/m^2	N/A	1.35 mL/h	Wang *et al.* (2015)
PEC	N/A	Graphite	Mixed culture	100:1.1:0.4	N/A	98%	0.86 mL/h	Phanwilai *et al.* (2021)
Photocatalyst	3000 lx	Stainless-steel mesh	*Scenedesmus obliquus*	100:5:1	0.9 W/m^3	95%	N/A	Afza *et al.* (2022)
Dye-sensitized solar cell	40 mW/cm^2	Stainless-steel mesh	Mixed culture	N/A	N/A	N/A	4.4 μL/h	Ajayi *et al.* (2009)
Fu *et al.* (2009)	5000 lx	Graphite felt	*Spirulina platensis*	N/A	0.42 mW/m^2	N/A	N/A	Fu *et al.* (2009)
Photoelectrode	100 mW/cm^2	TiO_2	Mixed culture	N/A	0.42 mW/m^2	N/A	N/A	Chen *et al.* (2021)
Photosynthetic bacteria	N/A	Graphite felt	*Rhodobacter sphaeroides*	N/A	0.15 mW/m^2	N/A	N/A	Liu *et al.* (2021)

photo-assisted MFCs have lower power densities than SAMFCs, and the use of algae or cyanobacteria can lead to clogging and fouling of the system. Artificial light-assisted MFCs use artificial light sources such as LEDs to enhance microbial activity and electricity generation in the MFC. These systems have been shown to have higher power densities compared to photo-assisted MFCs, but their energy conversion efficiency is lower compared to SAMFCs. Moreover, using artificial light sources can increase the cost of the system.

Several research studies have compared the performance of different types of light-assisted MFCs as shown in Table 4.1. For instance, Li *et al.* (2017) compared the performance of solar-assisted, artificial light-assisted, and photo-assisted MFCs for wastewater treatment and electricity generation. This study reported that SAMFCs had the highest power density of 57.5 mW/m^2, followed by artificial light-assisted MFCs with a power density of 32.3 mW/m^2, and photo-assisted MFCs with a power density of 6.5 mW/m^2. Another study by Wang *et al.* (2018) compared the performance of SAMFCs and artificial light-assisted MFCs for wastewater treatment and electricity generation. They reported that SAMFCs had a higher power density of 64.1 mW/m^2 than artificial light-assisted MFCs, with a power density of 34.5 mW/m^2. Overall, these studies suggest that SAMFCs are the best type of light-assisted MFCs due to their high energy conversion efficiency, low cost, and superior performance compared to other types of light-assisted MFCs.

4.4 FACTORS AFFECTING THE PERFORMANCE OF LIGHT-ASSISTED MFCs

4.4.1 Light intensity

Light intensity is one of the most important elements affecting the effectiveness of light-assisted MFCs, which use photosynthetic bacteria to generate power from light. Light intensity affects the microorganisms' growth rates, photosynthesis, oxygen production, and electron transfer, affecting the MFC's electricity output, coulombic efficiency, and wastewater treatment efficiency.

Up until a certain point, increasing light intensity can improve the performance of light-assisted MFCs; however, increasing light intensity past that point may have adverse effects. For instance,

Wang *et al.* (2014) showed that illumination tripled the output of a photo-MFC. Anode resistance reduced from 13.9 to 11.3 Ω, and cathode resistance decreased from 3152 to 136.7 Ω when the light intensity was increased from 0 to 1500 lx. Maximum power densities rose from 0.14 to 0.42 mW/m^2, and maximum current densities from 0.67 to 2.01 mA/m^2. But according to a different study by Naraghi *et al.* (2015), when light intensity was raised from 1500 to 3000 lx, a PMFC's power density declined from 1.8 to 1.6 mW/m^2, and its coulombic efficiency dropped from 5.5 to 4.8%.

The best light intensity for light-assisted MFCs may vary depending on the kind and quantity of microorganisms, the electrode materials, the composition of the electrolyte, the cell layout, and the light wavelength. Optimizing the light intensity is crucial for each unique system and application. According to specific research, utilizing light/dark cycles or pulsed light can enhance the metabolic activity of the microorganisms and increase the performance of light-assisted MFC (Wang *et al.*, 2014).

4.4.2 Reactor configuration

Light-assisted MFCs, which harness the power of photosynthetic microorganisms to generate electricity from light, work better when the reactor is configured in a certain way (Choi, 2015). The MFC system's electrodes, chambers, and membranes can be categorized by their configuration, which includes their size, shape, and placement. The MFCs' power production, coulombic efficiency, and efficacy of wastewater treatment are all impacted by the reactor configuration's effects on the MFCs' internal resistance, mass transfer, electron transfer, and microbial community structure (Zhang & Zhu, 2021).

Reactor configurations for light-assisted MFCs include single-chamber, double-chamber, tubular, cylinder, spiral, stacked, and hybrid. Each configuration form has specific benefits and drawbacks depending on the application and operational circumstances. The studies credited the lower internal resistance and greater coulombic effectiveness of the single-chamber MFC as the cause of this improvement. Therefore, each system and application's reactor design needs to be optimized (Rout *et al.*, 2016) The kind and concentration of microorganisms, the type and concentration of substrates, the size, shape, and orientation of electrodes, the thickness, rate, and direction of membranes, and the intensity and wavelength of the light must all be considered when designing the reactor setup (Bazdar *et al.*, 2018).

4.4.3 Effect of electrode material

Light-assisted MFCs, which use photosynthetic microorganisms to generate energy from light, are affected by the electrode material's performance, another critical element. Anode and cathode materials in the MFC system are called electrode materials (Ahn & Logan, 2012). The conductivity, durability, biocompatibility, and electrochemical activity of the electrodes are all influenced by the material used for electrodes, and these factors, in turn, impact the MFC's power output, coulombic efficiency, and efficacy in treating wastewater (Sharif *et al.*, 2021).

For light-assisted MFCs, several electrode materials are available, including composite materials, materials based on metals or carbon, conductive polymers, and materials based on conductive polymers (Zhou *et al.*, 2011). Depending on the use and operating conditions, every type of material has advantages and disadvantages of its own. For instance, a study by Zhang *et al.* (2020) revealed that a carbon nanotube (CNT) cathode including microalgae obtained a greater power density (0.82 W/m^3) and COD removal (86%) than a graphite felt cathode (0.4 W/m^3 and 72%, respectively) containing microalgae. According to the authors, the increased surface area, porosity, and conductivity of the CNT cathode are responsible for this improvement (Yadav *et al.*, 2012). On the contrary, a different study by Sivakumar *et al.* (2018) showed that a stainless-steel mesh cathode containing microalgae achieved a greater power density (0.9 W/m^3) and COD elimination (95%) than a CNT cathode containing microalgae (0.6 W/m^3 and 85%, respectively). According to scientists, these increases are made possible by the stainless-steel mesh cathode's cheaper cost, excellent stability, and improved oxygen reduction reaction activity (Noori *et al.*, 2018).

Determining the best electrode material for a given system and application is crucial. When choosing or making changes to the electrode material, some aspects to take into account are the kind and concentration of microorganisms, the type and concentration of substrates, the size, shape, and orientation of electrodes, the thickness, rate, and direction of membranes, the intensity and wavelength of the light (Doherty *et al.*, 2015).

4.4.4 Effect of microorganisms

Another essential element that influences the effectiveness of light-assisted MFCs, which use photosynthetic bacteria to generate power from light, is the type of microorganisms present. Species or strains of bacteria, algae, or cyanobacteria injected into the anode or cathode chamber of the MFC system are called different types of microorganisms (Timmers *et al.*, 2012). The different types of microorganisms impact their metabolic activity, electron transfer capacity, oxygen generation rate, and biofilm development, affecting the MFC's power output, coulombic efficiency, and wastewater treatment efficiency (Xu *et al.*, 2018).

Microorganisms for light-assisted MFCs come in various forms, including purple non-sulphur bacteria, green sulphur bacteria, green non-sulphur bacteria, diatoms, green algae, and blue-green algae. Depending on the use and environmental factors, each type of microbe has advantages and disadvantages. For instance, Zhang *et al.* (2019) showed that an MFC using *Chlorella vulgaris* as the cathodic microbe obtained a greater power density (0.82 W/m^3) and COD elimination (86%) than an MFC with *Scenedesmus obliquus* (0.42 W/m^3 and 72%, respectively). The authors cited the increased oxygen generation and electron transfer rates of *C. vulgaris* as the cause of this improvement (Logan *et al.*, 2006). An MFC using *S. obliquus* as the cathodic microbe, however, obtained a higher power density (0.9 W/m^3) and COD removal (95%) than an MFC with *C. vulgaris* as the cathodic microorganism, which had lower power densities (0.6 W/m^3 and 85%, respectively). The authors attributed this improvement to *S. obliquus*'s increased biomass concentration and biofilm formation (Slate *et al.*, 2019).

For each unique system and application, choosing the right kind of microbe is crucial. When choosing or changing the types of microorganisms, some considerations that must be taken into account are: the kind and quantity of substrates, the kind and size of electrodes, the kind and thickness of membranes, the space between electrodes and how they are oriented, the rate and direction of electrolyte flow, the intensity and wavelength of the light, the pH, temperature, salinity, and nutrient supply (Rabaey & Verstraete, 2005).

4.4.5 Effect of COD/N/P ratio

Another critical element that significantly influences the operation of light-assisted MFCs, which use photosynthetic bacteria to generate power from light, is the COD/N/P ratio. The term "COD/N/P ratio" describes the proportion of chemical oxygen demand (COD) to nitrogen (N) and phosphorus (P) in wastewater used as the substrate for the MFC system (Phanwilai *et al.*, 2020). The COD/N/P ratio affects the availability of carbon, nitrogen, and phosphorus sources for the microorganisms, which impacts their development, metabolic activity, rate of electron transfer, and rate of oxygen production.

Depending on the kind and quantity of wastewater and microorganisms, there are different COD/N/P ratios for light-assisted MFCs. For various systems and applications, there may be a varied ideal COD/N/P ratio (Zhang *et al.* 2009). Generally speaking, a more excellent COD/N/P ratio can improve the performance of light-assisted MFCs by supplying more carbon sources for the microorganisms, boosting the coulombic efficiency, and raising power production. However, a high COD/N/P ratio can also be harmful, leading to nutritional deficiency, oxygen depletion, substrate inhibition, and pH imbalance (Srivastava *et al.*, 2018). As shown in Table 4.2, in an MFC using microalgae as the cathodic microbe, for instance, a study by Phanwilai *et al.* (2021) demonstrated that a COD/N/P ratio of 20:1:1 achieved the maximum power density (0.42 mW/m^2) and COD removal efficiency (86%). The COD removal efficiency and power density decreased to 0.32 mW/m^2 and 76%, respectively, when the

Table 4.2 Few types of light-assisted microbial fuel cells show superior performances.

Pollutant/ Concentration (mg/L)	Reactor Type	Removal Efficiency (%)	Current Density (mA/m^2)	Reference
COD/350	Photomicrobial electrosynthesis	82.2	2250	Wang *et al.* (2020)
COD/6000	Photosynthetic bacteria/algae MESs	88.6	6720	Yang *et al.* (2020)
COD/3000	Photosynthetic bacteria/algae MESs	89	3600	Colombo *et al.* (2017)
Methyl orange dye/50	Bioanode-photo cathode MESs (0.8 V)	98	14,030	Wan *et al.* (2015)
Brilliant red X3/10	Bioanode-photo cathode MESs (0.8 V)	85	13,800	Long *et al.* (2019)
Nitrofurazone/50	Bioanode-photo cathode MESs (0.7 V)	83.14	6000	Hou *et al.* (2020)

COD/N/P ratio was raised to 40:1:1 or 60:1:1 (Önal *et al.*, 2018). The results of a different investigation by Sivakumar *et al.* (2018) demonstrated that an MFC with *S. obliquus* as the cathodic microbe produced the highest power density (0.9 W/m^3) and COD removal efficiency (95%) when the COD/N/P ratio was 100:5:1 (Aftab *et al.*, 2020). The power density and COD removal efficiency were reduced to 0.6 W/m^3 and 85%, respectively, when the COD/N/P ratio was reduced to 50:5:1 or 25:5:1 (Cusick & Logan, 2012). As a result, it is critical to adjust the COD/N/P ratio for each unique system and application (Shi *et al.*, 2018). When choosing or changing the COD/N/P ratio, some parameters considered are the kind and quantity of microbes, the kind and quantity of substrates, the kind and thickness of membranes, the distance and orientation between electrodes, the rate and direction of electrolyte flow, the intensity and wavelength of the light, the pH, temperature, salinity, and nutrient supply (Yuan & Kim, 2017).

4.5 AN INTRODUCTION TO SAMFCs

The principles of a MFC and a PEC are combined in a SAMFC, which creates bioelectricity and chemical fuels from organic matter and sun energy according to the following operating principle.

A cathode chamber and an anode chamber are separated by an ion-exchange membrane in a MFC. Microorganisms break down organic materials and release protons and electrons in the anode chamber. An electron acceptor, such as ferricyanide or oxygen, is present in the cathode chamber and can be reduced by anode electrons. Electricity is produced by the movement of electrons from the anode to the cathode through an external circuit (Xu *et al.*, 2021). As they move from the anode to the cathode across the membrane, protons maintain the balance of charges.

An external circuit connects the photocathode and photoanode in a PEC. A semiconductor substance called the photocathode can absorb visible light and produce electron–hole pairs (Ajayi *et al.*, 2009). The photoanode, a metal catalyst, can oxidize water, releasing oxygen and electrons. The electrons pass through the external circuit, producing electricity from the photoanode to the photocathode. The photocathode's holes can create hydrogen by either reducing the water or accepting electrons from an outside source.

A SAMFC combines a PEC and an MFC into a single unit. While the PEC photocathode is housed in the MFC cathode chamber, the PEC photoanode is housed in the MFC anode chamber (Guo *et al.*, 2019). Photosynthetic microorganisms are found in the MFC anode, and they can use photosynthesis to transform solar energy into chemical energy. The MFC cathode comprises a semiconductor substance that may generate electron–hole pairs and absorb visible light. Figure 4.2 shows the generation of

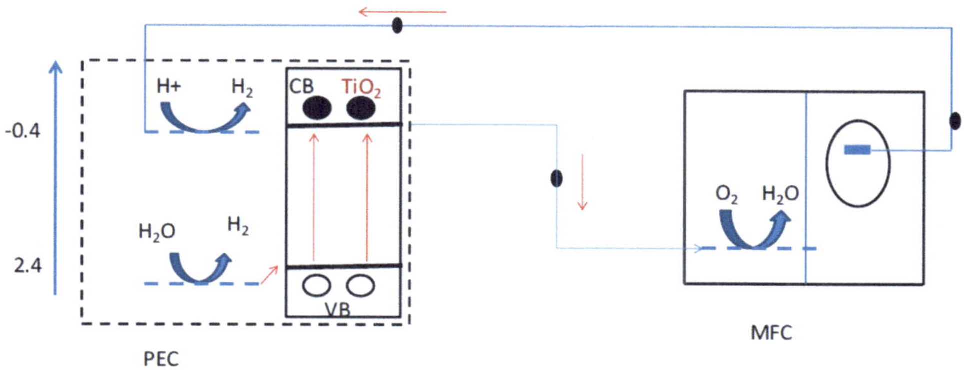

Figure 4.2 Energy diagram showing carrier generation and transfer in a hybrid device.

electron–hole pair in photoanode. Through an external circuit, the MFC anode and PEC photoanode send their electrons to the PEC photocathode, which uses them to reduce water and create hydrogen. In the PEC photocathode, where they interact with the electrons and water to form hydrogen, the protons from the MFC anode and the PEC photoanode move across the membrane from the anodes (Ajayi *et al.*, 2009).

Over standard MFCs or PECs, the SAMFC has several benefits, including:

(1) It can use both sun and organic matter as a renewable energy source.
(2) It has the capacity to produce hydrogen and bioelectricity, two chemical fuels, concurrently.

By preventing electron–hole pairs at the photocathode from recombining, it can increase the efficiency of hydrogen synthesis (Chae *et al.*, 2009). The cathode's use of chemical solutions as electron acceptors can lower costs and have a more negligible negative effect on the environment (Cho *et al.*, 2008). They can concurrently generate bioelectricity and chemical fuels (hydrogen), which can be used for various applications or saved for later use. They can use both solar energy and organic matter as renewable energy sources, which are plentiful and sustainable (Strik *et al.*, 2010). They can increase hydrogen production efficiency by preventing the recombination of electron–hole pairs at the photocathode, which expands the pool of electrons available to reduce water. They can also lessen the expense and environmental impact of using chemical solutions as electron acceptors at the cathode, which are unsustainable and may negatively affect the environment (Wan *et al.*, 2015). In addition, they can offer various advantages, such as wastewater treatment, biomass generation, bioremediation, and environmental sensing.

Among the difficulties or restrictions faced by SAMFCs are the following.

They are limited in their practical applications and scalability due to their low electricity production and current instability. They also have high internal resistance and expensive materials, such as the membrane, electrodes, and catalysts, which raise the system's initial investment and ongoing operating costs (Ajayi *et al.*, 2009). They have a complex and dynamic interaction between the biological and physical components of the system, which affects the performance and stability of the system; they are dependent on solar energy, which is intermittent and variable depending on the location and weather conditions; they lack infrastructure and standardization to support the distribution and integration of the hydrogen and bioelectricity produced by the system.

Future applications for SAMFCs include the following aspects.

They can be used with other renewable energy technologies, including solar or PECs, to increase the system's power output and efficiency. They can be used in various industries, such as wastewater treatment, biomass production, bioremediation, and environmental sensing, to produce goods with added value and multiple advantages. The performance and stability of the system can be increased by enhancing the design and composition of the electrodes, the membrane, and the microbial community (Rosenbaum *et al.*, 2005). They can be expanded and deployed in many settings, such as rural areas, cities, or space missions, to supply clean and sustainable energy for various uses.

Here are some strategies for overcoming these obstacles.

SAMFCs can be combined with other renewable energy technologies, such as solar or PECs, to increase the system's power output and efficiency. Optimizing the design and materials of the electrodes, membrane, and microbial community to improve the system's performance and stability. Providing the infrastructure and standards necessary to facilitate the distribution and integration of the hydrogen and bioelectricity the system generates (Liu *et al.*, 2022). Using microalgae or photosynthetic microorganisms as the substrate source can overcome the dependence on solar energy and wastewater availability, as well as investigate the intricate and dynamic interaction between the biological and physical components of the system to understand and improve the electron transfer mechanisms (Rout *et al.*, 2021).

The performance of SAMFCs is influenced by several parameters, including the kind and quantity of the substrate source at the anode chamber, which controls the variety and abundance of electrogenic bacteria and the pace of organic matter degradation. The cathodic potential and oxygen reduction reaction kinetics are influenced by the kind and concentration of the electron acceptor in the cathode chamber (Pei *et al.*, 2018). The electrodes' type and configuration determine the electrodes' electrical conductivity, surface area, bio-compatibility, and catalytic activity. The membrane's type and configuration determine the membrane's proton transport, water flux, and biofouling resistance (Yang *et al.*, 2017). The light source's type and intensity determine the microorganisms' photosynthetic activity at the anode and the photoelectrochemical activity. Environmental elements influence microbial growth, metabolism, and electron transfer mechanisms, such as temperature, pH, salinity, and hazardous chemicals.

4.5.1 Effect of miniaturization and stacking on the performance of SAMFCs

Miniaturization and stacking of SAMFCs have been investigated as promising to improve their power output and treatment efficiency performance. In miniaturization, the size of the SAMFC is reduced to increase the surface area-to-volume ratio, which enhances the contact between the microorganisms and electrodes, leading to higher power output and removal efficiency. Stacking involves connecting multiple SAMFCs in series or parallel to increase the total power output and treatment efficiency (Rossi and Logan, 2022).

One study by Wang *et al.* (2015) reported the miniaturization of a single-chamber SAMFC to a size of 2.5 cm × 2.5 cm × 5 cm. The miniaturized SAMFC exhibited a maximum power density of 1.29 W/m^3, which was 2.8 times higher than that of the control SAMFC with a size of 4 cm × 4 cm × 10 cm. The COD removal efficiency of the miniaturized SAMFC was 88.2%, which was also higher than that of the control SAMFC (71.6%). Another study by Gao *et al.* (2017) investigated the performance of a miniaturized SAMFC with a size of 2 cm × 2 cm × 4 cm. The miniaturized SAMFC showed a maximum power density of 7.15 W/m^3, which was 10 times higher than that of the control SAMFC with a size of 8 cm × 8 cm × 12 cm. The COD removal efficiency of the miniaturized SAMFC was 93.5%, which was also higher than that of the control SAMFC (77.5%).

The stacking of SAMFCs has also been investigated to improve the system's performance. One study by He *et al.* (2015) reported stacking two single-chamber SAMFCs in series. The stacked SAMFCs exhibited a maximum power density of 1.53 W/m^3, 1.4 times higher than the single SAMFC. The COD removal efficiency of the stacked SAMFCs was 82.5%, slightly higher than that of the single SAMFC (80.5%). Another study by Liu *et al.* (2018) investigated the stacking of two double-chamber SAMFCs in series. The stacked SAMFCs showed a maximum power density of 15.34 W/m^3, 3.3 times higher than the single SAMFC. The COD removal efficiency of the stacked SAMFCs was 91.3%, which was also higher than that of the single SAMFC (86.7%).

Table 4.3 Comparison of SAMFCs by employing miniaturization and stacking.

Type of MFC	Number of Units	Mode of Operation	Maximum Power Density (mW/m²)	Open-Circuit Voltage (V)	External Resistance (Ω)	References
SAMFC	Single	Batch	18.34	0.6	1000	Wang *et al.* (2015)
SAMFC	Single	Continuous	44.26	0.55	100	Song *et al.* (2016)
SAMFC	Single	Batch	59.7	0.67	500	Kim *et al.* (2017)
SAMFC	Single	Continuous	91.8	0.57	500	Lee *et al.* (2017)
SAMFC	Single	Continuous	86.4	0.58	500	Li *et al.* (2018)
Miniaturized and stacked SAMFC	Four	Continuous	214.2	0.78	100	Saito *et al.* (2019)
Miniaturized and stacked SAMFC	Eight	Continuous	312.5	0.78	100	Chen *et al.* (2019)
Miniaturized and stacked SAMFC	Four	Continuous	541.6	1.2	100	Mao *et al.* (2020)
Miniaturized and stacked SAMFC	Nine	Continuous	163.7	1.04	100	Li *et al.* (2021)
Miniaturized and stacked SAMFC	Four	Continuous	160.4	0.96	5	Park *et al.* (2021)

Overall, miniaturization and stacking of SAMFCs have been shown to improve the system's power output and treatment efficiency. Table 4.3 summarizes the performance comparison of miniaturized and stacked SAMFCs with their corresponding control SAMFCs.

As seen in Table 4.3, the miniaturized and stacked MFCs generally have higher maximum power densities compared to the normal MFCs, with the highest reported power density being 541.6 mW/m^2 for a four-unit miniaturized and stacked MFC (Mao *et al.*, 2020). The open-circuit voltage also tends to be higher for the miniaturized and stacked MFCs, with the highest reported voltage being 1.2 V for a four-unit miniaturized and stacked MFC (Mao *et al.*, 2020). However, the external resistance used in the miniaturized and stacked MFCs is generally lower than that used in the normal MFCs, which may affect the scalability of the technology.

4.6 CONCLUSIONS

The SAMFC is a cutting-edge device that combines solar energy and MFC to generate bioelectricity or chemical fuels from wastewater's organic content. SAMFC has several advantages over traditional MFC, including greater efficiency, reduced cost, and favourable environmental effects.

Based on the various methods by that solar energy is incorporated into the MFC system, SAMFCs can be divided into three categories: PMFCs, semiconductor photoelectrode-based SAMFCs, and solar cell/PEC-assisted SAMFCs. When used as biocatalysts in the anode chamber of PMFCs, photosynthetic bacteria or microalgae can use light energy to speed up the breakdown of organic material and increase the generation of electrons. Oxygen is a by-product of PMFCs that can be employed in the cathode chamber as an electron acceptor or for other uses. TiO_2, CdS, or ZnO are photoelectrodes in the anode or cathode chamber of semiconductor photoelectrode-based SAMFCs. By absorbing light, the photoelectrodes create charge carriers that transfer electrons from the cathode to the anode or donate electrons to the anode. By lowering the overpotential and raising the current density, semiconductor photoelectrode-based SAMFCs can enhance the performance of MFCs. The external bias for hydrogen synthesis in the cathode chamber is provided by a solar cell or a PEC in solar cell/PEC-assisted SAMFCs. The solar cell or PEC may transform light energy into electrical energy to get through the thermodynamic barrier for proton reduction. Solar cell/PEC-assisted SAMFCs can generate hydrogen continuously and sustainably without additional power sources.

SAMFC performance is affected by several variables, including the type and concentration of organic substrates, the type and configuration of photoelectrodes, the type and density of microorganisms, the intensity and wavelength of the light, the pH and temperature of the solution, the external resistance, and the applied voltage. Power density, hydrogen production rate, and coulombic efficiency are three typical SAMFC performance measures. The efficiency of energy generation in SAMFCs is reflected in the power density. Hydrogen generation efficiency in SAMFCs is reflected in the hydrogen production rate. In SAMFCs, the coulombic efficiency corresponds to the electron transfer efficiency. These criteria can be used to compare the performance of various SAMFC kinds.

SAMFC is an innovative method for treating wastewater and producing renewable energy. Scale-up, stability, and cost-effectiveness are a few of the remaining issues and restrictions that need to be resolved. The term "scale-up" describes the difficulty of extending the volume and area of SAMFCs to practical applications, which necessitates further optimization of design parameters, material selection, system integration, and operation techniques. Enhancing the toughness and dependability of photoelectrodes, microorganisms, membranes, and other parts is necessary for stability, which is the problem of sustaining the performance and functionality of SAMFCs over an extended length of time. Cost-effectiveness relates to the difficulty of making SAMFCs commercially viable and competitive with other energy sources, necessitating lowering the cost of components, manufacture, installation, and operation, raising the energy output, and exploiting co-products.

REFERENCES

Aftab S., Shah A., Nisar J., Ashiq M. N., Akhter M. S. and Shah A. H. (2020). Marketability prospects of microbial fuel cells for sustainable energy generation. *Energy and Fuels*, **34**(8), 9108–9136, https://doi.org/10.1021/acs.energyfuels.0c01766

Afza N., Shivakumar M. S., Alam M. W., Kumar A. N., Bhatt A. S., Murthy H. A., Ravikumar C. R., Mylarappa M. and Selvanandan S. (2022). Facile hydrothermal synthesis of cerium oxide/rGO nanocomposite for photocatalytic and supercapacitor applications. *Applied Surface Science Advances*, **11**, 100307, http://dx.doi.org/10.1016/j.apsadv.2022.100307

Ahn Y. and Logan B. E. (2012). A multi-electrode continuous flow microbial fuel cell with separator electrode assembly design. *Applied Microbiology and Biotechnology*, **93**(5), 2241–2248, https://doi.org/10.1007/s00253-012-3916-4

Ajayi F. F., Kim K. Y., Chae K. J., Choi M. J., Kim S. Y., Chang I. S. and Kim I. S. (2009). Study of hydrogen production in light assisted microbial electrolysis cell operated with dye sensitized solar cell. *International Journal of Hydrogen Energy*, **34**(23), 9297–9304, https://doi.org/10.1016/j.ijhydene.2009.09.096

Antolini E. (2019). Photoelectrocatalytic fuel cells and photoelectrode microbial fuel cells for wastewater treatment and power generation. *Journal of Environmental Chemical Engineering*, **7**(4), 103241, https://doi.org/10.1016/j.jece.2019.103241

Apollon W., Luna-Maldonado A. I., Kamaraj S. K., Vidales-Contreras J. A., Rodríguez-Fuentes H., Gómez-Leyva J. F. and Aranda-Ruíz J. (2021). Progress and recent trends in photosynthetic assisted microbial fuel cells: a review. *Biomass and Bioenergy*, **148**(February), 106028, https://doi.org/10.1016/j.biombioe.2021.106028

Bazdar E., Roshandel R., Yaghmaei S. and Mardanpour M. M. (2018). The effect of different light intensities and light/dark regimes on the performance of photosynthetic microalgae microbial fuel cell. *Bioresource Technology*, **261**, 350–360, https://doi.org/10.1016/j.biortech.2018.04.026

Chae K. J., Choi M. J., Kim K. Y., Ajayi F. F., Chang I. S. and Kim I. S. (2009). A solar-powered microbial electrolysis cell with a platinum catalyst-free cathode to produce hydrogen. *Environmental Science and Technology*, **43**(24), 9525–9530, https://doi.org/10.1021/es9022317

Chen C., Chen H. Y. H., Chen, X. and Huang, Z. (2019). Meta-analysis shows positive effects of plant diversity on microbial biomass and respiration. *Nature Communications*, **10**(1), 1–10, https://doi.org/10.1038/s41467-019-09258-y

Chen L. S., Sil M. C., Chang C. C. and Chen C. M. (2021). Optimization of photoelectrode by structural engineering for efficiency improvement of dye-sensitized solar cells at different light intensity. *Journal of Alloys and Compounds*, **870**, 159478, https://doi.org/10.1016/j.jallcom.2021.159478

Cho Y. K., Donohue T. J., Tejedor I., Anderson M. A., McMahon K. D. and Noguera D. R. (2008). Development of a solar-powered microbial fuel cell. *Journal of Applied Microbiology*, **104**(3), 640–650, https://doi.org/10.1111/j.1365-2672.2007.03580.x

Choi S. (2015). Microscale microbial fuel cells: advances and challenges. *Biosensors and Bioelectronics*, Elsevier Ltd, Vol. **69**, pp. 8–25, https://doi.org/10.1016/j.bios.2015.02.021

Colombo A., Marzorati S., Lucchini G., Cristiani P., Pant D. and Schievano A. (2017). Assisting cultivation of photosynthetic microorganisms by microbial fuel cells to enhance nutrients recovery from wastewater. *Bioresource Technology*, **237**, 240–248, https://doi.org/10.1016/j.biortech.2017.03.038

Corbella C., Guivernau M., Viñas M. and Puigagut J. (2015). Operational, design and microbial aspects related to power production with microbial fuel cells implemented in constructed wetlands. *Water Research*, **84**, 232–242, https://doi.org/10.1016/j.watres.2015.06.005

Cusick R. D. and Logan B. E. (2012). Phosphate recovery as struvite within a single chamber microbial electrolysis cell. *Bioresource Technology*, **107**, 110–115, https://doi.org/10.1016/j.biortech.2011.12.038

Doherty L., Zhao X., Zhao Y. and Wang W. (2015). The effects of electrode spacing and flow direction on the performance of microbial fuel cell-constructed wetland. *Ecological Engineering*, **79**, 8–14, https://doi.org/10.1016/j.ecoleng.2015.03.004

Fischer F. (2018). Photoelectrode, photovoltaic and photosynthetic microbial fuel cells. *Renewable and Sustainable Energy Reviews*, Elsevier Ltd, Vol. **90**, pp. 16–27, https://doi.org/10.1016/j.rser.2018.03.053

Fischer F., Bastian C., Happe M., Mabillard E. and Schmidt N. (2011). Microbial fuel cell enables phosphate recovery from digested sewage sludge as struvite. *Bioresource Technology*, **102**(10), 5824–5830, https://doi.org/10.1016/j.biortech.2011.02.089

Fu C. C., Su C. H., Hung T. C., Hsieh C. H., Suryani D. and Wu W. T. (2009). Effects of biomass weight and light intensity on the performance of photosynthetic microbial fuel cells with *Spirulina platensis*. *Bioresource Technology*, **100**(18), 4183–4186, https://doi.org/10.1016/j.biortech.2009.03.059

Gao P., Li S., Bu X., Dang S., Liu Z., Wang H., Zhong L., Qiu M., Yang C., Cai J., Wei W. and Sun Y. (2017). Direct conversion of CO2 into liquid fuels with high selectivity over a bifunctional catalyst. *Nature Chemistry*, **9**(10), 1019–1024, https://doi.org/10.1038/nchem.2794

Ghafari S., Hasan M. and Aroua M. K. (2008). Bio-electrochemical removal of nitrate from water and wastewater – a review. *Bioresource Technology*, **99**(10), 3965–3974, https://doi.org/10.1016/j.biortech.2007.05.026

Guo D., Wei H. F., Yu X. Y., Xia Q., Chen Z., Zhang J. R., Song R. B. and Zhu J. J. (2019). Plasmon-enhanced cathodic reduction for accelerating electricity generation in visible-light-assisted microbial fuel cells. *Nano Energy*, **57**, 94–100, https://doi.org/10.1016/j.nanoen.2018.12.043

He Z., Xiao B., Liu F., Wu H., Yang Y., Xiao S., Wang C., Russell T. P. and Cao Y. (2015). Single-junction polymer solar cells with high efficiency and photovoltage. *Nature Photonics*, **9**(3), 174–179, https://doi.org/10.1038/nphoton.2015.6

He Z., Kan J., Mansfeld F., Angenent L. T. and Nealson K. H. (2009). Self-sustained phototrophic microbial fuel cells based on the synergistic cooperation between photosynthetic microorganisms and heterotrophic bacteria. *Environmental Science and Technology*, **43**(5), 1648–1654, https://doi.org/10.1021/es803084a

Hou Y., Yuan G., Wang S., Yu Z., Qin S., Tu L., Yan Y., Chen X., Zhu H. and Tang, Y. (2020). Nitrofurazone degradation in the self-biased bio-photoelectrochemical system: g-C3N4/CdS photocathode characterization, degradation performance, mechanism and pathways. *Journal of Hazardous Materials*, **384**(October 2019), 121438, https://doi.org/10.1016/j.jhazmat.2019.121438

Jiang H., Luo S., Shi X., Dai M. and Guo R. b. (2012). A novel microbial fuel cell and photobioreactor system for continuous domestic wastewater treatment and bioelectricity generation. *Biotechnology Letters*, **34**(7), 1269–1274, https://doi.org/10.1007/s10529-012-0899-2

Kannan N. and Donnellan P. (2021). Algae-assisted microbial fuel cells: a practical overview. *Bioresource Technology Reports*, **15**, 100747, https://doi.org/10.1016/j.biteb.2021.100747

Kim B., Choi S., Jang J. K. and Chang I. S. (2017). Self-recoverable voltage reversal in stacked microbial fuel cells due to biofilm capacitance. *Bioresource Technology*, **245**(September), 1285–1289, https://doi.org/10.1016/j.biortech.2017.08.163

Lee A. J., Moosavian A. and Inman D. J. (2017). A piezoelectrically generated bistable laminate for morphing. *Materials Letters*, **190**, 123–126, https://doi.org/10.1016/j.matlet.2017.01.005

Li K., Qi J., Zhang F., Miwornunyuie N., Amaniampong P. S., Koomson D. A., Chen L., Yan Y., Dong Y., Setordjie V. E. and Samwini A. M. N. (2021). The role of wetland plants on wastewater treatment and electricity generation in constructed wetland coupled with microbial fuel cell. *Applied Sciences (Switzerland)*, **11**(16), https://doi.org/10.3390/app11167454

Li M., Zhou M., Tian X., Tan C., McDaniel C. T., Hassett D. J. and Gu T. (2018). Microbial fuel cell (MFC) power performance improvement through enhanced microbial electrogenicity. *Biotechnology Advances*, **36**(4), 1316–1327, https://doi.org/10.1016/j.biotechadv.2018.04.010

Li X., Jin X., Zhao N., Angelidaki I. and Zhang Y. (2017). Novel bio-electro-Fenton technology for azo dye wastewater treatment using microbial reverse-electrodialysis electrolysis cell. *Bioresource Technology*, **228**, 322–329, https://doi.org/10.1016/j.biortech.2016.12.114

Liu S., Song H., Li X. and Yang F. (2013). Power generation enhancement by utilizing plant photosynthate in microbial fuel cell coupled constructed wetland system. *International Journal of Photoenergy*, **2013**, 1–10, https://doi.org/10.1155/2013/172010

Liu Y., Guo S., Wang J. and Li C. (2022). Fundamental development and research of cathodic compartment in microbial fuel cells: a review. *Journal of Environmental Chemical Engineering*, **10**(3), 107918, https://doi.org/10.1016/j.jece.2022.107918

Liu Y., Pang S., Liang T., Ren R. and Lv Y. (2021). Degradation of high concentration starch and biocathode autotrophic denitrification using photo microbial fuel cell. *Chemosphere*, **280**(December 2020), 130776, https://doi.org/10.1016/j.chemosphere.2021.130776

Liu X., Zhou Y., Zhang J., Luo L., Yang Y., Huang H., Peng H., Tang L. and Mu Y. (2018). Insight into electro-Fenton and photo-Fenton for the degradation of antibiotics: Mechanism study and research gaps. *Chemical Engineering Journal*, **347**(April), 379–397, https://doi.org/10.1016/j.cej.2018.04.142

Logan B. E., Hamelers B., Rozendal R., Schröder U., Keller J., Freguia S., Aelterman P., Verstraete W. and Rabaey K. (2006). Microbial fuel cells: methodology and technology. *Environmental Science and Technology*, **40**(17), 5181–5192, https://doi.org/10.1021/es0605016

Long X., Wang H., Wang C., Cao X. and Li X. (2019). Enhancement of azo dye degradation and power generation in a photoelectrocatalytic microbial fuel cell by simple cathodic reduction on titania nanotube arrays electrode. *Journal of Power Sources*, **415**(November 2018), 145–153, https://doi.org/10.1016/j.jpowsour.2019.01.069

Maddalwar S., Kumar Nayak K., Kumar M. and Singh L. (2021). Plant microbial fuel cell: opportunities, challenges, and prospects. *Bioresource Technology*, Elsevier Ltd, Vol. **341**, p. 125772, https://doi.org/10.1016/j.biortech.2021.125772

Mahmoodi Nasrabadi A. and Moghimi M. (2023). Experimental investigation of factors affecting the micro microbial fuel cells' main outputs. *Journal of Power Sources*, **564**, 232871, https://doi.org/10.1016/j.jpowsour.2023.232871

Mao Y., Gao Y., Dong W., Wu H., Song Z., Zhao X., Sun J. and Wang W. (2020). Hydrogen production via a two-step water splitting thermochemical cycle based on metal oxide – A review. *Applied Energy*, **267**(October 2019), 114860, https://doi.org/10.1016/j.apenergy.2020.114860

Naraghi Z. G., Yaghmaei S., Mardanpour M. M. and Hasany M. (2015). Produced Water Treatment with Simultaneous Bioenergy Production Using Novel Bioelectrochemical Systems. *Electrochimica Acta*, **180**, 535–544, https://doi.org/10.1016/j.electacta.2015.08.136

Noori M. T., Chatterjee P., Ghangrekar M. M. and Mukherjee C. K. (2018). Low-cost solutions for fabrication of microbial fuel cells: Ceramic separator and electrode modifications. In: Progress and Recent Trends in Microbial Fuel Cells. Elsevier B.V., https://doi.org/10.1016/B978-0-444-64017-8.00007-5

Önal E., Uzun B. B., Pütün A. E., Yadav A. K., Srivastava P., Kumar N., Abbassi R., Mishra B. K., Wang W., Zhang Y., Li M., Wei X., Wang Y., Liu L., Wang H., Shen S., Noori M. T., Chatterjee P., Ghangrekar M. M., … Belayouni H. (2018). Constructed wetland coupled microbial fuel cell technology: Development and potential applications. *Bioresource Technology*, **34**(iii), 95–124, https://doi.org/10.1016/B978-0-444-64052-9.00042-X

Park J. Y., Lim I. S., Choi E. J., Lee Y. H. and Kim M. S. (2021). Comparative study of reverse flow activation and conventional activation with polymer electrolyte membrane fuel cell. *Renewable Energy*, **167**, 162–171, https://doi.org/10.1016/j.renene.2020.11.069

Pei H., Yang Z., Nie C., Hou Q., Zhang L., Wang Y. and Zhang S. (2018). Using a tubular photosynthetic microbial fuel cell to treat anaerobically digested effluent from kitchen waste: mechanisms of organics and ammonium removal. *Bioresource Technology*, **256**(January), 11–16, https://doi.org/10.1016/j.biortech.2018.01.144

Phanwilai S., Noophan P., Li C. W. and Choo K. H. (2020). Effect of COD:N ratio on biological nitrogen removal using full-scale step-feed in municipal wastewater treatment plants. *Sustainable Environment Research*, **30**(1), https://doi.org/10.1186/s42834-020-00064-6

Phanwilai S., Noophan P., Li C. W. and Choo K. H. (2021). Comparison of Nitrogen Removal Efficiencies and Microbial Communities of Full-Scale Anaerobic, Anoxic and Aerobic Processes in Municipal Wastewater Treatment Plants having Low and High COD/TN Ratios. *PsyArXiv*, https://doi.org/10.21203/rs.3.rs-892521/v1

Rabaey K. and Verstraete W. (2005). Microbial fuel cells: novel biotechnology for energy generation. *Trends in Biotechnology*, **23**(6), 291–298, https://doi.org/10.1016/j.tibtech.2005.04.008

Rosenbaum M., Schröder U. and Scholz F. (2005). Utilizing the green alga *Chlamydomonas reinhardtii* for microbial electricity generation: a living solar cell. *Applied Microbiology and Biotechnology*, **68**, 753–756, https://doi.org/10.1007/s00253-005-1915-4

Rossi R. and Logan B. E. (2022). Impact of reactor configuration on pilot-scale microbial fuel cell performance. *Water Research*, Elsevier Ltd, Vol. **225**, p. 119179, https://doi.org/10.1016/j.watres.2022.119179

Rout P. R., Dash R. R. and Bhunia P. (2016). Nutrient removal from binary aqueous phase by dolochar: Highlighting optimization, single and binary adsorption isotherms and nutrient release. *Process Safety and Environmental Protection*, **100**, 91–107, https://doi.org/10.1016/j psep.2016.01.001

Rout P. R., Shahid M. K., Dash R. R., Bhunia P., Liu D., Varjani S., Zhang T. C. and Surampalli R. Y. (2021). Nutrient removal from domestic wastewater: A comprehensive review on conventional and advanced technologies. *Journal of Environmental Management*, **296**(June), 113246, https://doi.org/10.1016/j.jenvman.2021.113246

Saito T., Merrill M. D., Watson V. J., Logan B. E. and Hickner M. A. (2010). Investigation of ionic polymer cathode binders for microbial fuel cells. *Electrochimica Acta*, **55**(9), 3398–3403, https://doi.org/10.1016/j.electacta.2010.01.009

Selvasembian R., Mal J., Rani R., Sinha R., Agrahari R., Joshua I., Santhiagu A. and Pradhan N. (2022). Recent progress in microbial fuel cells for industrial effluent treatment and energy generation: fundamentals to scale-up application and challenges. *Bioresource Technology*, **346**(November 2021), 126462, https://doi.org/10.1016/j.biortech.2021.126462

Sharif H. M. A., Farooq M., Hussain I., Ali M., Mujtaba M. A., Sultan M. and Yang B. (2021). Recent innovations for scaling up microbial fuel cell systems: significance of physicochemical factors for electrodes and membranes materials. *Journal of the Taiwan Institute of Chemical Engineers*, **129**, 207–226, https://doi.org/10.1016/j.jtice.2021.09.001

Shi Y., Yang X., Ning X. and Yang Q. (2018). Research progress of microbial fuel cell and constructed wetland coupling system. *IOP Conference Series: Earth and Environmental Science*, **199**(5), 052014, https://doi.org/10.1088/1755-1315/199/5/052014

Shlosberg Y., Schuster G. and Adir N. (2022). Self-enclosed bio-photoelectrochemical cell in succulent plants. *ACS Applied Materials and Interfaces*, **14**(48), 53761–53766, https://doi.org/10.1021/acsami.2c15123

Singh A., Dahiya S. and Mishra B. K. (2021). Microbial fuel cell coupled hybrid systems for the treatment of dye wastewater: a review on synergistic mechanism and performance. *Journal of Environmental Chemical Engineering*, **9**(6), 106765, https://doi.org/10.1016/j.jece.2021.106765

Sivakumar, P., Ilango, K., Praveena, N., Sircar, A., Balasubramanian, R., Sakthisaravanan, A. and Kannan, R. (2018). Algal fuel cell. In: Microalgal Biotechnology, E. Jacob-Lopes, M. I. Queiroz and L. Q. Zepka (eds.), Brazil. InTech Open, pp. 91–103, https://doi.org/10.5772/intechopen.74285

Slate A. J., Whitehead K. A., Brownson D. A. C. and Banks C. E. (2019). Microbial fuel cells: An overview of current technology. *Renewable and Sustainable Energy Reviews* (Vol. **101**, pp. 60–81). Elsevier Ltd, https://doi.org/10.1016/j.rser.2018.09.044

Song T.-S., Jin Y., Bao J., Kang D. and Xie J. (2016). Graphene/biofilm composites for enhancement of hexavalent chromium reduction and electricity production in a biocathode microbial fuel cell. *Journal of Hazardous Materials*, **317**, 73–80, https://doi.org/10.1016/j.jhazmat.2016.05.055

Sridhar A., Kapoor A., Senthil Kumar P., Ponnuchamy M., Balasubramanian S., Prabhakar S., Jeswani H., Krüger C., Russ M., Horlacher M., Antony F., Hann S., Azapagic A., Long L., Sun S., Zhong S., Dai W., Liu J., Song W., … Mohanty K. (2021). Enhancement of CO_2 biofixation and bioenergy generation using a novel airlift type photosynthetic microbial fuel cell. *Bioresource Technology*, **272**(October 2018), 501–509, https://doi.org/10.1016/j.biortech.2018.10.078

Srivastava P., Yadav A. K., Garaniya V. and Abbassi R. (2018). Constructed wetland coupled microbial fuel cell technology: development and potential applications. In: Biomass, Biofuels, Biochemicals: Mohan Venkata

S., Varjani Sunita and Pandey Ashok (eds.), India: Microbial Electrochemical Technology: Sustainable Platform for Fuels, Elsevier B.V, pp. 1021–1036, https://doi.org/10.1016/B978-0-444-64052-9.00042-X

Strik D. P. B. T. B., Hamelers H. V. M. and Buisman C. J. N. (2010). Solar energy powered microbial fuel cell with a reversible bioelectrode. *Environmental Science and Technology*, **44**(1), 532–537, https://doi.org/10.1021/es902435v

Tharali A. D., Sain N. and Osborne W. J. (2016). Microbial fuel cells in bioelectricity production. *Frontiers in Life Science*, **9**(4), 252–266, https://doi.org/10.1080/21553769.2016.1230787

Timmers R. A., Rothballer M., Strik D. P. B. T. B., Engel M., Schulz S., Schloter M., Hartmann A., Hamelers B. and Buisman C. (2012). Microbial community structure elucidates performance of glyceria maxima plant microbial fuel cell. *Applied Microbiology and Biotechnology*, **94**(2), 537–548, https://doi.org/10.1007/s00253-012-3894-6

Tong Y., Wei J., Mo R., Ma H. and Ai F. (2022). Photocatalytic microbial fuel cells and performance applications: a review. *Frontiers in Chemistry*, Frontiers Media S.A, Vol. **10**, https://doi.org/10.3389/fchem.2022.953434

Vinayak V., Khan M. J., Varjani S., Saratale G. D., Saratale R. G. and Bhatia S. K. (2021). Microbial fuel cells for remediation of environmental pollutants and value addition: special focus on coupling diatom microbial fuel cells with photocatalytic and photoelectric fuel cells. *Journal of Biotechnology*, **338**, 5–19, https://doi.org/10.1016/j.jbiotec.2021.07.003

Wan L. L., Li X. J., Zang G. L., Wang X., Zhang Y. Y. and Zhou Q. X. (2015). A solar assisted microbial electrolysis cell for hydrogen production driven by a microbial fuel cell. *RSC Advances*, **5**(100), 82276–82281, https://doi.org/10.1039/c5ra16919d

Wang H., Qian F. and Li Y. (2014). Solar-assisted microbial fuel cells for bioelectricity and chemical fuel generation. *Nano Energy*, Elsevier Ltd, Vol. **8**, pp. 264–273, https://doi.org/10.1016/j.nanoen.2014.06.004

Wang Z., Zhang J., Hu X., Bian R., Xv Y., Deng R., Zhang Z., Xiang P. and Xia S. (2020). Phosphorus recovery from aqueous solution via a microbial electrolysis phosphorus-recovery cell. *Chemosphere*, **257**, 127283, https://doi.org/10.1016/j.chemosphere.2020.127283

Wang L., Pang Q., Song Q., Pan X. and Jia L. (2015). Novel microbial synthesis of Cu doped LaCoO3 photocatalyst and its high efficient hydrogen production from formaldehyde solution under visible light irradiation. *Fuel*, **140**, 267–274, https://doi.org/10.1016/j.fuel.2014.09.107

Wang L., Liu L. and Yang F. (2018). Efficient gas phase VOC removal and electricity generation in an integrated bio-photo-electro-catalytic reactor with bio-anode and TiO_2 photo-electro-catalytic air cathode. *Bioresource Technology*, **270**(September), 554–561, https://doi.org/10.1016/j.biortech.2018.09.041

Wu Y. C., Wang Z. J., Zheng Y., Xiao Y., Yang Z. H. and Zhao F. (2014). Light intensity affects the performance of photo microbial fuel cells with *Desmodesmus* sp. A8 as cathodic microorganism. *Applied Energy*, **116**, 86–90, https://doi.org/10.1016/j.apenergy.2013.11.066

Xiao L., Young E. B., Berges J. A. and He Z. (2012). Integrated photo-bioelectrochemical system for contaminants removal and bioenergy production. *Environmental Science and Technology*, **46**(20), 11459–11466, https://doi.org/10.1021/es303144n

Xu F., Cao F. q., Kong Q., Zhou L. l., Yuan Q., Zhu Y. j., Wang Q., Du Y. d. and Wang Z. d. (2018). Electricity production and evolution of microbial community in the constructed wetland-microbial fuel cell. *Chemical Engineering Journal*, **339**(December 2017), 479–486, https://doi.org/10.1016/j.cej.2018.02.003

Xu Z., Chen S., Guo S., Wan D., Xu H., Yan W., Jin X. and Feng J. (2021). New insights in light-assisted microbial fuel cells for wastewater treatment and power generation: a win-win cooperation. *Journal of Power Sources*, **501**, 230000, https://doi.org/10.1016/j.jpowsour.2021.230000

Yadav A. K., Dash P., Mohanty A., Abbassi R. and Mishra B. K. (2012). Performance assessment of innovative constructed wetland-microbial fuel cell for electricity production and dye removal. *Ecological Engineering*, **47**, 126–131, https://doi.org/10.1016/j.ecoleng.2012.06.029

Yang Y., Zhao Y., Tang C., Xu L., Morgan D. and Liu R. (2020). Role of macrophyte species in constructed wetland-microbial fuel cell for simultaneous wastewater treatment and bioenergy generation. *Chemical Engineering Journal*, **392**(December 2019), 123708, https://doi.org/10.1016/j.cej.2019.123708

Yang Y., Qin M., Yang X. and He Z. (2017). Enhancing hydrogen production in microbial electrolysis cells by in situ hydrogen oxidation for self-buffering pH through periodic polarity reversal. *Journal of Power Sources*, **347**, 21–28, https://doi.org/10.1016/j.jpowsour.2017.02.046

Yi G., Wang B., Feng Y., Fang D., Yang L., Liu W., Zhang Y., Shao P., Pavlostathis S. G., Luo S., Luo X. and Wang A. (2022). The ins and outs of photo-assisted microbial electrochemical systems for synchronous wastewater

treatment and bioenergy recovery. *Resources, Conservation and Recycling*, **181**(September 2021), 106230, https://doi.org/10.1016/j.resconrec.2022.106230
Yuan P. and Kim Y. (2017). Increasing phosphorus recovery from dewatering centrate in microbial electrolysis cells. *Biotechnology for Biofuels*, **10**(1), 1–8, https://doi.org/10.1186/s13068-017-0754-8
Zhang X., Cheng S., Wang X., Huang X. and Logan B. E. (2009). Separator characteristics for increasing performance of microbial fuel cells. *Environmental Science and Technology*, **43**(21), 8456–8461, https://doi.org/10.1021/es901631p
Zhang M., Wang Y., Liang P., Zhao X., Liang M. and Zhou B. (2019). Combined photoelectrocatalytic microbial fuel cell (PEC-MFC) degradation of refractory organic pollutants and in-situ electricity utilization. *Chemosphere*, **214**, 669–678, https://doi.org/10.1016/j.chemosphere.2018.09.085
Zhang H., Yan Q., An Z. and Wen Z. (2022). A revolving algae biofilm based photosynthetic microbial fuel cell for simultaneous energy recovery, pollutants removal, and algae production. *Frontiers in Microbiology*, **13**, 990807, https://doi.org/10.3389/fmicb.2022.990807
Zhang F., Li J. and He Z. (2014). A new method for nutrients removal and recovery from wastewater using a bioelectrochemical system. *Bioresource Technology*, **166**, 630–634, https://doi.org/10.1016/j.biortech.2014.05.105
Zhang R. and Zhu Z. (2021). Microwave assisted hydrothermal conversion of waste cardboard. *Process Safety and Environmental Protection*, **156**, 209–218, https://doi.org/10.1016/j.psep.2021.10.005
Zhang Y., Jiang Q., Gong L., Liu H., Cui M. and Zhang J. (2020). In-situ mineral CO_2 sequestration in a methane producing microbial electrolysis cell treating sludge hydrolysate. *Journal of Hazardous Materials*, **394**(December 2019), 122519, https://doi.org/10.1016/j.jhazmat.2020.122519
Zhao J., Zhang H., Sun X., Hao S., Zhao P., Zhu X. and Dong S. (2022). Bifunctional bio-photoelectrochemical cells: a 'trading' platform for simultaneous production of electric power and hydrogen peroxide. *Journal of Materials Chemistry A*, **11**(2), 600–608, https://doi.org/10.1039/d2ta07761b
Zhou M., Chi M., Luo J., He H. and Jin T. (2011). An overview of electrode materials in microbial fuel cells. *Journal of Power Sources*, **196**(10), 4427–4435, https://doi.org/10.1016/j.jpowsour.2011.01.012

doi: 10.2166/9781789063813_0081

Chapter 5

Microbial fuel-cell technology for the treatment of dairy wastewater and concurrent generation of energy

*B. Monisha and Veena Gayathri Krishnaswamy**

Department of Biotechnology, Stella Maris College (Autonomous), Affiliated to University of Madras, Chennai, India
**Corresponding author: veenagayathri@stellamariscollege.edu.in*

ABSTRACT

Water pollution, which appears to be a global concern, is mostly caused by the discharge of several toxic compounds by various industries such as textiles, pulp and paper, and dairy sectors. Dairy wastewaters are difficult to manage because they include a wide range of contaminants. Wastewater treatment is an expensive and energy-intensive method that requires substantial energy to fulfil the energy demands of rising human population. Renewable energy-based wastewater treatment is a feasible option for overcoming these concerns. Microbial fuel cell (MFC) technology has shown promise as a sustainable strategy, combining energy and nutrient recovery to generate bioelectricity. MFCs can be employed in wastewater treatment, green power generation, biohydrogen synthesis, and ecologically friendly sewage treatment processes. This chapter focuses on different types of industrial wastewater and their physical, chemical, and biological treatment methods, as well as MFC technology and its techniques utilized for treating dairy wastewater, providing beneficial knowledge of combining it with currently employed conventional wastewater treatment procedures to achieve the degradation of various dairy contaminants. The chapter also discusses the types, processes, applications, challenges, and future prospects of wastewater treatment-related MFCs, with the goal of industrialization in the near future, leading to greener fuels and a more sustainable environment.

Keywords: dairy wastewater treatment, global issues, microbial fuel cells, pollutants, sustainability, generation of energy.

5.1 INTRODUCTION

Industrialization is a significant issue that leads to inadequate water resources and increased wastewater discharges. The emission of various hazardous chemicals by different industries is also the cause of water contamination, which seems to be a global problem (Malik *et al.*, 2022). However, one of the main sectors responsible for water contamination is the dairy industry and as the desire to enhance wastewater treatment standards has expanded, criteria have become significantly rigorous (Al-saned *et al.*, 2021). Due to its composition of lipids, proteins, and carbohydrates, dairy industry effluent is inherently complex. The Indian dairy industry produces 200 million tonnes of wastewater every year, largely from cleaning and washing operations, which is 2.5 times the overall quantity of milk produced worldwide (Elakkiya & Matheswaran, 2013). Dairy wastewater, with an excessive amount of pollutants dumped onto the land surface or into water, causes major environmental issues (Raghunath *et al.*, 2016). Aerobic and anaerobic biologic treatments, such as activated sludge method, trickling filter, aerated lagoons, and sequencing batch reactors (SBRs), two-phase anaerobic treatment systems, up-flow anaerobic sludge blanket (UASB), anaerobic digestion, membrane anaerobic reactor systems are frequently employed in treating dairy wastewater. Numerous disadvantages of these conventional systems exist, such as higher expenses, high energy consumption, and considerable sludge accumulation (Al-saned *et al.*, 2021). Microbial fuel cells (MFCs) are bioelectrochemical fuel cells that use active microorganisms as biocatalysts to produce bioenergy in anodic chambers, making them an environmentally friendly and sustainable method for dairy effluent removal (Dongre *et al.*, 2021). An MFC consists of a proton exchange membrane (PEM) that physically separates anode and cathode chambers (Ghasemi *et al.*, 2013). Organic substrates are oxidized by an active biocatalyst in the anode, which generates electrons and protons. Next, as the electrons are transported towards the cathode electrode through an external circuit, the PEM transports the protons to the cathode chamber (Mohyudin *et al.*, 2022). Due to the interaction of electrons and protons in the cathodic compartment, oxygen is reduced to water. A biocatalyst oxidizes carbon sources and generates electrons and protons, generating electric current through an external connection (Mostafa *et al.*, 2015). MFCs may also treat wastewater from domestic, agricultural, and dairy industries in addition to generating electricity (Mahato *et al.*, 2022). Depending on how electrons are conveyed from bacteria to the anode in MFCs, mediator MFCs and mediator-less MFCs can be differentiated. Mediator-less MFCs do not require mediators; a biofilm formed on the anode of mediator-less MFCs ferments complex organic compounds into simple products, which are then oxidized by electrochemically active organisms in the anode. However in mediator MFCs, electron shuttles or mediators are introduced into the system (Huang *et al.*, 2008; Mostafa *et al.*, 2015). Electrodes are crucial for exoelectrogenic biofilm growth and electrochemical reactions, enhancing the functionality and efficiency of MFCs. Granular activated carbon (GAC) or graphite granules (GGs) are the most common electrode materials used in MFCs, especially in large-scale systems because GAC has a high degree of micro-porosity and catalytic properties and GGs are more affordable with higher conductivity despite having a lower internal surface area (Huggins *et al.*, 2014). MFC technology has advanced significantly in recent years. Oxygen supply and consumption in the cathode chamber, substrate oxidation in the anode chamber, the electron shuttle from the anode compartment to the anode surface, and the permeability of the PEM are the primary factors which influence the functioning of an MFC (Mostafa *et al.*, 2015; Sharma & Li, 2010). The efficiency of an MFC was also enhanced using various electrodes such as activated carbon, carbon-based plain carbon paper, metal oxides, carbon cloth doped with nitrogen gas, and composites (Mohyudin *et al.*, 2022). MFCs provide sustainable electricity to remote populations, greatly lower the cost of water treatment, and reduce the pollution. To have a greater ecological impact, they may also generate electricity from waste materials such as wastewater and organic debris. This approach offers several ecological advantages. MFCs, similar to all fuel cells, have limitations such as limited power production capacity, high operation and material costs, and high maintenance and functionality requirements. The advantages and disadvantages of MFCs are listed in Table 5.1. This chapter discusses the effects of dairy effluents on the environment, MFC technology for the treatment of dairy wastewater, different types of MFCs, and their vital role in generating electricity.

Table 5.1 Advantages and disadvantages of MFCs.

Sl. No.	Advantages	Disadvantages
1	Eco-friendly (low carbon emission)	Low growth rate of microbes
2	Alternative source of fuel	Toxicity of the material
3	Low sludge producing	Requires high level of maintenance and functionality
4	Efficient conversion of substrates to electricity	High operation and material cost
5	Recovery of valuable compounds from substrates	Low power output

5.2 DIFFERENT INDUSTRIAL WASTEWATER TREATMENT METHODS

5.2.1 Textile industry

Textile processing is one of the most advanced and oldest industries. Due to a rise in textile mills and the wastewater they produce, global environmental issues have become significantly greater (Azanaw *et al.*, 2022). A large amount of extremely polluted water is released into the environment by the textile industry, which also consumes many types of synthetic colours and other chemicals. This dye-rich effluent has a substantial influence on the photosynthetic activity of aquatic plants and animals (Wei *et al.*, 2020). The use of synthetic colours may potentially influence human health as they contain compounds such as chlorine and heavy metals. There are numerous chemicals present in the wastewater generated from dyeing and printing processes, and their degradation is equally complicated. The structures of dyes, which are more complex and stable and primarily comprise aromatic rings such as benzene or naphthalene, are heterocyclic (Adane *et al.*, 2021). Although untreated effluent discharge into water bodies typically accounts for 80% of the textile industry's total emissions, this practice has been attributed to major damage to the environment (Wang, 2016). Biochemical oxygen demand (BOD), chemical oxygen demand (COD), total suspended solids (TSSs) and total dissolved solids (TDS), and temperature are all high in effluents, and contain significant salt concentrations and a high pH range (Azanaw *et al.*, 2022). Several chemical, biological, and physical techniques can be employed to treat textile effluents before it is reused in industrial and irrigation processes (Paździor *et al.*, 2019).

5.2.1.1 Physicochemical methods

Dye removal methods of physicochemical technology used for textile effluent wastewater include coagulation, ion exchange, filtration, and adsorption. The simplest and most common method is coagulation, which is used to remove contaminants from textile effluents. Alum and iron salts are frequently utilized as coagulants in this method (Queiroz *et al.*, 2020). Alum and activated carbon are the most efficient treatments, based on a study (Aleem *et al.*, 2020). The adsorption equilibrium separation technique is also used to remove dyes from wastewater. Although commercially available activated carbon is an effective adsorbent, it is expensive as well as difficult to recycle. Numerous studies have shown that low-cost absorbents are frequently used to remove dyes from wastewater. According to a study (Kumar & Ahmad, 2011), modified ginger waste is used as an adsorbent for the removal of crystal violet dye. Methylene blue and malachite green dyes were also removed using an agricultural waste from the powdered potato stems and leaves (Gupta *et al.*, 2016). For the biosorption method to remove dyes from textile wastewater effluents, a range of agricultural waste products has been employed, including wood apple shell (Jain & Jayaram, 2010), grapefruit peel (Saeed *et al.*, 2010), sugarcane bagasse ash (Kanawade & Gaikwad, 2011), *Capsicum annuum* seeds (Tunali Akar *et al.*, 2011), waste from palm trees (Belala *et al.*, 2011), and so on. To increase the effectiveness of wastewater treatment, filtration is used as a tertiary process. Nanofiltration is energy efficient as compared to other conventional separation techniques (Azanaw *et al.*, 2022). To soften hard water and with certain limitations in the removal of dyes, ion exchange treatments are used to eliminate cation and anion contaminants from effluents using synthetic resins. Salts, pollutants, and other impurities are removed from wastewater using the techniques of reverse osmosis and nanofiltration sheaths (Azanaw *et al.*, 2022).

5.2.1.2 Chemical treatment methods

Toxic contaminants from industry effluents are removed using chemical treatment technology. Effluents can be treated by chemical unit treatments such as chemical coagulation and precipitation, chemical oxidation, advanced oxidation processes (AOPs), chemical neutralization, and chemical stabilization (Samer, 2015). An ozonation process employs chemicals to physically and chemically eliminate synthetic colours from wastewater by using oxidizing agents such as ozone and hydrogen peroxide. Azo dye characteristic is a result of the integrated double bond being broken down by ozone gas during the ozonation process. It is beneficial because it may be used in a gaseous condition and does not produce solid waste. Even biodegradable dyes in wastewater have the potential to produce hazardous contaminants as a by-product (Miralles-Cuevas *et al.*, 2017). The incorporation of hydrogen peroxide and ultraviolet (UV) radiation in conjunction to remove dyes from wastewater is favourable owing to the formation of significant concentrations of hydroxyl radicals, which results in zero solid waste generation and a foul odour (Adane *et al.*, 2021). Hydroxyl radicals are extremely potent oxidizing agents, that are capable of rapidly reacting with dyes and inorganic and organic contaminants. The use of AOPs to remediate textile effluents is a promising strategy. AOPs involve Fenton's reagent and photocatalytic oxidation (Asghar *et al.*, 2015). Fenton's oxidation was an organic compound-based oxidation process that was catalysed by a metal. When ferrous sulphate ($FeSO_4$) and hydrogen peroxide (H_2O_2) are combined at a low pH, H_2O_2 is catalytically broken down by Fe^{2+}, which results in hydroxyl radicals with a strong oxidizing capacity that may rapidly oxidize organic molecules that are difficult to breakdown. As a result of the concurrent flocculation of reagents and dye molecules, the Fenton reaction develops iron sludge as a by-product (Aljuboury *et al.*, 2014). Under photocatalytic conditions, an innovative composite adsorbent consisting of calcium oxide loaded with silver nanoparticles had been employed to remove indigo carmine dyes (Alsohaimi *et al.*, 2020). According to a study on photocatalytic reaction, the combination of titanium dioxide (TiO_2) and H_2O_2 is the most effective method for treating textile effluents of organic contaminants (Garcia *et al.*, 2007).

5.2.1.3 Biological treatment methods

In biological processes of secondary treatment, the effluent is directed into a bioreactor where it is utilized by bacteria, algae, and fungus under either aerobic or anaerobic conditions (Samer, 2015). Anaerobic techniques are widely used for textile wastewater treatment because they endure more extensive organic loads, without the need for aeration, and generate lower sludge than aerobic procedures (Spagni *et al.*, 2012). Biological treatment methods provide a number of benefits over physical and chemical methods. They are economical, generate less sludge, and are ecologically sustainable (Singh *et al.*, 2019).

Aerobic treatments include activated sludge, oxidation ponds, aerobic digesters, and aeration lagoon, and anaerobic treatments include anaerobic digesters and septic tanks (Samer, 2015). Another aspect of the treatment is the removal of reactive colours from textile effluents using a fungal biomass, such as *Aspergillus fumigatus* (Kalaiarasi *et al.*, 2012). In another research, *Aspergillus niger*, a fungal culture, eliminated Remazol Brilliant Blue R and Acid Red 299 (NY1) (Benghazi *et al.*, 2014). In textile effluents, algae are also crucial for the adsorption of dyes. The ability of the brown macroalga *Stoechospermum marginatum* to absorb the colouring component Acid Orange II has been investigated (Kousha *et al.*, 2012). *Sargassum wightii* biomass is being utilized as a biosorbent to extract methylene blue dye from dyestuff (Kumar *et al.*, 2015). Dye degradation is a widespread microbiological process, and it is becoming more widespread. A study determined that the textile dye Reactive Orange 13 is degraded via enzymatic degradation by the bacterial stain *Alcaligenes faecalis* PMS-1 (Shah *et al.*, 2012). A 90% decolorizing activity against the dye Reactive Blue 19 was shown by the investigation to be possessed by *Enterobacter* sp. F (Holkar *et al.*, 2014). There are drawbacks of biological processes, such as drawn-out procedures, challenging-to-degrade dyes, and the presence of metals in polluted water can inhibit the growth of microbial species (Wei *et al.*, 2020).

5.2.2 Pulp and paper industry

The pulp and paper industry faces difficulties complying with environmental rules because it generates a significant volume of highly contaminated effluents. The third-largest source of wastewater is produced by the pulp and paper industry (Izadi *et al.*, 2018). The pulp and paper companies are extensive generators of natural resources, energy, and pollutants, all of which have a detrimental effect on the environment. Industries generate wastewater with inorganic and organic substances, elevated COD, toxic exposure, and limited biological degradation (Covinich *et al.*, 2014). Some of these contaminants are xenobiotics generated through pulp and paper manufacture, whereas others originate as naturally occurring contaminants. Even though the bleaching process creates substances that bioaccumulate, some of the substances are hazardous to fish and other aquatic life. BOD as well as COD levels in bleaching wastewater are significant, and they additionally include suspended matter, fatty acids, tannins, resins, and lignin (Ali & Sreekrishnan, 2001). Operations such as wood preparation, the pulping process, pulp washing, screening, bleaching, paper production, and coatings all affect the characteristics of the wastewater that has to be processed. The pulp and paper industry releases hazardous primary and secondary sludge that poses risks to both human health and the environment (Dixit *et al.*, 2020). Effluents should undergo treatment at wastewater treatment plants before being released into the environment (Dagar *et al.*, 2022).

5.2.2.1 Physicochemical methods

Due to their effectiveness in removing a range of pollutants from pulp and paper effluents, physicochemical treatment technologies such as coagulation, flocculation, reverse osmosis, adsorption, and oxidation have been widely utilized. The process of coagulation and flocculation is used to remove dissolved and suspended particles from wastewater (Izadi *et al.*, 2018). An investigation found that wet oxidation and coagulation together eliminated 51% of COD, 75% of lignin, and 83% of colour from effluents from pulp and paper mills (Buyukkamaci & Koken, 2010). Removing salt, colour, and adsorbable organic halogens is also effectively accomplished by a coupled membrane-based reverse osmosis and nanofiltration approach (Patel *et al.*, 2021). Coagulation has an impact on human health because it creates hazardous sludge and additional metals (Toczyłowska-Mamińska, 2017). Oxidation with ozone and hydrogen peroxide produces higher levels of pollutants (Covinich *et al.*, 2014). Due to membrane fouling, membrane-based technology also experiences flux reduction (Lin *et al.*, 2012). To remove harmful elements from the effluents produced from the pulp and paper industry, it is therefore vital to establish cost-effective and environmentally friendly approaches.

5.2.2.2 Chemical treatment methods

Treatment of paper machine filtrates, bleaching filtrates, and hazardous pulping effluents should be performed using chemical coagulation (Mehmood *et al.*, 2019). Aluminium chloride, alum, polyelectrolytes, and so on are among the most frequently used coagulants. Wastewater with a high concentration of suspended particles is treated using chemical coagulation following sedimentation. It was reported that ferric chloride, ferrous sulphate, aluminium chloride, and alum were shown to be effective in reducing the amount of carbon, colour, and turbidity in mechanical pulping effluents (Simonič & Vnučec, 2012). Using AOPs, the effluents from industries including petrochemical, textiles, and pulp and paper mills can possibly be treated. Ozonation, hydrogen peroxide, UV, Fenton's reagent, and photo-Fenton methods are applied to sterilize, disinfect, and remove colour and odour from the effluents produced by the paper and pulp manufacturing industry. Oxidant ozone may dissolve and remove pollutants with the aid of certain catalysts. The Fenton technique is chosen over other AOPs because it is more affordable and can function under typical solar radiation (Dixit *et al.*, 2020). AOPs have the ability to change resistant materials into inorganic compounds or partially mineral compounds into biodegradable materials. The use of hydroxyl radicals, which strike organic molecules for purpose of destruction, allow AOPs to remove both organic and inorganic contaminants (Covinich *et al.*, 2014).

5.2.2.3 Biological treatment methods

The utilization of bacteria, fungi, and enzymes in biological wastewater treatment processes allows for the removal of stray and delicate organic molecules from the environment. Microbes have found a way to reduce the load of organic compounds in effluents from pulp and paper manufacture, and this is a promising line of research (Patel *et al.*, 2021). Anaerobic treatments and activated sludge have been extensively employed in the great majority of pulp and paper mills worldwide. BOD, COD,

TSSs, and total organic carbon can be reduced by using activated sludge processes. An activated sludge process in a batch system has been shown to be capable of removing 74–95% COD from effluents of pulp and paper mills (Bengtsson *et al.*, 2008). Membrane bioreactor and moving-bed biofilm reactor techniques are used to effectively replace traditional biological treatments for the treatment of pulp and paper mill effluents (Iorhemen *et al.*, 2016). Anaerobic procedures are also used to treat wastewater from the pulp and paper industry. These processes include fluidized-bed reactors, rotating biological contractors, anaerobic sludge digestion, and UASB reactors. Fluidized-bed reactors have a greater effluent pollutant removal efficiency compared to UASB reactors, which requires less energy. Although anaerobic treatment generates biogas, it requires regular maintenance and monitoring. MFCs are a new field of research that can treat pulp and paper wastewater and complies with wastewater treatment standards for environmentally friendly treatment, but they must overcome challenges before large-scale application (Toczyłowska-Mamińska, 2017).

5.2.3 Dairy industry

Dairy production requires an extensive amount of water and produces effluent that is extremely polluted, making it one of the biggest contributors to wastewater in India. This effluent comprises organic materials including proteins, carbohydrates, and oil as well as undesirable variations in the pH level, temperature, and concentrations of both nitrate and phosphate. Untreated dairy effluent should not be released directly into water sources due to potential pollution issues such as dissolved oxygen reduction, volatile hazardous chemical release, loss of aquatic life, and increased biota sensitivity. Dairy companies require specialized procedures to control the pollutants in order to dispose of this effluent appropriately. Phytoremediation, aerobic and anaerobic treatments, reverse osmosis, electrochemical methods, and electrocoagulation can all be used to remediate dairy effluent. Additionally, natural coagulants could possibly be applied. The dairy industry treats contaminants using grease traps, oil–water separators, and flow equalization. To increase the amounts of phosphorus and nitrogen, reduce BOD, and keep organic carbon to a minimum, biological treatment uses both methods: anaerobic and aerobic. The key elements that comprise traditional aerobic treatment include activated sludge processes, rotating biological contactors, trickling filters, and aerated lagoons (Figure 5.1).

5.2.3.1 Different treatment techniques for dairy wastewater

A feasible technique for physico-chemically treating organic molecules in effluents is adsorption. The most often used adsorbent is activated carbon, while various alternatives can also be used, such as rice husk ash and coal-fly ash (Ahmaruzzaman & Gupta, 2011). To further treat dairy wastewater, chemical coagulants such as poly-aluminium chloride, ferric chloride, and aluminium sulphate might be used (Kaur, 2021). In accordance with further investigation, lactic acid bacteria that ferment lactose to lactic acid and degrade milk proteins may have the potential to naturally coagulate dairy effluents. COD was reduced by 65–78% and 49–82%, subsequently, when carboxymethyl cellulose and chitosan were employed (Seesuriyachan *et al.*, 2009). By using coagulants and flocculants to destabilize protein and lipid colloid particles, dissolved air flotation lowers organic loading. However, it necessitates more expensive chemical compounds which cause damage to the ecosystem and eliminate soluble materials. Before being disposed of, scum must be appropriately handled; however, if biopolymers are utilized, it can be used as a component in animal food (Kolev Slavov, 2017). The primary waste product from the manufacture of cheese is whey, which also has significant levels of organic material and mineral salts. Whey management has changed as a result of strict regulatory requirements for wastewaters and the high COD and BOD of whey. According to a study, utilizing nanofiltration to separate lactose from whey ultrafiltration permeate is effective (Cuartas-Uribe *et al.*, 2009).

Colloids and easily soluble pollutants are eliminated from the milk processing wastewater by chemical treatment. It comprises oxidation of reagents or pH level adjustment, and ferrous sulphate/hydrogen peroxide can remove as much as 80% of fats (Vlyssides *et al.*, 2012). To lessen negative effects, dairy wastewaters with extreme pH levels should be adjusted. To produce ideal coagulant conditions

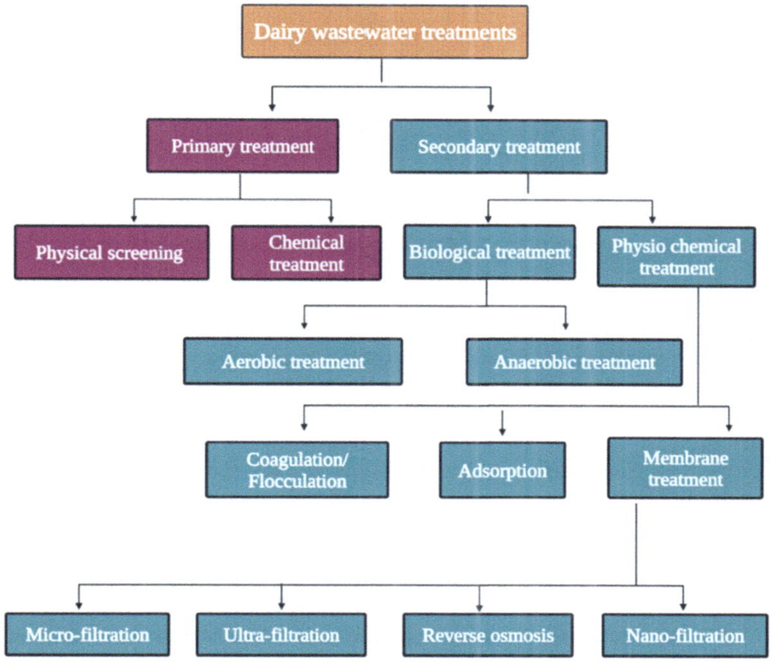

Figure 5.1 Flowchart of dairy wastewater treatment.

while using a dissolved air flotation, pH level must be maintained (Kolev Slavov, 2017). Through the process of electrocoagulation, pollutants from dairy effluents may be successfully removed using aluminium electrodes. During a 60-min interval at 60 V, electrocoagulation effectively removed >99.9% of bacterial markers as well as 98.84% of COD, 97 95% of BOD, and 97.75% of TSSs (Bazrafshan *et al.*, 2013). An efficient method for eliminating contaminants from the effluents of the manufacturing of ice cream has been demonstrated through research that combines electrocoagulation with sophisticated oxidation process technology. The outcomes have shown the efficacy of electrocoagulation, which can eliminate 40% of COD at a current density of 5 mA/cm^2 and increases efficiency to 25% when coupled with the Fenton procedure. By combining with the ozone processing, 30% more COD is additionally eliminated (Torres-Sánchez *et al.*, 2014).

On the basis of features, kinetic coefficient, and understanding of microorganisms, dairy wastewater may be treated by utilizing biological techniques such as active sludge processes. Organic waste from milk effluents has to be biologically degraded in order to be removed. The effluents from the production of milk and products correlated with milk are treated using the following methods: reduction of BOD and COD, total solids, volatile suspended solids, total Kjeldahl nitrogen, and total nitrogen. The elimination of phosphorus is less substantial, whereas nitrogen is reduced by 66–67% (Kaur, 2021). It was discovered that sludge thickening was caused by filamentous bacterial growth in an artificial effluent being treated in an anaerobic–anoxic–oxic system with a 7-day hydraulic retention time (HRT) and 20-day sludge age. Although total phosphorous was reliant on anoxic selector dimensions and the increase in nitrate, total nitrogen removal was maintained constant at 66%. However, a COD reduction of more than 90% was accomplished. Dairy wastewater with an elevated FOG (fats, oils, grease) concentration is treated using aerobic filters; however, the presence of excessive fats and biofilms could negatively impact productivity and result in biomass loss (Kolev Slavov, 2017). In SBRs, membrane-based methods effectively treat low-load dairy wastewater, obtaining higher BOD removal rates of over 97%, TSS-free wastewater, and total nitrogen reduction rates of 96% via assimilation (Bae *et al.*, 2003). Anaerobic systems are suitable for directly using dairy wastewater because they are less expensive than aerobic systems, do not generate disagreeable smell, and can be maintained adequately. High FOG degradation may be accomplished by employing synthetic milk wastewater in UASB reactors with FOG levels of 0.2, 0.6, and 1 g/L. Additionally, according to research, enzymatic pre-hydrolysis enhances the elimination of COD by 8.0% at the highest FOG levels (Kolev Slavov, 2017). In another study, significant amounts of BOD are efficiently eliminated from dairy effluents using membrane bioreactors; however, nitrogen is only partially removed due to membrane fouling, non-existence of anoxic zone, and expensive expenditures as well as operational expenses (Zhao *et al.*, 2020). An experiment utilizing mixed *Lactobacillus* and *Bacillus* in a fluidized bed bioreactor with low-density wood particles showed higher efficiency, obtaining 84% COD removal and 75% decreased BOD activity (Purushothaman *et al.*, 2020). In contrast to aerobic nitrogen, phosphorus, COD, and BOD removal using SBRs and up-flow anaerobic filters, dairy effluent treatment with membrane bioreactors and moving-bed bioreactors is not effective. As a consequence of SBR regarding excessive flow rates and ammonia, anaerobic treatment is inefficient. Combining aerobic and fermentative processes, which can replace conventional biological processes, would be a preferable strategy (Zhao *et al.*, 2020).

5.3 CHARACTERISTICS OF DAIRY WASTEWATER

Water is used extensively throughout the processing of milk for sanitation, washing, sterilization, heating, and cooling. Sanitary procedures account for 50–80% of the water utilized in the course of production, whereas the remaining 20–50% is clean water. The permitted wastewater generation in the dairy industry is 3 m^3/kL of milk processed, under the wastewater generation criteria (The Environment Protection rules, 1986). Effluents from the production of dairy products include starter cultures, by-products, pollutants, and washing of milk trucks, containers, equipment bottles, and reagents, as well as additives. Pollution of dairy effluent concentration differs based on an organization's operational description, capacity, manufacturing facilities, machinery, reuse of wastewater, and waste treatment. Carbohydrates, amino acids, and lipids can also be found in dairy effluents (Agrawal & Sarode, 2021). Upon hydrolysis, these organic compounds transform into sugars, acids, and fatty acids. The protein casein, milk sugar, fats, and inorganic salts are all found in dairy effluents, which results in significant BOD and COD. Dairy effluents have higher temperature, pH, TSSs, BOD, total nitrogen, phosphorus, fats, oil, and grease. The fermentation of milk sugar makes dairy waste whitish, somewhat alkaline, and acidic. The fine curd from the cheese waste gives it a suspended matter content, and the high oxygen demand imparts a polluting impact (Shete & Shinkar, 2013).

5.3.1 Sources of dairy industrial wastewater

5.3.1.1 Processing water

Through cooling and condensation, the drying of milk and whey generates the water needed for fermentation. These vapours, which could include volatile substances, whey, and milk droplets, can be collected or emitted with stormwater (Dongre *et al.*, 2021). They eradicate harmful substances and are reusable for employing in a variety of processes, including the production of steam, hot water, and membrane cleaning. For irrigation and room cleaning, water from the liquid cooling process utilized during pasteurization can be employed (Kolev Slavov, 2017).

5.3.1.2 Cleaning wastewater

Cleaning wastewater is frequently generated by machinery that has close association with dairy products, leaks, whey, clenching, defective machinery, and glitches in operation. More than 90% of the organic solids in effluents come from manufacturing by-products and dairy goods such as cheese, whey, butter, cream, yoghurt, fruit concentrates, and stabilizing agents (Dongre *et al.*, 2021). These effluents need further treatment because they are contaminated.

5.3.1.3 Sanitary wastewater

After secondary aerobic treatment, sanitary wastewater may provide nitrogen for dairy effluents and is utilized in restrooms, urban areas, and lavatories. When obtained individually from effluent streams, by-products of the production process which includes whey, milk, and whey permeates might be categorized into an entirely different group. In comparison to municipal wastewater, sanitary wastewater is usually routed straight to sewage treatment facilities (Kolev Slavov, 2017).

5.4 COMPOSITION OF DAIRY WASTEWATER

In comparison to sewage treatment facilities, dairy wastewater streams degrade biologically more quickly because they annually experience warmer temperatures of 17–25°C than municipal wastewater of 10–20°C. Milk and butter making factories have neutral pH, but dairy effluent has to be treated biologically at a pH of 6–9. Whey discharge causes the pH to reduce below 6.2. Although mineral acid coagulation makes acidic whey to have a pH of 4.3–4.6, the production of cheese formulates somewhat acidic whey with a pH of 5.9–6.6 (Venetsaneas *et al.*, 2009). Larger fat globules are extracted from milk by fat-rich foods such as cream, sour cream, and butter, which cause a variety of forms and concentrations of fat contaminants in the effluent. In comparison to other dairy effluents, the effluent from these products has considerably various kinds and concentrations of fat contaminants, and their removal by settling is more effective. The average FOG content in dairy effluents, which is mostly generated from high-fat products, is 0.2–0.4 g/L, with greater concentrations of 2.88 g/L by butter industries. The presence of lipids, proteins, and carbohydrates in quantities between 0.1 and 100 g/L contributes to the significant BOD and COD values in dairy effluents (Kolev Slavov, 2017). Lactose accounts for 90% of the absorption of BOD and COD. The amount of soluble COD increases by a significant solubility of lactose and is eliminated by biological processes. Because of the large COD weighs in milk and whey permeates, eliminating them directly into water bodies is not feasible (Wang *et al.*, 2009). With levels ranging from 0.8 to 77 g/L of COD and from 0.6 to 16 g/L of BOD, the organic matter contents in cheese whey effluents have elevated. Nitrogen and phosphorus contents in milk effluents vary accordingly (Carvalho *et al.*, 2013). The nitrogen level in urban dairies, dairies, and butter factories ranges from 4.2 to 6%, whereas cheese manufacturers account for 3.7% of BOD and have a phosphorus value of 0.6–0.7%. The amount of non-volatile suspended solids in primary as well as secondary sludge rises when industrial dairy effluents are highly salinized. Sodium, potassium, calcium, and chlorine ions are among the inorganic impurities, and they are produced in the highest concentrations in cheese and cottage cheese. The use of alkaline cleaning chemicals in milk facilities is indicated by increased sodium levels. It is important to take into account the wastewater pollutants caused by additives, cleaning agents, and other substances that enter sewer systems. Depending on the cleaning programme implemented, Clean-In-Place (CIP) technologies generate wastewater streams every 12–24 h with varying pH levels. The pH of the effluent is affected by CIP agents, which also provide less than 10% of BOD5 and COD load and up to 30% of the total flow rate of water to sanitation and sterilization processes (Kolev Slavov, 2017).

5.5 IMPACTS OF DAIRY EFFLUENTS ON THE ENVIRONMENT

On average, 2.5 L of wastewater are produced by the dairy sector for every litre of processed milk. Dairy effluent has a higher BOD and COD due to the presence of dairy product components such casein, inorganic salts, cleaning products, and disinfectants. As a result, environmental issues arise because these wastes are frequently disposed of into surrounding streams or land without being treated, exceeding the limitations established by the Indian Standard Institute and Bureau of Indian

Standards. The production of milk in the dairy sector has an effect on biodiversity, water quality, as well air quality. Additionally, the aquatic life and algae may be significantly impacted by high concentrations of polluted dairy effluents. The soluble organics, or suspended particles, in dairy effluents also aid in eutrophication (Shete & Shinkar, 2013). Even industrial activities may cause environmental pollution by releasing various types of atmospheric greenhouse gases such as carbon dioxide (CO_2), sulphur dioxide (SO_2), and nitrogen dioxide (NO_2). Although nitrous oxide (N_2O) was produced during the irrigation of effluents, methane (CH_4) emerges throughout the anaerobic waste treatment process. High particle emissions from stacks and other industrial operations can lead to dust, powdered form, and visual pollutants. Odour emissions must be taken into account at factories and manufacturing sites because waste treatment facilities might emit offensive odours (Raghunath *et al.*, 2016). Further research and the deployment of a variety of technologies are required to remediate dairy effluents in consideration of the significance of these emissions.

5.6 MICROBIOLOGY INVOLVED IN MFCs

MFCs employ the biocatalytic properties of surviving microbial organisms to transform the energy residing in chemical bonds into electrical current, minimizing the need for metal catalysts in the process. Bacterial organisms are capable of producing electrical power by utilizing organic material and biodegradable substances. Additionally, they have the ability to biodegrade or treat compounds that decompose, including dairy wastewater (Obileke *et al.*, 2021). MFC catalysts make use of organic-rich resources such as soil, freshwater sediment, marine sediment, polluted water, and activated sludge. Anaerobic substrate digestion frequently uses mixed cultures, although complex-mixed cultures are also permitted. Some single microorganisms produce electricity as a result of metabolic processes (Das & Mangwani, 2010). Mediating compounds affect microorganisms in open settings by transferring electrons from bacteria to electrodes. Numerous bacteria have been shown to be capable of generating electricity in MFCs by employing a mediator to speed up the flow of electrons between cells and anodic surfaces. In MFCs, the energy, hydrogen generation, and cost-effectiveness of *Geobacter* and *Shewanella* are investigated, with an emphasis on biological resilience and reactive stability. The electron transport to electrodes is carried out by exoelectrogens, microorganisms that release electrons exocellularly. Studies on dissimilatory metal-reducing bacteria such as *Geobacter* and *Shewanella* species show that outer-membrane cytochromes can be involved in this process. *Geobacter sulfurreducens* oxidizes organic substrates in order to transport electrons without the use of a mediator and to develop highly efficient nanowires. The recent discovery of nanowires introduces a whole new dimension to the study of extracellular electron transfer. The conductive, pili-like structures have been found in *G. sulfurreducens* PCA and *Shewanella oneidensis* MR-1. *G. sulfurreducens* was shown to be incapable of reducing insoluble electron acceptors when a pili gene was disrupted. However in *S. oneidensis*, deletion of MtrC- and OmcA-related genes resulted in poor conductivity of nanowires, decreased electrochemical activity, and inability to reduce insoluble electron acceptors (Logan & Regan, 2006). *Shewanella* species reduce their energy consumption by utilizing ferric iron or oxygen as terminal acceptors of electrons (Du *et al.*, 2007). However other studies have shown that *Geobacter metallireducens* uses the substrate acetate to oxidize organic materials in order to transfer electrons without the use of a mediator (Bond & Lovley, 2003). With an emphasis on temperature, mechanical pressure, pH level, and differences in the environment, the research process deals with the biological and reactive stability of microbes in MFCs. The results reveal a gradual rise in temperature and an abrupt drop in pH over time. Heat release and anaerobic microbial processes are held accountable for behavioural efficiency (Sahu, 2019). A dual-chambered MFC that used both aerobic and anaerobic anodic metabolism to generate energy from dairy industry wastewater had a maximum power density of 2.7 W/m^3 and a 91% COD removal efficiency. The most effective results were from anaerobic metabolism, which had a COD content of 1600 mg/L and a pH of 7, proving that MFCs are a reliable and significant source of bioenergy generation (Elakkiya &

Table 5.2 Standard norms of Central Pollution Control Board of India for dairy effluents.

Sl. No.	Industry	Parameter	Standards	
1	Dairy	Effluents	Concentration (mg/L) except for pH	Quantum per product processed
		pH	6.5–8.5	Nil
		BOD2 (3 days at 27°C)	100	Nil
		Suspended solids	150	Nil
		Oil and grease	10	Nil
		Wastewater generation	Nil	3 m^3/kL of milk

Source: Environment (Protection) Rules, 1986.

Table 5.3 Comparison of COD removal, coulombic efficiency, and power density generated in MFCs using dairy industry wastewater as substrate.

Sl. No.	MFC Type	Inoculum	COD	CODR (%)	Power Density	Columbic Efficiency (%)	References
1	Dual-chambered	Dairy wastewater	1600 mg/L	91	2.7 W/m^3	17	Elakkiya and Matheswaran (2013)
2	Single-chambered	Dairy wastewater	1487 mg/L	81	1.157 W/m^3	–	Mathuriya and Sharma (2010)
3	Single-chambered	Dairy wastewater	1000 mg/L	91	20.2 W/m^3	26.87	Mahdi Mardanpour *et al.* (2012)
4	Dual-chambered	Dairy wastewater	1.440–1.665 kg/m^3	53.50	144 mW/m^2	10.89	Parihar *et al.* (2022)

Matheswaran, 2013). On using a single-chambered MFC to generate electricity from dairy wastewater it had a maximum power density of 1.157 W/m^3 which was relatively a lower power generation when compared with other studies (Table 5.2) (Mathuriya & Sharma, 2010). However, an anode with graphite covering increases MFC efficiency, obtaining a maximum open circuit voltage of 810 mV and a power density of 20.2 W/m^3, substantially reducing COD and turbidity in dairy effluents. An annular single-chambered MFC also achieved a maximum columbic efficiency of 26.87% (Mahdi Mardanpour *et al.*, 2012). Parihar *et al.* (2022) revealed that using *Enterococcus faecalis*, which was isolated from dairy wastewater biofilms in a dual-chambered MFC, shows its potential for remediation of dairy wastewater and energy generation with a power density of 144 mW/m^2. Moreover, it demonstrated a COD removal efficiency of 53.5% and a coulombic efficiency of 10.89%. COD removal, coulombic efficiency, and power generation in MFCs using dairy industry wastewater as substrates in different studies are compared in Table 5.3.

5.7 MFCs IN DAIRY WASTEWATER TREATMENT AND ITS TYPES

MFCs are environmentally friendly bioelectrochemical devices that make use of bacteria for converting chemical energy into electrical energy. These green technologies enhance the handling of waste and bioelectricity generation. MFCs have anode and cathode chambers that are connected by a PEM and have active microorganisms on the anode side and abiotic material on the cathode. Exoelectrogens are microorganisms that function as biocatalysts, breaking down organic molecules to release electrons. Through an electric circuit, these electrons transfer to the cathode, where hydrogen ions combine with oxygen to generate water. As a result, MFCs have been proposed as a possible sustainable 'electricity' solution (Xu *et al.*, 2017). Due to its organic composition, which includes proteins and carbohydrates, dairy effluent is a good substrate for MFC applications. This helps remove COD and generate electricity. The biological degradation of colloidal organic materials, such as casein, may be the cause of the turbidity reduction in dairy effluent. It is a significant challenge to remove antibiotics from dairy effluent (Dongre *et al.*, 2021). The study highlights the effective functioning of a reactor as well as the production of energy using dual-chambered MFC technology and the treatment of dairy industry wastewater without the need of mediators or catalysts. However, higher power production is reported (Drisya & Manjunath, 2017). In the treatment of dairy wastewater, MFCs using graphite, carbon rods, aluminium rods, and stainless-steel electrodes yield elimination rates of 93.98, 90.63, and 57.52% for COD, BOD, and TDSs. This approach provides simultaneous effluent treatments and energy generation (Jyotishi & Rahi, 2021). A salt-bridge MFC with saline catholyte was used in laboratory research to treat dairy wastewater. The next generation of fuel-cell technology, two-chambered MFCs, attained maximum voltages and currents of 0.42 V, 0.46 V and 0.36 A, 0.42 A/L (Parkash *et al.*, 2015). A study examined 14 L reactors employing single-chambered air cathode MFCs for treating dairy wastewater with influent COD values ranging from 900 to 3830 mg/L. The system accomplished an organic elimination rate of 0.82 ± 0.11 kg COD/m^3/day, a mean organic removal efficiency of 67.9%, and an energy conversion efficiency of 17.4 J/kg COD removal (CODR). Small dairy companies may find that MFCs are a viable option for pre-treatment of effluents, achieving effluent discharge standards, and modifying to their specific capacity needs (Lóránt *et al.*, 2021). A dual-chambered MFC design for dairy wastewater treatment and bioelectricity production successfully removed 92.2% COD, 88.02% BOD, and 76.3% TDSs. Scaling up was possible as shown by the highest value of electricity generated, 644 mV (Sanjay & Udayashankara, 2019) (Figure 5.2).

As acetate substrate oxidized to carbon dioxide by microorganisms in the anodic chamber, electrons and protons are released. A direct contact with the electrochemically active carrier causes the diversion of electrons from the cell membrane through the electron mediator (Slate *et al.*, 2019). A further external integrated circuit transports these electrons to the cathode. To maintain charge neutrality, each transferred electron produces a proton that moves across a membrane that is permeable to proton to the cathodic chamber (Obileke *et al.*, 2021):

Anodic reaction:

$$CH_3COOH + H_2O \rightarrow 2CO_2 + 8H^+ + 8e^- \tag{5.1}$$

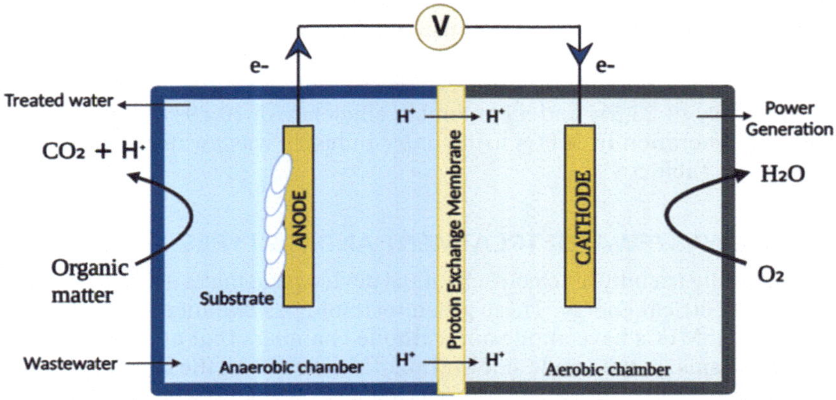

Figure 5.2 Schematic diagram of an MFC.

Electrons, beneficial by-products of the oxidation reaction, are produced at the cathode and are transported through a conducting wire to the water-filled cathodic chamber. Water molecules are generated when protons and oxygen combine with electrons (Slate *et al.*, 2019):

Cathodic reaction:

$$2O_2 + 8H^+ + 8e^+ \rightarrow 4H_2O \tag{5.2}$$

These molecules then pass across the PEM to the cathodic chamber where they are absorbed. For effective completion of the reaction, oxygen is provided. The overall reaction is: $CH_3COOH + 2O_2 \rightarrow 2H_2O + 2CO_2$ (Obileke *et al.*, 2021). The anode and cathode compartments are divided by a PEM, which allows protons to flow while restricting oxygen permeation. It hinders the exchange of electrons, oxygen, and substrate. Acetate is converted into carbon dioxide and water in the process, which also involves microbes (Barua *et al.*, 2019). There are two basic categories of MFCs: mediator MFCs and mediator-less MFCs.

5.7.1 Mediator MFCs

Chemical mediators such as neutral red, humic acid, and anthraquinone-2,6-disulphonate are used in mediator MFCs to transmit electrons from a cathode to an anode. These mediators are employed in effluent treatment methods and are known as 'electroactive metabolites'. Due to the presence of oxygen, which disrupts the mediator's function, anaerobic digestion is essential in mediator MFCs. Before presenting the final electron acceptor to the anode, the mediator receives electrons. The mediator deposits its electrons and then oxidizes back to its original condition (Flimban *et al.*, 2019). According to a study, bacteria either produce their mediator or transmit electrons to the electrode, allowing MFCs to function at elevated sustained activity levels (Obileke *et al.*, 2021).

5.7.2 Mediator-less MFCs

Without the need of intermediaries, some microorganisms can produce electricity. There are no additional external mediators introduced to the system in this sort of MFC. The majority of the bacteria in wastewater have a tendency to carry electrons to electrodes, which generate energy via lengthy extensions called nanowires (Logan *et al.*, 2006). A benefit of mediator-less MFCs over mediator MFCs is that it is non-toxic and more affordable.

Various two-compartment MFC designs are utilized in labs to maximize power outputs. Traditional rectangular, cylindrical, up-flow, and flat-plate MFCs are among these types; they may be used in batch or continuous modes of operation. To achieve the required power outputs, these designs are essential (Slate *et al.*, 2019).

5.8 APPLICATIONS OF MFCs

MFC technology has been advanced from its initial application of restricted wastewater treatment to handling several kinds of municipal, industrial, and agricultural wastewater. MFCs are cost-effective and stable, enabling the production of methane and bioelectricity during anaerobic digestion operations. It is a centuries-old method for treating dairy effluents, which is currently being used as a biosensor, as well as for bioelectricity generation, biohydrogen, nitrogen, and phosphorus recovery (Figure 5.3). They are capable of eliminating organic materials, sulphides, nitrites, phosphorus, and the salinity with up to 90% COD removal and 80% current efficiency (Dongre *et al.*, 2021). Wastewater is used as fuel by modern MFC technologies, which provide electricity for the generation of renewable energy. Alternative to generating energy, MFCs are capable of producing biohydrogen, resulting in hydrogen economy – a renewable supply for its needs. An excess voltage of 0.23 V or more must be applied to the anodic potential in order to produce hydrogen gas, and oxygen must be eliminated from the cathode chamber (Mostafa *et al.*, 2015). Pollutant analysis and process surveillance in biofuel cells are both possible with MFC technology. In addition to oxygen and hydrogen peroxide sensors, many

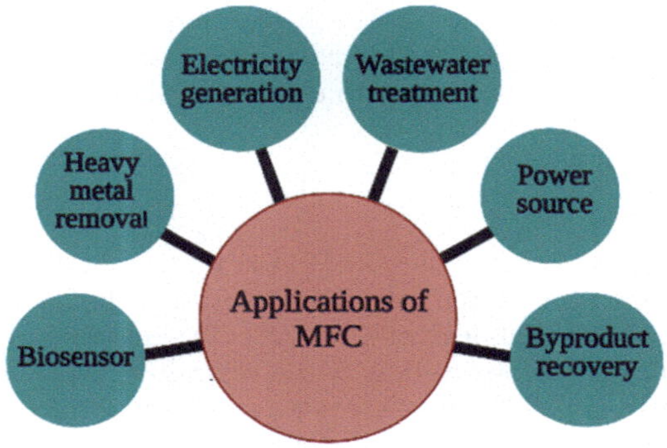

Figure 5.3 Applications of MFCs.

enzymatic glucose sensor types have been established. The elimination of hazardous substances such as phenols and petroleum compounds may also be achieved with MFCs, as well as the treatment of effluents from different industries and the production of renewable energy (Logan *et al.*, 2006; Obileke *et al.*, 2021). Additionally, they are able to manufacture biological electricity from trash produced by spacecraft. Using a biofilm electrode, MFC sensors monitor analyte concentration changes in aquatic environments, altering electrogenic microbes' metabolic activity and lowering necessity of external power (Yuan *et al.*, 2013).

5.9 GENERATION OF ENERGY USING MFCs

An appealing option for generating energy is MFC technology, which uses energy from microbial metabolism. MFCs are an effective, dependable process that uses non-toxic by-products and renewable energy sources. MFCs have demonstrated the ability to collect and transform chemical energy into electricity in situ, making them a feasible replacement for fossil fuels (Chaturvedi & Verma, 2016). A new photosynthetic MFC was constructed using dairy effluent as the anode chamber and *Synechococcus* sp. as the biocathode. The study revealed that a semi-continuous feed mode was advantageous for producing electricity and removing pollutants (Khodadi *et al.*, 2023). A dual-chambered MFC produced energy from dairy effluent using aerobic and anaerobic anodic metabolism. The co-modified polyoxometalate salt-based cathodes produced the most electricity, achieving a maximum power of 418.15 mW/m^2 after 96 h (Lachquer *et al.*, 2023). To treat dairy, municipal, and paper mill wastewater, a dual-chambered MFC with aluminium as a cathode and anode was used and in 10 days of operation, a power level of 12.10 mA was attained using three distinct types of MFCs (Vandana *et al.*, 2023). In MFCs, electrodes are made of graphite, carbon rods, aluminium rods, and stainless steel. According to a study, an MFC with stainless steel as the electrode and an agar–sodium chloride salt bridge as a PEM provides 37.651 μW of power and 0.0677 W-s of electrical energy over the 6-h detention period (Drisya & Manjunath, 2017). Further research reveals that the agar, sodium chloride salt bridge, when used as a PEM bridge, and graphite, carbon rods, aluminium rods, and stainless steel were utilized as electrodes, as is a more effective design, yielding energy with a voltage range of 359–1106 mV following a 10-day detention period (Jyotishi & Rahi, 2021) (Table 5.4). Also, utilizing activated sludge as a microbial consortium to explore the use of dairy effluent as a substrate for generating power in MFC technology and its performance were evaluated, which reveals a maximum power density of 40 mW/m^2, proving that it could effectively treat dairy effluent (Roh *et al.*, 2012). Dairy and household wastewater were used as feedstock in another study, which employed barium titanate nanoparticles as a cathode material in MFCs. With the maximum power densities rising from 64 to 698 mW/m^2, the

Table 5.4 Comparison of generated power density levels in MFCs using different types of electrodes.

Sl. No.	MFC Type	Electrode	Power Level	References
1	Dual-chambered	Polyoxometalate salt-based cathodes	418.15 mW/m²	Lachquer *et al.* (2023)
2	Dual-chambered	Aluminium as a cathode and anode	12.10 mA	Vandana *et al.* (2023)
3	MFCs	Stainless steel	37.651 μW	Drisya and Manjunath (2017)
4	MFCs	Graphite, carbon rods, aluminium rods, and stainless steel	359–1106 mV	Jyotishi and Rahi (2021)
5	MFCs	Barium titanate nanoparticles as a cathode	64–698 mW/m²	Touach *et al.* (2023)

energy performance improved under illuminating conditions. Water is also purified by this sustainable energy generation technique (Touach *et al.*, 2023).

5.10 CURRENT STRATEGIES AND FUTURE PROSPECTS

Due to many advantages, including environmental sustainability, MFCs play a significant role in the production of electricity and the removal of contaminants from effluents. The generation of electricity from organic materials, particularly from organic waste, is a promising application of MFC technology. However, they must also deal with challenges such as output design and economics. MFCs do not pose a serious threat to renewable energy sources or wastewater treatment systems. They can adapt to different chemical substrates and use biological processes to transform chemical energy into electrical energy. The primary disadvantage of using MFCs is inadequate power generation. Future barrier to this technique is the high cost of cathode catalysts, electrode components, and membranes. Future research should focus on developing an electrode with a larger area of surface that will help enhance the power production. It is essential to lower the operating costs and boost power output in order to make MFC technology commercially viable. However, by isolating effective bacteria or developing modified strains utilizing rDNA technologies, its low power density may be overcome. It is possible for bacterial strains to develop mediators that effectively transmit electrons to the anode, and finding a novel mediator which can enhance the efficacy of MFC technology. Further research should be conducted to minimize limiting factors and figure out metabolic processes in order to identify highly electrochemically active microbes. Researchers working on MFCs face additional challenges as a result of the prolonged time consumption. However, it is currently difficult for them to build large-scale MFCs that have both high-power generation and reliable performance.

5.11 CONCLUSION

MFC technology is environmentally friendly and can successfully remove pollutants from dairy wastewater. The environmental effects of dairy effluents, various types of MFCs, their vital role in energy production, and MFC technology for treating dairy wastewater have all been outlined in this chapter. Latest developments in MFC technologies have been comprehensively covered, including improved structural design, the use of innovative biocatalysts and materials for the anode, cathode, and biocathode, an extensive microbial community, substrate selection, and pollution removal. MFCs are widely acknowledged as an excellent and efficient solution for treating wastewater with the added benefit of producing electricity. However, because of their low power output and difficulty in scaling up, microorganisms and produced waste have been used to generate electricity until now. Therefore, it is possible that MFCs are becoming more and more viable for generating energy based on advances in power densities, the removal of COD, the degradation of pollutants, and increasing demands for electricity produced without CO_2 emissions. Given this, it is possible to conclude that MFCs might be commercialized in large-scale businesses by increasing power density and overall efficacy, reducing resource costs, and facilitating endless advancements in the environmentally friendly treatment of wastewater and electricity production.

ACKNOWLEDGEMENT

We would like to extend our gratitude to the Management of Stella Maris College (Autonomous) for their constant support.

REFERENCES

Adane T., Adugna A. T. and Alemayehu E. (2021). Textile industry effluent treatment techniques. *Journal of Chemistry*, **2021**, 5314404, https://doi.org/10.1155/2021/5314404

Agrawal V. and Sarode D. D. D. (2021). An overview on dairy industry wastewater and its Indian scenario. *International Journal of Engineering Applied Sciences and Technology*, **6**(1), 363–368, https://doi.org/10.33564/ijeast.2021.v06i01.055

Ahmaruzzaman M. and Gupta V. K. (2011). Rice husk and its ash as low-cost adsorbents in water and wastewater treatment. *Industrial and Engineering Chemistry Research*, **50**(24), 13589–13613, https://doi.org/10.1021/ie201477c

Al-saned A. J. O., Kitafa B. A. and Badday A. S. (2021). Microbial fuel cells (MFC) in the treatment of dairy wastewater. *IOP Conference Series: Materials Science and Engineering*, **1067**(1), 012073, https://doi.org/10.1088/1757-899x/1067/1/012073

Aleem M., Cao J., Li C., Rashid H., Wu Y., Nawaz M. I., Abbas M. and Akram M. W. (2020). Coagulation- and adsorption-based environmental impact assessment and textile effluent treatment. *Water, Air, and Soil Pollution*, Springer, **231**(2), 1–8, https://doi.org/10.1007/s11270-020-4400-x

Ali M. and Sreekrishnan T. R. (2001). Aquatic toxicity from pulp and paper mill effluents: a review. *Advances in Environmental Research*, **5**(2), 175–196, https://doi.org/10.1016/S1093-0191(00)00055-1

Aljuboury A., Palaniandy P., Bin H. and Aziz A. (2014). A review on the Fenton process for wastewater treatment. *Journal of Innovative Engineering*, **2**(3).

Alsohaimi I. H., Nassar A. M., Seaf Elnasr T. A. and Cheba B. (2020). A novel composite silver nanoparticles loaded calcium oxide stemming from egg shell recycling: a potent photocatalytic and antibacterial activities. *Journal of Cleaner Production*, **248**, 119274, https://doi.org/10.1016/j.jclepro.2019.119274

Asghar A., Raman A. A. A. and Daud W. M. A. W. (2015). Advanced oxidation processes for in-situ production of hydrogen peroxide/hydroxyl radical for textile wastewater treatment: a review. *Journal of Cleaner Production*, **87**(1), 826–838, https://doi.org/10.1016/j.jclepro.2014.09.010

Azanaw A., Birlie B., Teshome B. and Jemberie M. (2022). Textile effluent treatment methods and eco-friendly resolution of textile wastewater. *Case Studies in Chemical and Environmental Engineering*, **6**, 100230, https://doi.org/10.1016/j.cscee.2022.100230

Bae T. H., Han S. S. and Tak T. M. (2003). Membrane sequencing batch reactor system for the treatment of dairy industry wastewater. *Process Biochemistry*, **39**(2), 221–231, https://doi.org/10.1016/S0032-9592(03)00063-3

Barua E., Hossain M. S., Shaha M., Islam E., Zohora F. T., Protity A. T., Mukharjee S. K., Sarker P. K., Salimullah M. and Hashem A. (2019). Generation of electricity using microbial fuel cell (MFC) from sludge. *Bangladesh Journal of Microbiology*, **35**(1), 23–26, https://doi.org/10.3329/bjm.v35i1.39800

Bazrafshan E., Moein H., Kord Mostafapour F. and Nakhaie S. (2013). Application of electrocoagulation process for dairy wastewater treatment. Journal of Chemistry, **2013**, 640139, https://doi.org/10.1155/2013/640139

Belala Z., Jeguirim M., Belhachemi M., Addoun F. and Trouvé G. (2011). Biosorption of basic dye from aqueous solutions by date stones and palm-trees waste: kinetic, equilibrium and thermodynamic studies. *Desalination*, **271**(1–3), 80–87, https://doi.org/10.1016/j.desal.2010.12.009

Benghazi L., Record E., Suárez A., Gomez-Vidal J. A., Martínez J. and de la Rubia T. (2014). Production of the phanerochaete flavido-alba laccase in *Aspergillus niger* for synthetic dyes decolorization and biotransformation. *World Journal of Microbiology and Biotechnology*, **30**(1), 201–211, https://doi.org/10.1007/s11274-013-1440-z

Bengtsson S., Werker A. and Welander T. (2008). Production of polyhydroxyalkanoates by glycogen accumulating organisms treating a paper mill wastewater. *Water Science and Technology*, **58**(2), 323–330, https://doi.org/10.2166/wst.2008.381

Bond D. R. and Lovley D. R. (2003). Electricity production by *Geobacter sulfurreducens* attached to electrodes. *Applied and Environmental Microbiology*, **69**(3), 1548–1555, https://doi.org/10.1128/AEM.69.3.1548-1555.2003

Buyukkamaci N. and Koken E. (2010). Economic evaluation of alternative wastewater treatment plant options for pulp and paper industry. *Science of the Total Environment*, **408**(24), 6070–6078, https://doi.org/10.1016/j.scitotenv.2010.08.045

Carvalho F., Prazeres A. R. and Rivas J. (2013). Cheese whey wastewater: characterization and treatment. *Science of the Total Environment*, **445–446**, 385–396, https://doi.org/10.1016/j.scitotenv.2012.12.038

Chaturvedi V. and Verma P. (2016). Microbial fuel cell: a green approach for the utilization of waste for the generation of bioelectricity. *Bioresources and Bioprocessing*, **3**(1), 1–14, https://doi.org/10.1186/s40643-016-0116-6

Covinich L. G., Bengoechea D. I., Fenoglio R. and Area M. C. (2014). Advanced oxidation processes for wastewater treatment in the pulp and paper industry: a review. *American Journal of Environmental Engineering*, **4**(3), 56–70, https://doi.org/10.5923/j.ajee.20140403.03

Cuartas-Uribe B., Alcaina-Miranda M. I., Soriano-Costa E., Mendoza-Roca J. A., Iborra-Clar M. I. and Lora-García J. (2009). A study of the separation of lactose from whey ultrafiltration permeate using nanofiltration. *Desalination*, **241**(1–3), 244–255, https://doi.org/10.1016/j.desal.2007.11.086

Dagar S., Singh S. K. and Gupta M. K. (2022). Economics of advanced technologies for wastewater treatment: evidence from pulp and paper industry. *Frontiers in Environmental Science*, **10**, 960639, https://doi.org/10.3389/fenvs.2022.960639

Das S. and Mangwani N. (2010). Recent developments in microbial fuel cells: a review. *Journal of Scientific and Industrial Research*, **69**(10), 727–731.

Dixit M., Liu H., Luo J. and Shukla P. (2020). Effluents detoxification from pulp and paper industry using microbial engineering and advanced oxidation techniques. *Journal of Hazardous Materials*, **398**, 122998, https://doi.org/10.1016/j.jhazmat.2020.122998

Dongre A., Sogani M., Sonu K., Syed Z. and Sharma G. (2021). Treatment of dairy wastewaters: evaluating microbial fuel cell tools and mechanism. In: Environmental Issues and Sustainable Development, S. Sarvajayakesavalu and P. Charoensudjai (eds.), IntechOpen, Rijeka, Croatia, https://doi.org/10.5772/intechopen.93911

Drisya C. M. and Manjunath N. T. (2017). Dairy wastewater treatment and electricity generation using microbial fuel cell. *International Research Journal of Engineering and Technology*, **4**(8), 1293–1296.

Du Z., Li H. and Gu T. (2007). A state of the art review on microbial fuel cells: a promising technology for wastewater treatment and bioenergy. *Biotechnology Advances*, **25**(5), 464–482, https://doi.org/10.1016/j.biotechadv.2007.05.004

Elakkiya E. and Matheswaran M. (2013). Comparison of anodic metabolisms in bioelectricity production during treatment of dairy wastewater in microbial fuel cell. *Bioresource Technology*, **136**, 407–412, https://doi.org/10.1016/j.biortech.2013.02.113

Flimban S. G. A., Ismail I. M. I., Kim T. and Oh S. E. (2019). Overview of recent advancements in the microbial fuel cell from fundamentals to applications: design, major elements, and scalability. *Energies*, **12**(17), 1–20, https://doi.org/10.3390/en12173390

Garcia J. C., Oliveira J. L., Silva A. E. C., Oliveira C. C., Nozaki J. and de Souza N. E. (2007). Comparative study of the degradation of real textile effluents by photocatalytic reactions involving $UV/TiO_2/H_2O_2$ and $UV/Fe^{2+}/H_2O_2$ systems. *Journal of Hazardous Materials*, **147**(1–2), 105–110, https://doi.org/10.1016/j.jhazmat.2006.12.053

Ghasemi M., Wan Daud W. R., Ismail M., Rahimnejad M., Ismail A. F., Leong J. X., Miskan M. and Ben Liew K. (2013). Effect of pre-treatment and biofouling of proton exchange membrane on microbial fuel cell performance. *International Journal of Hydrogen Energy*, **38**(13), 5480–5484, https://doi.org/10.1016/j.ijhydene.2012.09.148

Gupta N., Kushwaha A. K. and Chattopadhyaya M. C. (2016). Application of potato (*Solanum tuberosum*) plant wastes for the removal of methylene blue and malachite green dye from aqueous solution. *Arabian Journal of Chemistry*, **9**, S707–S716, https://doi.org/10.1016/j.arabjc.2011.07.021

Holkar C. R., Pandit A. B. and Pinjari D. V. (2014). Kinetics of biological decolorisation of anthraquinone based Reactive Blue 19 using an isolated strain of *Enterobacter* sp. F NCIM 5545. *Bioresource Technology*, **173**, 342–351, https://doi.org/10.1016/j.biortech.2014.09.108

Huang L., Zeng R. J. and Angelidaki I. (2008). Electricity production from xylose using a mediator-less microbial fuel cell. *Bioresource Technology*, **99**(10), 4178–4184, https://doi.org/10.1016/j.biortech.2007.08.067

Huggins T., Wang H., Kearns J., Jenkins P. and Ren Z. J. (2014). Biochar as a sustainable electrode material for electricity production in microbial fuel cells. *Bioresource Technology*, **157**, 114–119, https://doi.org/10.1016/j.biortech.2014.01.058

Iorhemen O. T., Hamza R. A. and Tay J. H. (2016). Membrane bioreactor (MBR) technology for wastewater treatment and reclamation: membrane fouling. *Membranes*, **6**(2), 33, https://doi.org/10.3390/membranes6020033

Izadi A., Hosseini M., Najafpour Darzi G., Nabi Bidhendi G., Pajoum Shariati F. and Mosaddeghi M. R. (2018). Perspectives on membrane bioreactor potential for treatment of pulp and paper industry wastewater: a critical review. *Journal of Applied Biotechnology Reports*, **5**(4), 139–150, https://doi.org/10.29252/JABR.05.04.02

Jain S. and Jayaram R. V. (2010). Removal of basic dyes from aqueous solution by low-cost adsorbent: wood apple shell (*Feronia acidissima*). *Desalination*, **250**(3), 921–927, https://doi.org/10.1016/j.desal.2009.04.005

Jyotishi P. and Rahi D. C. (2021). Microbial Fuel Cell: An Application for Dairy Wastewater Treatment and Electricity Generation, Springer, Singapore, **77**, 1083–1089, https://doi.org/10.1007/978-981-15-5195-6_81

Kalaiarasi K., Lavanya A., Amsamani S. and Bagyalakshmi G. (2012). Decolourization of textile dye effluent by non-viable biomass of *Aspergillus fumigatus*. *Brazilian Archives of Biology and Technology*, **55**(3), 471–476, https://doi.org/10.1590/S1516-89132012000300019

Kanawade S. M. and Gaikwad R. W. (2011). Removal of dyes from dye effluent by using sugarcane bagasse ash as an adsorbent. International Journal of Chemical Engineering and Applications, **2**, 203–206, https://doi.org/10.7763/ijcea.2011.v2.103

Kaur N. (2021). Different treatment techniques of dairy wastewater. *Groundwater for Sustainable Development*, **14**, 100640, https://doi.org/10.1016/j.gsd.2021.100640

Khodadi S., Karbassi A., Tavakoli O., Baghdadi M. and Zare Z. (2023). Simultaneous dairy wastewater treatment and bioelectricity production in a new microbial fuel cell using photosynthetic *Synechococcus*. International Microbiology, **26**, 741–756, https://doi.org/10.1007/s10123-023-00328-2

Kolev Slavov A. (2017). Dairy wastewaters – general characteristics and treatment possibilities – a review. *Food Technology and Biotechnology*, **55**(1), 14–28, https://doi.org/10.17113/ftb.55.01.17.4520

Kousha M., Daneshvar E., Sohrabi M. S., Jokar M. and Bhatnagar A. (2012). Adsorption of acid orange II dye by raw and chemically modified brown macroalga *Stoechospermum marginatum*. *Chemical Engineering Journal*, **192**, 67–76, https://doi.org/10.1016/j.cej.2012.03.057

Kumar R. and Ahmad R. (2011). Biosorption of hazardous crystal violet dye from aqueous solution onto treated ginger waste (TGW). *Desalination*, **265**(1–3), 112–118, https://doi.org/10.1016/j.desal.2010.07.040

Kumar P. S., Pavithra J., Suriya S., Ramesh M. and Kumar K. A. (2015). *Sargassum wightii*, a marine alga is the source for the production of algal oil, bio-oil, and application in the dye wastewater treatment. *Desalination and Water Treatment*, **55**(5), 1342–1358, https://doi.org/10.1080/19443994.2014.924032

Lachquer F., Touach N., Benzaouak A., Oulmekki A., Lotfi E. M., El Mahi M., Hernández-Fernández F. J. and Toyir J. (2023). New application of polyoxometalate salts as cathode materials in single chamber MFC using wastewater for bioenergy production. *Processes*, **11**(3), 836, https://doi.org/10.3390/pr11030836

Lin H., Gao W., Meng F., Liao B. Q., Leung K. T., Zhao L., Chen J. and Hong H. (2012). Membrane bioreactors for industrial wastewater treatment: a critical review. *Critical Reviews in Environmental Science and Technology*, **42**(7), 677–740, https://doi.org/10.1080/10643389.2010.526494

Logan B. E. and Regan J. M. (2006). Microbial fuel cells – challenges and applications. *Environmental Science and Technology*, **40**(17), 5172–5180, https://doi.org/10.1021/es0627592

Logan B. E., Hamelers B., Rozendal R., Schröder U., Keller J., Freguia S., Aelterman P., Verstraete W. and Rabaey K. (2006). Microbial fuel cells: methodology and technology. *Environmental Science and Technology*, **40**(17), 5181–5192, https://doi.org/10.1021/es0605016

Lóránt B., Gyalai-Korpos M., Goryanin I. and Tardy G. M. (2021). Application of air cathode microbial fuel cells for energy efficient treatment of dairy wastewater. *Periodica Polytechnica Chemical Engineering*, **65**(2), 200–209, https://doi.org/10.3311/PPch.16695

Mahato J., Miah M., Shovon M. S., Roy N., Easmin M. S. and Sharma S. C. D. (2022). Electricity production by microbial fuel cell using cheese whey wastewater of the dairy industry in Rajshahi, Bangladesh. *Chemical and Biochemical Engineering Quarterly*, **35**(4), 421–430, https://doi.org/10.15255/CABEQ.2021.1922

Mahdi Mardanpour M., Nasr Esfahany M., Behzad T. and Sedaqatvand R. (2012). Single chamber microbial fuel cell with spiral anode for dairy wastewater treatment. *Biosensors and Bioelectronics*, **38**(1), 264–269, https://doi.org/10.1016/j.bios.2012.05.046

Malik S., Kishore S., Prasad S. and Shah M. P. (2022). A comprehensive review on emerging trends in industrial wastewater research. *Journal of Basic Microbiology*, **62**(3–4), 296–309, https://doi.org/10.1002/jobm.202100554

Mathuriya S. A. and Sharma V. N. (2010). Bioelectricity production from various wastewaters through microbial fuel cell technology. *Journal of Biochemical Technology*, **2**(2009), 133–137, http://www.jbiochemtech.com/index.php/jbt/article/view/JBT213

Mehmood K., Rehman S. K. U., Wang J., Farooq F., Mahmood Q., Jadoon A. M., Javed M. F. and Ahmad I. (2019). Treatment of pulp and paper industrial effluent using physicochemical process for recycling. *Water (Switzerland)*, **11**(11), 2393, https://doi.org/10.3390/w11112393

Miralles-Cuevas S., Oller I., Agüera A., Llorca M., Sánchez Pérez J. A. and Malato S. (2017). Combination of nanofiltration and ozonation for the remediation of real municipal wastewater effluents: acute and chronic toxicity assessment. *Journal of Hazardous Materials*, **323**, 442–451, https://doi.org/10.1016/j.jhazmat.2016.03.013

Mohyudin S., Farooq R., Jubeen F., Rasheed T., Fatima M. and Sher F. (2022). Microbial fuel cells a state-of-the-art technology for wastewater treatment and bioelectricity generation. *Environmental Research*, **204**, 112387, https://doi.org/10.1016/j.envres.2021.112387

Mostafa R., Arash A., Soheil D., Alireza Z. and Sang-Eun O. (2015). Microbial fuel cell as new technology for bioelectricity generation: a review. *Alexandria Engineering Journal*, **54**, 745–756.

Obileke K. C., Onyeaka H., Meyer E. L. and Nwokolo N. (2021). Microbial fuel cells, a renewable energy technology for bio-electricity generation: a mini-review. *Electrochemistry Communications*, **125**, 107003, https://doi.org/10.1016/j.elecom.2021.107003

Parihar P. S., Keshavkant S. and Jadhav S. K. (2022). Electrogenic potential of *Enterococcus faecalis* DWW1 isolated from the anodic biofilm of a dairy wastewater fed dual chambered microbial fuel cell. *Journal of Water Process Engineering*, **45**, 102503, https://doi.org/10.1016/j.jwpe.2021.102503

Parkash A., Aziz S., Nazir I. and Soomro S. A. (2015). Utilizing dairy wastewater for electricity generation using environment-friendly double chambered microbial fuel cell. *NUST Journal of Engineering Sciences*, **8**(1), 44–50.

Patel K., Patel N., Vaghamshi N., Shah K., Duggirala S. M. and Dudhagara P. (2021). Trends and strategies in the effluent treatment of pulp and paper industries: a review highlighting reactor options. *Current Research in Microbial Sciences*, **2**, 100077, https://doi.org/10.1016/j.crmicr.2021.100077

Paździor K., Bilińska L. and Ledakowicz S. (2019). A review of the existing and emerging technologies in the combination of AOPs and biological processes in industrial textile wastewater treatment. *Chemical Engineering Journal*, **376**, 120597, https://doi.org/10.1016/j.cej.2018.12.057

Purushothaman K. and Jena H. M. (2020). Biological treatment of synthetic dairy wastewater in FBBR. *Journal of the Indian Chemical Society*, **97**(12b), 2847–2853.

Queiroz R. de C. S. de, Lôbo I. P., Ribeiro V. de S., Rodrigues L. B. and Almeida Neto J. A. de. (2020). Assessment of autochthonous aquatic macrophytes with phytoremediation potential for dairy wastewater treatment in floating constructed wetlands. *International Journal of Phytoremediation*, **22**(5), 518–528, https://doi.org/10.1080/15226514.2019.1686603

Raghunath B. V., Punnagaiarasi A., Rajarajan G., Irshad A., Elango A. and Mahesh Kumar G. (2016). Impact of dairy effluent on environment – a review. Environmental Science and Engineering (Subseries: Environmental Science), 239–249, https://doi.org/10.1007/978-3-319-27228-3_22

Roh S. H., Lee S. W., Kim K. R. and Kim S. I. L. (2012). Electricity generation from dairy wastewater using microbial fuel cell. *Applied Chemistry for Engineering*, **23**(3), 297–301.

Saeed A., Sharif M. and Iqbal M. (2010). Application potential of grapefruit peel as dye sorbent: kinetics, equilibrium and mechanism of crystal violet adsorption. *Journal of Hazardous Materials*, **179**(1–3), 564–572, https://doi.org/10.1016/j.jhazmat.2010.03.041

Sahu O. (2019). Sustainable and clean treatment of industrial wastewater with microbial fuel cell. *Results in Engineering*, **4**, 100053, https://doi.org/10.1016/j.rineng.2019.100053

Samer M. (2015). Biological and chemical wastewater treatment processes. Wastewater Treatment Engineering, **150**, 61250, https://doi.org/10.5772/61250

Sanjay S. and Udayashankara T. H. (2019). Dairy wastewater treatment with bio-electricity generation using dual chambered membrane-less microbial fuel cell. *Materials Today: Proceedings*, **35**, 308–311, https://doi.org/10.1016/j.matpr.2020.01.533

Seesuriyachan P., Kuntiya A., Sasaki K. and Techapun C. (2009). Biocoagulation of dairy wastewater by *Lactobacillus casei* TISTR 1500 for protein recovery using micro-aerobic sequencing batch reactor (micro-aerobic SBR). *Process Biochemistry*, **44**(4), 406–411, https://doi.org/10.1016/j.procbio.2008.12.006

Shah P. D., Dave S. R. and Rao M. S. (2012). Enzymatic degradation of textile dye reactive orange 13 by newly isolated bacterial strain *Alcaligenes faecalis* PMS-1. *International Biodeterioration and Biodegradation*, **69**, 41–50, https://doi.org/10.1016/j.ibiod.2012.01.002

Sharma Y. and Li B. (2010). The variation of power generation with organic substrates in single-chamber microbial fuel cells (SCMFCs). *Bioresource Technology*, **101**(6), 1844–1850, https://doi.org/10.1016/j.biortech.2009.10.040

Shete B. S. and Shinkar N. P. (2013). Dairy industry wastewater sources, characteristics & its effects on environment department of civil engineering. *International Journal of Current Engineering and Technology*, **3**(5), 1611–1615.

Simonič M. and Vnučec D. (2012). Coagulation and UF treatment of pulp and paper mill wastewater in comparison. *Central European Journal of Chemistry*, **10**(1), 127–136, https://doi.org/10.2478/s11532-011-0121-8

Singh R. P., Singh P. K., Gupta R. and Singh R. L. (2019). Treatment and Recycling of Wastewater from Textile Industry, Springer, Singapore, 225–266, https://doi.org/10.1007/978-981-13-1468-1_8

Slate A. J., Whitehead K. A., Brownson D. A. C. and Banks C. E. (2019). Microbial fuel cells: an overview of current technology. *Renewable and Sustainable Energy Reviews*, **101**, 60–81, https://doi.org/10.1016/j.rser.2018.09.044

Spagni A., Casu S. and Grilli S. (2012). Decolourisation of textile wastewater in a submerged anaerobic membrane bioreactor. *Bioresource Technology*, **117**, 180–185, https://doi.org/10.1016/j.biortech.2012.04.074

Toczyłowska-Mamińska R. (2017). Limits and perspectives of pulp and paper industry wastewater treatment – a review. *Renewable and Sustainable Energy Reviews*, **78**, 764–772, https://doi.org/10.1016/j.rser.2017.05.021

Torres-Sánchez A. L., López-Cervera S. J., de la Rosa C., Maldonado-Vega M., Maldonado-Santoyo M. and Peralta-Hernández J. M. (2014). Electrocoagulation process coupled with advance oxidation techniques to treatment of dairy industry wastewater. *International Journal of Electrochemical Science*, **9**(11), 6103–6112.

Touach N., Benzaouak A., Toyir J., El Hamdouni Y., El Mahi M., Lotfi E. M., Labjar N., Kacimi M. and Liotta L. F. (2023). $BaTiO_3$ functional perovskite as photocathode in microbial fuel cells for energy production and wastewater treatment. *Molecules*, **28**(4), 1894, https://doi.org/10.3390/molecules28041894

Tunali Akar S., Gorgulu A., Akar T. and Celik S. (2011). Decolorization of Reactive Blue 49 contaminated solutions by *Capsicum annuum* seeds: batch and continuous mode biosorption applications. *Chemical Engineering Journal*, **168**(1), 125–133, https://doi.org/10.1016/j.cej.2010.12.049

Vandana Tiwari N. K. and Kumar R. (2023). Electricity production from different effluent using microbial fuel cell. Green Energy and Technology, 253–261, https://doi.org/10.1007/978-981-99-2279-6_22

Venetsaneas N., Antonopoulou G., Stamatelatou K., Kornaros M. and Lyberatos G. (2009). Using cheese whey for hydrogen and methane generation in a two-stage continuous process with alternative pH controlling approaches. *Bioresource Technology*, **100**(15), 3713–3717, https://doi.org/10.1016/j.biortech.2009.01.025

Vlyssides A. G., Tsimas E. S., Barampouti E. M. P. and Mai S. T. (2012). Anaerobic digestion of cheese dairy wastewater following chemical oxidation. *Biosystems Engineering*, **113**(3), 253–258, https://doi.org/10.1016/j.biosystemseng.2012.09.001

Wang D. M. (2016). Environmental Protection in Clothing Industry, World Scientific, 729–735, https://doi.org/10.1142/9789814749916_0076

Wang S., Chandrasekhara Rao N., Qiu R. and Moletta R. (2009). Performance and kinetic evaluation of anaerobic moving bed biofilm reactor for treating milk permeate from dairy industry. *Bioresource Technology*, **100**(23), 5641–5647, https://doi.org/10.1016/j.biortech.2009.06.028

Wei F., Shahid M. J., Alnusairi G. S. H., Afzal M., Khan A., El-Esawi M. A., Abbas Z., Wei K., Zaheer I. E., Rizwan M. and Ali S. (2020). Implementation of floating treatment wetlands for textile wastewater management: a review. *Sustainability (Switzerland)*, **12**(14), 5801, https://doi.org/10.3390/su12145801

Xu X., Zhao Q., Wu M., Ding J. and Zhang W. (2017). Biodegradation of organic matter and anodic microbial communities analysis in sediment microbial fuel cells with/without Fe(III) oxide addition. *Bioresource Technology*, **225**, 402–408, https://doi.org/10.1016/j.biortech.2016.11.126

Yuan Y., Zhou S., Liu Y. and Tang J. (2013). Nanostructured macroporous bioanode based on polyaniline-modified natural loofah sponge for high-performance microbial fuel cells. *Environmental Science and Technology*, **47**(24), 14525–14532, https://doi.org/10.1021/es404163g

Zhao K., Wu Y. W., Young S. and Chen X. J. (2020). Biological Treatment of Dairy Wastewater: A Mini Review. *Journal of Environmental Informatics Letters*, **4**(1), 22–31, https://doi.org/10.3808/jeil.202000036

doi: 10.2166/9781789063813_0101

Chapter 6

Bioenergy recovery from industrial wastewater through sustainable bioelectrochemical systems

Vishnuprasad Selvaraj[1], *Prasanna Kumar S. Mural*[2], *Abdul Rehman*[3] *and Karthick Senthilkumar*[3]*

[1]*Department of Chemical Engineering, Saintgits College of Engineering, Kottayam 686532, India*
[2]*Department of Metallurgical Engineering & Materials Science, Indian Institute of Technology Bombay, Maharashtra 400076, India*
[3]*Department of Chemical Engineering, Motilal Nehru National Institute of Technology, Allahabad 211004, India*
**Corresponding author: karthicks@mnnit.ac.in*

ABSTRACT

The rapid growth of industrialization has heightened the demand for energy, leading to increased pressure on finite petroleum resources. Consequently, research efforts have intensified to explore renewable and sustainable energy sources. Microbial fuel cell (MFC) technology has emerged as a promising bioelectrochemical platform, capable of generating bioelectricity while utilizing microorganisms to degrade organic contaminants found in wastewater. However, the successful scaling-up of MFCs remains a significant technical challenge, hindering their practical application. This chapter delves into the power generation potential of MFCs using various industrial wastewater substrates. It underscores the influence of crucial factors on cell performance, including substrate type, quantity, pH levels, and temperature regulation within the chambers. Despite its numerous advantages, this technology also presents certain challenges and potential outcomes, particularly concerning energy recovery from the effluents processed within MFCs.

Keywords: microbial fuel cell, wastewater, bioelectricity, green energy, bioelectrochemical system.

6.1 INTRODUCTION

The world is currently experiencing exponential population growth, coupled with a simultaneous expansion of industries striving to meet the demands of progress and development (Chen *et al.*, 2012; Yazdi *et al.*, 2015). This surge in human activities and industrialization places immense strain on non-renewable energy resources, prompting an urgent need to explore alternative and sustainable energy sources. The adverse environmental impacts of conventional energy production are becoming increasingly evident, necessitating a transition towards more environmentally friendly solutions. Recently, wastewater has emerged as an exceptionally valuable resource which not only offers a potential remedy for water scarcity but also represents a largely untapped source of energy (Xu *et al.*, 2021). Within wastewater lies a treasure trove of valuable materials such as alginate, bioplastics, cellulose, fibres, and metals, as well as essential nutrients such as phosphorus and nitrogen. The ability to harness the latent energy concealed within wastewater presents a dual opportunity: efficient waste management and the sustainable generation of energy. Microbial electrochemical technologies, particularly microbial fuel cells (MFCs) and microbial electrolysis cells (MECs), have been developed as promising tools to achieve dual objectives of waste management and energy recovery (Ren *et al.*, 2022; Sharma & Chhabra, 2023). These innovative technologies enable the biological treatment of wastewater while simultaneously recovering energy in the form of electricity (Gao *et al.*, 2021; Srivastava *et al.*, 2022). This convergence of wastewater treatment and energy recovery represents an environmentally sustainable approach to addressing the growing energy crisis and the escalating challenges of wastewater management (Hernández-Fernández *et al.*, 2015).

MFCs, in particular, have garnered attention in recent decades due to their capability to bridge the gap between wastewater treatment and energy conservation. The concept of using microorganisms to produce electricity in MFCs traces its roots back to 1911, laying the foundation for extensive research and development in the 20th century (Dong *et al.*, 2015; Velvizhi & Venkata Mohan, 2011). MFCs function by utilizing the metabolic processes of microorganisms, which feed on the organic substances found in wastewater and subsequently convert a portion of the energy into electricity. The surplus energy generated can be utilized for various applications, rendering MFCs an attractive option for sustainable energy production (Mohyudin *et al.*, 2022; Naina Mohamed *et al.*, 2018).

It was observed that wastewater contains approximately 10 times greater energy than that is required to treat them with adequate discharge standards (Ajiboye *et al.*, 2021; Pannell *et al.*, 2016). If this surplus energy can be harnessed efficiently, it has the potential to substantially reduce our reliance on non-renewable energy sources and mitigate the environmental impact of wastewater discharges. Over the years, extensive research has explored various aspects of MFCs, including cell designs, electrode materials, electron transport mechanisms, and membrane types. These efforts have aimed to optimize MFC performance and make them a practical and cost-effective solution for both wastewater treatment and energy generation (Neoh *et al.*, 2016; Pandey *et al.*, 2016). In this chapter, we will delve deeper into the world of MFCs, exploring their principles, applications, challenges, and potential future developments. By harnessing the power of microorganisms, MFCs offer a unique and promising avenue for simultaneously addressing environmental and energy-related concerns.

6.2 EMERGENCE OF MFCs

The global landscape of energy generation is undergoing a transformation, driven by the imperative need for sustainable solutions to meet the ever-increasing demand for power (Kumar *et al.*, 2013; Ortiz-Martínez *et al.*, 2015). Simultaneously, environmental conservation has taken centre stage in this evolving narrative. In this context, wastewater effluent originating from industrial, municipal, and various other sources emerges as a prime candidate for both energy harvesting and bioremediation (Lin *et al.*, 2023).

Microbial fuel technology, encapsulated within the realm of MFCs, stands as a beacon of innovation in the age-old quest for effective wastewater management. An MFC is generally a bioelectrochemical system (BES) which can be used as a power source. The basic principle of an MFC involves the use of bacteria or other microorganisms to break down organic substrates, producing electrons and protons as by-products. These electrons can then be captured and used as electrical energy (Mukherjee *et al.*, 2021; Varanasi *et al.*, 2019). The working principle of MFCs is shown in Figure 6.1. MFCs represent a convergence of microbiology, electrochemistry, and environmental science, offering an ideal solution to the long-standing challenges posed by wastewater treatment and energy production (Ieropoulos *et al.*, 2012; Sun *et al.*, 2017).

6.2.1 Understanding the parameters of MFC efficiency

The efficacy of MFC technology hinges on three fundamental parameters that serve as barometers of its efficiency.

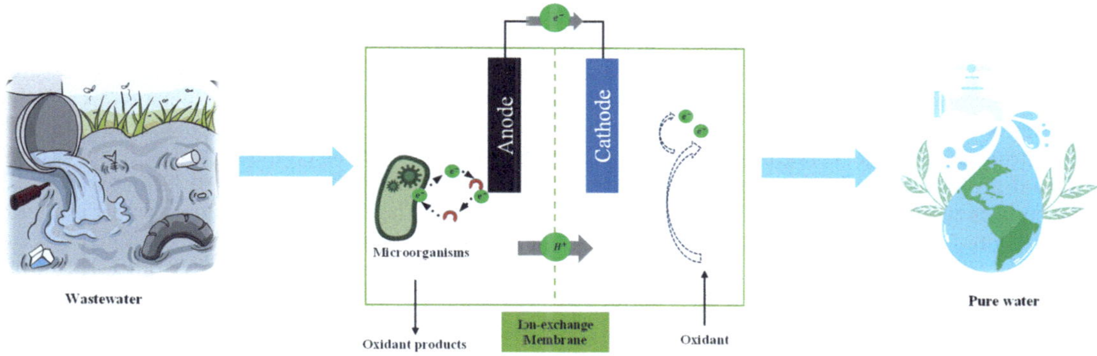

Figure 6.1 Working diagram of an MFC for sewage treatment and power generation.

6.2.1.1 Coulombic efficiencies: gauging electron transfer

Coulombic efficiency assumes a crucial role in evaluating MFC performance. It provides insights into the efficiency of electron transfer within the system, a critical factor in optimizing energy generation (He *et al.*, 2005). Understanding how efficiently electrons are harnessed during the degradation of organic matter is pivotal for enhancing MFCs' energy production capabilities (Christgen *et al.*, 2015; Rabaey *et al.*, 2005). Eslami *et al.* examined the coulombic efficiency in order to determine the performance of MFCs using recyclable membrane polypropylene. In that study, very low coulombic efficiency of about 0.02% was attained; the low efficiency was due to the low current attainment and the resistance occurred internally was so high. Also the consumption of organic substances in a substrate prevent the high generation of electricity in MFCs (Eslami *et al.*, 2023). In another investigation, Wang discussed how the nitrite concentration affects the coulombic efficiency in an MFC and found that the efficiency was reduced to 5.4% at 60 ppm of nitrite concentration in the influent whereas the output current was maximum for the same concentration of nitrite (Wang *et al.*, 2020). These studies proved that based on the type of substrate and organic substances the coulombic efficiency varied.

6.2.1.2 Chemical oxygen demand: a window into pollution mitigation

Chemical oxygen demand (COD) serves as a vital metric in assessing effectiveness of MFCs. It offers a quantifiable measure of the reduction in organic pollutants during the process of bioelectricity generation. As MFCs concurrently address wastewater treatment and energy production, COD measurements illuminate their role in mitigating environmental pollution (Cai *et al.*, 2018; He *et al.*, 2006; Pandey *et al.*, 2016). Hence this statement was confirmed by Ni *et al.*, (2009) and it showed that the concentration of substrate directly affected the removal of COD and voltage output. A sudden decrement in the substrate concentration decreased the voltage of an MFC (Cell & Swine, 2020). Vinasse was the major residue generated from various industries through the fermentation of alcohol. Vinasse is a liquid residue from the sugarcane-based ethanol industry. After sugarcane juice fermentation by yeast, ethanol concentration in the fermented broth is no more than 10% v/v (due to its toxicity). During distillation, the ethanol is recovered and everything left is called vinasse. It is produced in high volumes (12–15 L for each litre of ethanol) and is rich in minerals (Neto *et al.*, 2019). This liquid contained high organic matter that was used as a substrate in MFCs. The organic compound in the vinasse of about 6760 ppm produced 0.81 V with 83% COD removal and 0.61 V was generated from vinasse contained 10,604 ppm organic matter with a COD removal of about 93%. Therefore, below 6000 ppm concentration in MFCs led to achieve higher output voltage and COD removal was considered to be a saturated level of organic content in the vinasse to operate effectively in MFCs (López *et al.*, 2017). Thus, the concentration of substrate in wastewater and COD percentage directly affect the output voltage.

6.2.2 Diverse sources of wastewater for MFC applications

The adaptability of MFCs becomes apparent when exploring the diverse range of wastewater sources. Some of the sources are brewery wastewater, domestic wastewater, meat-packing wastewater, hydrogen fermentation reactor effluents, food-processing wastewater, and swine wastewater. Table 6.1 presents the diverse wastewater types employed as substrates within MFC systems. This extensive testing underscores the versatility of MFCs in addressing the unique challenges posed by various industrial and municipal wastewater streams (Fornero *et al.*, 2010; Oh *et al.*, 2010; Shi *et al.*, 2017). Some studies were discussed in this chapter that represented the sources of influents and effluents which affected the output voltage. Yang *et al.* explained that using double-chambered MFCs, the molasses wastewater was treated that produced output voltage of about 209 mV. In addition to that, the voltage rose to 535 mV after 10 days of operation (Yang *et al.*, 2022). In another study, the domestic

Table 6.1 Various types of wastewater used as substrates in MFC systems.

Sl. No.	Anode	Cathode	Substrate Type	COD (%)	P_{dmax}(mW/m^3)
1	Carbon cloth	Carbon cloth	D-Glucose	93	2160
2	Carbon cloth	Carbon cloth	D-Fructose	88	1810
3	Carbon cloth	Carbon cloth	L-Rhamnose	90	1320
4	Carbon cloth	Carbon cloth	D-Arabinose	93	2030
5	Carbon cloth	Carbon cloth	D-Galacturonic acid	80	1480
6	Carbon cloth	Carbon cloth	D-Gluconic acid	93	2050
7	Carbon cloth	Carbon cloth	Mannitol	91	1490
8	Carbon cloth	Carbon cloth	Arabitol	91	2030
9	Carbon cloth	Carbon cloth	Xylitol	91	2110
10	Carbon cloth	Carbon cloth	L-Asparagine	93	595
11	Carbon cloth	Carbon cloth	L-Glutamic acid	95	686
12	Carbon cloth	Carbon cloth	L-Lysine	93	592
13	Carbon cloth	Carbon cloth	L-Arginine	91	727
14	Graphite fibre brushes	Wet-proofed carbon cloth	Formic acid	NA	62
15	Graphite fibre brushes	Wet-proofed carbon cloth	Succinic acid	NA	444
16	Toray carbon paper	Carbon paper	Butyrate	>98	305
17	Graphite felt	Graphite felt	Propionate	NA	115.6
18	Toray carbon paper	Carbon paper	Ethanol	NA	488
19	Carbon paper	Pt-coated carbon paper	Pyridine	NA	142.1
20	Carbon paper	Pt-coated carbon paper	Indole	95	228
21	Graphite plate	Carbon cloth + graphite plate	Tetrathionate	NA	13.9
22	Plain carbon paper	Carbon paper with Pt	Sulphide	NA	13
23	Carbon cloth	Platinum cathode	Sulphate	NA	>10 μW/cm^2
24	Graphite plate	Graphite plate	Naphthalene	67	132
25	Graphite plate	Graphite plate	Fluorene	74	139

Source: Cha *et al.* (2010); Choudhury *et al.* (2017); Do *et al.* (2018); Muga and Mihelcic (2008); Oh *et al.* (2010); and Thulasinathan *et al.* (2022).

wastewater was used in up-flow MFCs where activated carbon was used as cathode material with output voltage of 5.2 V (Joel & Okabe, 2020). Abu-reesh *et al.* utilized petroleum refinery wastewater that contained a high amount of chemical energy. A dual-chambered fuel cell was used for treating this type of wastewater to recover 762 mV of voltage (Abu-reesh *et al.*, 2022). From these literature studies it was observed that the amount of voltage produced was varied with respect to different types of wastewater. Hence the source of wastewater also played a major role in voltage output.

6.2.3 Challenges of complex organic substrates

However, the path to harnessing the potential of wastewater is not without obstacles. Complex organic substrates, often residues from different sources, may introduce challenges interfering with electricity generation (Ajiboye *et al.*, 2021; Potter, 1911; Rahimnejad *et al.*, 2011). These challenges include potential toxicity due to elevated ammonia concentrations or the production of volatile acids during hydrolysis and substrate fermentation. Sonawane *et al.* reported that when different substrates were used, such as acetic acid and glucose, in an open circuit the voltage produced was 0.3 and 0.8 V (Sonawane *et al.*, 2022). Similarly, fruit waste was used as a substrate in an MFC and different voltages were produced, for example 0.259, 0.255, and 0.32 V when it was operated at different temperatures at 40, 50, and 60°C (Rahman *et al.*, 2021). Table 6.2 displays the physicochemical characteristics of wastewater. Addressing these complexities necessitates innovative strategies and a deep understanding of microbial interactions within MFCs.

6.2.4 Role of microbiology in BESs

Microbiology serves as the linchpin in the interdisciplinary realm of BESs. These innovative systems encompass a diverse array of technologies, with MFCs occupying a prominent position. MFCs are aided by enzymatic fuel cells, microbial desalination cells, microbial solar cells, MECs, microbial electro-synthesis cells (MESCs), and microbial reverse electro-dialysis cells (MRCs). Together, they represent a tapestry of innovation reshaping our approach to energy generation, wastewater treatment, and pollutant removal (Aiyer, 2020; Logan, 2010; Logan & Regan, 2006; Ucar *et al.*, 2017). A BES in the treatment of azo dyes was discussed by Sun *et al.* It clearly analysed the performance of a BES with respect to different parameters such as external resistance, application of potential, microorganisms concentration, azo dye concentration, and so on. It revealed that a predominant role was played by the microorganisms to degrade the organic pollutant in an efficient manner when compared to other parameters (Sun *et al.*, 2022). The major advantages of BESs are, it can treat even

Table 6.2 Physicochemical parameters of municipal textile and tannery wastewater.

Sl. No.	Parameters	Values		
		Textile Wastewater	Tannery Wastewater	Municipal Wastewater
1	Colour	Black	Greyish	Greyish black
2	pH	10.03 ± 0.01	6.36 ± 0.01	7.9 ± 0.02
3	BOD	465 ± 1	50 ± 0	140 ± 0.5
4	DO	1.7 ± 0.1	0.9 ± 0	4.35 ± 0.05
5	EC	2597 ± 2.64	18.56 ± 0.01	795 ± 1
6	Salinity	8.13 ± 0.05	10.98 ± 0.01	0.39 ± 0.005
7	Turbidity	19.9 ± 0.05	18.5 ± 0.11	3.7 ± 0.1
8	COD	854 ± 0	1134 ± 0	983 ± 0
9	TDS	7152 ± 2.08	9322 ± 1	398 ± 1.73

Source: Ajiboye *et al.* (2021); Neoh *et al.* (2016); Pannell *et al.* (2016) Ren *et al.* (2012); and Sikder and Rahman (2023). BOD: biological oxygen demand; DO: dissolved oxygen; EC: electrical conductivity; COD: chemical oxygen demand; TDS: total dissolved solids.

multiple pollutants and also less power and little concentration of substances are required to treat the wastewater (Srikanth *et al.*, 2014). One of the most efficient BES methods to produce electricity is using MFCs. Even valuable products can be produced from BESs using MECs (Horv *et al.*, 2023).

6.2.5 Mechanisms of MFCs

To unlock the full potential of MFCs, a deeper understanding of their operational mechanisms is paramount. These systems employ microorganisms as biocatalysts to oxidize organic matter, liberating electrons during degradation (Liu *et al.*, 2014; Schröder, 2007; Song *et al.*, 2019). These liberated electrons embark on a journey, traversing conductive electrode materials until they reach a terminal electron acceptor, ultimately resulting in the generation of bioelectricity. The typical electrode reactions ((6.1)–(6.3)) that occur in an MFC when an organic substrate such as glucose is used are shown below:

Reaction at the anode:

$$C_6H_{12}O_6 + 6H_2O \rightarrow 6CO_2 + 24H^+ + 24e^- \quad (6.1)$$

Reaction at the cathode:

$$6O_2 + 24e^- + 24H^+ \rightarrow 12H_2O \quad (6.2)$$

Overall reaction:

$$C_6H_{12}O_6 + 6O_2 \rightarrow 6CO_2 + 6H_2O \quad (6.3)$$

However, the realm of electrogenic microorganisms is diverse, showcasing various mechanisms for electron transfer through electrodes in different MFCs. This diversity underscores the adaptability of MFC technology to a wide array of organic substrates (Qiao *et al.*, 2008; Song *et al.*, 2019; Xu *et al.*, 2015b).

6.2.6 Versatile substrates for bioelectricity generation

In the pursuit of sustainable energy solutions, MFCs are successful as versatile platforms capable of generating bioelectricity from a myriad of organic substrates (Mateo *et al.*, 2018). Carbohydrates, starch, chitin, cellulose, organic acids, proteins, amino acids, pathogenic pollutants, and other toxic waste chemicals have all been harnessed to produce bioelectricity within MFCs (Pandey *et al.*, 2016; Pant *et al.*, 2010). This flexibility renders MFCs suitable for diverse scenarios, offering hope for a greener future.

6.2.7 MFCs in diverse wastewater applications

As we conclude our exploration of MFCs, we bear witness to their extensive applications across diverse wastewater-related domains. These applications extend from household and municipal settings to the bustling landscapes of the brewery and distillery operations, food industry, paper mills, swine farms, textile factories, metal-contaminated environments, mining industry sites, and even marine sediments (Pandey *et al.*, 2016; Thulasinathan *et al.*, 2022). MFCs present sustainable solutions that address a myriad of wastewater-related challenges, heralding a new era of greener and more efficient energy generation and waste management.

6.3 INCREASING INDUSTRIAL WASTEWATER AS AN AREA OF GLOBAL CONCERN CREATING POTENTIAL AREA FOR MFCs

In today's contemporary world, the challenges posed by wastewater production and its far-reaching societal impacts have become increasingly pronounced. Nations are actively advancing their developmental agendas, with sectors such as energy, healthcare, transportation, and infrastructure experiencing rapid growth (Kim *et al.*, 2016; Shah *et al.*, 2019). Consequently, the production of wastewater has surged, resulting in the contamination of both surface and groundwater sources. However, it is worth noting that wastewater, often viewed as a problem, also holds immense potential as an energy reservoir (Ajiboye *et al.*, 2021; Deepika *et al.*, 2015; Prabu *et al.*, 2011; Thulasinathan *et al.*, 2022).

The concept of harnessing renewable energy from wastewater is still in its nascent phase. Traditional thermal power plants, reliant on fossil fuels, continue to be a dominant source of energy generation (Kassouri *et al.*, 2022). Unfortunately, these facilities release substantial volumes of carbon dioxide (CO_2) along with other greenhouse gases, contributing significantly to climate change. With industrialization and economic expansion persisting at an unremitting pace, curtailing the generation of wastewater remains a challenging endeavour (Kazemi *et al.*, 2015; Srikanth *et al.*, 2014).

The causes of water pollution are multifaceted, stemming from residential areas, industrial processes, mining activities, and infiltration (Tansini *et al.*, 2022). However, one of the most noteworthy contributors to this predicament is the extensive water consumption by industries on a massive scale. Wastewater is commonly classified into various categories, which encompass rainfall runoff from non-absorbent surfaces, household wastewater, industrial wastewater, and agricultural wastewater. Various approaches to modification of MFCs for higher power generation are shown in Figure 6.2. Industrial wastewater encompasses a range of sources, from cooling water to variable composition washing effluents and biodegradable and potentially hazardous manufacturing or process water (Hammoudeh *et al.*, 2014).

Wastewaters exhibit substantial distinctions when compared to sources of drinking water, which are typically rivers, lakes, or reservoirs (Ahiahonu *et al.*, 2022; Srikanth *et al.*, 2014). They exhibit higher levels of contaminants, particularly from industrial sources. The indiscriminate discharge of wastewater from diverse origins inflicts harm on ecosystems, leading to detrimental effects such as hypertrophication, hypoxia, and algal blooms (Aryal *et al.*, 2017). Furthermore, the toxicity of these wastewaters varies according to their composition, dictated by their respective sources. The presence

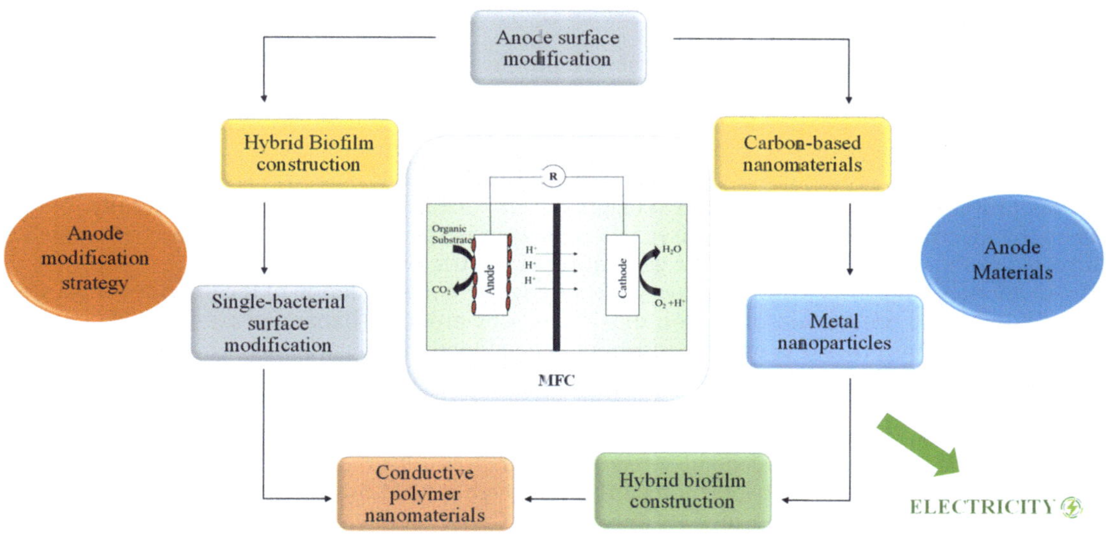

Figure 6.2 Various approaches in MFC modification for higher power generations.

of a multitude of pollutants in wastewater originating from various sources makes wastewater treatment a complex and multifaceted challenge.

Conventional methods for treating wastewater depend on a combination of chemical, physical, and biological techniques to eliminate solids, colloids, organic substances, nutrients, and soluble contaminants such as heavy metals and organics (Li *et al.*, 2008). Each of these treatment methods has its unique constraints, including considerations such as expenses, practicality, dependability, effectiveness, ecological consequences, operational demands, sludge production, preliminary treatment requirements, and the possibility of producing detrimental secondary substances (Liang *et al.*, 2018; Min *et al.*, 2005).

Current wastewater treatment approaches are well-known for their expensive set-up and significant energy demands. Additionally, these systems typically do not generate income or enhance the value of the treated wastewater (Ali *et al.*, 2015; Bird *et al.*, 2022). The idea of sustainable wastewater treatment is emerging as an encouraging answer to confront the urgent challenges of energy scarcity, resource depletion, and pollution. Sustainable treatment methods aim to reduce resource usage, operate with no net energy consumption, sustain consistent treatment efficiency, produce top-quality effluents suitable for water recycling, maintain a balance between investments and economic benefits, support social fairness, and minimize environmental consequences (Ge & He, 2016; Kim *et al.*, 2010). Zhang *et al.* utilized a double-chambered fuel cell to treat domestic wastewater. In that system, COD was removed up to 91.7% with a density of power up to 2.05 W/m^3 (Zhang *et al.*, 2016). In case of treating real wastewaters from textile industries, the COD removal percentage achieved was 98% with 123.2 W/m^3 power density (Mario *et al.*, 2017).

Effluent pollutants are usually eliminated through a combination of physicochemical and/or biological techniques, with research concentrating on cost-efficient and effective system combinations or innovative alternatives. Among these alternatives, MFCs and MECs emerge as more economically viable choices. These systems employ microorganisms as biocatalysts, enabling the breakdown of organic substances and the transfer of electrons to the anode's surface, ultimately leading to the generation of bioelectricity (Bhowmick *et al.*, 2019; Zhuang *et al.*, 2012).

6.4 PHENOMENA INVOLVED IN UTILIZING INDUSTRIAL WASTEWATER IN ELECTRICITY GENERATION

MFCs harness the potential of microorganisms to produce electrical energy through the biochemical breakdown of organic compounds. This intricate process involves essential components: a proton-exchange membrane (PEM) that facilitates separation, a cathode, and an anode (Firdous *et al.*, 2018). Inside an MFC, as organic substances undergo decomposition, it results in the generation of electrons and protons within the cell. These electrons travel through an external circuit to reach the cathode, whereas protons move through a PEM (Huggins *et al.*, 2013; Jiang *et al.*, 2011). Typically, oxygen acts as the electron acceptor, but in cases where oxygen is scarce within a cell, these electrons are redirected to the cathode, leading to the generation of electricity.

In an anodic chamber of MFCs, microorganisms engage in the oxidation of the provided substrates, which leads to the production of carbon dioxide as a by-product (Babanova *et al.*, 2020; Li *et al.*, 2015). Crucially, this process does not contribute to a net increase in carbon emissions, as the carbon originates from renewable biomass sources. Moreover, the bacteria in MFCs can be utilized to generate electricity from organic wastewater and domestic sewage. Figure 6.3 illustrates the advantages of an MFC system for wastewater treatment.

The generation of electricity depends on an external circuit for the movement of electrons within an anodic chamber. Concurrently, protons diffuse through a PEM into the cathodic chamber, where they combine with oxygen to form water molecules. Consequently, microorganisms in the anodic chamber extract electrons and protons from the organic substrates, employing them in a dissimilative manner (Oh *et al.*, 2010).

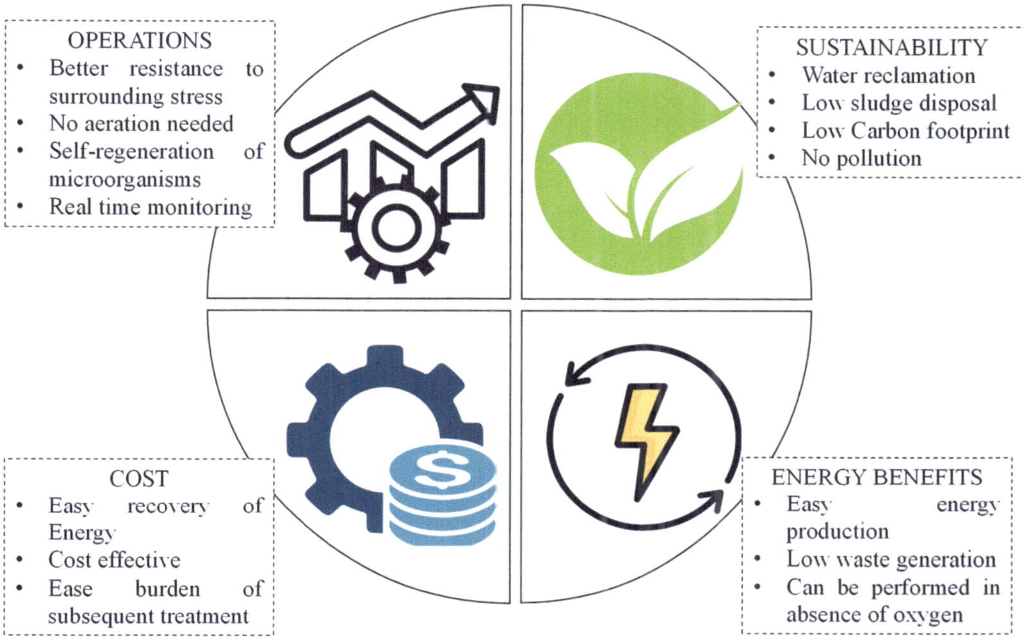

Figure 6.3 Pros of an MFC system for wastewater treatment.

In terms of its fundamental operating principle, MFCs exploit the metabolic capabilities of microorganisms to break down organic substrates and generate electricity through the transfer of electrons from the cell to the circuit. Inside the anodic chamber, bacteria decompose organic substrates, producing protons and electrons in the absence of oxygen. These electrons then traverse the electron transport chain to reach the final electron acceptor, which can be oxygen, nitrate, or Fe(III) (Liang *et al.*, 2021). However, in the absence of an electron acceptor, electrons may be used for electron transfer to the anode. One of the highest water utilizing industries is the dairy industry. It can be treated by MFCs with single- and double chambers connected in series or parallel. Mansoorian *et al.* produced the density of power up to 621.13 W/m^3 from a double-chambered MFC operated using a membrane in the absence of mediators. This system also reduced the level to 90.46% present in the wastewater (Mansoorian, 2016).

6.5 ROLE OF MICROBIAL COMMUNITY IN PHYTOREMEDIATION OF HEAVY METALS FROM INDUSTRIAL WASTEWATER

A variety of bacterial genera, such as *Bacillus*, *Mycobacterium*, *Paenibacillus*, *Alcaligenes*, *Acidovorax*, *Rhodococcus*, and *Pseudomonas*, have extensively examined for their roles in phytoremediation (Choudhury *et al.*, 2017; Ghasemi *et al.*, 2013). Additionally, microalgae have emerged as effective agents for environmentally friendly and practical bioremediation of heavy metals (Barbosa *et al.*, 2017; Mohan *et al.*, 2014). Significantly, the leftover microalgal biomass following bioremediation can be efficiently repurposed for the production of valuable nutraceuticals and metabolites. Enhancing the rhizosphere microbiota of plants to boost metal solubility facilitates the synthesis of organic acids and polysaccharides, providing a competitive advantage in phytoremediation (Ajiboye *et al.*, 2021; Lin *et al.*, 2023).

The interaction between plants and microorganisms, coupled with the uptake of heavy metals, encompasses a range of biochemical processes. These processes involve translocation, chelation, immobilization, solubilization, precipitation, volatilization, and complexation. For example, different elemental pollutants enter plants through nutrient transport systems. In phytodegradation, organic contaminants undergo metabolism within plant cells with the assistance of specific enzymes. Dehalogenase and nitro reductase enzymes play vital roles in breaking down anilines, pesticides, and chlorinated compounds, aiding in the degradation of nitro aromatic compounds (McAnulty *et al.*, 2017; Slate *et al.*, 2019). Table 6.3 displays the microorganisms utilized for the generation of bioelectricity in MFCs.

Metals exhibit a high degree of metallic activity, whereas non-metals lack metallic properties. Metalloids, on the contrary, occupy an intermediate position between metals and non-metals. Industrial areas often experience elevated concentrations of metals and metalloids due to continuous human activities. These pollutants pose significant health risks, including the development of diseases such as cancer. Therefore, it is imperative to raise awareness about the health hazards associated with toxic chemicals present in wastewater, especially concerning the absorption of metals into soil, water, and edible plants from contaminated sites (Li *et al.*, 2014; Mohanakrishna *et al.*, 2018).

Industrial wastewater frequently contains hazardous pollutants, such as manganese (Mn), nickel (Ni), cadmium (Cd), copper (Cu), arsenic (As), chromium (Cr), lead (Pb), zinc (Zn), and other heavy metals, which present challenges for growing herbs in affected fields. Moreover, certain plant species, including *Cistus salviifolius*, *Mentha suaveolens*, *Phytolacca americana*, *Digitalis purpurea*, *Agrostis castellana*, *Hypochaeris radicata*, *Pteridium aquilinum*, and *Pinus pinaster*, demonstrate potential for detoxifying metals and metalloids such as zinc (Zn), tungsten (W), antimony (Sb), and arsenic (As), making them viable options for wastewater treatment (Estrada-Arriaga *et al.*, 2018; Pannell *et al.*,

Table 6.3 Microorganisms utilized in MFCs for the power generation.

Sl. No.	Microorganism	Substrate/ co-substrate	Power density (mW/m^2)	Mediator
1	*Shewanella oneidensis* strain 14063	Sodium pyruvate	>40 for acid orange 7 (AO7); initial conc. is 70 mg/L	1-Amino-2-naphthol, one of the metabolites of AO7 reductive decolourization
2	*S. oneidensis*	Lactate	24	Anthraquinone-2,6-disulphonate (AQDS)
3	*Klebsiella pneumoniae*	Glucose	NA	HNQ as mediator biomineralized manganese as electron acceptor
4	*Rhodoferax ferrireducens*	Glucose, xylose sucrose, maltose	158	Mediator-less MFCs
5	*K. pneumoniae* strain L17	Glucose	34.77	Mediator-less MFCs
6	*Nocardiopsis* sp. KNU (S strain)	CMC	162	Mediator-less MFCs
7	*Streptomyces enissocaesilis* KNU (K strains)	CMC	145	Mediator-less MFCs
8	*Pseudomonas* species	Luria–Bertani (LB) medium	NA	Phenazine-1-carboxamide
9	*Pseudomonas* sp.	Peptone	979	Methylene blue
10	*Escherichia coli* strain K-12	Sucrose	215	Mediator-less MFCs

Source: Arun *et al.* (2020) and Chaturvedi and Verma (2016).

2016; Xu *et al.*, 2015a). These types of heavy metals were commonly treated using MFCs. But the role of microorganisms in treating heavy metals was highlighted in some studies. Wu *et al.* explained that *Gammaproteobacteria* was an enriched species found in the biofilm present in the anode of a double-chambered MFC and was majorly responsible for the removal of chromium metal and produced electricity simultaneously (Wu *et al.*, 2015). Three heavy metals such as copper, chromium, and cadmium were removed by Amanze *et al.*, through *Castellaniella* species. Nearly 100% removal efficiency was attained with more electric current output of about 320 mW/m^2 and a COD removal of about 91.15% (Amanze *et al.*, 2022).

6.6 FUTURE PERSPECTIVE OF EMPLOYING MFCs IN BIOELECTRICITY GENERATION

MFCs have the potential to address numerous challenges associated with conventional wastewater treatment methods. This rapidly evolving technology offers the unique advantage of simultaneously treating wastewater and harnessing power. MFCs represent a promising avenue for generating electricity from biomass through the utilization of bacteria and wastewater. Ongoing global research endeavours are dedicated to transforming this field into a viable energy producer, with recent findings revealing that wastewater contains approximately 5–10 times more energy than needed for its treatment. MFCs are gaining recognition for their role in constructing energy-efficient wastewater treatment systems, a process referred to as bioelectricity production, where organisms' metabolic processes lead to the production of electrons (Grattieri & Minteer, 2018; Waller & Trabold, 2013; Zabihallahpoor *et al.*, 2015).

There are several compelling advantages of using MFCs over other available options for generating energy from organic matter:

(1) *Enhanced conversion efficiency*: MFCs excel in converting the energy from substrates into electrical energy, resulting in an increased output and high conversion efficiency.
(2) *Ambient operation*: MFCs operate under ambient conditions and are capable of functioning at lower temperatures, distinguishing them from current bioenergy processes.
(3) *Minimal gas treatment*: In contrast to some alternative methods, MFCs do not require the treatment of off-gases, as they primarily consist of carbon dioxide with no other toxic residues.
(4) *Self-aeration*: MFCs offer the added advantage of self-aeration, eliminating the need for a separate aeration facility, as aeration can be integrated into the cathode itself.
(5) *Versatility and wide applicability*: MFCs demonstrate a broad range of advantages, particularly in industries reliant on electrical instruments and appliances. Their successful use in fuel applications significantly contributes to their appeal in meeting various energy requirements.

6.7 CONCLUSION

MFCs stand at the forefront of cutting-edge technology for generating electricity through microbial metabolism. However, several noteworthy limitations hinder their widespread adoption. A significant drawback is the low power density attained when utilizing xenobiotics and waste materials compared to purer carbon sources such as glucose. This limitation restricts their effectiveness in waste management and their capacity to generate electricity for everyday needs. Additionally, the cost of pure carbon sources can be prohibitive for routine electricity generation. Moreover, as MFCs are scaled-up for practical applications, there is an urgent need for further research aimed at improving their overall efficiency. Potential solutions to address these challenges include the isolation of highly efficient microorganisms and the utilization of genetically engineered strains. Additionally, the exploration of advanced techniques such as air cathodes, stacked reactors, and cloth electrode assemblies holds significant promise in surmounting practical barriers to the implementation of MFCs. Although MFCs offer a ground-breaking approach to sustainable energy generation and wastewater treatment, ongoing innovation and research are crucial to unlock their full potential and make them a viable solution for a wide range of applications.

COMPETING INTERESTS

The authors declare no competing interests.

REFERENCES

Abu-reesh I. M., Kunju A. and Sevda S. (2022). Performance of microbial fuel cells in treating petroleum refinery wastewater. *Journal of Water Process Engineering*, **49**, 103029, https://doi.org/10.1016/j.jwpe.2022.103029

Ahiahonu E. K., Anku W. W., Roopnarain A., Green E., Serepa-Dlamini M. H. and Poomani Govender P. (2022). Exploring indigenous freshwater chlorophytes in integrated biophotovoltaic system for simultaneous wastewater treatment, heavy metal biosorption, CO_2 biofixation and biodiesel generation. *Bioelectrochemistry*, **147**, 108208, https://doi.org/10.1016/j.bioelechem.2022.108208

Aiyer K. S. (2020). How does electron transfer occur in microbial fuel cells? *World Journal of Microbiology and Biotechnology*, **36**, 1–9, https://doi.org/10.1007/s11274-020-2801-z

Ajiboye T. O., Oyewo O. A. and Onwudiwe D. C. (2021). Simultaneous removal of organics and heavy metals from industrial wastewater: a review. *Chemosphere*, **262**, 128379, https://doi.org/10.1016/j.chemosphere.2020.128379

Ali A. E. H., Gomaa Ola M., Fathey R., El Kareem H. A. and Zaid M. A. (2015). Optimization of double chamber microbial fuel cell for domestic wastewater treatment and electricity production. *Journal of Fuel Chemistry and Technology*, **43**, 1092–1099, https://doi.org/10.1016/s1872-5813(15)30032-3

Amanze C., Zheng X., Man M., Yu Z., Ai C., Wu X., Xiao S., Xia M., Yu R., Wu X. and Shen L. (2022). Recovery of heavy metals from industrial wastewater using bioelectrochemical system inoculated with novel *Castellaniella* species. *Environmental Research*, **205**, 112467, https://doi.org/10.1016/j.envres.2021.112467

Arun S., Sinharoy A., Pakshirajan K. and Lens P. N. L. (2020). Algae based microbial fuel cells for wastewater treatment and recovery of value-added products. *Renewable and Sustainable Energy Reviews*, **132**, 110041, https://doi.org/10.1016/j.rser.2020.110041

Aryal N., Halder A., Zhang M., Whelan P. R., Tremblay P.-. L., Chi Q. and Zhang T. (2017). Freestanding and flexible graphene papers as bioelectrochemical cathode for selective and efficient CO_2 conversion. *Scientific Reports*, **7**, 1–8, https://doi.org/10.1038/s41598-017-09841-7

Babanova S., Jones J., Phadke S., Lu M., Angulo C., Garcia J., Carpenter K., Cortese R., Chen S., Phan T. and Bretschger O. (2020). Continuous flow, large-scale, microbial fuel cell system for the sustained treatment of swine waste. *Water Environment Research*, **92**, 60–72, https://doi.org/10.1002/wer.1183

Barbosa S. G., Peixoto L., Ter Heijne A., Kuntke P., Alves M. M. and Pereira M. A. (2017). Investigating bacterial community changes and organic substrate degradation in microbial fuel cells operating on real human urine. *Environmental Science: Water Research & Technology*, **3**, 897–904, https://doi.org/10.1039/c7ew00087a

Bhowmick G. D., Das S., Ghangrekar M. M., Mitra A. and Banerjee R. (2019). Improved wastewater treatment by combined system of microbial fuel cell with activated carbon/TiO_2 cathode catalyst and membrane bioreactor. *Journal of the Institution of Engineers (India): Series A*, **100**, 675–682, https://doi.org/10.1007/s40030-019-00406-7

Bird H., Heidrich E. S., Leicester D. D. and Theodosiou P. (2022). Pilot-scale microbial fuel cells (MFCs): a meta-analysis study to inform full-scale design principles for optimum wastewater treatment. *Journal of Cleaner Production*, **346**, 131227, https://doi.org/10.1016/j.jclepro.2022.131227

Cai L., Zhang H., Feng Y., Wang Y. and Yu M. (2018). Sludge decrement and electricity generation of sludge microbial fuel cell enhanced by zero valent iron. *Journal of Cleaner Production*, **174**, 35–41, https://doi.org/10.1016/j.jclepro.2017.10.300

Cell F. and Swine U. (2020). Effects of concentration variations on the performance and microbial community in microbial.

Cha J., Choi S., Yu H., Kim H. and Kim C. (2010). Directly applicable microbial fuel cells in aeration tank for wastewater treatment. *Bioelectrochemistry*, **78**, 72–79, https://doi.org/10.1016/j.bioelechem.2009.07.009

Chaturvedi V. and Verma P. (2016). Microbial fuel cell: a green approach for the utilization of waste for the generation of bioelectricity. *Bioresources and Bioprocessing*, **3**, 1–4, https://doi.org/10.1186/s40643-016-0116-6

Chen Q., Li J., Li X., Huang K., Zhou B., Cai W. and Shangguan W. (2012). Visible-light responsive photocatalytic fuel cell based on WO_3/W photoanode and Cu_2O/Cu photocathode for simultaneous wastewater treatment and electricity generation. *Environmental Science & Technology*, **46**, 11451–11458, https://doi.org/10.1021/es302651q

Choudhury P., Uday U. S. P., Mahata N., Tiwari O. N., Ray R. N., Bandyopadhyay T. K. and Bhunia B. (2017). Performance improvement of microbial fuel cells for waste water treatment along with value addition: a review on past achievements and recent perspectives. *Renewable and Sustainable Energy Reviews*, **79**, 372–389, https://doi.org/10.1016/j.rser.2017.05.098

Christgen B., Scott K., Dolfing J., Head I. M. and Curtis T. P. (2015). An evaluation of the performance and economics of membranes and separators in single chamber microbial fuel cells treating domestic wastewater. *PLoS ONE*, **10**, 1–13, https://doi.org/10.1371/journal.pone.0136108

Deepika J., Meignanalakshmi S. and Thilagaraj R. W. (2015). The optimization of parameters for increased electricity production by a microbial fuel cell using rumen fluid. *International Journal of Green Energy*, **12**, 333–338, https://doi.org/10.1080/15435075.2013.847838

Do M. H., Ngo H. H., Guo W. S., Liu Y., Chang S. W., Nguyen D. D., Nguyem L. D. and Ni B. J. (2018). Challenges in the application of microbial fuel cells to wastewater treatment and energy production: a mini review. *Science of the Total Environment*, **639**, 910–920, https://doi.org/10.1016/j.scitotenv.2018.05.136

Dong Y., Feng Y., Qu Y., Du Y., Zhou X. and Liu J. (2015). A combined system of microbial fuel cell and intermittently aerated biological filter for energy self-sufficient wastewater treatment. *Scientific Reports*, **5**, 1–8, https://doi.org/10.1038/srep18070

Eslami S., Bahrami M., Zandi M., Fakhar J., Gavagsaz-Ghoachani R., Noorollahi Y., Phattanasak M. and Nahid-Mobarakeh B. (2023). Performance investigation and comparison of polypropylene to Nafion117 as the membrane of a dual-chamber microbial fuel cell. *Cleaner Materials*, **8**, 100184, https://doi.org/10.1016/j.clema.2023.100184

Estrada-Arriaga E. B., Hernández-Romano J., García-Sánchez L., Guillén Garcés R. A., Bahena-Bahena E. O., Guadarrama-Pérez O. and Moeller Chavez G. E. (2018). Domestic wastewater treatment and power generation in continuous flow air-cathode stacked microbial fuel cell: effect of series and parallel configuration. *Journal of Environmental Management*, **214**, 232–241, https://doi.org/10.1016/j.jenvman.2018.03.007

Firdous S., Jin W., Shahid N., Bhatti Z. A., Abbasi U., Mahmood Q. and Ali A. (2018). The performance of microbial fuel cells treating vegetable oil industrial wastewater. *Environmental Technology & Innovation*, **10**, 143–151, https://doi.org/10.1016/j.eti.2018.02.006

Fornero J. J., Rosenbaum M. and Angenent L. T. (2010). Electric power generation from municipal, food, and animal wastewaters using microbial fuel cells. *Electroanalysis*, **22**, 832–843, https://doi.org/10.1002/elan.200980011

Gao Y., Wang S., Yin F., Hu P., Wang X., Liu Y. and Liu H. (2021). Enhancing sensitivity of microbial fuel cell sensors for low concentration biodegradable organic matter detection: regulation of substrate concentration. *Journal of Environmental Sciences*, **101**, 227–235, https://doi.org/10.1016/j.jes.2020.08.020

Ge Z. and He Z. (2016). Long-term performance of a 200 liter modularized microbial fuel cell system treating municipal wastewater: treatment, energy, and cost. *Environmental Science: Water Research & Technology*, **2**, 274–281, https://doi.org/10.1039/c6ew00020g

Ghasemi M., Daud W. R. W., Hassan S. H. A., Oh S. E., Ismail M., Rahimnejad M. and Jahim J. Md (2013). Nano-structured carbon as electrode material in microbial fuel cells: a comprehensive review. *Journal of Alloys and Compounds*, **580**, 245–255, https://doi.org/10.1016/j.jallcom.2013.05.094

Grattieri M. and Minteer S. D. (2018). Microbial fuel cells in saline and hypersaline environments: advancements, challenges and future perspectives. *Bioelectrochemistry*, **120**, 127–137, https://doi.org/10.1016/j.bioelechem.2017.12.004

Hammoudeh S., Nguyen D. K. and Sousa R. M. (2014). Energy prices and CO_2 emission allowance prices: a quantile regression approach. *Energy Policy*, **70**, 201–206, https://doi.org/10.1016/j.enpol.2014.03.026

He Z., Minteer S. D. and Angenent L. T. (2005). Electricity generation from artificial wastewater using an upflow microbial fuel cell. *Environmental Science & Technology*, **39**, 5262–5267, https://doi.org/10.1021/es0502876

He Z., Wagner N., Minteer S. D. and Angenent L. T. (2006). An upflow microbial fuel cell with an interior cathode: assessment of the internal resistance by impedance spectroscopy. *Environmental Science & Technology*, **40**, 5212–5217, https://doi.org/10.1021/es060394f

Hernández-Fernández F. J., Pérez De Los Ríos A., Salar-García M. J., Ortiz-Martínez V. M., Lozano-Blanco L. J., Godínez C., Tomás-Alonso F. and Quesada-Medina J. (2015). Recent progress and perspectives in microbial fuel cells for bioenergy generation and wastewater treatment. *Fuel Processing Technology*, **138**, 284–297, https://doi.org/10.1016/j.fuproc.2015.05.022

Horv N., Szuhaj M. and Kov L. (2023). Bioelectrochemical systems (BES) for biomethane production – review.

Huggins T., Fallgren P. H., Jin S. and Ren Z. J. (2013). Energy and performance comparison of microbial fuel cell and conventional aeration treating of wastewater. *Journal of Microbial & Biochemical Technology*, **6**, 1–5, https://doi.org/10.4172/1948-5948.s6-002

Ieropoulos I., Greenman J. and Melhuish C. (2012). Urine utilisation by microbial fuel cells; energy fuel for the future. *Physical Chemistry Chemical Physics*, **14**, 94–98, https://doi.org/10.1039/c1cp23213d

Jiang D., Curtis M., Troop E., Scheible K., McGrath J., Hu B., Suib S., Raymond D. and Li B. (2011). A pilot-scale study on utilizing multi-anode/cathode microbial fuel cells (MAC MFCs) to enhance the power production in wastewater treatment. *International Journal of Hydrogen Energy*, **36**, 876–884, https://doi.org/10.1016/j.ijhydene.2010.08.074

Joel N. D. and Okabe S. (2020). Domestic wastewater treatment and energy harvesting by serpentine up- flow MFCs equipped with PVDF-based activated carbon air-cathodes and a low voltage booster. *Chemical Engineering Journal*, **380**, 122443, https://doi.org/10.1016/j.cej.2019.122443

Kassouri Y., Bilgili F. and Kuşkaya S. (2022). A wavelet-based model of world oil shocks interaction with CO_2 emissions in the US. *Environmental Science & Policy*, **127**, 280–292, https://doi.org/10.1016/j.envsci.2021.10.020

Kazemi M., Biria D. and Rismani-Yazdi H. (2015). Modelling bio-electrosynthesis in a reverse microbial fuel cell to produce acetate from CO_2 and H_2O. *Physical Chemistry Chemical Physics*, **17**, 12561–12574, https://doi.org/10.1039/c5cp00904a

Kim J. R., Premier G. C., Hawkes F. R., Rodríguez J., Dinsdale R. M. and Guwy A. J. (2010). Modular tubular microbial fuel cells for energy recovery during sucrose wastewater treatment at low organic loading rate. *Bioresource Technology*, **101**, 1190–1198, https://doi.org/10.1016/j.biortech.2009.09.023

Kim K. Y., Yang W., Evans P. J. and Logan B. E. (2016). Continuous treatment of high strength wastewaters using air-cathode microbial fuel cells. *Bioresource Technology*, **221**, 96–101, https://doi.org/10.1016/j.biortech.2016.09.031

Kumar G. G., Sarathi V. G. S. and Nahm K. S. (2013). Recent advances and challenges in the anode architecture and their modifications for the applications of microbial fuel cells. *Biosensors and Bioelectronics*, **43**, 461–475, https://doi.org/10.1016/j.bios.2012.12.048

Li Z., Zhang X. and Lei L. (2008). Electricity production during the treatment of real electroplating wastewater containing Cr^{6+} using microbial fuel cell. *Process Biochemistry*, **43**, 1352–1358, https://doi.org/10.1016/j.procbio.2008.08.005

Li W. W., Yu H. Q. and He Z. (2014). Towards sustainable wastewater treatment by using microbial fuel cells-centered technologies. *Energy & Environmental Science*, **7**, 911–924, https://doi.org/10.1039/c3ee43106a

Li W., Niu Q., Zhang H., Tian Z., Zhang Y., Gao Y., Li Y-. Y., Nishimura O. and Yang M. (2015). UASB treatment of chemical synthesis-based pharmaceutical wastewater containing rich organic sulfur compounds and sulfate and associated microbial characteristics. *Chemical Engineering Journal*, **260**, 55–63, https://doi.org/10.1016/j.cej.2014.08.085

Liang P., Duan R., Jiang Y., Zhang X., Qiu Y. and Huang X. (2018). One-year operation of 1000-L modularized microbial fuel cell for municipal wastewater treatment. *Water Research*, **141**, 1–8, https://doi.org/10.1016/j.watres.2018.04.066

Liang B., Zhang X., Zhong M., Lv C. and Li K. (2021). Transition metal (Fe, Co, Ni) and sulfur codoped nitrogen-enriched hydrothermal carbon as high-performance cathode catalyst for microbial fuel cell. *Journal of Power Sources*, **506**, 230178, https://doi.org/10.1016/j.jpowsour.2021.230178

Lin C., Liang S., Yang X. and Yang Q. (2023). Toxicity monitoring signals analysis of selenite using microbial fuel cells. *Science of the Total Environment*, **862**, 160801, https://doi.org/10.1016/j.scitotenv.2022.160801

Liu X. W., Chen J. J., Huang Y. X., Sun X. F., Sheng G. P., Li D. B., Xiong L., Zhang Y. Y., Zhao F. and Yu H. Q. (2014). Experimental and theoretical demonstrations for the mechanism behind enhanced microbial electron transfer by CNT network. *Scientific Reports*, **4**, 1–7, https://doi.org/10.1038/srep03732

Logan B. E. (2010). Scaling up microbial fuel cells and other bioelectrochemical systems. *Applied Microbiology and Biotechnology*, **85**, 1665–1671, https://doi.org/10.1007/s00253-009-2378-9

Logan B. E. and Regan J. M. (2006). Electricity-producing bacterial communities in microbial fuel cells. *Trends in Microbiology*, **14**, 512–518, https://doi.org/10.1016/j.tim.2006.10.003

López Velarde Santos M., Rodríguez Valadéz F. J., Mora Solís V., González Nava C., Cornejo Martell A.J. and Hensel O. (2017). Performance of a microbial fuel cell operated with vinasses using different cod concentrations. *Revista Internacional de Contaminación Ambiental*, **33**, 521–528, https://doi.org/10.20937/RICA.2017.33.03.14

Mansoorian H. J. (2016). Evaluation of dairy industry wastewater treatment and simultaneous bioelectricity generation in a catalyst-less and mediator-less membrane microbial fuel cell. *Journal of Saudi Chemical Society*, **20**, 88–100, https://doi.org/10.1016/j.jscs.2014.08.002

Mario P., Urquizo G., Kadier A. and Logro W. (2017). Single chamber microbial fuel cell (SCMFC) with a cathodic microalgal biofilm: a preliminary assessment of the generation of bioelectricity and biodegradation of real dye textile wastewater. *Chemosphere*, **176**, 378–388, https://doi.org/10.1016/j.chemosphere.2017.02.099

Mateo S., Cañizares P., Rodrigo M. A. and Fernandez-Morales F. J. (2018). Driving force behind electrochemical performance of microbial fuel cells fed with different substrates. *Chemosphere*, **207**, 313–319, https://doi.org/10.1016/j.chemosphere.2018.05.100

McAnulty M. J., Poosarla V. G., Kim K. Y., Jasso-Chávez R., Logan B. E. and Wood T. K. (2017). Electricity from methane by reversing methanogenesis. *Nature Communications*, **8**, 15419, https://doi.org/10.1038/ncomms15419

Min B., Kim J. R., Oh S. E., Regan J. M. and Logan B. E. (2005). Electricity generation from swine wastewater using microbial fuel cells. *Water Research*, **39**, 4961–4968, https://doi.org/10.1016/j.watres.2005.09.039

Mohan S. V., Srikanth S., Chiranjeevi P., Arora S. and Chandra R. (2014). Algal biocathode for in situ terminal electron acceptor (TEA) production: synergetic association of bacteria-microalgae metabolism for the functioning of biofuel cell. *Bioresource Technology*, **166**, 566–574, https://doi.org/10.1016/j.biortech.2014.05.081

Mohanakrishna G., Abu-Reesh I. M., Kondaveeti S., Al-Raoush R. I. and He Z. (2018). Enhanced treatment of petroleum refinery wastewater by short-term applied voltage in single chamber microbial fuel cell. *Bioresource Technology*, **253**, 16–21, https://doi.org/10.1016/j.biortech.2018.01.005

Mohyudin S., Farooq R., Jubeen F., Rasheed T., Fatima M. and Sher F. (2022). Microbial fuel cells a state-of-the-art technology for wastewater treatment and bioelectricity generation. *Environmental Research*, **204**, 112387, https://doi.org/10.1016/j.envres.2021.112387

Muga H. E. and Mihelcic J. R. (2008). Sustainability of wastewater treatment technologies. *Journal of Environmental Management*, **88**, 437–447, https://doi.org/10.1016/j.jenvman.2007.03.008

Mukherjee A., Zaveri P., Patel R. and Shah M. T. (2021). Optimization of microbial fuel cell process using a novel consortium for aromatic hydrocarbon bioremediation and bioelectricity generation. *Journal of Environmental Management*, **298**, 113546, https://doi.org/10.1016/j.jenvman.2021.113546

Naina Mohamed S., Thota Karunakaran R. and Manickam M. (2018). Enhancement of bioelectricity generation from treatment of distillery wastewater using microbial fuel cell. *Environmental Progress & Sustainable Energy*, **37**, 663–668, https://doi.org/10.1002/ep.12734

Neoh C. H., Noor Z. Z., Mutamim N. S. A. and Lim C. K. (2016). Green technology in wastewater treatment technologies: integration of membrane bioreactor with various wastewater treatment systems. *Chemical Engineering Journal*, **283**, 582–594, https://doi.org/10.1016/j.cej.2015.07.060

Neto C. J. D., Letti L. A. J., Karp S. G., Vítola F. M. D. and Soccol C. R. (2019). Production of biofuels from algae biomass by fast pyrolysis. In: Biofuels from algae, Elsevier, A. Pandey, D-J. Lee, Y. Chisti and C.R. Soccol (eds.), pp. 461–473.

Ni B. J., Xie W-M., Liu S-G., Yu H-Q., Wang Y-Z., Wang G. and Dai X-L. (2009). Granulation of activated sludge in a pilot-scale sequencing batch reactor for the treatment of low-strength municipal wastewater, **43**(3), 751–761, https://doi.org/10.1016/J.WATRES.2008.11.009

Oh S. T., Kim J. R., Premier G. C., Lee T. H., Kim C. and Sloan W. T. (2010). Sustainable wastewater treatment: How might microbial fuel cells contribute. *Biotechnology Advances*, **28**, 871–881, https://doi.org/10.1016/j.biotechadv.2010.07.008

Ortiz-Martínez V. M., Salar-García M. J., de los Ríos A. P., Hernández-Fernández F. J., Egea J. A. and Lozano L. J. (2015). Developments in microbial fuel cell modeling. *Chemical Engineering Journal*, **271**, 50–60, https://doi.org/10.1016/j.cej.2015.02.076

Pandey P., Shinde V. N., Deopurkar R. L., Kale S. P., Patil S. A. and Pant D. (2016). Recent advances in the use of different substrates in microbial fuel cells toward wastewater treatment and simultaneous energy recovery. *Applied Energy*, **168**, 706–723, https://doi.org/10.1016/j.apenergy.2016.01.056

Pannell T. C., Goud R. K., Schell D. J. and Borole A. P. (2016). Effect of fed-batch vs. continuous mode of operation on microbial fuel cell performance treating biorefinery wastewater. *Biochemical Engineering Journal*, **116**, 85–94, https://doi.org/10.1016/j.bej.2016.04.029

Pant D., Van Bogaert G., Diels L. and Vanbroekhoven K. (2010). A review of the substrates used in microbial fuel cells (MFCs) for sustainable energy production. *Bioresource Technology*, **101**, 1533–1543, https://doi.org/10.1016/j.biortech.2009.10.017

Potter M. C. (1911). Electrical effects accompanying the decomposition of organic compounds the decomposition electrical effects accompanying of organic the fermentative activity of yeast and other organisms. *Cultures of Proceedings of the Royal Society*, **84**, 260–276.

Prabu S. L., Suriyaprakash T. N. K. and Kumar J. A. (2011). Wastewater treatment technologies: a review. *Pharma Times*, **43**, 9–13.

Qiao Y., Li C. M., Bao S. J., Lu Z. and Hong Y. (2008). Direct electrochemistry and electrocatalytic mechanism of evolved *Escherichia coli* cells in microbial fuel cells. *Chemical Communications*, **11**, 1290–1292, https://doi.org/10.1039/b719955d

Rabaey K., Clauwaert P., Aelterman P. and Verstraete W. (2005). Tubular microbial fuel cells for efficient electricity generation. *Environmental Science & Technology*, **39**, 8077–8082, https://doi.org/10.1021/es050986i

Rahimnejad M., Ghoreyshi A. A., Najafpour G. and Jafary T. (2011). Power generation from organic substrate in batch and continuous flow microbial fuel cell operations. *Applied Energy*, **88**, 3999–4004, https://doi.org/10.1016/j.apenergy.2011.04.017

Rahman W., Suzana Y. and Aida Mohammad S. N. A. (2021). Screening of fruit waste as substrate for microbial fuel cell (MFC). 020003.

Ren H., Lee H. S. and Chae J. (2012). Miniaturizing microbial fuel cells for potential portable power sources: promises and challenges. *Microfluidics and Nanofluidics*, **13**, 353–381, https://doi.org/10.1007/s10404-012-0986-7

Ren Z., Ji G., Liu H., Li P., Huang J. and Lichtfouse E. (2022). Higher performances of open vs. closed circuit microbial fuel cell sensor for nitrate monitoring in water. *Journal of Environmental Chemical Engineering*, **10**, 107807, https://doi.org/10.1016/j.jece.2022.107807

Schröder U. (2007). Anodic electron transfer mechanisms in microbial fuel cells and their energy efficiency. *Physical Chemistry Chemical Physics*, **9**, 2619–2629, https://doi.org/10.1039/b703627m

Shah S., Venkatramanan V. and Prasad R. (2019). Microbial fuel cell: sustainable green technology for bioelectricity generation and wastewater treatment. In: Shah, S., Venkatramanan, V., Prasad, R. (eds) Sustainable Green Technologies for Environmental Management. Springer-Singapore, pp. 199–218, https://doi.org/10.1007/978-981-13-2772-8_10

Sharma A. and Chhabra M. (2023). The versatility of microbial fuel cells as tools for organic matter monitoring. *Bioresource Technology*, **377**, 128949, https://doi.org/10.1016/j.biortech.2023.128949

Shi X., Leong K. Y. and Ng H. Y. (2017). Anaerobic treatment of pharmaceutical wastewater: a critical review. *Bioresource Technology*, **245**, 1238–1244, https://doi.org/10.1016/j.biortech.2017 08.150

Sikder S. and Rahman M. (2023). Efficiency of microbial fuel cell in wastewater (municipal, textile and tannery) treatment and bioelectricity production. *Case Studies in Chemical and Environmental Engineering*, **8**, 100421, https://doi.org/10.1016/j.cscee.2023.100421

Slate A. J., Whitehead K. A., Brownson D. A. C. and Banks C. E. (2019). Microbial fuel cells: an overview of current technology. *Renewable and Sustainable Energy Reviews*, **101**, 60–81, https://doi.org/10.1016/j.rser.2018.09.044

Sonawane J. M., Mahadevan R., Pandey A. and Greener J. (2022). Recent progress in microbial fuel cells using substrates from diverse sources. *Heliyon*, **8**, e12353, https://doi.org/10.1016/j.heliyon.2022.e12353

Song H. L., Zhu Y. and Li J. (2019). Electron transfer mechanisms, characteristics and applications of biological cathode microbial fuel cells – a mini review. *Arabian Journal of Chemistry*, **12**, 2236–2243, https://doi.org/10.1016/j.arabjc.2015.01.008

Srikanth S., Maesen M., Dominguez-Benetton X., Vanbroekhoven K. and Pant D. (2014). Enzymatic electrosynthesis of formate through CO_2 sequestration/reduction in a bioelectrochemical system (BES). *Bioresource Technology*, **165**, 350–354, https://doi.org/10.1016/j.biortech.2014.01.129

Srivastava R. K., Sarangi P. K., Vivekanand V., Pareek N., Shaik K. B. and Subudhi S. (2022). Microbial fuel cells for waste nutrients minimization: recent process technologies and inputs of electrochemical active microbial system. *Microbiological Research*, **265**, 127216, https://doi.org/10.1016/j.micres.2022.127216

Sun H., Luo S., Jin R. and He Z. (2017). Ensemble engineering and statistical modeling for parameter calibration towards optimal design of microbial fuel cells. *Journal of Power Sources*, **356**, 288–298, https://doi.org/10.1016/j.jpowsour.2017.02.051

Sun L., Mo Y. and Zhang L. (2022). A mini review on bio-electrochemical systems for the treatment of azo dye wastewater: state-of-the-art and future prospects. *Chemosphere*, **294**, 133801, https://doi.org/10.1016/j.chemosphere.2022.133801

Tansini A., Pavlovic J. and Fontaras G. (2022). Quantifying the real-world CO_2 emissions and energy consumption of modern plug-in hybrid vehicles. *Journal of Cleaner Production*, **362**, 132191, https://doi.org/10.1016/j.jclepro.2022.132191

Thulasinathan B., Jayabalan T., Arumugam N., Kulanthaisamy M. R., Kim W., Kumar P., Govarthanan M. and Alagarsamy A. (2022). Wastewater substrates in microbial fuel cell systems for carbon-neutral bioelectricity generation: an overview. *Fuel*, **317**, 123369, https://doi.org/10.1016/j.fuel.2022.123369

Ucar D., Zhang Y. and Angelidaki I. (2017). An overview of electron acceptors in microbial fuel cells. *Frontiers in Microbiology*, **8**, 1–14, https://doi.org/10.3389/fmicb.2017.00643

Varanasi J. L., Veerubhotla R., Pandit S. and Das D. (2019). Biohydrogen Production Using Microbial Electrolysis Cell: Recent Advances and Future Prospects. Microbial electrochemical technology. Elsevier B.V. (Netherlands), pp. 843–869. https://doi.org/10.1016/B978-0-444-64052-9.00035-2

Velvizhi G. and Venkata Mohan S. (2011). Biocatalyst behavior under self-induced electrogenic microenvironment in comparison with anaerobic treatment: evaluation with pharmaceutical wastewater for multi-pollutant removal. *Bioresource Technology*, **102**, 10784–10793, https://doi.org/10.1016/j.biortech.2011.08.061

Waller M. G. and Trabold T. A. (2013). Review of microbial fuel cells for wastewater treatment: large-scale applications, future needs and current research gaps. In International Conference on Fuel Cell Science, Engineering and Technology, American Society of Mechanical Engineers. **55522**, 1–7, https://doi.org/10.1115/FuelCell 2013-18185

Wang R., Wang X. and Zhou X. (2020). Effect of anolytic nitrite concentration on electricity generation and electron transfer in a dual-chamber microbial fuel cell. *Environmental Science and Pollution Research*, **27** (2020), 9910–9918, https://doi.org/10.1007/s11356-019-07323-z

Wu X., Zhu X., Song T., Zhang L., Jia H. and Wei P. (2015). Effect of acclimatization on hexavalent chromium reduction in a biocathode microbial fuel cell. *Bioresource Technology*, **180**, 185–191, https://doi.org/10.1016/j.biortech.2014.12.105

Xu B., Ge Z. and He Z. (2015a). Sediment microbial fuel cells for wastewater treatment: challenges and opportunities. *Environmental Science: Water Research & Technology*, **1**, 279–284, https://doi.org/10.1039/c5ew00020c

Xu L., Zhang G. Q., Yuan G. E., Liu H. Y., Liu J. D. and Yang F. L. (2015b). Anti-fouling performance and mechanism of anthraquinone/polypyrrole composite modified membrane cathode in a novel MFC-aerobic MBR coupled system. *RSC Advances*, **5**, 22533–22543, https://doi.org/10.1039/c5ra00735f

Xu M., Li J., Liu B., Yang C., Hou H., Hu J., Yang J., Xiao K., Liang S. and Wang D. (2021). The evaluation of long term performance of microbial fuel cell based Pb toxicity shock sensor. *Chemosphere*, **270**, 129455, https://doi.org/10.1016/j.chemosphere.2020.129455

Yang J., Cao X., Sun Y., Yang G. and Yi W. (2022). Recovery of microbial fuel cells with high COD molasses wastewater and analysis of the microbial community. *Biomass and Bioenergy*, **161**, 106450, https://doi.org/10.1016/j.biombioe.2022.106450

Yazdi H., Alzate-Gaviria L. and Ren Z. J. (2015). Pluggable microbial fuel cell stacks for septic wastewater treatment and electricity production. *Bioresource Technology*, **180**, 258–263, https://doi.org/10.1016/j.biortech.2014.12.100

Zabihallahpoor A., Rahimnejad M. and Talebnia F. (2015). Sediment microbial fuel cells as a new source of renewable and sustainable energy: present status and future prospects. *RSC Advances*, **5**, 94171–94183, https://doi.org/10.1039/c5ra15279h

Zhang G., Lee D. and Cheng F. (2016). Treatment of domestic sewage with anoxic/oxic membrane-less microbial fuel cell with intermittent aeration. *Bioresource Technology*, **218**, 680–686, https://doi.org/10.1016/j.biortech.2016.07.006

Zhuang L., Zheng Y., Zhou S., Yuan Y., Yuan H. and Chen Y. (2012). Scalable microbial fuel cell (MFC) stack for continuous real wastewater treatment. *Bioresource Technology*, **106**, 82–88, https://doi.org/10.1016/j.biortech.2011.11.019

doi: 10.2166/9781789063813_0119

Chapter 7

Application of microbial desalination cell technology to treat industrial wastewater

Abhimanyu Sharma[1], *Karan Singh*[1]*, *HarKamal Singh*[1], *Monika Verma*[1], *Tamanna Thakur*[2], *Mandeep Kaur*[3] *and Dharmendra*[4]

[1]*School of Civil Engineering, Lovely Professional University, Phagwara, Punjab, India 144411*
[2]*Postdoctoral Scientist at EMPA, Zurich 8600, Switzerland*
[3]*Chitkara College of Pharmacy, Chitkara University, Punjab, India 140401*
[4]*Department of Civil Engineering, National Institute of Technology, Hamirpur, H.P., India 177005*
**Corresponding author: karans72@gmail.com*

ABSTRACT

The high salinity and complicated chemical makeup of industrial effluent create a serious threat to the environment. Traditional treatment procedures are not always successful in eradicating water impurities, whereas desalination operations can be both expensive and requires a lot of energy usage. To deal with that, microbial desalination cell (MDC) technology has emerged as a viable option in recent years for the purification of industrial wastewater. The purpose of this chapter is to introduce readers to MDCs and their potential uses in industrial wastewater treatment by discussing their basic concepts, essential parts, performance evaluation, and potential in the future. In addition, the challenges and prospective strategies for enhancing MDC effectiveness and commercial viability are discussed.

7.1 INTRODUCTION

The term 'industrial wastewater' refers to the large volume of wastewater produced by industries of all types. Complex mixtures of organic and inorganic contaminants, metals, salts, and salts are common in this type of wastewater. Consequently, damage to ecosystems and human health can result from the unregulated release of industrial effluent. Thus, it is critical to protect ecosystems and human health by properly treating industrial effluent. Manufacturing, textile processing, chemical production, metal processing, petroleum refining, and food processing are just a few of the various industries that generate industrial wastewater. The chemical composition of industrial effluent varies according to the specific type of industry and practices adopted by that industry. Some typical examples of pollutants found in industrial wastewater are organic compounds, inorganic contaminants, and salinity because they are all by-products of manufacturing processes (Manna & Sen, 2023).

There are many factors to consider while attempting to treat industrial wastewater, including the water's heterogeneous makeup, high salinity, toxicity and dangerous compounds, sludge management, cost, and energy intensity. Emerging technologies and breakthroughs, such as advanced oxidation processes (AOPs), membrane-based technologies, electrochemical technologies, and biological treatment systems, have been developed to address the challenges of treating industrial wastewater.

Desalination technologies are also vital in combating water scarcity and addressing rising demand for freshwater supplies. The principles of operation and obstacles faced by various desalination methods are described which includes thermal-based and membrane-based systems. Multi-stage flash distillation (MSF) and multi-effect distillation (MED) are two examples of thermal desalination processes that use heat to purify potable water from seawater. Semipermeable membranes are used in membrane-based procedures like reverse osmosis (RO) and electrodialysis (ED) to purify water by removing salt and other contaminants. However, there are drawbacks to desalination systems, such as their high capital and operational costs, environmental effect, and difficulties in disposing of brine and fouling of membranes. To maintain water security and sustainability, widespread application of desalination technology needs to overcome these obstacles and creating sustainable desalination systems. To address these issues and to enhance the efficiency, affordability, and environmental sustainability of desalination processes, microbial desalinization seems to be a viable option (Parsa, 2023; Singh and Dharmendra, 2020).

Microbial desalination cell (MDC) technology has emerged as a promising approach for the treatment of industrial wastewater, offering potential advantages over conventional treatment methods. MDCs combine microbial electrochemical processes and desalination techniques to simultaneously treat wastewater and generate electrical energy. This study presents the basic working principles of MDCs which include the ion-selective membranes, and use of exoelectrogenic microorganisms. This technology has an edge on conventional techniques in better removal of impure organic matter, energy generation and concurrent desalination. MDCs provide a way to reduce the energy consumption and environmental impact of conventional remediation methods (Sophia *et al.*, 2016). Not only it effectively treats the effluent and removes salts, but also reduces energy consumption significantly. There is denying that this research approach still requires little more exploration to exploit the potential of MDC technology completely, even then implementation of this technology is promising for the more sustainable and cost-effective treatment of industrial wastewater. This would also prove to be helpful for the safeguard of environment and to conserve water supplies.

7.2 MICROBIAL DESALINATION CELLS

Electrochemical MDCs are an innovative methodology for the treatment of wastewater, concomitantly addressing the issue of salt removal from the water. This is achieved by integrating microbial processes with several desalination methodologies. The performance and efficiency of MDCs are greatly influenced by their design and constituent components. In this section of the discourse, we will provide an overview of the construction process of MDCs and delineate the key constituents that constitute these edifices (Sophia *et al.*, 2016).

7.2.1 Electrode configuration

The majority of MDCs consists of two distinct chambers, each equipped with its own ion-selective membrane. The anode and cathode electrodes are each allocated to distinct locations. The utilization of either a single-chamber or a dual-chamber configuration is contingent upon the precise requirements of the task and the desired outputs.

7.2.1.1 Anode

Microorganisms are accountable for the generation of electrons and the oxidation of substances at the anode. The material is commonly composed of a conductive substance possessing a substantial surface area, such as graphite or carbon cloth, with the primary objective of facilitating the adherence of microorganisms and generation of the biofilm. In the electrochemical reactions that are taking place inside the cell, the anode acts as a conduit for the passage of electrons to and from their respective destinations.

7.2.1.2 Cathode

Reduction processes are carried out at the cathode, and these reactions often include the reduction of oxygen or other electron acceptors. Platinum, carbon, and cathodes that are referred to as 'air-breathing' are all examples of common cathode materials. By eliminating electrons that were produced at the anode, the cathode is an essential component in the process of keeping the redox equilibrium of the MDC stable.

7.2.2 Ion-selective membrane

The ion-selective membrane plays an important function in the process because it enables selective ion transport between the anode and the cathode compartments of the battery. The movement of ions from the salty water (at the anode) to the clean water (at the cathode) is made possible by it, which is essential for the process of desalination. Because of the particular ion transport properties that they possess, cation exchange membranes (CEMs) and anion exchange membranes (AEMs) are two types of membranes that are extensively used in a variety of applications.

7.2.3 Proton exchange membrane

Proton exchange membranes (PEMs), as opposed to ion-selective membranes, are utilized in the construction of some MDC designs. The PEM functions as a gate, allowing protons to pass through but excluding all other ions from the system. Because of the way that this system is set up, it is feasible to generate a proton gradient, which improves both the efficiency of energy production and the desalination process.

7.2.4 External circuit and load

Electrons can travel from the anode to the cathode in an MDC because of the connection to an external circuit. Capacitors, resistors, and possibly even other electrical components could be incorporated into the circuit in order to regulate the flow of current and increase the amount of energy that is produced. The load that is connected to the circuit has the ability to collect the electrical energy that is produced by the MDC.

7.2.5 Monitoring and control systems

Monitors and controllers, such as pH and temperature sensors, flow metres, and automation systems, are some examples of the types of devices that might be included in a more advanced MDC system. Because of the nature of these components, the system's parameters can be monitored in real time, which makes it possible for more accurate optimization and more reliable operation.

The MDC system is dependent on the use of components that have been meticulously planned out and produced using high-quality materials in order to achieve the best possible performance. The efficacy of MDCs is evaluated based on many key factors, namely the surface area of the electrodes, the membrane's selectivity, the spacing between the electrodes, and the arrangement of the electrodes.

7.3 OPERATIONAL PRINCIPLE

The notion of concurrently harnessing renewable energy and purifying saline water is a captivating concept. The fundamental principles behind the operation of microbial electrochemical cells (MECs) are rooted in the electrochemical processes and energy generation that occur between bacteria and electrodes. The fundamental objective of a membrane distillation column (MDC), commonly referred to as an MDC, is to facilitate the desalination of saltwater through the elimination of salt ions and other potential contaminants. The desalination of water via a membrane distillation technique requires the simultaneous occurrence of ion migration and electrostatic adsorption mechanisms. Ions, exemplified by Na^+ and Cl^-, traverse an ion-selective membrane throughout their transit towards the electrodes, where they are ultimately accumulated. Simultaneously, the phenomenon of electrostatic adsorption induces the attraction of ions towards the electrodes' surfaces that possess a positive charge (Sophia *et al.*, 2016).

7.3.1 Anode reactions

After exoelectrogenic bacteria have oxidized organic materials, the electrons that are produced are delivered into the electrode at the location known as the anode. This process is known as extracellular electron transfer (EET). The typical anode reactions include the oxidation of organic compounds present in the wastewater, such as glucose or acetate, by microorganisms, releasing electrons and protons. The electrons generated from this oxidation process are transferred to the anode, creating an electric current.

7.3.2 Cathode reactions

The cathode compartment of the MDC is responsible for the reduction reactions. The generated electrons at the anode travel through an external circuit to the cathode, where they combine with protons and reduce an electron acceptor (e.g., oxygen, nitrate). In some MDC configurations, alternative cathode reactions, such as hydrogen evolution or sulphate reduction, can also take place. These reactions enable the flow of current, which can be harvested for energy production.

7.3.3 Various microbial processes involved in MDCs

MDCs utilize the metabolic activities of microorganisms to achieve desalination and energy production. There are two processes: (1) exoelectrogenesis, and (2) electromethogenesis.

(1) **Exoelectrogenesis:** Exoelectrogenesis, the process of microorganisms releasing electrons during the oxidation of organic matter, forms the foundation of MDC operation. In exoelectrogenesis, exoelectrogenic microorganisms (also known as exoelectrogens) possess unique mechanisms that enable them to transfer electrons extracellularly. The success of MDCs relies on the presence and activity of exoelectrogens. These mechanisms include:

   (a) **Direct electron transfer (DET):** Some microorganisms, such as *Geobacter* and *Shewanella* species, are capable of directly transferring electrons to the anode electrode surface through specialized outer membrane proteins called microbial nanowires or conductive pili. This direct contact facilitates efficient electron transfer from the microbial cells to the anode.

   (b) **Mediated electron transfer (MET):** Other microorganisms, such as certain strains of *Pseudomonas* and *Rhodoferax* species, produce redox mediators (e.g., flavins) that shuttle electrons from the cells to the anode. The redox mediators act as electron shuttles, accepting electrons from the cells and transferring them to the electrode surface.

   Microbial desalination in MDCs involves both biological and electrochemical mechanisms. The primary biological desalination mechanism is driven by the ion transport properties of microorganisms. As the anode biofilm bacteria such as *Geobacter, Shewanella,* and *Pseudomonas* species, consume organic matter by oxidation, they release protons, leading to a decrease in pH. In contrast, the cathode biofilm harbours electroactive microorganisms, including oxygen-reducing bacteria, denitrifiers, and sulphate-reducing bacteria, depending on the electron acceptor used. The generated protons migrate towards the cathode through a CEM, creating an ion gradient. This gradient promotes the migration of cations (e.g., sodium) from the saline compartment to the desalination chamber, reducing the salinity of the water. Simultaneously, anions (e.g., chloride) migrate in the opposite direction to maintain charge balance. The electrochemical mechanism contributes to desalination by ED, where charged ions are attracted or repelled by the applied electric field, facilitating their transport across selective ion exchange membranes.

(2) **Electromethogenesis:** Electromethogenesis, a distinct microbial process within MDCs, involves the production of methane gas through electrochemical reactions. Electromethogenesis occurs

through the reduction of carbon dioxide (CO_2) using electrons derived from the oxidation of organic matter at the anode. The following mechanisms are involved:

(a) **Electromethanogenesis:** Electromethanogenic microorganisms, such as *Methanobacterium* and *Methanococcus* species, utilize the released electrons from the anode to reduce CO_2 and produce methane (CH_4) as a metabolic by-product. This process occurs in the cathode chamber of the MDC.

(b) **Hydrogenotrophic methanogenesis:** In some MDC configurations, hydrogenotrophic methanogens, such as *Methanococcus* species, utilize molecular hydrogen (H_2) generated at the cathode as an electron donor, along with CO_2, to produce methane.

Microbial pathways: Electromethanogenesis in MDCs involves two primary microbial pathways:

(a) **Acetate pathway:** Some electromethanogens can directly convert acetate, an intermediate product of organic matter oxidation, into methane. Acetoclastic methanogens, such as *Methanosarcina* species, play a significant role in this pathway.

(b) **Hydrogen pathway:** Electromethanogens utilizing the hydrogen pathway consume hydrogen gas and carbon dioxide to produce methane. These hydrogenotrophic methanogens can use H_2 generated at the cathode or dissolved H_2 in the system.

7.3.4 Energy generation, optimization, and efficiency

The operational principle of an MDC allows for the simultaneous desalination of saline water and generation of electrical energy. The microbial metabolism involved in the anode reactions generates electrons, which can be harvested as an electric current. This electrical energy can be utilized for various applications, including powering small devices or contributing to the overall energy grid. Efforts are ongoing to optimize MDC performance and improve energy efficiency. Factors such as the choice of electrode materials, membrane selection, microbial community composition, and system configuration all play a significant role in the overall efficiency and desalination capacity of the MDC (Parsa, 2023).

7.4 CASE STUDIES: APPLICATION OF MDCs IN INDUSTRIAL WASTEWATER TREATMENT

7.4.1 Textile industry wastewater treatment

Textile industries are huge consumers of high-quality water and further manufacturing processes result in release of highly polluted water. Bhatt and Rani (2013) reported that out of total industrial effluent; 17–20% is generated from textile wastewater. Two major processes involved in textile production are dry and wet processes. In dry processing majorly solid waste is generated. Wet treatment processes involved in textile industry are sizing, scouring, bleaching, dyeing/printing and finishing. Dye-laden wastewater has adverse impact on aquatic life as well as on humans due to the presence of hazardous chemicals like heavy metals, NaOH, acid, starch, and so on (Adane *et al.*, 2021; Holkar *et al.*, 2016) which is not easily degradable by conventional treatment methods. Therefore, it is highly essential to treat effluent from textile industry before discharging into the water body. Vineta *et al.* (2014) bifurcated the textile wastewater treatment process in three parts: *Primary* (screening, homogenization, sedimentation, neutralization, chemical coagulation, mechanical flocculation), *Secondary* (aerobic treatment, aerated lagoons, anaerobic treatment, activated sludge process, oxidation ditch, trickling filtration, and pond) and *Tertiary* (membrane technologies, oxidation technique, adsorption, electrolytic precipitation, electrochemical processes, foam fractionation, ion exchange method, thermal evaporation, and photo catalytic degradation).

Coagulation is a known physiochemical technique for the removal of pollutants from effluent water. Alum and iron salts are majorly used coagulants. Flocculation and coagulation are less efficient to dye removal and by-product sludge is generated. Chemical treatment methods like AOP and chemical oxidation are widely used for dye, toxic metal and odour removal from effluent. Oxidants like ClO_2, Cl,

H_2O_2, and O_3 are utilized for remedial operations (Asghar *et al.*, 2015). In the ozonation process, the conjugate double bond in azo dyes (responsible for imparting colour in dyes) is broken (Forgacs *et al.*, 2004). Limitation of this method is involvement of high cost (Gosavi & Sharma, 2014) and production of highly toxic by-product (Miralles-Cuevas *et al.*, 2017). Sorption technology has also gained much attention due to its removing capability of different dyes from wastewater effluent. Activated carbon is a suitable adsorbent for removal of dye due to adsorption capability and large surface area. Few researchers performed sorption studies using other adsorbents also like modified wheat residue, modified ginger waste, potato plant waste and so on (Charola *et al.*, 2018; Kumar & Ahmad, 2011; Zhong *et al.*, 2011). The high costs involved and difficulty in recycling are limitations to this technique.

Discussion on various limitations and disadvantages of treating textile industry effluent by the above-mentioned methods, directs the use of hybrid method like MDC.

7.4.2 Petroleum refinery effluent treatment

Petroleum refinery effluent (PRE) are generated during manufacturing fuels, lubricants, refining crude oil (Harry, 1995) composed of oil and grease along with other highly toxic organic compounds such as hydrocarbons, phenol, and dissolved minerals (Basheer *et al.*, 2011; Mi-Seon *et al.*, 2008; Wake, 2005). Manufacturing of petroleum product requires plenty of water and consequently large amounts of wastewater is generated (Coelho *et al.*, 2006; Doggett & Rascoe, 2009; Saien & Nejati, 2007). PRE is naturally oxidized to some extent into bi-products that are extremely toxic (Kavitha & Palanivelu 2004) which remain in the environment for longer durations due to polycyclic aromatics (Mrayyana & Battikhi, 2005).

PRE treatment processes mainly involve coagulation, adsorption, chemical oxidation and biological techniques (Abdelwahab *et al.*, 2009; El-Naas *et al.*, 2009; Jou & Huang, 2003; Ma *et al.*, 2009; Serafim, 1979). Membranes and micro-wave-assisted catalytic air oxidation are also utilized for cleansing operations (Rahman & Al-Malack, 2006). However, the efficiency and reaction rates of such methods are proven to be very low particularly for large volumes of effluent (Kuyukina *et al.*, 2009; Rahman & Al-Malack, 2006). Heterogeneous photocatalytic degradation is also a utilized process due to its potential for eliminating organic substrates (Akpan & Hameed, 2009; Rajeshwar *et al.*, 2008). The process results in complete mineralization and produces no sludge (Lin, 2005; Wang *et al.*, 1999). But the process is based upon high running cost (Kavitha & Palanivelu, 2004).

Biological anaerobic oxidation (special case of MDC) technique is found to be excellent in remediation of organic material by electro-active bacteria (Jaroo *et al.*, 2021). They utilized air cathode MDC's capability to remediate oil refinery wastewater. Sevda *et al.* (2016) studied PRE treatment with MDC and the effect of salt concentration on the process. The treatment efficiency was found to be dependent on salt concentration as reduced internal resistance enhances MDC performance.

7.4.3 Paper and pulp mill wastewater treatment

The paper and pulp industry utilizes large amounts of water, around 20,000 and 60,000 gallons per ton of product (Nemerow & Dasgupta, 1991) and processing results in a large amount of wastewater effluent. Major processes resulting in effluent generation are pulp washing, screening, bleaching, paper machine, and coating operation. Paper and pulp effluent has some adverse effects such as scum formation, chlorinated phenol effluent, thermal impacts, slime growth, toxicity in discharged water, odour and aesthetic issues (Berube & Kahmark, 2001; Mohamed *et al.*, 1989; Pokhrel & Viraraghavan, 2004).

Physiochemical treatment processes involving sedimentation, screening, adsorption, oxidation, coagulation, electrolysis, and so on are utilized for paper and pulp effluent treatment (De Pinho *et al.*, 2000; Korhonen *et al.*, (2000); Murthy *et al.*, 1991; Thompson *et al.* 2001; Zamora *et al.*, 1998). But these methods either prove to be very costly or not that effective (Freire *et al.* 2000; Laari *et al.* 2000).

Biological treatment methods like activated sludge process, aerated lagoons, aerobic biological reactors were used intensively to remove organic pollutants, chlorinated acetic acids (Bryant *et al.*, 1997; Mohamed *et al.*, 1989; Norris *et al.*, 2000; Schnell *et al.*, 1992; Thompson *et al.*, 2001). Further treatment process was necessary to fully treat the effluent (Schnell *et al.*, 1992). Therefore, MDC can prove to be an efficient technology to treat the paper and pulp mill effluent.

7.4.4 Food processing wastewater treatment

Although food processing industry is essential for food supply chain and plays a major role in sustainable development goals (Nguegan & Mafini, 2017) but is a major fresh water consumer. Studies show the food processing industry is one of the major consumers of industrial water (Compton *et al.*, 2018). Major effluents of food processing industry are nutrients such as nitrogen and phosphorous, fatty acids, minerals such as phosphate, calcium, magnesium, iodine, various ions and solvents, hydrocarbons through the pesticides, high level of biological oxygen demand, and chemical oxygen demand, suspended loads (Shahid *et al.*, 2020; Udugama *et al.*, 2020).

Major treatment processes involved are: coagulation, filtration, algal treatment, and sequential batch reactor (SBR) for meat, fruit, and vegetable industry (Cristóvão *et al.*, 2014); adsorption, ultra-filtration, electrochemical peroxidation for edible oil industry (Azmi & Yunos, 2014; Sharma & Simsek, 2019); nano-filtration, gravitational methods, AOPs, adsorption for dairy industry effluent (Bazrafshan *et al.*, 2016; Bruguera-Casamada *et al.*, 2019; Falahati *et al.*, 2018); electrochemical treatment for bakery industry discharge (De Santana *et al.*, 2018). Although most of the treatment process result in issues like fouling (Li *et al.*, 2020), high investment cost and require further studies for efficient development of treatment models (Rezvani *et al.*, 2019). MDC can prove to be most efficient and economical method for treatment process of various ions, minerals, hydrocarbons generated by food processing industry.

7.5 CONCLUSION

An overview of working operation, principle, electron transfer mechanism, advantage and limitations of MDC is discussed. A comprehensive review of MDC for treating effluent generated through various industries is also presented. Limitations of treating effluent with traditional approach bring the necessity to utilize MDC technology for industrial wastewater treatment. From the numerous literature it was observed that MDC has better treatment efficiency than traditional techniques, such as coagulation, filtration, algal treatment, sequential batch reactor, activated sludge process, aerated lagoons, aerobic biological reactors, and so on. However, practical implementation of MDC is still deficient for large-scale treatment, and further studies for efficient treatment and cost-effectiveness need to be carried out.

REFERENCES

Abdelwahab O., Amin N. K. and El-Ashtoukhy E.-S. Z. (2009). Electrochemical removal of phenol from oil refinery wastewater. *Journal of Hazardous Materials*, **163**, 711–716, https://doi.org/10.1016/j.jhazmat.2008.07.016

Adane T., Adugna A. T. and Alemayehu E. (2021). Textile industry effluent treatment techniques. *Journal of Chemistry*, **2021**, 1–14, https://doi.org/10.1155/2021/5314404

Akpan U. G. and Hameed B. H. (2009). Parameters affecting the photocatalytic degradation of dyes using TiO_2-based photocatalysts: a review. *Journal of Hazardous Materials*, **170**, 520–529, https://doi.org/10.1016/j.jhazmat.2009.05.039

Asghar A., Abdul Raman A. A. and Wan Daud W. M. A. (2015). Advanced oxidation processes for in-situ production of hydrogen peroxide/hydroxyl radical for textile wastewater treatment: a review. *Journal of Cleaner Production*, **87**(1), 826–838, https://doi.org/10.1016/j.jclepro.2014.09.010

Azmi N. S. and Yunos K. F. M. (2014). Wastewater treatment of palm oil mill effluent (POME) by ultrafiltration membrane separation technique coupled with adsorption treatment as pre-treatment. *Agriculture and Agricultural Science Procedia*, **2**, 257–264, https://doi.org/10.1016/j.aaspro.2014.11.037

Basheer D., Wan D. and Abdul Aziz A. R. (2011). Treatment technologies for petroleum refineries effluents: a review. *Process Safety and Environmental Protection*, **89**, 95–105, https://doi.org/10.1016/j.psep.2010.11.003

Bazrafshan E., Kord Mostafapour F., Alizadeh M. and Farzadkia M. (2016). Dairy wastewater treatment by chemical coagulation and adsorption on modified dried activated sludge: a pilot-plant study. *Desalination and Water Treatment*, **57**, 8183–8193, https://doi.org/10.1080/19443994.2015.1018331

Berube P. R. and Kahmark K. A. (2001). Pulp and paper mill effluents. *Water Environment Research*, **73**(5), 1–36.

Bhatt P. and Rani A. (2013). Textile dyeing and printing industry: an environmental hazard. *Asian Dyer*, **10**(6), 51–54.

Bruguera-Casamada C., Araujo R. M., Brillas E. and Sirés I. (2019). Advantages of electro-Fenton over electrocoagulation for disinfection of dairy wastewater. *Chemical Engineering Journal*, **376**, 119975, https://doi.org/10.1016/j.cej.2018.09.136

Bryant C. W., Barkley W. A., Garett R. M. and Gardner F. D. (1997). Biological nitrification of kraft wastewater. *Water Science and Technology*, **35**(2–3), 147–153, https://doi.org/10.2166/wst.1997.0504

Charola S., Yadav R., Das P. and Maiti S. (2018). Fixed-bed adsorption of reactive orange 84 dye onto activated carbon prepared from empty cotton flower agro-waste. *Sustainable Environment Research*, **28**(6), 298–308, https://doi.org/10.1016/j.serj.2018.09.003

Coelho A., Castro A. V., Dezotti M. and Sant'Anna Jr. G. L. (2006). Treatment of petroleum refinery sourwater by advanced oxidation processes. *Journal of Hazardous Materials*, **137**, 178–184, https://doi.org/10.1016/j.jhazmat.2006.01.051

Compton M., Willis S., Rezaie B. and Humes K. (2018). Food processing industry energy and water consumption in the Pacific northwest. *Innovative Food Science & Emerging Technologies*, **47**, 371–383, https://doi.org/10.1016/j.ifset.2018.04.001

Cristóvão R. O., Botelho C. M., Martins R. J., Loureiro J. M. and Boaventura R. A. (2014). Primary treatment optimization of a fish canning wastewater from a Portuguese plant. *Water Resources and Industry*, **6**, 51–63, https://doi.org/10.1016/j.wri.2014.07.002

De Pinho M. N., Minhalma M., Rosa M. J. and Taborda F. (2000). Integration of flotation/ultrafiltration for treatment of bleached pulp effluent. *Pulp Pap Can*, **104**(4), 50–54.

De Santana M. M., Zanoelo E. F., Benincá C. and Freire F. B. (2018). Electrochemical treatment of wastewater from a bakery industry: experimental and modeling study. *Process Safety and Environmental Protection*, **116**, 685–692, https://doi.org/10.1016/j.psep.2018.04.001

Doggett T. and Rascoe A. (2009). Global Energy Demand Seen up 44 Percent by 2030. http://www.reuters.com/articles/GCAGreenBusiness/idUSN2719528620090527 (accessed 22 September 2009)

El-Naas M. H., Al-Zuhair S. and Alhaija M. A. (2009). Reduction of COD in refinery wastewater through adsorption on date-Pit activated carbon. *Journal of Hazardous Materials*, **173**, 750–757, https://doi.org/10.1016/j.jhazmat.2009.09.002

Falahati F., Baghdadi M. and Aminzadeh B. (2018). Treatment of dairy wastewater by graphene oxide nanoadsorbent and sludge separation, using in situ sludge magnetic impregnation (ISSMI). *Pollution*, **4**, 29–41.

Forgacs E., Cserhati T. and Oros G. (2004). Removal of synthetic dyes from wastewaters: a review. *Environment International*, **30**(7), 953–971, https://doi.org/10.1016/j.envint.2004.02.001

Freire R. S., Kunz A. and Duran N. (2000). Some chemical and toxicological aspects about paper mill effluent treatment with ozone. *Environmental Technology*, **21**, 717–721, https://doi.org/10.1080/09593332108618088

Gosavi V. and Sharma S. (2014). A general review on various treatment methods for textile wastewater. *Journal of Environmental Science, Computer Science and Engineering & Technology*, **3**(1), 29–39.

Harry M. F. (1995). Industrial Pollution Handbook. McGraw-Hill. Inc., New York.

Holkar C. R., Jadhav A. J., Pinjari D. V., Mahamuni N. M. and Pandit A. B. (2016). A critical review on textile wastewater treatments: possible approaches. *Journal of Environmental Management*, **182**, 351–366, https://doi.org/10.1016/j.jenvman.2016.07.090

Jaroo S. S., Jumaah G. F. and Abbas T. R. (2021). The operation characteristics of air cathode microbial desalination cell to treat oil refinery wastewater. *IOP Conference Series: Earth and Environmental Science*, **877**, 012002, https://doi.org/10.1088/1755-1315/877/1/012002

Jou C. G. and Huang G. (2003). A pilot study for oil refinery wastewater treatment using a fixed film bioreactor. *Advances in Environmental Research*, **7**, 463–469, https://doi.org/10.1016/S1093-0191(02)00016-3

Kavitha V. and Palanivelu K. (2004). The role of ferrous ion in Fenton and photo-Fenton processes for the degradation of phenol. *Chemosphere*, **55**, 1235–1243, https://doi.org/10.1016/j.chemosphere.2003.12.022

Korhonen S. M., Metsarinne S. E. and Tuhakanen T. A. (2000). Removal of ethylenediaminenetraacetic acid (EDTA) from pulp mill effluents by ozonation. *Ozone: Science & Engineering*, **22**, 279–286, https://doi.org/10.1080/01919510008547211

Kumar R. and Ahmad R. (2011). Bio sorption of hazardous crystal violet dye from aqueous solution onto treated ginger waste (TGW). *Desalination*, **265**(1–3), 112–118, https://doi.org/10.1016/j.desal.2010.07.040

Kuyukina M. S., Ivshina I. R., Serebrennikova M. K., Krivorutchko A. B., Podorozhko E. A., Ivanov R. V. and Lozinsky V. I. (2009). Petroleum-contaminated water treatment in a fluidized-bed bioreactor with immobilized *Rhodococcus* cells. *International Biodeterioration & Biodegradation*, **63**, 427–432, https://doi.org/10.1016/j.ibiod.2008.12.001

Laari A., Korhonen S., Kallas J. and Tuhkanen T. (2000). Selective removal of lipophilic wood extractives from paper mill water circulations by ozonation. *Ozone: Science & Engineering*, **22**, 585–605, https://doi.org/10.1080/01919510009408801

Li C., Sun W., Lu Z., Ao X. and Li S. (2020). Ceramic nanocomposite membranes and membrane fouling: a review. *Water Research*, **175**, 115674, https://doi.org/10.1016/j.watres.2020.115674

Lin H. T. (2005). Photocatalysis in a Novel Semiconducting Optical Fiber Monolithic Reactor for Wastewater Treatment. PhD Thesis, Louisiana State University.

Ma F., Guo J. B., Zhao L. J., Chang C. C. and Cui D. (2009). Application of bioaugmentation to improve the activated sludge system into the contact oxidation system treating petrochemical wastewater. *Bioresource Technology*, **100**, 597–602, https://doi.org/10.1016/j.biortech.2008.06.066

Manna M. and Sen S. (2023). Advanced oxidation process: a sustainable technology for treating refractory organic compounds present in industrial wastewater. *Environmental Science and Pollution Research*, **30**(10), 25477–25505, https://doi.org/10.1007/s11356-022-19435-0

Miralles-Cuevas S., Oller I., Agüera A., Llorca M., Sanchez J. A. and Malato S. (2017). Combination of nanofiltration and ozonation for the remediation of real municipal wastewater effluents: acute and chronic toxicity assessment. *Journal of Hazardous Materials*, **323**, 442–451, https://doi.org/10.1016/j.jhazmat.2016.03.013

Mi-Seon J., Eldon R., Soo-Hong K. and Hung-Suck P. (2008). An analysis of synergistic and antagonistic behavior during BTEX removal in a batch system using response surface methodology. *Journal of Hazardous Materials*, **152**, 1276–1284, https://doi.org/10.1016/j.jhazmat.2007.08.002

Mohamed M., Matayun M. and Lim T. S. (1989). Chlorinated organics in tropical hardwood kraft pulp and paper mill effluents and their elimination in an activated sludge treatment system. *Pertanika*, **2**(3), 387–394.

Mrayyana B. and Battikhi M. N. (2005). Biodegradation of total organic carbons (TOC) in Jordanian petroleum sludge. *Journal of Hazardous Materials*, **120**, 127–134, https://doi.org/10.1016/j.jhazmat.2004.12.033

Murthy B. S. A., Sihorwala T. A., Tilwankar H. V. and Killedar D. J. (1991). Removal of colour from pulp and paper mill effluents by sorption technique – a case study. *Indian Journal of Environmental Protection*, **11**(5), 360.

Nemerow N. L. and Dasgupta A. (1991). Industrial and Hazardous Waste Management. Van Nostrand Reinhold, New York.

Nguegan C. A. and Mafini C. (2017). Supply chain management problems in the food processing industry: implications for business performance. *Acta Commercii*, **17**, 1–15.

Norris P., Marshall R. and Richard M. (2000). High temperature on activated sludge treatment performance and sludge quality in a recycle mill. Tappi International Environmental Conference and Exhibit, Denver, CO, vol. 1. Norcross, GA, 30092, USA, Technical Association for Pulp and Paper Industry (TAPPI); 383–387.

Parsa S. M. (2023). Mega-scale desalination efficacy (reverse osmosis, electrodialysis, membrane distillation, MED, MSF) during COVID-19: evidence from salinity, pretreatment methods, temperature of operation. *Journal of Hazardous Materials Advances*, **9**, 100217, https://doi.org/10.1016/j.hazadv.2022.100217

Pokhrel D. and Viraraghavan T. (2004). Treatment of pulp and paper mill wastewater – a review. *Science of the Total Environment*, **333**, 37–58, https://doi.org/10.1016/j.scitotenv.2004.05.017

Rahman M. M. and Al-Malack M. H. (2006). Performance of a crossflow membrane bioreactor (CF–MBR) when treating refinery wastewater. *Desalination*, **191**, 16–26, https://doi.org/10.1016/j.desal.2005.05.022

Rajeshwar K., Osugi M. E., Chanmanee W., Chenthamarakshan C. R., Zanoni M. V. B., Kajitvichyanukul P. and Krishnan-Ayer R. (2008). Heterogeneous photocatalytic treatment of organic dyes in air and aqueous media. *Journal of Photochemistry and Photobiology C: Photochemistry Reviews*, **9**, 171–192, https://doi.org/10.1016/j.jphotochemrev.2008.09.001

Rezvani F., Sarrafzadeh M. H., Ebrahimi S. and Oh H. M. (2019). Nitrate removal from drinking water with a focus on biological methods: a review. *Environmental Science and Pollution Research*, **26**, 1124–1141, https://doi.org/10.1007/s11356-017-9185-0

Saien J. and Nejati H. (2007). Enhanced photocatalytic degradation of pollutants in petroleum refinery wastewater under mild conditions. *Journal of Hazardous Materials*, **148**, 491–495, https://doi.org/10.1016/j.jhazmat.2007.03.001

Schnell A., Hall E. R. and Skog S. (1992). Anaerobic and aerobic treatability of high-yield sulphate spent liquor. *Water Quality Research Journal*, **27**(3), 601–620, https://doi.org/10.2166/wqrj.1992.038

Serafim A. J. (1979). Solid Retention Time on Carbon Adsorption of Organics in Secondary Effluents from Treatment of Petroleum Refinery Waste. PhD Thesis, Texas A&M University.

Sevda S., Abu-Reesh I. M., Yuan H. and He Z. (2016). Bioelectricity generation from treatment of petroleum refinery wastewater with simultaneous seawater desalination in microbial desalination cells. *Energy Conversion and Management*, **141**, 101–107, https://doi.org/10.1016/j.enconman.2016.05.050

Shahid M., Khalid S., Murtaza B., Anwar H., Shah A. H., Sardar A., Shabbir Z. and Niazi N. K. (2020). A critical analysis of wastewater use in agriculture and associated health risks in Pakistan. *Environmental Geochemistry and Health*, **45**, 5599–5618. Advance online publication. https://doi.org/10.1007/s10653-020-00702-3

Sharma S. and Simsek H. (2019). Treatment of canola-oil refinery effluent using electrochemical methods: a comparison between combined electrocoagulation + electrooxidation and electrochemical peroxidation methods. *Chemosphere*, **221**, 630–639, https://doi.org/10.1016/j.chemosphere.2019.01.066

Singh K. and Dharmendra (2020). Performance of a dual chamber microbial fuel cell using sodium chloride as catholyte. *Pollution*, **6**(1), 79–86.

Sophia A. C., Bhalambaal V. M., Lima E. C. and Thirunavoukkarasu M. (2016). Microbial desalination cell technology: contribution to sustainable wastewater treatment process, current status and future applications. *Journal of Environmental Chemical Engineering*, **4**(3), 3468–3478, https://doi.org/10.1016/j.jece.2016.07.024

Thompson G., Swain J., Kay M. and Forster C. F. (2001). The treatment of pulp and paper mill effluent: a review. *Bioresource Technology*, **77**(3), 275–286, https://doi.org/10.1016/S0960-8524(00)00060-2

Udugama I. A., Petersen L. A., Falco F. C., Junicke H., Mitic A., Alsina X. F., Mansouri S. S. and Gernaey K. V. (2020). Resource recovery from waste streams in a water-energy-food nexus perspective: toward more sustainable food processing. *Food and Bioproducts Processing*, **119**, 133–147, https://doi.org/10.1016/j.fbp.2019.10.014

Vineta S., Silvana Z., Sanja R. and Golomeova S. (2014). Methods for wastewaters treatment in textile industry. International Scientific Conference.

Wake H. (2005). Oil refineries: a review of their ecological impacts on the aquatic environment. *Estuarine, Coastal and Shelf Science*, **62**, 131–140, https://doi.org/10.1016/j.ecss.2004.08.013

Wang K.-H., Hsieh Y.-H., Ko R.-C. and Chang C.-C. (1999). Photocatalytic degradation of wastewater from manufactured fiber by titanium dioxide suspensions in aqueous solution. *Environment International*, **25**, 671–676, https://doi.org/10.1016/S0160-4120(99)00035-5

Zamora P. P., Esposito E., Pelegrini R., Groto R. and Duran N. (1998). Effluent treatment of pulp and paper, and textile industries using immobilised horseradish peroxidase. *Environmental Technology*, **19**, 55–63, https://doi.org/10.1080/09593331908616655

Zhong Q. Q., Yue Q. Y., Li Q., Xu and Gao B. Y. (2011). Preparation, characterization of modified wheat residue and its utilization for the anionic dye removal. *Desalination*, **267**(2–3), 193–200, https://doi.org/10.1016/j.desal.2010.09.025

doi: 10.2166/9781789063813_0129

Chapter 8

Application of different catholytes in microbial fuel cells and their application for industrial wastewater

Karan Singh[1]*, Abhimanyu Sharma[1], Jaspreet Singh[1], Amol Sharma[1], Dharmendra[2], Mandeep Kaur[3] and Nidhi Selwal[4]

[1]School of Civil Engineering, Lovely Professional University, Phagwara, Punjab 144411, India
[2]Department of Civil Engineering, National Institute of Technology, Hamirpur, HP 177005, India
[3]Chitkara College of Pharmacy, Chitkara University, Punjab 140401, India
[4]School of Bioengineering and Biosciences, Lovely Professional University, Phagwara, Punjab 144411, India
*Corresponding author: karans72@gmail.com

ABSTRACT

Utilizing different catholytes in microbial fuel cells (MFCs) has demonstrated considerable potential in the remediation of industrial effluents. The choice of a catholyte is crucial in enhancing the performance and efficiency of MFCs, which in turn has a significant influence on the overall wastewater treatment process. Several catholytes, such as oxygen, ferricyanide, and other redox mediators, have been discovered and utilized to enhance electron transfer kinetics and optimize power production in MFCs. The selection of an appropriate catalyst is determined by factors such as the distinctive composition of wastewater, the desired level of treatment effectiveness, and the economic viability. The highest mean power output was attained by combining ferricyanide with a multilayer structure, such as a hexacyanoferrate cathode. Utilizing several catholytes in MFCs is an innovative and encouraging approach for treating industrial wastewater. The potential of MFCs to provide sustainable and efficient solutions for industrial wastewater treatment is increasing as catalyst selection, system design, and operating procedures continue to progress. Fuel cells are anticipated to operate within the mesophilic temperature range, which is optimal for wastewater purification.

Keywords: microbial fuel cell, industrial wastewater, COD removal, catholyte, ferricyanide.

8.1 INTRODUCTION

In the future, demand for renewable energy may account for a sizable share of both global energy production and consumption (Li *et al.*, 2018; Mohan *et al.*, 2008). Non-renewable energy sources, such as oil and coal, are unlikely to have a substantial impact on the future of global energy (Chaturvedi & Katoch, 2020). As non-renewable energy supplies are being depleted at a far more rapid pace than in the past, it is necessary to develop technologies that generate renewable energy at a reasonable price. Biomass, also known as organic matter, is a diversified renewable energy source in India that plays a significant role in the country's energy landscape. Recent statistics show that biomass contributes substantially to India's renewable energy output. According to India's Ministry of New and Renewable Energy, biomass still accounts for around 32% of total primary energy consumption in the country, and more than 70% of the country's population relies on it. As of 2013, power generating capacity in India surpassed 2670 GW, with renewable energy accounting for approximately 10.5%. This also signifies the significant contribution of renewable energy sources to the country's overall power generation landscape. Biomass-derived renewable energy generation accounts for around 12.83% of overall energy output (Kumar *et al.*, 2015). This emphasizes the abundant availability of biomass as a suitable substrate for the functioning of microbial fuel cells (MFCs). As previously stated, MFC technology has enormous potential for improving self-sufficiency and resourcefulness in wastewater treatment operations (Aryal *et al.*, 2017; Bhowmick *et al.*, 2019; Chakraborty *et al.*, 2020; Du *et al.*, 2007; Pandey *et al.*, 2016; Slate *et al.*, 2019). Several crucial components are involved in an MFC. These consist of an anode and a cathode separated by a proton, cation, or anion membrane. There is also a salt bridge that connects the anodic and cathodic zones. At the cathode, oxygen acts as an electron acceptor, allowing water to be produced as a by-product. A biofilm on the anode's surface works as a catalyst, allowing biochemical energy to be converted into electrons (Bhowmick *et al.*, 2019; Das & Ghangrekar, 2018; Singh & Dharmendra, 2020). Recent studies (Das & Ghangrekar, 2019; Das *et al.*, 2019a; Li *et al.*, 2011; Pant *et al.*, 2010; Zhou *et al.*, 2011) indicate that MFCs offer an environmentally friendly approach to power generation. The materials used in MFCs play a crucial role in enhancing both the proliferation of microorganisms and the efficiency of reactions, thus positively impacting the overall effectiveness of MFC technology while minimizing the environmental impact. Obtaining the maximum feasible levels of columbic efficiency and power output from MFCs is one of the most difficult aspects of these devices. There are additional challenges to address in the realm of MFCs, including the necessity to reduce costs and develop an inherently scalable architecture for MFC systems. These challenges must be resolved to ensure wider adoption and deployment of MFC technology on a larger scale (Das *et al.*, 2019b; Goglio *et al.*, 2019; Logan, 2008; Sevda *et al.*, 2015). This chapter explores the many different elements of MFC performance, including its potential uses, drawbacks, and the roles that substrates and microorganisms play in the process.

8.2 ELECTRON TRANSFER MECHANISM

In the process of transferring electrons from organic matter to the anode electrode of an MFC, two primary methods are commonly employed: explicit electron transfer and facilitated electron transfer. These mechanisms play crucial roles in facilitating the movement of electrons within an MFC system. Pycocyanin is a notable illustration of a self-produced mediator that bacteria use to facilitate the transfer of electrons to the anode's surface. This pigment, found in certain bacterial species, plays a vital role in mediating the electron transfer process within MFCs. By employing pycocyanin, bacteria are able to efficiently transfer electrons from their metabolic reactions to the anode surface, enhancing the overall electrical output of MFC systems (Logan, 2008; Rabaey *et al.*, 2004). Pycocyanin is produced by *Pseudomonas aeruginosa*. *Shewanella oneidensis*, *Geothrix fermentans*, and other bacteria similar to them need to interact with external mediators in order to generate energy. MFCs have had certain chemical mediators added to them in order to improve the efficiency with which microorganisms such as yeast, glucose, acetate, and others transmit electrons. According to Chaudhuri and Lovley's research in 2003, anode electrodes are used to establish a biofilm in order to increase energy production and facilitate the flow of electrons. When electrons reach the surface of an electrode, they are released within the anode, causing an electrochemical process to begin. When electrons are liberated at the anode, a number of chemical processes occur that contribute to the general functioning of the system. In the case of a direct electron transfer, the outer membrane is not necessary. Certain bacteria, such as *Shewanella putrefaciens*, *Geobacter sulfurreducens*, and *Rhodoferax ferrireducens*, utilize a direct electron transfer mechanism. On the contrary, an indirect electron transfer occurs when a third party, either naturally occurring or intentionally inserted, helps transfer electrons to a cathode. Electron shuttles play a crucial role in transporting electrons between the surfaces of the electrodes and the microorganisms involved in the process.

Figure 8.1 shows both the insistent and alternate MFC methods. Anaerobic reactions convert organic material into electrons (e^-) and hydrogen ions (H^+) in the anode compartment. Hydrogen ions, denoted by the symbol H^+, were moved through a membrane and into the cathode compartment, whereas electrons, denoted by the symbol e^-, were transmitted *via* an external circuit. In the cathode compartment, oxygen performs the role of an electron acceptor. It does this by bringing together hydrogen ions (H^+) and electrons (e^-) to produce water. Equations (8.1) and (8.2) show the reactions that take place when glucose ($C_6H_{12}O_6$) is the anolyte in an MFC. These reactions take place when oxygen (O_2) is used as an electron acceptor at the anode of an MFC.

At the cathode:

$$O_2 + 4e^- + 4H^+ \rightarrow 2H_2O \tag{8.1}$$

At the anode:

$$C_6H_{12}O_6 + 6H_2O \rightarrow 6CO_2 + 24e^- + 24H^+ \tag{8.2}$$

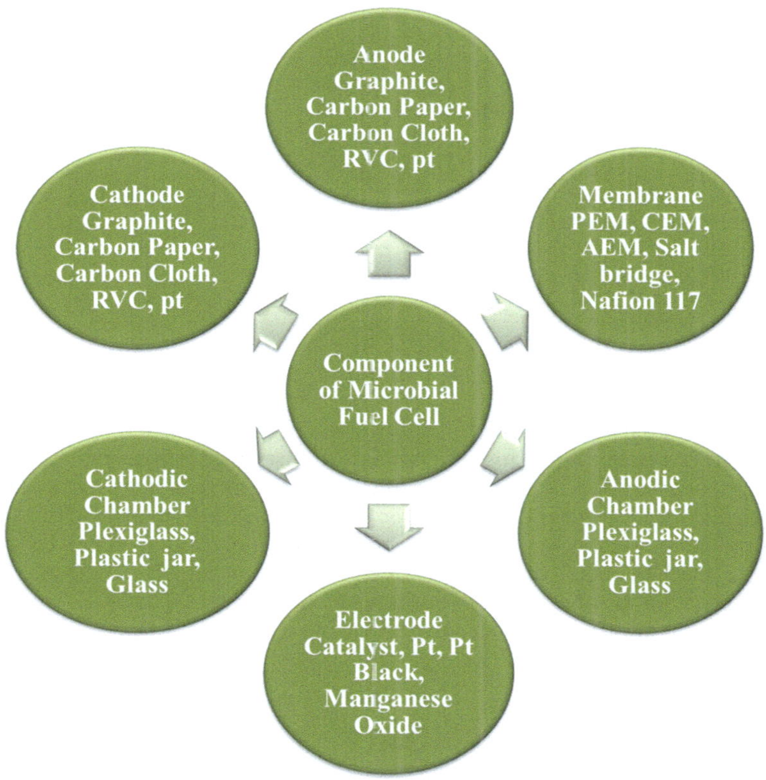

Figure 8.1 Components of an MFC.

8.3 DIFFERENT CATHOLYTES USED IN MFCs

8.3.1 Algae as biocatholyte

The redox processes that take place during photosynthesis are among the most complex and involved throughout life. Solar energy and carbon dioxide (CO_2) are the only two inputs required for this process to result in the production of oxygen, carbohydrates, and other molecules such as pigments, proteins, and lipids (Khoo *et al.*, 2023). The growth of autotrophic algae is conceivable under controlled circumstances such as those observed in a laboratory. The cultivation of algae and the accumulation of algal biomass are often accomplished *via* the use of photobioreactors, algal ponds, and lagoons. Microalgae are exceptional in that they are able to grow heterotrophically even in the absence of light by using a wide variety of carbon substrates. This feature allows them to thrive in environments with little to no light. This heterotrophic type of algal development, on the contrary, is vulnerable to pollution and the growth inhibition caused by additional microorganisms (Muthuraman & Kasianantham, 2023). The cultivation of phytoplankton in photobioreactors takes place in a mixotrophic mode, which combines autotrophic and heterotrophic processes. The mixotrophic nature of algae, together with its other properties such as CO_2 absorption and oxygen production, as well as its logarithmic growth pattern, makes it a good candidate for use as an MFC biocatholyte. In a bacterial–algal MFC, the bacteria in the anolyte are responsible for oxidizing the substrate, whereas the algae in the catholyte are responsible for producing oxygen. During photosynthesis, this excess oxygen acts as an electron acceptor, which leads to a net decrease in the amount of carbon dioxide in the atmosphere (Ahirwar *et al.*, 2023). This process also results in the creation of water and additional algal biomass. The following chemical equations present the photosynthetic processes that take place inside an algal-MFC cathodic chamber:

Dark reaction:

$$C_6H_{12}O_6\ (\text{biomass}) + 6O_2 \rightarrow 6CO_2 + 6H_2O \tag{8.3}$$

Light reaction:

$$6CO_2 + 12H^+ + 12e^- \rightarrow C_6H_{12}O_6\ (\text{biomass}) + 3O_2 \tag{8.4}$$

Cathodic reaction:

$$O_2 + 4H^+ + 4e^- \rightarrow 2H_2O \tag{8.5}$$

A cathodic compartment is responsible for maintaining the cycle of self-replicating algae growth. Holdup algae have the potential to produce a biofilm on the electrode as well as the hollow exteriors. The algal biofilm acts as a passive electron acceptor, enabling electrons to enter the algal cell body as they move through the MFC circuit. This is possible because the biofilm is composed of algal cells. When working with suspensions of planktonic algae, mediators are required to facilitate the flow of electrons from the cathode into the algae. After having removed electrons from algal cells, these oxidized and degraded mediators are then returned to the electrolyte where they were originally found. It has been established that the amount of dissolved oxygen present inside the algal biofilm has a significant impact on the effectiveness of the electrochemical technique. The development of a biofilm on the cathode facilitates direct electron transmission, which lowers the ohmic and charge transfer resistance-related MFC losses.

8.3.2 Bacterial microflagellate catholytes

Bacterio-algal MFCs are distinct from normal MFCs owing to the fact that the cathodic chamber of the former contains algae. Both the biomass and pigments that the algae produce aid in the oxygenation of the cathodic chamber. The cathode function is served by the oxygen that is generated by photosynthesis in algae; this oxygen receives electrons. The ability of algae to withstand changes in the concentration of cations is due to the fact that diverse ions may flow across the membrane from the anode chamber. There are a variety of possible uses for algal biomass, including the production of animal feed and energy, as well as the development of bioproducts *via* anaerobic digestion and other forms of bioprocessing. In batch reactions for the generation of biomass, higher hydraulic retention time (HRT) periods result in larger algal yields. This is in contrast to the situation in a continuous system, where shorter HRTs lead to lower yields. This disadvantage of continuous operation may be mitigated in a number of ways, one of which is by transferring algal biomass between the photobioreactor and the cathodic compartment of an MFC. There is a bottleneck in the process of producing electricity because of the pace at which oxygen is transformed into water in the cathode compartment. One can obtain more oxygen reduction by using a catalyst at the cathode or by adding a strong reducing salt such as potassium ferricyanide to the catholyte. Both of these strategies have the potential to be successful. In addition to rotating the electrode, sparging the cathode chamber with air or pure oxygen and rotating the chamber itself are two other helpful ways to boost the quantity of oxygen that is accessible. On the contrary, the algae cultures in the cathode chamber had the potential to deliver oxygen in a reliable and consistent manner. When there are algal biofilms on the cathode surface of MFCs, oxygen transfer is likely to be slowed down. As a result, MFCs will not work as well. The production of algal biofilms and the thickness of such biofilms are both influenced by the surface textures of the electrodes. The formation of thick biofilms is encouraged on surfaces that are extremely porous and/or rough, whereas the formation of thin biofilms is encouraged on surfaces that are smooth. It is hard to track algae growth by looking at the optical density in the cathode chamber because algae can grow in three different ways: as algal biofilms on the electrode and chamber walls, as suspended algal aggregates, and as suspended biomass in solution. Monitoring the development of algae may also be accomplished in a number of different ways, such as by conducting an analysis of proteins or by collecting measurements of dry biomass. Both of these methods have advantages and disadvantages. Bacterio-algal membrane fuel cells, more often referred to as MFCs, have made use of a broad range of catholytes. These catholytes include, but are not limited to, Bold's basal medium, which contains 300 mg/dm^3 of algal biomass, 1 g of NH_4Cl, and 0.13 g of dipotassium phosphate in tap water, in addition to a huge number of other catholytes. Because many species of phytoplankton were utilized in each research project, there was a wide range of power generation that could be accounted for by this fact. This phytoplankton required a wide variety of growing conditions and exhibited their own unique set of growth properties. The solubility restrictions placed on ferricyanide, in contrast to oxygen, do not affect its concentration (Rhoads *et al.*, 2005). Because of this, ferricyanide is widely utilized as an electron donor in MFC research. Aelterman *et al.* (2006), and Rabaey *et al.* (2003) all came to the same conclusion: the standard redox potential of ferricyanide, which is given in Equation (8.2), is not as high as that of oxygen. However, the significantly lower overpotential of ferricyanide led to significantly faster reaction rates and a much higher power output. Ferricyanide with a carbon electrode generated 50–80 percentage points more power than oxygen with a platinum (Pt)–carbon cathode, according to Oh *et al.* (2004), who attributed this difference to greater mass transfer efficiencies and a larger cathode potential. It is widely known that the manufacturing of potassium ferrocyanide cannot be maintained continuously. This is true despite the fact that potassium ferrocyanide is an efficient electron acceptor for the generation of electricity. It is risky because the chemicals used are difficult to recycle or repurpose. As a result of this, ferricyanide is normally only used in controlled environments, such as laboratories (Logan *et al.*, 2006). Ferricyanide is still an important cathodic electron acceptor for lab demonstrations of several basic ideas owing to how stable it is and how well it works in a system. Aelterman *et al.* (2006) conducted studies comparing series and parallel arrangements to evaluate the performance of MFCs. Their findings found that using a multilayer structure with a hexacyanoferrate cathode resulted in the highest average power generation of 258 W/m^3/h. This chapter demonstrates the usefulness of this particular configuration in maximizing power generation in MFC systems. Ferricyanide, known for its catalytic properties, has been extensively employed in the assessment of diverse electrode materials. This compound has proven useful in evaluating the performance and characteristics of various electrodes used in different applications. Recent research has focused on determining the effectiveness of three unique MFC procedures in removing nitrogen and carbon from wastewater. The continuously running MFC with ferricyanide performed the best out of the three different types of MFC that were investigated (type 1, continuous operation, and type 3, continuous operation with oxygen). This superiority stems from its exceptional capability to remove current, carbon, and nitrogen. Specifically, readings of 0.83 V for the ferricyanide current and 0.58 V for the oxygen current have been recorded. Notably, when compared to oxygen, ferricyanide exhibits a 36-percentage-point increase in carbon removal and a 9-percentage-point increase in nitrogen removal (Zain *et al.*, 2015). These findings underscore the enhanced effectiveness of ferricyanide in the removal of carbon and nitrogen compounds.

8.3.3 Nitrogen species

Nitrate is one of the most common forms of nitrogen found in rivers; it is also one of the most harmful compounds to both people and animals, leading to a wide range of environmental and health issues (Demirel *et al.*, 2014). Nitrate is one of the most deadly forms of nitrogen. Exposure to nitrate can have severe consequences for living organisms, as it can disrupt vital biological processes and lead to detrimental health effects. The potent and deadly nature of nitrate demands utmost caution and careful handling to mitigate potential risks and safeguard both human and environmental well-being. According to Shen *et al.*'s (2009) research, the maximum allowable concentration of nitrate in drinking water in the United States is 44.43 mg/L, whereas in the European Union it is 50 mg/L. Biocatalysts have made it possible for nitrate to function as an electron acceptor in membrane fuel cells, which are used in the process of denitrification and the generation of energy. Clauwaert *et al.* (2007) presented that nitrate might function well as a cathodic electron acceptor in MFCs. Their work also revealed that denitrification (the process of removing nitrates) could be accomplished by microorganisms in a tubular reactor without the need for external energy input. These findings highlight the inherent capability of microorganisms to perform denitrification, showcasing their potential for sustainable and energy-efficient nitrogen removal processes. During the same time period, Lefebvre *et al.* examined a cathodic process that was analogous to this one in an MFC with two chambers (Lefebvre *et al.*, 2008). Through the process of electrochemical denitrification, they were able to convert 73.4% of the total nitrogen into N_2 gas in their experiment. Acetate being an anodic substrate, made it possible for nitrate to be taken away from the cathode (Equations (8.3)–(8.6)). The maximal cell potential at 1000 Ω external resistance was just 0.095 V, which is a significant amount lower than the values that have been previously recorded for oxygen. It is possible that this is due to the low redox potential of nitrate, which is 0.74 V. Using an MFC and the technique of aerobic nitrification, Virdis *et al.* (2008) developed a novel method for extracting carbon and nitrogen from water. This process was successful in their demonstration. It is possible that this may open a door for more research as well as an extension of applications. Ammonium- and organic matter-containing effluents were originally transferred into the anode compartment. Here, the organic matter was oxidized and electrons become free as a result. To convert ammonium to nitrate, an external aerobic nitrification tank was used. This tank was supplied with effluents from the anode. After being enriched with nitrate, this stream was sent to the cathode compartment of the MFC, which is where the nitrate was broken down by electrons. Nitrate, an electron acceptor, received the electrons produced at the beginning of the process. This particular system had a nitrogen removal rate of kg chemical oxygen demand (COD)/m^3 net cathodic compartment (NCC) (Virdis *et al.*, 2008), and its volumetric power density was 34.61 W/mC. The fact that ammonia may flow through a cation-exchange membrane (CEM) from an anode to a cathode as part of this process is one of the drawbacks of the procedure. This results in insufficient nitrogen removal from effluents. To find a solution to this issue, Virdis and colleagues merged the processes of nitrification and denitrification called simultaneous nitrification and denitrification (SND) into a single chamber known as the cathode (Virdis *et al.*, 2010). Even though there is a larger concentration of dissolved oxygen in the water than there would be in a conventional

SND system, denitrification may still take place. Denitrifying bacteria were able to thrive on the porous surface of graphite granules, which led to this finding (Virdis *et al.*, 2010). This observation was based on the fact that denitrifying bacteria were able to thrive. Significant progress has been made in both the research and development of technologies for achieving SND in MFCs. Yu *et al.* (2011) suggested combining the membrane-aerated biofilm process with the MFC process in order to achieve simultaneous nitrification, denitrification, and organic carbon removal in a two-chambered MFC system. By combining the two processes, this would be possible. When using this technique, the effectiveness of removing total carbon was 97%, whereas the efficiency of removing nitrogen was 52%. To concurrently remove carbon dioxide and nitrogen, Xie *et al.* (2011) designed an oxic/anoxic biocathode system. Oxic and anoxic biocathodes will remove ammonium and nitrate from a system at the same time that COD is oxidizing at the anode. With this set-up, we were able to achieve the maximum oxic biocatalytic power density of 14 W/m^3 while also achieving the maximum anaerobic biocatalytic power density of 7.2 W/m^3. The highest possible eradication rates for COD were 99.8%, 94.7%, and 97.3%, respectively. Research has been conducted focusing on the processes of cathode nitrate denitrification as well as abiotic cathode nitrate reduction. According to Fang *et al.* (2011), nitrate levels in the cathode compartment may decrease from 49 to 25 mg/L with a power density of up to 7.2 mW/m^2 at a resistance of 470 Ω. This is possible with a certain amount of power. Ammonia was the predominant by-product of reducing nitrates, accounting for 51.8% of the reaction, whereas nitrite was only produced in very small quantities (0.6%). Research on nitrite, a crucial intermediate product of nitrate reduction, has not been carried out to the same extent as research on nitrate reduction at the cathode, which has been the subject of a great deal of investigation. Virdis *et al.* (2008) reveal that a carbon-to-nitrogen ratio could reduce if nitrite worked better as a terminal electron acceptor at the cathode of an MFC. Exoelectrogenic bacteria are found to be capable of using both nitrate and nitrite for nitrogen reduction (Puig *et al.*, 2011). However, the generation of electricity is impacted by the oxidation of nitrite in the presence of oxygen at the cathode, which may take place as a result of either biological or electrochemical processes (Puig *et al.*, 2011). In the most current research on the effectiveness of MFCs in removing nitrate, field applications were shown to be practical. The energy that was generated by the organic pollutants in river Xin Romania was transferred to the nitrate in the river, which functioned as a receiver for the energy. A multifunction cleaner with a single compartment was able to attain a power density of 88 mW/m^2 and a current density of 310 mA/m^2. According to Cucu *et al.* (2016), the removal efficiency for both organic contaminants and nitrates was 97%. Equations (8.2) through (8.6) illustrate the denitrification process, in which nitrous oxide is a crucial intermediate component. It is of the utmost importance to cut down emissions of N_2O, which are among the most powerful greenhouse gases. The laws of thermodynamics point to N_2O as perhaps the most promising electron acceptor among the oxidized nitrogen intermediates that are produced during the denitrification process. Desloover *et al.* (2011) found that the cathode chamber has the capacity to absorb N_2O between 0.76 and 1.83 kg N/m^3 NCC.

8.3.4 Persulphate

Persulphate has a wide range of uses, some of which include the elimination of contaminants in soil and groundwater, the bleaching of hair, and the micro-etching of copper-printed circuit boards, the measurement of total organic carbon, and the cleaning of swimming pools. According to Li *et al.* (2009), persulphate is a hazardous waste because of its oxidizing properties. But persulphate can be used in MFCs because it has a good oxidation–reduction potential of 2.12 V, which is higher than that of many electron acceptors used in these devices, such as permanganate. Moreover, persulphate is advantageous for MFC applications due to its sulphate composition. Because of its many favourable characteristics, persulphate was chosen to serve as the electron acceptor in this experiment. The power density in the MFC rose by a factor of 2 when persulphate was used instead of $K_3Fe(CN)_6$ as the reducing agent, from 83.9 to 166.7 mW/m^2. There is a possibility that a $K_2S_2O_8$-containing MFC as well as a $K_3Fe(CN)_6$-containing MFC will not function at medium-to-high current densities. This behaviour was attributed to the ferricyanide solution having quicker electron reduction kinetics on the surface of the carbon electrode, as stated by Li *et al.* (2009).

8.3.5 Permanganate

Under both acidic and alkaline conditions, Equation (8.6) shows the transformation of permanganate into manganese dioxide (MnO_2) by only requiring the donation of three electrons. Permanganate has a property that allows it to receive electrons into its structure. As the oxidation potential of permanganate is higher under conditions that are more acidic than in settings that are more alkaline, it stands to reason that the power output of permanganate would be greater in the former. In the tests that were conducted (You *et al.*, 2006), different pH values were used in order to investigate the effectiveness of permanganate in MFCs.

$$MnO^{4-} + 4H^+ + 3e^- \rightarrow MnO_2 + 2H_2O \tag{8.6}$$

The electron acceptor was permanganate, which produced a power density of 115.60 mW/m^2 at a current density of 0.017 mA/cm^2. This was 4.5 and 11.3 times higher than the power densities produced by hexacyanoferrate (25.62 mW/m^2) and oxygen (10.2 mW/m^2), respectively. In the same research, a bushing MFC that used permanganate as an electron acceptor was able to generate the most possible power density, which was 3986.72 mW/m^2 at a current density of 0.59 mA/cm^2. You *et al.* (2006) have shown that permanganate works well as a cathodic electron acceptor in metal–organic framework reactors. This is an extremely important point to keep in mind. The experiment, on the contrary, has a number of errors. During the process of producing electricity, the depletion of permanganate, similar to the depletion of other soluble electron acceptors, requires regular liquid replenishment. Because the cathode potential is particularly pH-sensitive, the authors assume that only small-scale power supplies may profit from the pH management of the solution. This is due to the fact that the cathode has potential. The fact that the system does not include a catalytic phase, on the contrary, is beneficial (You *et al.*, 2006). A recent study focused on the ideal concentration of permanganate from the perspective of energy generation. Eliato *et al.* (2016) found that the maximum power density of potassium permanganate at 400 mM is 93.13 mW/m^2, and that at this power density, the current density is 0.03 mA/cm^2.

8.3.6 Manganese dioxide

Manganese dioxide has proven to be effective in batteries and alkaline fuel cells as a cathode material and catalysts, according to Li *et al.* (2010a). By using the redox pair that MnO_2 and Mn^{2+} form, the cathode is able to transfer electrons to an electron acceptor. It is more effective to use electron mediators between the cathode and oxygen due to the challenges associated with direct oxygen consumption (i.e. limited solubility). These challenges make it difficult to use oxygen directly. Rhoads *et al.* (2005) focused on the possibility of biomineralizing manganese oxides. A layer of manganese dioxide first develops on the cathode, where an input of electrons from the anode reduces it. According to Liew *et al.* (2015), lower price of manganese dioxide makes it a competitive option to platinum when it comes to serving as a cathode catalyst. In addition to this, it is advantageous for the mechanism known as the electron mediator. The greatest volumetric anode density ever recorded for a tube MFC was 3,773,347 mW/m^3 when manganese dioxide was utilized as a catalyst. Notably, the low price of MnO_2 makes it a feasible alternative to Pt in applications (Zhang *et al.*, 2009).

8.3.7 Mercury (Hg)

According to Wang *et al.* (2011), either nicotinamide adenine dinucleotide + hydrogen (NADH) or nicotinamide adenine dinucleotide (NAD^+) can substitute mercury, which has a redox potential of roughly 320 mV (Hg^{2+}), as an electron acceptor. The use of mercury in MFCs provides the dual benefit of removing mercury from rivers and also generating electricity from the element. According to Wang *et al.*, using this approach resulted in the production of elemental mercury on the surface of the cathode and dimercury dichloride (Hg_2Cl_2) deposits on the bottom of the cathode chamber. This allowed for a peak power density of 433.1 mW/m^2 to be achieved (Wang *et al.*, 2011).

8.3.8 Iron (Fe)

Capacity of iron to act as an electron mediator enables it to be used in a way that makes it possible to enhance the efficiency of cathodes. The redox pair Fe^{3+}/Fe^{2+} is the one that is used most often in MFCs. By using Equation (8.7), the ferric iron that is present in a cathode chamber may be transformed into ferrous iron:

$$Fe^{3+} + e^{-} \rightarrow Fe^{2+} \tag{8.7}$$

This reversible electron transfer mechanism offers a number of benefits, some of which are listed below: fast reactions, high standard potentials, biological degradability (Heijne *et al.*, 2006), and the release of essential chemicals such as phosphate. When this redox was paired with a bipolar membrane and a graphite electrode, the greatest power density that could be achieved was 0.86 W/m^2 at a current density of 4.5 A/m^2. The coulombic efficiency and energy recovery ranged from 80% to 95% and from 18% to 29%, respectively (Heijne *et al.*, 2006). An oxidation reaction is required to drive the Fe^{3+}/Fe^{2+} redox cycle to a successful conclusion. In 2007, Heijne and his colleagues focused on how well an MFC worked with continuous ferrous iron oxidation. They performed this by breaking down the iron with *Acidithiobacillus ferrooxidans*, an acidophilic chemolithoautotrophic microbe. They found that changing ferrous iron to ferric iron enhanced power generation by 38%, bringing it to 1.2 W/m^2 and 4.4 A/m^2, respectively. This was in contrast to the research they had conducted earlier. Iron has the potential to act as both an electron acceptor and a moderator of the flow of electrons. In a different study, iron was employed in the form of ferric phosphate ($FePO_4$), which is a phosphate. The substance $FePO_4$ may be found in sewage sludge, and this substance serves not only as a source of orthophosphate but also as an electron acceptor owing to the high concentration of Fe^{3+} that it contains. Phosphorus is a component that is necessary for a great number of manufacturing and agricultural procedures. According to Cordell *et al.*'s (2009) findings, this essential chemical will run out somewhere between 50 and 100 years from now. According to Usharani and Lakshmanaperumalsamy (2010), the recovery of phosphate is more important than the dispersion of phosphate because it is one of the key drivers of eutrophication. In comparison to the power densities shown by other electron acceptors and mediators, ferric iron demonstrated rather high levels of activity. For ferric iron MFCs, a bipolar membrane is needed rather than a CEM, which is often used. Because CEMs are capable of transporting a wide variety of cations, including protons, one cannot utilize them to control the pH of a cathode chamber. According to Heijne *et al.* (2006), use of ferric iron necessitates either a bipolar membrane or the presence of an acid. The high purity of the phosphate that is produced as a result of this process is the fundamental benefit of using it. As a result, phosphate may be easily separated from iron and other components that might possibly be harmful. A pH value of <2.5 is necessary in order to retain solubility of ferric iron intact. This is due to the fact that, at higher pH values, ferric iron tends to precipitate as ferric iron hydroxides. It has been discovered that the presence of these precipitates is deleterious to the membrane's function. In addition, according to Fischer *et al.* (2011), in order to transfer electrons and protons to Fe^{3+}, a cathodic mediator, such as methylene blue, is necessary. This may limit its ability to find widespread use. Iron has traditionally been known for its role as an electron acceptor; however, recent studies (Nguyen *et al.*, 2016; Santoro *et al.*, 2016) have shown that it also has the potential to be used in the production of highly effective catalysts.

8.3.9 Copper

Copper is one of the many heavy metals that may be found in sediment and water; its presence there is mostly the consequence of emissions from mining and metallurgical processes. Copper is a micronutrient that is required for life, but at high concentrations, it may be poisonous to living beings (Alaoui-Sossé *et al.*, 2004). Because of this, eliminating copper is very necessary. Heijne *et al.* (2010) and Tao *et al.* (2011b) focused on the possibility of two-compartment MFCs being able to produce energy while also recovering copper. Through the use of a bipolar membrane, Heijne *et al.* (2010) were able to segregate pH and recover copper using MFCs. It was possible to attain a removal efficiency of more than 99.88% by operating at a maximum power density of 0.80 W/m^2 and a current density of 3.2 A/m^2. Although copper(II) oxide (CuO) and copper(I) oxide (Cu_2O) were not found, it was noted that pure copper crystals were growing on the surface of the cathode. According to studies that these same researchers have conducted earlier, the pH in the cathode portion of the bipolar membrane remained rather near neutral. Tao *et al.* (2011b) carried out an experiment to observe what would happen when a proton exchange membrane (PEM) and a solution of cupric sulphate were used as a catalyst to speed up the reduction of Cu^{2+} in a membrane-fault current cell. The highest power density of a glucose-fed MFC was reported to be 339 mW/m^3, and this was established when the starting copper concentration was 6412.5267 mg Cu^{2+}/L. With an external resistance of 15 Ω and a copper concentration of 0.4 mg Cu^{2+}/L at the start of the experiment, it was possible to remove more than 99% of copper. To further reduce the cost of fabrication associated with this strategy, Tao *et al.* (2011a) developed a lab-scale membrane-free buffled MFC. According to Tao *et al.* (2011a), after 144 h of exposure to a copper concentration of 500 mg/L, 70% of was removed from the solution. Copper, on the contrary, is an excellent electron acceptor, in contrast to oxygen (Tao *et al.*, 2011b). The use of cathodic copper reduction has therefore resulted in an increase in the potential for MFCs. It is possible to vary the amount of copper to be lowered and the amount of energy to be generated depending on the design and characteristics of reactors. Investigations into multiple batch cycle operations for the production of electricity using a variety of cathode materials have been carried out recently (Wu *et al.*, 2016a). Carbon rods, titanium sheets, and stainless-steel woven mesh were all put together based on their paces as potential candidates for the role of cathode material in copper removal processes. It has been determined that stainless-steel woven mesh is a cathode material that is both the most effective and the most cost-effective. When copper removal is necessary in an MFC, the amount of copper deposited on the cathode has a substantial impact on both the amount of copper removed and the power density. This method is in its infancy; thus, more research into the catalytic behaviour of copper for oxygen reduction at the cathode and the construction of a more effective reactor are necessary. According to research conducted by Wu *et al.* (2016b), the power density required to remove copper from an MFC may attain 33.6 W/m^3 depending on the kind of reactor, electron source, anode, and cathode materials used.

8.3.10 Chromium

Chromium has been shown to be capable of performing the role of an electron acceptor in a number of investigations (Li *et al.*, 2009; Sahinkaya *et al.*, 2016, among others). When wastewaters containing natural or synthetic chromium were treated in MFCs, the process resulted in a simultaneous decrease of chromium and the generation of energy. This reduction process is theoretically conceivable with a redox potential of 1.33 V, which is important to note from a thermodynamic point of view. Using a synthetic effluent with 200 mg Cr(IV)/L, both the highest power density of 150 mW/m^2 (0.04 mA/cm^2) and the highest open-circuit voltage (OCV) of 0.91 V were possible to obtain. Both of these results are impressive. According to the findings of this experiment (Wang *et al.*, 2008), a decrease in pH has a beneficial impact on the lowering of Cr(VI). Li *et al.* (2009) examined the same procedure but using actual electroplating effluent that contained Cr(VI). According to the results of this investigation, the material used for the electrode is an essential factor in the removal of Cr(VI). Electroplating wastewater, which included 204 mg Cr(VI)/L, was treated using either graphite paper or graphite plates as a cathode material. However, the graphite paper produced better results (a power density of 1600 mW/m^2 and a chromium removal rate of 99.5%). In their 2009 study, Y. Li and his colleagues focused on the reduction of Cr(VI) in an MFC photoelectrochemical cell-linked system, which is different from how Cr(VI) is usually reduced in MFCs. In an experiment by Y. Li *et al.*, 97% of Cr(IV) that had been present at an initial concentration of 26 mg/L had removed after 26 h of exposure to light (Li *et al.*, 2009). The maximum potential was 0.80 V under light irradiation, but it was only 0.55 V in the absence of light irradiation. Researchers were able to transform the energy from the sun into a single-MFC unit and clean waste using a cathode that was covered with rutile. According to research carried out by Li Y. and his colleagues in 2009, synergies between a biocatalysed anode and a rutile-coated cathode increased power output and decreased Cr(IV) (Li *et al.*, 2009). Recent studies have shown that raising the amount of microbes in the cathode chamber may boost the efficiency of the chromium reduction process. An exoelectrogenic biofilm was placed near the anode so that it could be converted into a biocathode. This was done in order to accomplish this goal. According to Wu *et al.*'s (2015) research, cutting-edge technology is 2.9% more successful in lowering Cr(VI) than the more traditional biocatalysts. Other recent research on reducing Cr(VI) has focused on using self-assembled graphene biocathodes and changing the materials of electrodes (Song *et al.*, 2016; Wu *et al.*, 2016a). The cathode material for chromium reduction is often the subject of attention in the field of contemporary research. A membrane fuel cell (MFC) was used to test how well a carbon nanofibre electrode coated with alumina (AA) and nickel (Ni) nanoparticles (NPs) could remove Cr(VI). The newly designed electrode has the capacity to reduce 100 mg Cr(VI)/L at a rate of 2.12 g/m^3-h when subjected to a power density of 1540 mW/m^2. Gupta *et al.* concluded that the effectiveness of Columbia (cathodic columbic efficiency) was 93%. After treatment with an abiotic cathode for 45 h, alkaline Cr(VI) wastewater showed a decrease in chromium concentration of 10 mg/L and a power density of 21.4 mW/m^2 (Gupta *et al.*, 2017; Xafenias *et al.*, 2015).

8.3.11 Triiodide

A iodide/iodine redox couple has the same potential as the Fe^{3+}/Fe^{2+} redox couple mentioned in Section 8.3.9 on copper to act as an electron mediator. Using this redox pair at the cathode provides a number of distinct benefits. There is no loss of triiodide (I_3) in the catholyte because it may be replaced when it is used up. The iodide anion combines with an iodine molecule in water, which then leads to the production of triiodide (I_3), which then causes the cycle to continue. Additionally, it indicates the viability of using carbonaceous materials as cathodes in an efficient manner. Both fundamental and non-fundamental surroundings do not affect the inertness of triiodide. A iodide/iodine redox dyad has the right properties to act as a mediator for electrons in a cathode chamber, which is where electrons are received and moved. Li J. and his colleagues were able to show in 2010 that this redox dyad could work as the electron acceptor or mediator by using a two-chambered MFC (Li *et al.*, 2010b). According to Li *et al.* (2010a), the combination of 1.2 mM I_3 and 0.2 mM KI produced the highest power density of 484.0 mW/m^2. Because the majority of ongoing research is concentrated on H-type MFCs, there is still a need for improvements in the design of reactors in order to make it possible to generate substantial amounts of electricity. Microorganisms that are actively engaged in electrochemistry face a risk while they are in the anodic chamber because of the I_3 ion. According to Li J. and colleagues, while developing a new configuration, it is vital to take this constraint into consideration in order to improve performance (Li *et al.*, 2010b).

8.3.12 Carbon dioxide

The cathode voltage that results from the reduction of carbon dioxide is also modestly low as a result of its low redox potential. The CO_2 reduction potential is 0.420 V when the pH of the solution is 7. However, in order for there to be a generation of electricity, the cathode potential must be higher than the anode potential. According to Cao *et al.*'s (2009) research, in order to bring down CO_2 levels, it is necessary to bring in more energy. Cao *et al.* (2009) provided evidence that a biocathode MFC may harness the power of the sun to cut down carbon dioxide emissions. This allows for both the creation of biomass and a decrease in carbon dioxide emissions. Because of this bioreaction, it is possible to sequester carbon dioxide. According to Villano *et al.* (2010), cathodic chambers have the ability to convert carbon dioxide into methane as well. The oxidation of organic compounds results in the production of electrons and carbon dioxide, both of which may contribute to the synthesis of methane. In 2010, Villano *et al.* (2010) deployed a two-chambered MFC to illustrate the practicability of this approach in real-world applications. Utilizing this approach may bring beneficial results. First, the process of oxidizing organic matter is maintained separately from the process of generating methane. This keeps any inhibitors in the effluent away from the methanogens. This is accomplished by separating the two processes. The temperature of the cathode need not be maintained at a steady level, which means that this process utilizes less energy. This technique is also responsible for regulating later processes in the sequence, such as anaerobic digestion and the creation of methane. According to Villano *et al.*'s (2010) research, the system continues to operate normally even when just trace amounts of the substrate are present. The cathode reduction of carbon dioxide into biofuels or commercial chemicals using microbes and energy supplied from the outside has attracted a lot of attention in recent years. Researchers in this field have opened a new avenue for the generation of biofuels or chemicals by eliminating barriers to natural photosynthesis processes. The linked processes are known as microbial electrosynthesis or, more recently, artificial photosynthesis. These methods have the potential to create a number of carbon-based energy-storage compounds, such as acetate (Patil *et al.*, 2015), acetic acid (Gildemyn *et al.*, 2015), butyrate (Ganigué *et al.*, 2015), and ethanol (Gildemyn *et al.*, 2015; Pant *et al.*, 2010; Patil *et al.*, 2015).

8.3.13 Perchlorate

Perchlorate is an interesting contaminant in drinking water due to its high mobility and thyroid-inhibiting impact (Cetin *et al.*, 2015; Ucar *et al.*, 2017). These findings were published in two separate studies. According to Ucar *et al.* (2015a), biological reduction is more cost-effective when compared to other treatment methods. According to research by Butler *et al.* (2010), a very active population of microbes that reduce perchlorate was used in the cathode chamber to turn perchlorate into chloride. The current output by an MFC operating on acetate and perchlorate was typically measured at 0.28 mA. At this point in time, the perchlorate elimination rate that was the highest was 24 mg/L. It is possible to get rid of perchlorate that is commonly present in groundwater by using this technique. However, the levels of perchlorate that are generally found in sedimentary fluids are in the parts per million (ppm) range, making it challenging to generate power. Groundwater may often be found to have unacceptable levels of nitrate content. As a result, MFCs may be used for the removal of both perchlorate and nitrate. Recent studies have concentrated on finding ways to get rid of nitrate and perchlorate in a biocathode that is responsible for autotrophic denitrification. An MFC that was fed with acetate was able to cut the amount of input nitrate by 87.05% while also reducing the amount of perchlorate by 53.14%. Research conducted by Jiang *et al.* (2017) in 2017 found that a ratio of 1:1 between NO_3 and ClO_4 is the most effective combination. According to research carried out and published in 2017 by Lian *et al.* (2017), decreasing levels of perchlorate and nitrate using acetate is the most efficient use of electron donors. Because of its high solubility in water, acetate has a significant chance of being wasted while treating potable water for nitrate contamination. This is despite the fact that acetate is an efficient organic electron source that may reduce both nitrate and perchlorate. Under these conditions, sulphur and other inorganic electron sources could be more desirable than organic ones. When inorganic electron donors are utilized, for example sulphur, effluent may become acidic and sulphate may be formed. This is one of the potential drawbacks of using inorganic electron donors. Recent studies (Ucar *et al.*, 2015b, 2017) have demonstrated that both these processes may be utilized to remove nitrate and perchlorate from drinkable water as well as groundwater.

8.3.14 Chloroethenes

Chlorinated aliphatic hydrocarbons, often known as CAHs, present a considerable risk owing to the fact that they are both poisonous and carcinogenic. These chemicals are commonly employed as solvents and degreasing agents. According to Holliger and Schraa (1994), some anaerobic microbes have the ability to dechlorinate CAHs by using an external electron donor or external voltage. One alternate use of this method is to make use of insoluble electrodes in order to provide electrons to communities that are engaged in dechlorination. Aulenta *et al.* (2007) conducted research with two different communities: a mixed culture of bacteria that could dechlorinate and a pure culture of *Geobacter lovleyi*. They found that trichloroethylene (TCE) could be dechlorinated in a mixed culture when it was fed acetate. The results of the dechlorination were *cis*-Dichloroethene (DCE) (83.9% by molecular weight), vinyl chloride, ethene, and ethane. On the basis of the results of research that used a mixture of cultures, a polarized carbon paper electrode could be the only electron donor needed to completely dechlorinate TCE. It is likely that unexpected reactions could take place and by-products would be produced if external electron donors were brought into the polluted region. As bacterial oxidation takes place at the anode, use of a solid electrode in an MFC is beneficial because it prevents any external organic matter from being transferred to the site (Aulenta *et al.*, 2007).

8.3.15 2-Chlorophenol

On the basis of the research conducted on chloroethene, it is advisable to utilize a solid electrode as the primary electron source instead of relying on soluble electron donors. Strycharz *et al.* (2010) report that electrodes have a lot of potential for bioremediation of chlorinated contaminants and metals because they can provide electrons needed to reduce pollutants. In this context, the commonly employed species *Geobacter* plays a vital role. Strycharz *et al.* (2010) have also shown that *Anaeromyxobacter dehalogenans* can provide electrons to 2-chlorophenol, which causes it to lose its chlorine and turn into phenol. During the experiment, 10 mM of acetate was used as the substrate for *A. dehalogenans*, and 80 mM of 2-chlorophenol was used as the electron acceptor. Strycharz *et al.* (2010) suggest that utilizing electrodes as an electron source for bioremediation shows promising potential. The study recorded the highest dechlorination rates of 40 M Cl/day (in 200 mL), indicating the feasibility of achieving such rates. In 2012, Akbulut *et al.* (2012) conducted a study on the dechlorination of 2-chlorophenol, utilizing a primitive form of the laccase enzyme. According to their findings, under optimal dechlorination conditions, the enzyme successfully removed 1.5 μM of 2-chlorophenol. According to research conducted by Strycharz *et al.* in 2010, using solid electrodes as an electron source for the reduction of chloroethene has a number of benefits. Firstly, it offers the theoretical possibility of efficiently delivering electrons to microorganisms for the purpose of pollutant reduction. Secondly, direct interaction between contaminants and the electrode can help prevent undesirable reactions. This is facilitated by the ease with which the electrode functions as an electron donor on-site, simplifying its implementation. Once this process takes place, the accumulated metal impurities can be effectively eliminated from the surface of an electrode (Strycharz *et al.*, 2010).

8.3.16 Oxygen

Oxygen has historically been thought of as an electron acceptor with the greatest promise owing to its high redox potential in an oxygen reduction reaction (ORR) process, its availability, and its inexpensive cost. Harnisch and Schroder (2010) have been thought of as the electron acceptor with the greatest promise owing to the high redox potential it has in the ORR process, its availability, and its inexpensive cost. Harnisch and Schroder (2010) provided in-depth examinations of the ORR's underpinnings that may be found in other sources. Increasing the effective surface area of a cathode (Erable *et al.*, 2009; Freguia *et al.*, 2007), increasing the pressure to reduce concentration loss (Fornero *et al.*, 2008), utilizing a membrane–cathode assembly to improve proton transfer, and employing catalysts to accelerate the reaction kinetics are some of the methods that can be utilized to improve ORR performance. Pt was the most desirable catalyst for ORRs due to its strong affinity for oxygen and minimal activation loss. The limiting influence that cathode reactions have on the overall performance of MFCs was established by Logan *et al.* (2005). They performed this by showing that a Pt-based MFC could yield a fivefold increase in power output when compared to a standard carbon cathode MFC. However, the usefulness of catalysts based on platinum was restricted due to its expensive cost, its rarity in nature, and its toxicity. Lower Pt loadings down to 0.1 mg/cm^2 were studied in order to limit the use of Pt. Comparable performance was achieved to that of 2 mg/cm^2 (Cheng *et al.*, 2006), which suggests that this method might be used to lower the capital cost for Pt-based MFCs. However, scalable MFCs still needed catalysts that did not include any noble metals and were even inexpensive. Metal macrocycles (Cheng *et al.*, 2006; Kim *et al.*, 2011; Yuan *et al.*, 2011; Zhao *et al.*, 2005, 2006), metal oxides (PbO_2; Morris *et al.*, 2007; Fe/Fe_2O_3; Zhuang *et al.*, 2010; Mn_xO_2), and nitrate-treated Vulcan XC macrocycle catalysts excelled as ORR catalysts, demonstrating low internal resistance and high OCVs in addition to high power outputs (60–80% of Pt-based MFCs as control experiment correspondingly). However, because of the intrinsic toxicity of heavy metals, their practical uses were severely restricted; hence, additional transition metals were examined. However, intermediate electron acceptors with high OCVs and maximum power density (MPD) might be explored further as potential feasible replacements for metal oxides due to their excellent performance. The findings of these investigations could lead to the development of innovative ORR treatment methods. Enzymes and microorganisms have been popular choices for use as biocathodes in recent years (He & Angenent, 2006). When a mediator was present, the highest power density of MFCs with the highly efficient catalyst laccase (*Trametes versicolor*) was 10 times that of control Pt-based MFCs of the same design, and the OCV in these cells achieved 1.1 V (Schaetzle *et al.*, 2009). However, enzyme-catalysed MFCs had a critical drawback that made them unsuitable for continuous use: the activity of the enzyme decreased with time. The display of enzymes on the surface of the host cell, as was observed for anode microorganisms (Fishilevich *et al.*, 2009), is one possible solution to this issue that might be implemented.

Microorganisms are an additional kind of biocatalyst that have self-generational economics and a straightforward approach to cleaning polluted environments, which shows promise but has not yet been proven. Rhoads *et al.* (2005) used bacteria capable of oxidizing manganese as a biocatalyst, whereas Ter Heijne *et al.* (2007) focused on ferrous-oxidizing bacteria, also known as FOB. The subsequent step was to use various inoculums in mixed cultures as biocathodes (Chen *et al.*, 2008, 2010; You *et al.*, 2009). Also, the ability of isolated strains to speed up ORRs was examined, and it was found that these strains could be added to MFCs (Carbajosa *et al.*, 2010; Cournet *et al.*, 2010a, 2010b; Freguia *et al.*, 2010). Rabaey *et al.* (2008) established that isolated populations are unable to generate the same amounts of electricity and current as mixed populations. This mismatch could be caused by a number of factors, such as insufficient pH control for isolated strains, bacterial densities that have not been checked, and the possibility of heterotrophic or autotrophic growth due to organic crossover. However, biocathodes had a possible drawback in the sense that they were commonly utilized in aqueous cathode compartments. This meant that the running expenses for aeration in large-scale MFCs might increase as a consequence of the use of biocathodes. When comparing catalysts for MFC air cathodes, chemical catalysts worked well, lasted for a long time, and were relatively priced. Enzyme catalysts, on the contrary, performed well but were the most expensive and lasted for a least amount of time. Microorganisms performed well, lasted for the longest time, were the least expensive to operate, and cost the least to maintain; yet, they needed a great deal of maintenance and were the most expensive to run. There were positives and negatives associated with each of the three different kinds. It is required to study new methodologies in order to fully benefit from each kind of MFC (e.g. miniaturized MFCs with chemical catalysts, disposable MFCs with enzyme catalysts, and stacked or scalable MFCs with biocatalysts) (Rabaey *et al.*, 2008).

8.4 MFCs IN DIFFERENT INDUSTRIAL WASTEWATER

MFCs have the potential to be exploited as pre-treatment systems for wastewater, creating energy in the process, in contrast to conventional wastewater treatment methods, which require significant amounts of energy. It is possible for nutrient-rich wastewater to sustain a large biofuel production capacity while also integrating smoothly with the infrastructure already in place for treating wastewater. A good indicator of the quality of treated effluents is the degree to which it reduces its COD. Biodegradation is the process by which organic substrates are broken down by anaerobic consortia in an anodic chamber of an MFC. Studies that utilized MFCs revealed COD removal efficiencies of up to 90% in certain cases. It has been revealed that bacterio-algal MFCs have a high capacity for the elimination of COD. These results show that using microbial consortia in MFCs is a good way to get rid of COD in wastewater treatment processes. Although a cathodic chamber was responsible for removing 58% of COD, an anodic chamber was able to remove 74% of it. COD removal in bacterio-algal MFCs is an exciting invention that has the potential to dramatically expand the effect of the technology as well as its commercial use, despite the fact that it has received relatively little attention in recent years. The methanogenesis reaction's use of electrons, the cathodic biofilm's aerobic respiration, and oxygen crossing are just a few of the factors that could slow down the removal of COD. Traditional activated sludge systems are responsible for an indirect environmental burden in the form of sludge disposal, which accounts for half of the operating cost of a traditional wastewater treatment (WWT). It is necessary to perform tertiary treatment in a wastewater treatment plant (WWTP) in addition to the use of phytoplankton for the purpose of biological nutrient removal. When compared to bacterial systems, thermophilic and acidophilic algae strains are capable of producing two times as much biomass and generating 20% more net energy than their bacterial counterparts. An algal system can remove 1 kg of biochemical oxygen demand (BOD), which then provides enough biomass for anaerobic digestion to generate 1 kW-h of methane-powered energy. In research that compared the environmental implications of bioenergy from algae to those of more traditional crops such as maize, canola, and switchgrass, Clarens and Colosi (2012) discovered that the generation of algae from wastewater had positive effects on the surrounding ecosystem. It is not possible to compare the environmental and economic impacts of micro-algal biofuels to those of traditional fossil fuels because this would require the use of fertilizers that are both energy-intensive and expensive.

8.5 APPLICATIONS OF MFCs

8.5.1 Industrial applications of MFCs

In recent years, membrane filtration and concentration has emerged as a potentially useful technology for the purification of wastewater containing simple to complex carbohydrates as well as industrial effluents. In addition to the standard effluent treatment capabilities, this system also provides choices for pre-treatment and post-treatment of wastewater. As can be observed in Figure 8.2, it also offers a workable replacement option for traditional biological processes that are used in wastewater treatment plants. It has been shown that an MFC is beneficial to several types of industrial effluents, including those from the textiles, distillery, brewery, molasses, leather, pulp and paper, dairy, food processing, and agro-waste sectors, amongst others, in both laboratory and commercial settings (Das & Ghangrekar, 2018; Singh & Dharmendra, 2020). Following the failure of various field applications over the course of the last two decades, a number of well-established as well as newly emerging industrial enterprises have started providing MFCs for the on-site treatment of industrial wastewater. The leading innovators of MFCs, Israeli small- and medium-sized enterprises, developed a test run of 16 modules of MFC units using carbon felt electrodes and a permeable filter as the membrane. This was performed to treat industrial wastewater. Lebone Solutions, which is situated in the United States, suggests making use of MFCs in order to manage manure and graphite textile electrodes in order to harvest energy from soil. The Dutch company Plant-e B.V. came up with tubular plant MFCs and tubular electrode assemblies for use in plant-MFC applications. Several corporations from all over the world, such as IntAct Labs LLC (United States), Hy-SyEnce (United States), Indian Oil Corporation Ltd. (India), Tata Consultancy Services (India), and a few others, are working on the research, development, and improvement of MFCs by utilizing novel redox catalysts and materials.

8.5.2 Electricity generation and wastewater treatment

The activation of sediment by means of electrical energy is a substantial consumer of power throughout the process of commercial wastewater treatment. If biological waste remediation is monitored and examined, there is a possibility that MFCs might be used more effectively. A biosensor is created when the ratio of the amount of organic matter present in effluent to the amount of MFCs produced. These completely integrated wastewater treatment facilities are responsible for recovering energy and decreasing the creation of surplus sediment without significantly compromising the mineralization of organic matter or the rest of the process. Additionally, these facilities are designed to reduce the development of excess sediment. The expenses of the procedure, however, need to be brought down before it can be considered economically viable (Das *et al.*, 2019a). Either omitting the membrane altogether or making use of a cationic membrane, the production of which is more cost-effective, may be carried out to achieve this goal. If aerobic biomass is available, it is not necessary to use the expensive cathode catalysts. Additionally, MFCs may be run in wastewater treatment facilities, which further reduce operational expenses.

8.5.3 Nutrient removal

Microbial oxidation is the primary method used at the anode in the process of removing organic compounds. Since the discovery of biocathodes and the relative reduction phenomenon at the cathode, an MFC has been given a new lease of life as a treatment method for wastewater. This made it possible to get rid of a number of harmful pollutants, such as perchlorate, nitrate, chlorinated compounds, copper, iron, nitrate, and mercury. The results of the first study on nitrate denitrification in MFCs, which was published in 2007, have been confirmed by the fact that full denitrification can occur at

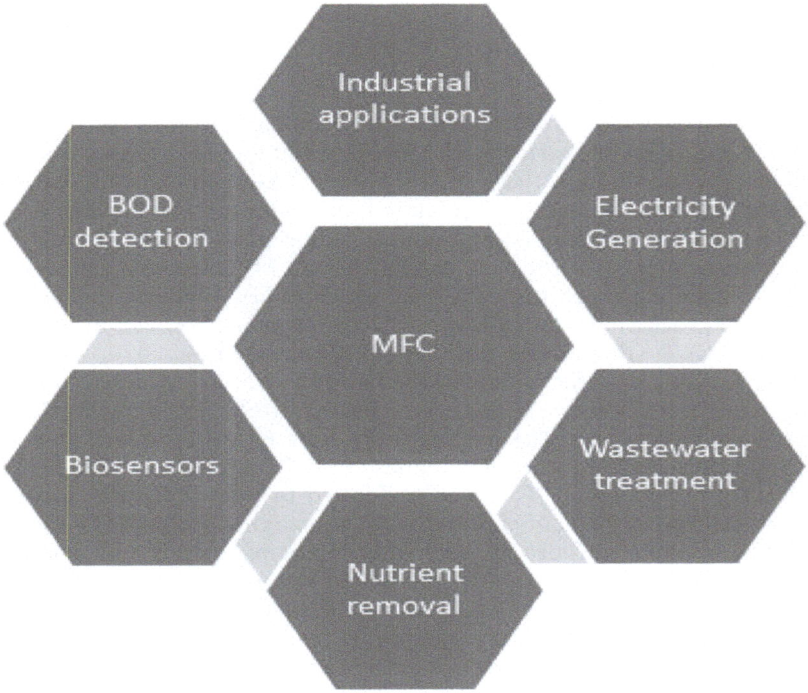

Figure 8.2 Applications of MFCs.

the cathode in a tubular reactor without adding an additional donor source. Getting rid of carbon and nitrogen at the same time using a unique approach is possible by combining MFCs with aerobic nitrification (Das *et al.*, 2019b; Singh & Dharmendra, 2020).

8.5.4 Real-time applications and sensor-based applications

In addition to its application in the treatment of wastewater and the elimination of toxins from industrial effluents, MFCs may also be used as a source of power for a wide variety of electrical devices and sensors. An MFC is able to measure a wide range of ambient water quality monitoring parameters, including BOD and COD, dissolved oxygen concentration, heavy metal dose, volatile fatty acids, toxicity detection, gas detection, and microbial activity estimates. These parameters may be measured online or on-site. In addition to supplying energy to electronic devices, micro-form-factor connections are able to provide an external power source to smaller electronic sensors. MFC technology has a lot of potential applications in the robotics sector because it can simulate the opportunistic behaviour of robots and derive electricity from the organic matter of substrates *via* electrochemical redox reactions. When gastrobots were first developed, they obtained their power supply from *Escherichia coli* fuel cells. This power supply was then utilized to charge the robot's onboard batteries, which in turn powered the robot's many controllers and actuators. Ecobot-I was the first robot to be successfully shown that was powered only by stacking MFC units, but Ecobot-II is a more autonomously behaved, more complicated robot that utilizes the same amount of MFC units but has an operational cycle that lasts for 12 days. For space applications, underground conditions, and isolated places within the constraints of their working surroundings, it is presumed that the aforementioned examples of autonomous robots driven by stacked MFCs are capable of being self-sufficient in terms of both their energy needs and their ability to maintain themselves (Bhowmick *et al.*, 2019; Pant *et al.*, 2010).

8.5.5 Biosensors

Electrodes may be used to immobilize bacteria, and a membrane can be used to prevent germs from moving to the other compartment during the construction of biosensors. The diffusion of any potentially harmful component *via* a sensor may be measured by the voltage differential that exists between the electrodes (Das & Ghangrekar, 2019). Biosensors have a wide variety of uses, including the detection of pesticides in rivers, the conduct of research in polluted regions, the measurement and indication of pollution, and the determination of unlawful waste disposal. At the entrances to wastewater treatment plants, biosensors have been installed specifically for this purpose. Recent research has shown that MFCs may be successfully used as biosensors for the purpose of detecting and measuring levels of the cocaine metabolite benzoylecgonine that are present in human urine.

8.5.5.1 MFCs as self-powered biosensors versus traditional whole-cell biosensors

Because of their remarkable sensitivity to a wide range of biological and environmental variables, MFCs are also capable of performing the role of biosensors. Self-powered devices, which do not need an external source of electricity to work, can be fabricated by combining the energy production and sensing functions of MFCs. These devices can detect and measure the needed parameters on their own remotely. This allows the devices to detect and quantify the required parameters from a greater distance. MFCs have been put through a number of different tests to observe how well they perform in their unconventional roles as electrical biosensors. When a biological recognition element comes in contact with an electrochemical transduction element, e.g. the anode or cathode electrodes in MFCs, specific analytical data are obtained. The meaning of the term 'electrical biosensor' as used by the IUPAC is that enzymes, fluorescent proteins, and many other molecules that emit fluorescence or pigment are the kinds of biological components that are often put to use in traditional biosensors for the purpose of identifying analytes. However, the fact that typical biosensors need an external power source to run the transducer and the other electronic components that are responsible for signal conversion, amplification, and transmission significantly restricts the range of applications for which these sensors may be used to monitor the environment remotely. Microorganisms, on the contrary, are the biological sensing elements in an MFC. The reaction of microorganisms is quantified *via* the anode and cathode electrodes of an MFC in the form of an electrical current output. There is no need for a transducer because the analytical output signal of MFC-based biosensors is already measured as an electric current. Therefore, biosensors based on MFCs are more expensive and energy efficient than traditional biosensors while also continuing to be recyclable and beneficial to the environment. In conclusion, it would seem that biosensors that are based on MFCs are most useful for monitoring microbial activity and the related parameters such as BOD or detecting inhibitors and toxicants. MFC-based biosensors are considered the next generation of biosensing technology because of the promising applications they have for the rapid monitoring of environmental factors such as pH, temperature, composition, and concentration of organic matter, as well as other parameters related to the quality of water effluents (Sun *et al.*, 2015).

8.5.5.2 MFC-based biosensors for BOD detection

BOD is a key indication of water quality and is also one of the most frequently utilized indicators. To be precise, the content of biodegradable organics in wastewater is commonly tested *via* the BOD measurement, which is typically performed over 5 days (BOD5). Before and after subjecting a sample of water to incubation at 20°C for 5 days, the quantity of dissolved oxygen in the water is determined. When utilizing this procedure, obtaining reliable results requires a significant investment of both time and effort. As a result, biosensing using MFCs looks to be an alternative to a traditional BOD test that lasts for 5 days that is more efficient, more accurate, and requires less effort on the part of the user. Over the course of the last several years, MFCs have gained a significant amount of traction as a preferred method for BOD measurement. Current efficiency, which is also called coulombic or Faraday efficiency, is proportional to the amount of oxidizable organic matter in the feedstock or fuel of a membrane fuel cell (MFC), which is another name for this type of fuel cell. The most notable benefits that an MFC-based BOD biosensor offers in comparison to traditional ones are its accuracy and repeatability, response speed (quick monitoring), long-term operational stability, and minimal maintenance and servicing needs. All of these factors make it superior to conventional biosensors. For instance, Kim *et al.* (2011) created an MFC-based system for remote BOD detection that is capable of providing a steady current over a period of more than 5 years. Observations have shown that changes in temperature, pH, conductivity, and the concentration of inorganic solids have a substantial impact on the sensor's sensitivity. However, the stability and sensitivity of MFC-based biosensors for BOD monitoring are dependent on a wide variety of other factors. These other factors include the concentration of dissolved oxygen in the cathode chamber, the external resistance, the effluent flow rate, and anything else that has an effect on the electricity generation of an MFC. The main drawbacks of MFC-based BOD biosensors are their slow metabolic rates and the limited variety of microorganisms and chemical substances that they can detect. Researchers have examined the possibility of creating electrogenic bacteria that are highly active and adaptable and can oxidize a wide range of organic compounds efficiently. This would improve the effectiveness and adaptability of MFC-based sensors for detecting BOD. This will allow sensors to be used more effectively and more widely. Regulation of BOD is particularly relevant to the research of microorganisms and pollutants in groundwater. MFC-based BOD sensors provide options that are quick, easy, non-invasive, and affordable in comparison to conventional monitoring techniques, which may be time-consuming and expensive. Zhang *et al.* (2015) proposed a novel design for a submersible MFC so that they could achieve their goal. The cell acted as a sensor for real-time measurement of BOD and microbiological activity in groundwater, and a battery was not required. The cell was able to detect microbial activity and BOD in authentically polluted groundwater in less than 3 h, with fluctuations of 15–22% and 6–16%, respectively. Kharkwal *et al.* (2017) have only recently developed a reliable BOD sensor based on an MFC by making use of MnO_2 catalysts. The use of manganese dioxide as a cathode was one strategy that was used to lower the cost of producing MFCs for potential widespread application. The cell was examined using a synthetic sodium acetate solution in addition to effluent that had not been treated. According to the results, the expected range for the BOD levels was between 3% and 12% of the BOD5 readings, which is indicative of agreement. It has been established that the applicable MFC-based BOD biosensor has a stability that is greater than 1.5 years.

8.5.5.3 MFC-based biosensors for water toxicity detection

The detection of water toxicity is vital for defining the measures necessary to supply water that is safe for the sustenance of human beings, animals, and plants within the required quality range. Safety for humans, animals, and plant life depends on meeting the quality requirements. MFCs seem to be the systems with the highest level of sensitivity because the metabolic activity of microorganisms and, as a result, the rate at which they take in substrates are altered by the presence of any toxicant in the aqueous feedstock. As a result, the presence of harmful substances in circulating water, as well as the amount of such substances, may be quickly identified by analysing the disruptions in the electric current that are created by MFCs (Yu *et al.*, 2017).

8.5.5.4 Techno-economic assessment of MFCs

The cost of the reactor's electrodes, membrane (if one is required), electrical connections, manufacture (making or buying), coating (if one is needed) of the electrodes, maintenance, and other related expenses are all factors that should be taken into account financially. The specifics include the efficiency of the treatment, duration of the procedure, the amount of energy that is created, the value of by-products, and so on. The technologies are compared and contrasted with one another using few criteria, and their potential in the market is assessed. When determining the effectiveness of a process, it is a common practice to employ both linear and circular methodologies. Sewage treatment facilities that have the capability of releasing cleaned water into a nearby stream are able to use linear models. The therapy process involves both aerobic and anaerobic components. Aerobic therapy necessitates the circulation of oxygen, which makes it an inefficient kind of treatment. On the contrary, anaerobic therapy calls for no additional energy input but takes much more time. This construction showcases environmentally friendly methods if it is equipped with MFC systems. In this scenario, MFC systems regulate the anaerobic treatment and generate electricity without the need for further purification of methane gas that a typical anaerobic digester would create. This is because MFC systems utilize a membrane film reactor. The solid waste from a traditional treatment plant is used as a fuel to power the MFC, which creates energy. Because of this, the quantity of garbage that may be considered significant is cut down, which in turn lowers the cost of disposal. It is possible for the capital cost of MFCs to be 30 times higher than the cost of wastewater treatment technologies that are more traditionally used in certain circumstances. In spite of the fact that efforts have been made to run MFCs on a pilot scale in order to evaluate whether or not it would be feasible to run MFCs on a large scale, the development of MFCs has been slowed down owing to difficulties in both operational and economic aspects. An MFC's performance may be evaluated using a variety of technical parameters, including COD elimination, energy generation, coulombic efficiency, HRT, output voltage, and power density. The effectiveness of MFCs may be affected by a variety of factors, including the composition of electrodes and substrates, the volume of the reactor, and the kind of membrane used. The present demand cannot be satisfied with a single MFC; however, stacking MFCs may provide a route towards more ecologically responsible wastewater treatment. When a large number of MFC modules are stacked, the cost of construction has the potential to increase if alternative low-cost materials that do not compromise MFC performance are not investigated. Because the internal resistance of a system rise as its volume grows, stacking may result in complexity in design and operation, leakage, maintenance, very high costs, and even a loss in power production. Optimization studies are absolutely necessary steps before releasing the product to a broader populace. Microorganisms play an important role, and it is vital for them to come into contact with an anode. The cost of constructing a treatment facility is directly proportional to the amount of reactor or plant capacity that is required. When treated at a smaller scale (1200 m^3/day, COD of 610–1116 mg/L), effluents from paper pulp cost \$312/m^3 with

>85% COD removal; however, when scaled-up to 100,000 m^3/day, the cost reduced to \$39/m^3 while maintaining the same level of effectiveness. MFC technique demands a bigger initial investment, with an estimated cost of Rs. 329,000 (\$4386) for a capacity of 1.5 m^3 that can treat septage effluents and generate energy; this kind of toilet is known as a 'bioelectric toilet'. The fact that the technology has a 10-year payback period demonstrates its viability in the long run (Mehmood *et al.*, 2019).

8.6 CONCLUSIONS

The use of various catholytes in MFCs has shown significant promise for the treatment of industrial wastewater. The selection of a catholyte is critical to improving the performance and efficiency of MFCs, thereby impacting the total wastewater treatment process. Various catholytes have been identified and used to improve electron transfer kinetics and maximize power generation in MFCs, including oxygen, ferricyanide, and other redox mediators. The choice of a suitable catalyst is influenced by criteria such as the unique wastewater composition, required treatment efficiency, and economic feasibility. Maximum average power output was achieved by combining ferricyanide with a multilayer structure, such as a hexacyanoferrate cathode. The use of various catholytes in MFCs is a unique and promising method for industrial wastewater treatment.

MFCs' promise towards sustainable and efficient solutions for industrial wastewater treatment is growing due to continual advances in catalyst selection, system design, and operating techniques. The use of NPs, which speed up the electron transfer process; genetically altered microbes; regulated or pre-treated inoculum; and decreasing the amount of time required for MFC commencement are all approaches that may be used to improve the effectiveness of MFCs. It is expected that fuel cells would work at temperatures in the mesophilic range, which are ideal for the purification of wastewater.

8.7 FUTURE PERSPECTIVES

Prior to the turn of the century, an MFC was considered a potentially effective approach for cleaning wastewater and probing new frontiers in the remediation of industrial pollutants. In addition, it was thought to be a good idea to try increasing the energy output of an MFC by experimenting with new types of materials, redox catalysts, and architectural strategies. The amount of energy that MFCs can harvest increases slightly with an increase in the anodic chamber's size. The scalability of MFCs has to change towards modularity and the stacking of small components in order to accommodate advancements in the use of electrical circuits. Additionally, as bioreactors were modular and stacked, there was a voltage reversal and ionic cross-conduction in electrolyte solutions. The hope that had been lost in relation to the potential of this bioelectrochemical technology has been reignited thanks to a select group of promising experimental projects using MFCs for wastewater treatment, robotics, and biosensors. In spite of its relatively high initial cost, MFC technology has a payback time of 20 years, making it comparable with other technologies that are being used for wastewater treatment. The integration of MFC technology with conventional treatment systems may result in the removal of additional pollutants and the recovery of more resources from wastewater. The utilization of MFCs for the treatment of industrial wastewater as well as the emergence of organizations prepared to combine biofuel cells into ground demonstrations signal the viable use of this technology.

REFERENCES

Aelterman P., Rabaey K., Verstraete W., Pham H. T. and Boon N. (2006). Continuous electricity generation at high voltages and currents using stacked microbial fuel cells. *Environmental Science & Technology*, **40**, 3388–3394, https://doi.org/10.1021/es0525511

Ahirwar A., Das S., Das S., Yang Y. H., Bhatia S. K., Vinayak V. and Ghangrekar M. M. (2023). Photosynthetic microbial fuel cell for bioenergy and valuable production: a review of circular bio-economy approach. *Algal Research*, **70**, 102973, https://doi.org/10.1016/j.algal.2023.102973

Akbulut M., Aytar P., Gedikli S., Ünal A., Çabuk A. and Kolankaya N. (2012). Dechlorination of chlorine from 2-chlorophenol with *Trametes versicolor* crude laccase. *Journal of Engineering and Natural Sciences*, **30**, 19.

Alaoui-Sossé B., Genet P., Vinit-Dunand F., Toussaint M. L., Epron D. and Badot P. M. (2004). Effect of copper on growth in cucumber plants (*Cucumis sativus*) and its relationships with carbohydrate accumulation and changes in ion contents. *Plant Science*, **166**, 1213–1218, https://doi.org/10.1016/j.plantsci.2003.12.032

Aryal N., Ammam F., Patil S. A. and Pant D. (2017). An overview of cathode materials for microbial electrosynthesis of chemicals from carbon dioxide. *Green Chemistry*, **19**(24), 5748–5760, https://doi.org/10.1039/C7GC01801K

Aulenta F., Pera A., Rossetti S., Petrangeli Papini M. and Majone M. (2007). Relevance of side reactions in anaerobic reductive dechlorination microcosms amended with different electron donors. *Water Research*, **41**, 27–38, https://doi.org/10.1016/j.watres.2006.09.019

Bhowmick G. D., Das S., Adhikary K., Ghangrekar M. M. and Mitra A. (2019). Using rhodium as a cathode catalyst for enhancing the performance of microbial fuel cells. *International Journal of Hydrogen Energy*, **44**(39), 22218–22222, https://doi.org/10.1016/j.ijhydene.2019.06.063

Butler C. S., Clauwaert P., Green S. J., Verstraete W. and Nerenberg R. (2010). Bioelectrochemical perchlorate reduction in a microbial fuel cell. *Environmental Science & Technology*, **44**, 4685–4691, https://doi.org/10.1021/es901758z

Cao X., Huang X., Liang P., Boon N., Fan M., Zhang L. and Zhang X. (2009). A completely anoxic microbial fuel cell using a photo-biocathode for cathodic carbon dioxide reduction. *Energy & Environmental Science*, **2**, 498–501, https://doi.org/10.1039/b901069f

Carbajosa S., Malki M., Caillard R., Lopez M. F., Palomares F. J., Martin-Gago J. A., Rodriguez N., Amils R., Fernandez V. M. and De Lacey A. L. (2010). Electrochemical growth of *Acidithiobacillus ferrooxidans* on a graphite electrode for obtaining a biocathode for direct electrocatalytic reduction of oxygen. *Biosensors & Bioelectronics*, **26**, 877–880, https://doi.org/10.1016/j.bios.2010.07.037

Cetin U., Goncu B. and Ucar D. (2015). Perchlorate removal with inorganic electron donors. *Sigma Journal of Engineering and Natural Sciences*, **33**, 286–296.

Chakraborty I., Das S., Dubey B. K. and Ghangrekar M. M. (2020). Novel low-cost proton exchange membrane made from sulphonated biochar for application in microbial fuel cells. *Materials Chemistry and Physics*, **239**, 122025, https://doi.org/10.1016/j.matchemphys.2019.122025

Chaturvedi N. K. and Katoch S. S. (2020). Remedial technologies for aniline and aniline derivatives elimination from wastewater. *Journal of Health and Pollution*, **10**(25), 200302, https://doi.org/10.5696/2156-9614-10.25.200302

Chaudhuri S. K. and Lovley D. R. (2003). Electricity generation by direct oxidation of glucose in mediatorless microbial fuel cells. *Nature Biotechnology*, **21**(10), 1229–1232, https://doi.org/10.1038/nbt867

Chen G. W., Choi S. J., Lee T. H., Lee G. Y., Cha J. H. and Kim C. W. (2008). Application of biocathode in microbial fuel cells: cell performance and microbial community. *Applied Microbiology and Biotechnology*, **79**, 379–388, https://doi.org/10.1007/s00253-008-1451-0

Chen G. W., Cha J. H., Choi S. J., Lee T. H. and Kim C. W. (2010). Characterization of an open biocathode microbial fuel cell for electricity generation and effluent Polish. *Korean Journal of Chemical Engineering*, **27**, 828–835, https://doi.org/10.1007/s11814-010-0142-6

Cheng S., Liu H. and Logan B. E. (2006). Power densities using different cathode catalysts (Pt and CoTMPP) and polymer binders (Nafion and PTFE) in single chamber microbial fuel cells. *Environmental Science & Technology*, **40**, 364–369, https://doi.org/10.1021/es0512071

Clarens A. and Colosi L. (2012). Life cycle assessment of algae-to-energy systems. In: Advanced Biofuels and Bioproducts. J. W. Lee (ed.), Springer New York, New York, NY, pp. 759–778.

Clauwaert P., Van der Ha D., Boon N., Verbeken K., Verhaege M., Rabaey K. and Verstraete W. (2007). Open air biocathode enables effective electricity generation with microbial fuel cells. *Environmental Science & Technology*, **41**(21), 7564–7569.

Cordell D., Drangert J.-O. and White S. (2009). The story of phosphorous: global food security and food for thought. *Global Environmental Change*, **19**, 292–305, https://doi.org/10.1016/j.gloenvcha.2008.10.009

Cournet A., Berge M., Roques C., Bergel A. and Delia M. L. (2010a). Electrochemical reduction of oxygen catalyzed by *Pseudomonas aeruginosa*. *Electrochimica Acta*, **55**, 4902–4908, https://doi.org/10.1016/j.electacta.2010.03.085

Cournet A., Delia M. L., Bergel A., Roques C. and Berge M. (2010b). Electrochemical reduction of oxygen catalyzed by a wide range of bacteria including Gram-positive. *Electrochemistry Communications*, **12**, 505–508, https://doi.org/10.1016/j.elecom.2010.01.026

Cucu A., Tiliakos A., Tanase I., Serban C. E., Stamatin I., Ciocanea A. and Nichita C. (2016). Microbial fuel cell for nitrate reduction. *Energy Procedia*, **85**, 156–161, https://doi.org/10.1016/j.egypro.2015.12.286

Das S. and Ghangrekar M. M. (2018). Value added product recovery and carbon dioxide sequestration from biogas using microbial electrosynthesis. *Indian Journal of Experimental Biology*, **56**(7), 470–478.

Das S. and Ghangrekar M. M. (2019). Tungsten oxide as electrocatalyst for improved power generation and wastewater treatment in microbial fuel cell. *Environmental Technology*, **41**, 1–8.

Das I., Das S., Chakraborty I. and Ghangrekar M. M. (2019a). Bio-refractory pollutant removal using microbial electrochemical technologies: a short review. *Journal of the Indian Chemical Society*, **96**(4), 493–497.

Das S., Das S. and Ghangrekar M. M. (2019b). Quorum-sensing mediated signals: a promising multi-functional modulators for separately enhancing algal yield and power generation in microbial fuel cell. *Bioresource Technology*, **294**, 122138, https://doi.org/10.1016/j.biortech.2019.122138

Demirel S., Uyanık İ., Yurtsever A., Çelikten H. and Uçar D. (2014). Simultaneous bromate and nitrate reduction in water using sulfur-utilizing autotrophic and mixotrophic denitrification processes in a fixed bed column reactor. *Clean: Soil, Air, Water*, **42**, 1185–1189, https://doi.org/10.1002/clen.201300475

Desloover J., Puig S., Virdis B., Boeckx P., Verstraete W. and Boon N. (2011). Biocathodic nitrous oxide removal in bioelectrochemical systems. *Environmental Science & Technology*, **45**, 10557–10566, https://doi.org/10.1021/es202047x

Du Z., Li H. and Gu T. (2007). A state of the art review on microbial fuel cells: a promising technology for wastewater treatment and bioenergy. *Biotechnology Advances*, **25**(5), 464–482, https://doi.org/10.1016/j.biotechadv.2007.05.004

Eliato T. R., Pazuki G. and Majidian N. (2016). Potassium permanganate as an electron receiver in a microbial fuel cell. *Energy Sources, Part A: Recovery, Utilization and Environmental Effects*, **38**, 644–651, https://doi.org/10.1080/15567036.2013.818079

Erable B., Duteanu N., Kumar S. M. S., Feng Y. J., Ghangrekar M. M. and Scott K. (2009). Nitric acid activation of graphite granules to increase the performance of the non-catalyzed oxygen reduction reaction (ORR) for MFC applications. *Electrochemistry Communications*, **11**, 1547–1549, https://doi.org/10.1016/j.elecom.2009.05.057

Fang C., Min B. and Angelidaki I. (2011). Nitrate as an oxidant in the cathode chamber of a microbial fuel cell for both power generation and nutrient removal purposes. *Applied Biochemistry and Biotechnology*, **164**, 464–474, https://doi.org/10.1007/s12010-010-9148-0

Fischer F., Bastian C., Happe M., Mabillard E. and Schmidt N. (2011). Microbial fuel cell enables phosphate recovery from digested sewage sludge as struvite. *Bioresource Technology*, **102**, 5824–5830, https://doi.org/10.1016/j.biortech.2011.02.089

Fishilevich S., Amir L., Fridman Y., Aharoni A. and Alfonta L. (2009). Surface display of redox enzymes in microbial fuel cells. *Journal of the American Chemical Society*, **131**, 12052–12053, https://doi.org/10.1021/ja9042017

Fornero J. J., Rosenbaum M., Cotta M. A. and Angenent L. T. (2008). Microbial fuel cell performance with a pressurized cathode chamber. *Environmental Science & Technology*, **42**, 8578–8584, https://doi.org/10.1021/es8015292

Freguia S., Rabaey K., Yuan Z. and Keller J. (2007). Non-catalyzed cathodic oxygen reduction at graphite granules in microbial fuel cells. *Electrochimica Acta*, **53**, 598–603, https://doi.org/10.1016/j.electacta.2007.07.037

Freguia S., Tsujimura S. and Kano K. (2010). Electron transfer pathways in microbial oxygen biocathodes. *Electrochimica Acta*, **55**, 813–818, https://doi.org/10.1016/j.electacta.2009.09.027

Ganigué R., Puig S., Batlle-Vilanova P., Balaguer M. D. and Colprim J. (2015). Microbial electrosynthesis of butyrate from carbon dioxide. *Chemical Communications*, **51**, 3235–3238, https://doi.org/10.1039/c4cc10121a

Gildemyn S., Verbeeck K., Slabbinck R., Andersen S. J., Prévoteau A. and Rabaey K. (2015). Integrated production, extraction, and concentration of acetic acid from CO_2 through microbial electrosynthesis. *Environmental Science & Technology Letters*, **2**, 325–328, https://doi.org/10.1021/acs.estlett.5b00212

Goglio A., Marzorati S., Rago L., Pant D., Cristiani P. and Schievano A. (2019). Microbial recycling cells: first steps into a new type of microbial electrochemical technologies, aimed at recovering nutrients from wastewater. *Bioresource Technology*, **277**, 117–127, https://doi.org/10.1016/j.biortech.2019.01.039

Gupta S., Yadav A. and Verma N. (2017). Simultaneous Cr(VI) reduction and bioelectricity generation using microbial fuel cell based on alumina–nickel nanoparticles-dispersed carbon nanofiber electrode. *Chemical Engineering Journal*, **307**, 729–738, https://doi.org/10.1016/j.cej.2016.08.130

Harnisch F. and Schroder U. (2010). From MFC to MXC: chemical and biological cathodes and their potential for microbial bioelectrochemical systems. *Chemical Society Reviews*, **39**, 4433–4448, https://doi.org/10.1039/c003068f

He Z. and Angenent L. T. (2006). Application of bacterial biocathodes in microbial fuel cells. *Electroanalysis*, **18**, 2009–2015, https://doi.org/10.1002/elan.200603628

Heijne A., Hamelers H. V. M., De Wilde V., Rozendal R. A. and Buisman C. J. N. (2006). A bipolar membrane combined with ferric iron reduction as an efficient cathode system in microbial fuel cells. *Environmental Science & Technology*, **40**, 5200–5205, https://doi.org/10.1021/es0608545

Heijne A. T., Liu F., van der Weijden R., Weijma J., Buisman C. J. N. and Hamelers H. V. M. (2010). Copper recovery combined with electricity production in a microbial fuel cell. *Environmental Science & Technology*, **44**, 4376–4381, https://doi.org/10.1021/es100526g

Holliger C. and Schraa G. (1994). Physiological meaning and potential for application of reductive dechlorination by anaerobic bacteria. *FEMS Microbiology Reviews*, **15**, 297–305, https://doi.org/10.1111/j.1574-6976.1994.tb00141.x

Jiang C., Yang Q., Wang D., Zhong Y., Chen F., Li X., Zeng G., Li X. and Shang M. (2017). Simultaneous perchlorate and nitrate removal coupled with electricity generation in autotrophic denitrifying biocathode microbial fuel cell. *Chemical Engineering Journal*, **308**, 783–790, https://doi.org/10.1016/j.cej.2016.09.121

Kharkwal S., Tan Y. C., Lu M. and Ng H. Y. (2017). Development and long-term stability of a novel microbial fuel cell BOD sensor with MnO_2 catalyst. *International Journal of Molecular Sciences*, **18**(2), 276, https://doi.org/10.3390/ijms18020276

Khoo K. S., Ahmad I., Chew K. W., Iwamoto K., Bhatnagar A. and Show P. L. (2023). Enhanced microalgal lipid production for biofuel using different strategies including genetic modification of microalgae: a review. *Progress in Energy and Combustion Science*, **96**, 101071, https://doi.org/10.1016/j.pecs.2023.101071

Kim J. R., Kim J. Y., Han S. B., Park K. W., Saratale G. D. and Oh S. E. (2011). Application of Co-naphthalocyanine (CoNPc) as alternative cathode catalyst and support structure for microbial fuel cells. *Bioresource Technology*, **102**, 342–347, https://doi.org/10.1016/j.biortech.2010.07.005

Kumar A., Kumar N., Baredar P. and Shukla A. (2015). A review on biomass energy resources, potential, conversion and policy in India. *Renewable and Sustainable Energy Reviews*, **45**, 530–539, https://doi.org/10.1016/j.rser.2015.02.007

Lefebvre O., Al-Mamun A. and Ng H. Y. (2008). A microbial fuel cell equipped with a biocathode for organic removal and denitrification. *Water Science and Technology*, **58**, 881–885, https://doi.org/10.2166/wst.2008.343

Li J., Fu Q., Liao Q., Zhu X., Ye D. D. and Tian X. (2009). Persulfate: a self-activated cathodic electron acceptor for microbial fuel cells. *Journal of Power Sources*, **194**, 269–274, https://doi.org/10.1016/j.jpowsour.2009.04.055

Li J., Fu Q., Zhu X., Liao Q., Zhang L. and Wang H. (2010a). A solar regenerable cathodic electron acceptor for microbial fuel cells. *Electrochimica Acta*, **55**, 2332–2337, https://doi.org/10.1016/j.electacta.2009.11.064

Li X., Hu B., Suib S., Lei Y. and Li B. (2010b). Manganese dioxide as a new cathode catalyst in microbial fuel cells. *Journal of Power Sources*, **195**, 2586–2591, https://doi.org/10.1016/j.jpowsour.2009.10.084

Li W. W., Sheng G. P., Liu X. W. and Yu H. Q. (2011). Recent advances in the separators for microbial fuel cells. *Bioresource Technology*, **102**(1), 244–252, https://doi.org/10.1016/j.biortech.2010.03.090

Li S., Chen G. and Anandhi A. (2018). Applications of emerging bioelectrochemical technologies in agricultural systems: a current review. *Energies*, **11**(11), 2951, https://doi.org/10.3390/en11112951

Lian J., Tian X., Li Z., Guo J., Guo Y., Yue L., Ping J. and Duan L. (2017). The effects of different electron donors and electron acceptors on perchlorate reduction and bioelectricity generation in a microbial fuel cell. *International Journal of Hydrogen Energy*, **42**, 544–552, https://doi.org/10.1016/j.ijhydene.2016.11.027

Liew K. B., Daud W. R. W., Ghasemi M., Loh K. S., Ismail M. and Lim S. S. (2015). Manganese oxide/functionalised carbon nanotubes nanocomposite as catalyst for oxygen reduction reaction in microbial fuel cell. *International Journal of Hydrogen Energy*, **40**, 11625–11632, https://doi.org/10.1016/j.ijhydene.2015.04.030

Logan B. E. (2008). Microbial Fuel Cells. John Wiley & Sons.

Logan B. E., Murano C., Scott K., Gray N. D. and Head I. M. (2005). Electricity generation from cysteine in a microbial fuel cell. *Water Research*, **39**, 942–952, https://doi.org/10.1016/j.watres.2004.11.019

Logan B. E., Hamelers B., Rozendal R., Schröder U., Keller J., Freguia S., Alterman P., Verstraete W. and Rabaey K. (2006). Microbial fuel cells: methodology and technology. *Environmental Science & Technology*, **40**, 5181–5192, https://doi.org/10.1021/es0605016

Mehmood K., Rehman S. K. U., Wang J., Farooq F., Mahmood Q., Jadoon A. M., … and Ahmad I. (2019). Treatment of pulp and paper industrial effluent using physicochemical process for recycling. *Water*, **11**(11), 2393, https://doi.org/10.3390/w11112393

Mohan S. V., Saravanan R., Raghavulu S. V., Mohanakrishna G. and Sarma P. N. (2008). Bioelectricity production from wastewater treatment in dual chambered microbial fuel cell (MFC) using selectively enriched mixed microflora: effect of catholyte. *Bioresource Technology*, **99**(3), 596–603, https://doi.org/10.1016/j.biortech.2006.12.026

Morris J. M., Jin S., Wang J. Q., Zhu C. Z. and Urynowicz M. A. (2007). Lead dioxide as an alternative catalyst to platinum in microbial fuel cells. *Electrochemistry Communications*, **9**, 1730–1734, https://doi.org/10.1016/j.elecom.2007.03.028

Muthuraman V. S. and Kasianantham N. (2023). Valorization opportunities and adaptability assessment of algae based biofuels for futuristic sustainability. *Process Safety and Environmental Protection*, **174**, 694–721.

Nguyen M. T., Mecheri B., Iannaci A., D'Epifanio A. and Licoccia S. (2016). Iron/polyindole-based electrocatalysts to enhance oxygen reduction in microbial fuel cells. *Electrochimica Acta*, **190**, 388–395, https://doi.org/10.1016/j.electacta.2015.12.105

Oh S., Min B. and Logan B. E. (2004). Cathode performance as a factor in electricity generation in microbial fuel cells. *Environmental Science & Technology*, **38**, 4900–4904, https://doi.org/10.1021/es049422p

Pandey P., Shinde V. N., Deopurkar R. L., Kale S. P., Patil S. A. and Pant D. (2016). Recent advances in the use of different substrates in microbial fuel cells toward wastewater treatment and simultaneous energy recovery. *Applied Energy*, **168**, 706–723, https://doi.org/10.1016/j.apenergy.2016.01.056

Pant D., Van Bogaert G., Diels L. and Vanbroekhoven K. (2010). A review of the substrates used in microbial fuel cells (MFCs) for sustainable energy production. *Bioresource Technology*, **101**(6), 1533–1543, https://doi.org/10.1016/j.biortech.2009.10.017

Patil S. A., Arends J. B. A., Vanwonterghem I., Van Meerbergen J., Guo K., Tyson G. W., and Rabaey K. (2015). Selective enrichment establishes a stable performing community for microbial electrosynthesis of acetate from CO_2. *Environmental Science & Technology*, **49**, 8833–8843, https://doi.org/10.1021/es506149d

Puig S., Serra M., Vilar-Sanz A., Cabré M., Bañeras L., Colprim J. and Balaguer M. D. (2011). Autotrophic nitrite removal in the cathode of microbial fuel cells. *Bioresource Technology*, **102**, 4462–4467, https://doi.org/10.1016/j.biortech.2010.12.100

Rabaey K., Lissens G., Siciliano S. D. and Verstraete W. (2003). A microbial fuel cell capable of converting glucose to electricity at high rate and efficiency. *Biotechnology and Bioengineering*, **25**, 1531–1535, https://doi.org/10.1023/A:1025484009367

Rabaey K., Boon N., Siciliano S. D., Verhaege M. and Verstraete W. (2004). Biofuel cells select for microbial consortia that self-mediate electron transfer. *Applied and Environmental Microbiology*, **70**(9), 5373–5382, https://doi.org/10.1128/AEM.70.9.5373-5382.2004

Rabaey K., Read S. T., Clauwaert P., Freguia S., Bond P. L., Blackall L. L. and Keller J. (2008). Cathodic oxygen reduction catalyzed by bacteria in microbial fuel cells. *ISME Journal*, **2**, 519–527, https://doi.org/10.1038/ismej.2008.1

Rhoads A., Beyenal H. and Lewandowski Z. (2005). Microbial fuel cell using anaerobic respiration as an anodic reaction and biomineralized manganese as a cathodic reactant. *Environmental Science & Technology*, **39**, 4666–4671, https://doi.org/10.1021/es048386r

Sahinkaya E., Yurtsever A. and Ucar D. (2016). A novel elemental sulfur-based mixotrophic denitrifying membrane bioreactor for simultaneous Cr(VI) and nitrate reduction. *Journal of Hazardous Materials*, **324**, 15–21, https://doi.org/10.1016/j.jhazmat.2016.02.032

Santoro C., Serov A., Stariha L., Kodali M., Gordon J., Babanova S., Bretschger O., Artyushkov K. and Atanassov P. (2016). Iron based catalysts from novel low-cost organic precursors for enhanced oxygen reduction reaction in neutral media microbial fuel cells. *Energy & Environmental Science*, **9**, 2346–2353, https://doi.org/10.1039/C6EE01145D

Schaetzle O., Barriere F. and Schroder U. (2009). An improved microbial fuel cell with laccase as the oxygen reduction catalyst. *Energy & Environmental Science*, **2**, 96–99, https://doi.org/10.1039/B815331K

Sevda S., Chayambuka K., Sreekrishnan T. R., Pant D. and Dominguez-Benetton X. (2015). A comprehensive impedance journey to continuous microbial fuel cells. *Bioelectrochemistry*, **106**, 159–166, https://doi.org/10.1016/j.bioelechem.2015.04.008

Shen J., He R., Han W., Sun X., Li J. and Wang L. (2009). Biological denitrification of high-nitrate wastewater in a modified anoxic/oxic-membrane bioreactor (A/O-MBR). *Journal of Hazardous Materials*, **172**, 595–600, https://doi.org/10.1016/j.jhazmat.2009.07.045

Singh K. and Dharmendra (2020). Performance of a dual chamber microbial fuel cell using sodium chloride as catholyte. *Pollution*, **6**(1), 79–86.

Slate A. J., Whitehead K. A., Brownson D. A. and Banks C. E. (2019). Microbial fuel cells: an overview of current technology. *Renewable and Sustainable Energy Reviews*, **101**, 60–81, https://doi.org/10.1016/j.rser.2018.09.044

Song T., Jin Y., Bao J., Kang D. and Xie J. (2016). Graphene/biofilm composites for enhancement of hexavalent chromium reduction and electricity production in a biocathode microbial fuel cell. *Journal of Hazardous Materials*, **317**, 73–80, https://doi.org/10.1016/j.jhazmat.2016.05.055

Strycharz S. M., Gannon S. M., Boles A. R., Franks A. E., Nevin K. P. and Lovley D. R. (2010). Reductive dechlorination of 2-chlorophenol by *Anaeromyxobacter dehalogenans* with an electrode serving as the electron donor. *Environmental Microbiology Reports*, **2**, 289–294, https://doi.org/10.1111/j.1758-2229.2009.00118.x

Sun J. Z., Peter Kingori G., Si R. W., Zhai D. D., Liao Z. H., Sun D. Z., … and Yong Y. C. (2015). Microbial fuel cell-based biosensors for environmental monitoring: a review. *Water Science and Technology*, **71**(6), 801–809, https://doi.org/10.2166/wst.2015.035

Tao H. C., Li W., Liang M., Xu N., Ni J. R. and Wu W. M. (2011a). A membrane-free baffled microbial fuel cell for cathodic reduction of Cu(II) with electricity generation. *Bioresource Technology*, **102**, 4774–4778, https://doi.org/10.1016/j.biortech.2011.01.057

Tao H. C., Liang M., Li W., Zhang L. J., Ni J. R. and Wu W. M. (2011b). Removal of copper from aqueous solution by electrodeposition in cathode chamber of microbial fuel cell. *Journal of Hazardous Materials*, **189**, 186–192, https://doi.org/10.1016/j.jhazmat.2011.02.018

Ter Heijne A., Hamelers H. V. M. and Buisman C. J. N. (2007). Microbial fuel cell operation with continuous biological ferrous iron oxidation of the catholyte. *Environmental Science & Technology*, **41**, 4130–4134, https://doi.org/10.1021/es0702824

Ucar D., Cokgor E. U. and Şahinkaya E. (2015a). Evaluation of nitrate and perchlorate reduction using sulfur-based autotrophic and mixotrophic denitrifying processes. *Water Science and Technology*, **91**, 1–11, https://doi.org/10.2166/ws.2015.129

Ucar D., Cokgor E. U. and Sahinkaya E. (2015b). Simultaneous nitrate and perchlorate reduction using sulfur-based autotrophic and heterotrophic denitrifying processes. *Journal of Chemical Technology & Biotechnology*, **91**, 1471–1477, https://doi.org/10.1002/jctb.4744

Ucar D., Cokgor E. U., Sahinkaya E., Cetin U., Bereketoglu C., Calimlioglu B., Goncu B. and Yurtsever A. (2017). Simultaneous nitrate and perchlorate removal from groundwater by heterotrophic-autotrophic sequential system. *International Biodeterioration & Biodegradation*, **116**, 83–90, https://doi.org/10.1016/j.ibiod.2016.10.01

Usharani K. and Lakshmanaperumalsamy P. (2010). Studies on the removal efficiency of phosphate from wastewater using *Pseudomonas* sp. YLW-7 and *Enterobacter* sp. KLW-2. *Global Journal of Environmental Research*, **4**, 83–89.

Villano M., Aulenta F., Ciucci C., Ferri T., Giuliano A. and Majone M. (2010). Bioelectrochemical reduction of CO_2 to CH_4 via direct and indirect extracellular electron transfer by a hydrogenophilic methanogenic culture. *Bioresource Technology*, **101**, 3085–3090, https://doi.org/10.1016/j.biortech.2009.12.077

Virdis B., Rabaey K., Yuan Z. and Keller J. (2008). Microbial fuel cells for simultaneous carbon and nitrogen removal. *Water Research*, **42**, 3013–3024, https://doi.org/10.1016/j.watres.2008.03.017

Virdis B., Rabaey K., Rozendal R. A., Yuan Z. and Keller J. (2010). Simultaneous nitrification, denitrification and carbon removal in microbial fuel cells. *Water Research*, **44**, 2970–2980, https://doi.org/10.1016/j.watres.2010.02.022

Wang G., Huang L. and Zhang Y. (2008). Cathodic reduction of hexavalent chromium [Cr(VI)] coupled with electricity generation in microbial fuel cells. *Biotechnology Letters*, **30**, 1959–1966, https://doi.org/10.1007/s10529-008-9792-4

Wang Z., Lim B. and Choi C. (2011). Removal of Hg^{2+} as an electron acceptor coupled with power generation using a microbial fuel cell. *Bioresource Technology*, **102**, 6304–6307, https://doi.org/10.1016/j.biortech.2011.02.027

Wu X., Zhu X., Song T., Zhang L., Jia H. and Wei P. (2015). Effect of acclimatization on hexavalent chromium reduction in a biocathode microbial fuel cell. *Bioresource Technology*, **180**, 185–191, https://doi.org/10.1016/j.biortech.2014.12.105

Wu D., Huang L., Quan X. and Li Puma G. (2016a). Electricity generation and bivalent copper reduction as a function of operation time and cathode electrode material in microbial fuel cells. *Journal of Power Sources*, **307**, 705–714, https://doi.org/10.1016/j.jpowsour.2016.01.022

Wu X., Tong F., Yong X., Zhou J., Zhang L., Jia H., and Wei P. (2016b). Effect of NaX zeolite-modified graphite felts on hexavalent chromium removal in biocathode microbial fuel cells. *Journal of Hazardous Materials*, **308**, 303–311, https://doi.org/10.1016/j.jhazmat.2016.01.070

Xafenias N., Zhang Y. and Banks C. J. (2015). Evaluating hexavalent chromium reduction and electricity production in microbial fuel cells with alkaline cathodes. *International Journal of Environmental Science and Technology*, **12**, 2435–2446, https://doi.org/10.1007/s13762-014-0651-7

Xie S., Liang P., Chen Y., Xia X. and Huang X. (2011). Simultaneous carbon and nitrogen removal using an oxic/anoxic-biocathode microbial fuel cells coupled system. *Bioresource Technology*, **102**, 348–354, https://doi.org/10.1016/j.biortech.2010.07.046

You S., Zhao Q., Zhang J., Jiang J. and Zhao S. (2006). Short communication a microbial fuel cell using permanganate as the cathodic electron acceptor. *Journal of Power Sources*, **162**, 1409–1415, https://doi.org/10.1016/j.jpowsour.2006.07.063

You S. J., Ren N. Q., Zhao Q. L., Wang J. Y. and Yang F. L. (2009). Power generation and electrochemical analysis of biocathode microbial fuel cell using graphite fibre brush as cathode material. *Fuel Cells*, **9**, 588–596, https://doi.org/10.1002/fuce.200900023

Yu C.-P., Liang Z., Das A. and Hu Z. (2011). Nitrogen removal from wastewater using membrane aerated microbial fuel cell techniques. *Water Research*, **45**, 1157–1164, https://doi.org/10.1016/j.watres.2010.11.002

Yu D., Bai L., Zhai J., Wang Y. and Dong S. (2017). Toxicity detection in water containing heavy metal ions with a self-powered microbial fuel cell-based biosensor. *Talanta*, **168**, 210–216, https://doi.org/10.1016/j.talanta.2017.03.048

Yuan Y., Ahmed J. and Kim S. (2011). Polyaniline/carbon black composite-supported iron phthalocyanine as an oxygen reduction catalyst for microbial fuel cells. *Journal of Power Sources*, **196**, 1103–1106, https://doi.org/10.1016/j.jpowsour.2010.08.112

Zain S. M., Ching N. L., Jusoh S. and Yunus S. Y. (2015). Different types of microbial fuel cell (MFC) systems for simultaneous electricity generation and pollutant removal. *Jurnal Teknologi*, **74**, 13–19, https://doi.org/10.11113/jt.v74.4544

Zhang B., Zhao H., Shi C., Zhou S. and Ni J. (2009). Simultaneous removal of sulfide and organics with vanadium(V) reduction in microbial fuel cells. *Journal of Chemical Technology & Biotechnology*, **84**, 1780–1786, https://doi.org/10.1002/jctb.2244

Zhang Y., Wang Y. and Angelidaki I. (2015). Alternate switching between microbial fuel cell and microbial electrolysis cell operation as a new method to control H_2O_2 level in bioelectro-Fenton system. *Journal of Power Sources*, **291**, 108–116, https://doi.org/10.1016/j.jpowsour.2015.05.020

Zhao F., Harnisch F., Schroder U., Scholz F., Bogdanoff P. and Herrmann I. (2005). Application of pyrolysed iron(II) phthalocyanine and CoTMPP based oxygen reduction catalysts as cathode materials in microbial fuel cells. *Electrochemistry Communications*, **7**, 1405–1410, https://doi.org/10.1016/j.elecom.2005.09.032

Zhao F., Harnisch F., Schrorder U., Scholz F., Bogdanoff P. and Herrmann I. (2006). Challenges and constraints of using oxygen cathodes in microbial fuel cells. *Environmental Science & Technology*, **40**, 5193–5199, https://doi.org/10.1021/es060332p

Zhou M., Chi M., Luo J., He H. and Jin T. (2011). An overview of electrode materials in microbial fuel cells. *Journal of Power Sources*, **196**(10), 4427–4435, https://doi.org/10.1016/j.jpowsour.2011.01.012

Zhuang L., Zhou S. G., Li Y. T., Liu T. L. and Huang D. Y. (2010). In situ Fenton-enhanced cathodic reaction for sustainable increased electricity generation in microbial fuel cells. *Journal of Power Sources*, **195**, 1379–1382, https://doi.org/10.1016/j.jpowsour.2009.09.011

doi: 10.2166/9781789063813_0155

Chapter 9

Value-added gaseous fuel recovery from industrial wastewater and sludge using microbial electrochemical technology

*Isha Raval, Upexa Patel, Ekta Patel and Ambika Arkatkar**

Department of Biotechnology, Veer Narmad South Gujarat University, Surat, Gujarat, India
**Corresponding author: asarkatkar@vnsgu.ac.in*

ABSTRACT

Synthesis of value-added gases using biological systems has been accepted as one of the important scientific endeavors for many decades. The generation of electricity and the production of acetic acid, formic acid, methane, hydrogen, and other value-added products can all be done with the help of recent technological advancements that use wastewater as a raw material. These optimized biosynthesis processes which are environmentally friendly and cost effective can be considered for scale-up level experiments. Microbial electrochemical technology combines microbes, electrochemistry, and material science that can be amalgamated to efficiently convert the chemical energy contained in the organic molecules. The wastewater and sludge effluent from industrial sector can be used as a probable nutrient source for the microbes. Factors such as nutrient availability, toxicity, variation in microbial population, operating parameters, and so on, have great influence on the yield and quality of the gaseous fuel. This chapter summarizes recent updates on microbial electrochemical systems being utilized for gaseous recovery and parameters that influence productivity of gases.

Keywords: gaseous fuel, microbial electrochemical technology, industrial wastewater, wastewater toxicity, value-added product

9.1 INTRODUCTION

9.1.1 Today's scenario of energy

Energy is the foundation of contemporary society and a requirement for long-term climate control (Islam *et al.*, 2014). The population of the planet is growing every day. Due to this population growth and energy demand is also increasing exponentially. Over the last decades fossil fuels (coal, oil, natural gas) have been the primary source of energy and they are meeting up to 95% of global energy needs (Liu, 2015). The present scenario is that it takes a very long time to form fossil fuels naturally (non-renewable energy source), its consumption is on the higher side, but we have very limited natural resources. It has been proved that when fossil fuels are burned, harmful greenhouse gases (GHGs) are emitted. Carbon dioxide emissions are one of the primary reasons for climate change.

9.1.2 Other major resources of energy

Renewable energy is produced from natural resources that can be renewed more quickly than the rate of their consumption. We have many renewable energy sources around us that can be utilized for generating energy. Compared to burning fossil fuels, this will produce significantly fewer emissions. The most effective way to combat the climate catastrophe is to switch from fossil fuels, which now account for the majority of emissions, to renewable energy.

New long-lasting energy sources that have negligible environmental impact is the demand of our era. It is necessary to find an alternate method of generating energy in the form of electricity. To make a sustainable and easy future, studies on locating opportunities for renewable and unexplored assets to replace traditional fuels, fight climate adjustments, and decrease the rate of GHG emissions are increasing exponentially. Recently, researchers are putting their effort into the use of renewable electricity assets, which include solar electricity, wind strength, geothermal electricity, hydropower, ocean energy, bioenergy, hydrogen, and biogas, rather than fossil fuels (Table 9.1).

9.1.3 Waste generation

Large quantities of waste generation and water scarcity are different transnational problems because of population. Now, there is a huge strain on water delivery and sewer structures, because of their energy-driven maintenance. Huge amounts of waste from different sectors, which include business,

Table 9.1 Renewable energy sources.

Sr. No.	Source of Energy	Remarks	References
1.	Solar	Solar energy is obtained from sunlight. The energy is intercepted by the Earth 10 000 times faster than human consumption. It is affordable and the cheapest source of electricity. A solar panel that converts solar energy to electricity has a lifespan of 30 years.	Shaikh *et al.* (2017)
2.	Wind	It uses the kinetic energy of moving air. It has vast potential for deployment, especially in remote regions.	Daut *et al.* (2012)
3.	Geothermal	Thermal energy, operating from last hundred years is trapped inside the earth's interior is the source for geothermal energy.	Salazar *et al.* (2017)
4.	Hydropower	This makes use of the force of water flowing downhill from higher altitudes. It can be generated from, reservoirs (rely on stored water) and free flowing river (rely on flow of the river).	Kaunda *et al.* (2012)
5.	Tidal	Tidal energy is generated by ocean waves during tides. It can be derived from: tidal streams, barrages, and tidal lagoons.	Khare and Bhuiyan (2022)
6.	Bioenergy	Many different organic materials can be used to create bioenergy. Burning biomass creates GHG emissions.	Gupta *et al.* (2014)
7.	Hydrogen	Hydrogen-derived energy is equivalent to gasoline energy. There are various forms of hydrogen: (1) Gray hydrogen – produced from fossil fuels (2) Blue hydrogen – produced by storing or reusing exhausted CO_2 for other uses (3) Green hydrogen – produced from renewable energy sources	Rahimi (2022)

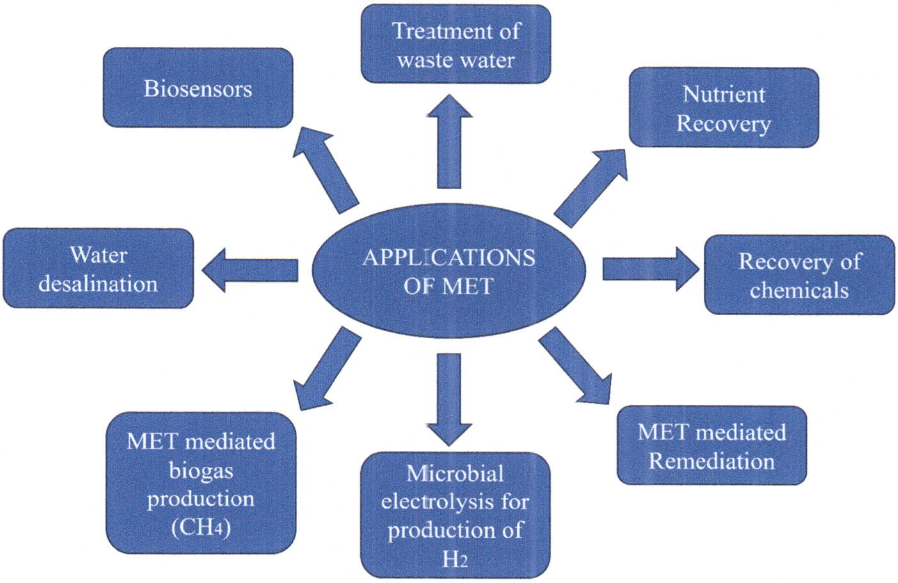

Figure 9.1 Applications of MET.

home, husbandry, and industries are accrued, and transferred each day to different waste treatment plants (WTPs). Chemical precipitation, coagulation, flocculation, froth flotation, adsorption, membrane technology, ion exchange, aerobic activated sludge treatment, aerobic digester, and so on are one of the few methods used in WTPs for treatment of wastewater (Rahimi, 2022). The technologies have their own advantages as well as disadvantages. These technologies may have inordinate electricity demand, high cost, and may not be environmentally friendly. Scientists have therefore developed a keen interest in sustainable waste management techniques that can also serve as a source of energy generation from waste. Organic matters are present in wastewater, and they can be used as energy resources. For the utilization of this energy source, microbial electrochemical techniques (MET) are used in the waste-treatment process as well as in the creation of useful products (Figure 9.1).

9.1.4 Microbial electrochemical technology

It uses microorganisms as electrochemical catalysts, merging the microbial metabolism with electrochemical processes to supply bioelectricity, biofuels, hydrogen, and other treasured chemical compounds. Microbial electrochemical systems (MESs) are the devices used in METs.

A system is commonly composed of two chambers: an anode and a cathode chamber separated via a selectively permeable proton/cation-exchange membrane (CEM) or a salt bridge. In a MES, the anode chamber carries microbes that act as biocatalysts under anaerobic conditions and the cathode chamber consists of the electron acceptors (e.g., oxygen). Electrons generated from the oxidation of natural compounds are conveyed to the anode. Electrons produced with the aid of the microbes are transferred to the anode immediately via 'nanowires' or outer-membrane proteins, or indirectly through the use of electron shuttling retailer. Those electrons attain the cathode across an external circuit, and for each electron conducted, protons react on the cathode to complete the reaction and sustain the electric current (Arkatkar *et al.*, 2020).

Table 9.2 Different types of METs.

Sr. No.	Type	Remarks	End-Product
1.	MFC	Chemical energy to electric energy	Electricity
2.	MEC	External supply of electricity	Hydrogen, acetate, ethanol, butyrate
3.	Microbial electrosynthesis (MES)	Biogas production	Methane
4.	Microbial remediation cell (MRC)	Reduced toxicity of contaminants	Contaminant remediation and power generation
5.	MDC	Uses electric potential difference among the electrodes to perform desalination	Sustainable drinking water
6.	Microbial solar cell (MSC)	Photosynthetic reaction	Water, hydrogen, methane, ethanol

Source: Rout *et al.* (2021).

There are different types of METs. However extensively, all of them have the similar working principle. Various designs and configurations are being experimented to optimize the assembly of three simple elements (anode, cathode, and electrode) in a functioning device (Table 9.2).

9.2 ESSENTIALS FOR MESs

9.2.1 Components

The MES system encompasses two separate chambers, the anode and the cathode. The cathode compartment is taken into consideration for electrodes to attract oxygen. To stop oxygen from entering the anode chamber, a proton-exchange membrane is employed to separate the chambers.

Electrodes: The anode and cathode are solid electron conductors that convey the electric current into the MES. The electrode factors such as the electrode material, conductivity, mass transfer, surface area, cost, scalability, biocompatibility affect the performance of MES. Graphite sheet, carbon felt, carbon fabric, and carbon mesh are mostly used as the anode, whereas graphite and platinum are mostly used as the cathode (Rahimi, 2022).

Membrane: Two kinds of membranes are used in MES, semipermeable and pass through membranes. The semipermeable membrane appoints charge transfer between the chambers. However, pass through membranes can be applied when both charge and oxygen crossover is required. Sometimes there may be no need to apply an ion-exchange membrane (Rahimi, 2022).

Substrate: Substrate is the waste that is targeted as energy resource. The substrate's nature, components, and concentration have an effect on the microbial community and the power recuperation, together with electricity or hydrogen production during the MES operation. Organic substrates like glucose, sodium acetate, cellulose, agro-manner wastes such as residuals, oil refinery waste, and dairy and vegetable wastes are used as a substrate in MES. They provide strength for bacterial cellular growth. In most instances, acetate is considered as first-rate substrate as it is the simplest substrate to breakdown and has good energy contents compared to other substrates (Rahimi, 2022).

9.2.1.1 How does it work?

The important contributor in METs working principle is the innate functionality of special microbes, referred to as electroactive microorganisms, they can link their metabolism with the delivery of electrons inside as well as outside their cellular system. In the natural environment, minerals have metals that are insoluble in nature. These minerals can act as electron donors or acceptors. Thus, under natural conditions the electroactive microbes donate electrons to these minerals. In METs electrodes act as electron acceptors. In anaerobic surroundings, synergic consortia of fermentative and bio electrogenic microbes work together. The fermentative microbes break down complicated organic compounds into simpler structures, such as hydrogen, gasoline, acetate, polymers, ethanol, and other lengthy-chain fatty acids. Those molecules are oxidized by bioelectrogenic microbes. In MES, the electrons thus generated by transfer from electron donor to terminal electron acceptor outside the cell are in an insoluble form. The travelling of these electrons across the electrodes leads to the generation of electrical energy. The colonies of microbes in favorable conditions cause the formation of biofilm, whose capacity is to conduct electrons on the electrodes (Louro *et al.*, 2018).

9.3 VALUABLE PRODUCTS GENERATED USING MESs

Production of valuable products from organic matter present in waste is one of the most important applications of MET. Value-added products generated by MET include hydrogen, hydrogen peroxide, methane, volatile fatty acids (VFA), alcohols, metals, and so on (Kong *et al.*, 2020). In this section, as per the context of the chapter, production of valuable gases will be discussed.

Hydrogen: Hydrogen, a promising energy source, can be easily produced from organic materials and does not release any pollutants during its combustion. It has been shown that the utilization of organic matter as well as the generation of hydrogen can be significantly improved by combining MET with dark fermentation (DF). This system incorporates the complementary strengths of MET and DF and overcome their drawbacks. This mixture will accelerate the creation of hydrogen while promoting the breakdown of solid or liquid waste (Kong *et al.*, 2020).

Methane: Methane is one of the common sources of energy. Methane can be produced at cathode in two different ways, directly by conversion of carbon to methane (Eq. 9.1) or indirectly by incorporating ready hydrogen in the reaction (Eqs. 9.2 and 9.3)

$$\text{Directly} \quad CO_2 + 8H^+ + 8e^- \rightarrow CH_4 + 2H_2O \tag{9.1}$$

$$\text{Indirectly} \quad 8H^+ + 8e^- \rightarrow 4H_2 \tag{9.2}$$

$$4H_2 + CO_2 \rightarrow CH_4 + 2H_2O \tag{9.3}$$

Waste has gained attention for its role in the bioelectrosynthesis of methane, and by improving operational conditions, more methane can be produced. The formation of biocathodic methane and extracellular electron transfer (EET) by bacteria at the biocathode are closely connected processes. Anaerobic digestion (AD) and methanogenesis have been studied in combination to treat complicated wastes, such as solid waste. In contrast to MET, which can eliminate VFAs, speed up methanogenesis, and raise the methane content in the biogas, AD can enhance the hydrolysis and acidity of wastes. The hydrolysis can be made better using the AD-MET integrated system.

9.4 WASTE UTILIZATION IN MESs

There are mainly two categories of waste source which are used in MET (Kong *et al.*, 2020), namely wastewater and solid waste. The category differs in their source and components (Table 9.3). The difference in components directly associated with the nature and complexity of the substrate. The nature and complexity of substrate can hugely impact the parameters like chemical oxygen demand (COD), biological oxygen demand (BOD), total dissolved solid (TDS), volatile suspended solids (VSS). The change in the said parameters can be associated with biological factors such as microbial growth, requirement of metabolic processes, possible by-product, time duration of utilization, and so on.

Table 9.3 List of the wastes which are used in MET as a substrate.

Category	Waste	Source	Components	References
Wastewater	Domestic wastewater	Household wastewater like, toilet, kitchen sink, dish washer, bath/shower, clothes washer, miscellaneous, and so on	Feces, food particles, different metals, organic compounds, grease, organisms, and so on	Geary (1998)
	Industrial wastewater	Industries such as cement works, lime kilns, pharmaceuticals, food, rubber, textiles, paper pulp, organics compost, meat processing, breweries, chemical plants, and so on.	Heavy metals, chlorinated solvents, acids, detergents, oil, flammables, sulfide, salt, acetone, methyl ethyl ketone, isopropanol, and so on.	Sathya *et al.* (2022)
	Agro-industrial wastewater	Industries like, dairy, brewery, winey, palm oil, agriculture products processing, agriculture residues, livestock, and so on.	Whey, cream, sugar, alcohol, metals, phenolic and organic compounds, acids, nutrients, oil, soil particles, manure, and so on.	Bolognesi *et al.* (2020)
Solid waster	Municipal waste	Households, office buildings, institutions, small businesses, garden waste, street sweeping, market cleansing, and so on.	Paper, food, green waste, plastics, rubbers, wood, tires, tin cans, jars, clothing, paints, and so on.	Zhou *et al.* (2014)
	Biomass waste	Animal, industrial, agriculture and forestry residues, sewage sludge, municipal solid waste, and so on.	Corn stover, corncob, wheat shoot, soya stalk, bagasse, seaweed, coconut husk, spirulina powder, sawdust, wood chips, and so on.	Kalak (2023)

9.5 FACTORS THAT AFFECT PRODUCTION OF VALUABLE GASES

Between electroactive microorganisms and extracellular substrates or partners, EET is a typical process where the electrons are exchanged. Since it drives the biogeochemical cycles of vital elements and promotes bioremediation and bioenergy recovery, the EET process is extremely important for ecosystems. It also contributes to sustainable waste management. There are various factors of a MES set up that affect the production of valuable gases like type of microorganisms, substrate, inoculum, nature of wastewaters, temperature, pH, conductivity, conductive materials, external resistance, type and composition of electrode material, and reactor design, membrane type and so on (Figure 9.2).

9.5.1 Effect of microorganism and their metabolism

Inorganic and organic waste materials contain chemical energy that can be converted to electrical energy by microorganisms equipped with EET systems. Microorganisms such as *Shewanella oneidensis, Geobacter sulfurreducens, Haloferax volcanii, Natrialba magadii, Cyanobacteria, Synechococcus elongates* are majorly used (Das *et al.*, 2019). Enzymes that are secreted from cells, cell components, or a whole microbial cell enhance the rate of reaction. Bacterial biofilm can interact with materials such as electrodes and exchange electrons. Flavins, cytochromes, cell surface protrusions, and simple metabolite reactions can all contribute to this electron exchange in a bacterial biofilm.

In the anode chamber, organic compounds are oxidized, resulting in the generation of protons and electrons. While electrons go to the cathode over the external circuit, protons are delivered to it via the membrane. Eventually, protons and electrons run out in the cathode chamber where oxygen is converted to water. The electron acceptor for the cathode process in MES is usually oxygen.

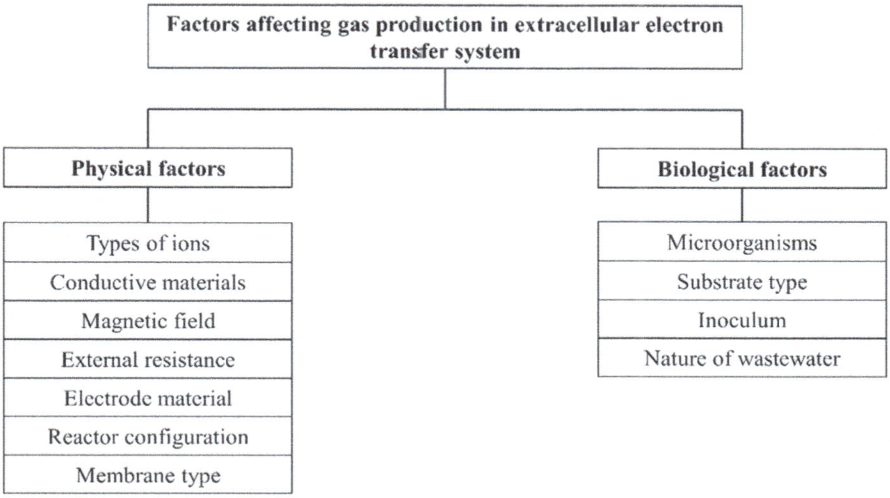

Figure 9.2 Factors that affect production of valuable gases in METs.

Cyanobacteria have been used in the cathode chamber as oxygen sources. In addition to being able to directly harvest algal biomass for use as fuel feedstock in the anode, algal cathodes offer the advantages of not requiring mechanical cathode air supply, which results in lower running costs and total CO_2 emissions from the anodic bacterial respiration (Harnisch *et al.*, 2011).

9.5.2 Effect of substrate type

Gas production depends significantly on the type of wastewater used and microbial population. The presence of bacteriocins or high concentrations of organic acids in particular wastewaters have a major impact on H_2 production. Bacteriocins found in lactic acid bacteria-rich wastewaters may prevent H_2 generation. Thus, to increase the production of H_2, wastewater is often subjected to pretreatment like filtration, sedimentation, coagulation, aerobic and anaerobic process, adsorption, ion exchange, extraction, and so on (Li & Chen, 2018).

The three principal types of wastewaters that have frequently been fed to or processed in METs are household wastewater, swine wastewater, and landfill leachate. Each type of water affects the efficiency of treatment, hydrogen production, and generation of power in METs. Household wastewater was found to be the perfect source for MET generation and production of power and hydrogen, while METs employing landfill leachate as a substrate had the lower power densities and production rates of hydrogen. It is interesting to note that reactors treating swine wastewater have demonstrated higher treatment efficiency, as evidenced by COD elimination.

A high ionic strength typically lowers internal resistance, which causes the anode potential to drop and the cathode potential to rise, thus increasing the overall circuit voltage. Although swine wastewater and landfill leachate have significant levels of biodegradable organics and good ionic strength, the presence of relatively large amounts of harmful chemicals such ammonium and heavy metals negatively impact microbial development. Reduced microbial growth ultimately affects the performance of reactors when these types of wastewaters were used to feed METs (Li & Chen, 2018).

9.5.3 Effect of inoculum

For the production of H_2, mixed cultures are widely used because they are reliable, manageable, and less contaminated (Jack, 2016). Such type of cultures can be found in many places and can make use of different kinds of organic material in wastewater. A typical source of mixed culture for the generation of H_2 is the sludge from anaerobic digesters. Pretreatment time and temperature affects the amount of H_2 produced during heating treatment. On the other side, acid treatments result in poor H_2 generation efficiency. To treat anaerobic sludge, variety of inoculum pretreatments were evaluated by the researchers, such as heat treatment at 90°C for 15 min, ultra sonification at 20 kHz, alkaline treatment in pH 10 for 15 min, and a fresh hydrodynamic cavitation pretreatment technique (Prakash *et al.*, 2018).

9.5.4 Effect of temperature

EET is temperature-sensitive mechanism. As the working temperature changes, there are significant variations in the capability of power generation in METs and the efficiency of hydrogen production in reactors. In general, temperatures between 30°C and 40°C are suitable for the performance of these reactors. Power generation is inhibited above 45°C or below 15°C and hydrogen production in METs. It has been found that METs can effectively eliminate the most COD when the temperature is between 25°C and 35°C. Studies have shown that in a mix culture condition electrogenic bacteria predominate and the electrochemically active biofilms grow and function better at ideal temperatures around 30°C. Reactors can function in real-world applications for wastewater treatment and can withstand temperature variations. This enables the electroactive microbes to adapt to unfavorable temperatures. It should be noted that the ideal temperatures for producing power and hydrogen differ from those for removing COD (Li & Chen, 2018).

9.5.5 Effect of pH

The performance of MES is significantly impacted by the pH in the anode chamber. Proton transfer via the membrane is considered as one of the rate-limiting factors in any MES system. While significant production of hydrogen was found at pH levels between 5 and 6, pH ranges between 6.5 and 7.5 were reported to have higher power densities (up to ~1200 mW/m^2). Both types of reactors exhibit significant pH effects on COD removal. In comparison to pH values that are incredibly low or high, at the optimum pH values, the elimination of COD is approximately three times higher. This information indicates that activity of microbial cells is slower at a suboptimal pH than at an ideal pH. It is possible that proton transfer becomes poor, and this affects the power generation. The gradient in proton concentration speeds up the proton transfer at relatively low pH levels, increasing the availability of proton in the cathode chamber, these protons will be utilized for the synthesis of biohydrogen. Thus, hydrogen generation in METs prefers a pH that is considerably lower (pH 6 to 3).

However, as the pH falls below 3, the reactor's performance suffers because majority of electrogenic microbes can no longer survive in the acidic environment. Even under such harsh low-pH circumstances, some species have been reported to work effectively in METs, producing power densities of 20–55 mW/m^2. In the case of hydrogen production, it should be noted that biohydrogen production cannot be achieved without a proton in the cathode chamber (Li & Chen, 2018).

9.5.6 Effect of conductivity of electrolytes and electrodes

The METs primarily use two types of media: aqueous media, which is used during the treatment of wastewater, and solid media, which is used during the remediation of soil and sediment. The electron transmission in the two different types of media is affected by the aqueous concentration of ions, sediment and soil moisture levels, salinity, porosity, and other factors. Anaerobic bacteria that can drive more electrons from organic pollutants can more easily accelerate their metabolic reaction rate thanks to the anode in MES, which serves as a constant source of electron acceptors. It has been reported that in systems that involves soil and sediments, the large internal resistance of system or the low electrical conductivity acts as a limiting factor. This leads to low coulombic efficiency as well as the suppression of microorganism metabolism. Substrates degradation and the electricity generation in a MES are determined by the microbial metabolism and EET efficiency (Stefanova *et al.*, 2018).

The conductivity of electrolyte is governed by the ion concentration. Directly reducing internal resistance and increasing electron transmission, leading to a rise in ion concentration. Certain metal ions, including Na^+, K^+, Cu^{2+}, and Zn^{2+} are required for the growth of microorganisms. These substances may encourage the growth of microorganisms when they reach a specific concentration, but if the content increases above a certain level, the environment becomes toxic to microorganisms and can even make these individuals nonviable (Gadkari *et al.*, 2020; Zhang *et al.*, 2019).

According to an examination of the effects of varying NaCl concentrations (0.5–2.5%) on the anaerobic breakdown of petroleum hydrocarbons in a MES, the efficiency of the MES improved (power density increased to 1.06 mW/m^2) after the concentration of NaCl increased until the value reached 1%. Another illustration is related to the basic coenzyme formation by the core ion Cu^{2+} in many biological processes. However, when Cu^{2+} concentrations are too high, active oxides are produced that bind with biological macromolecules like proteins and impair their ability to function normally. Studies have also demonstrated that the performance of electron transport is improved by the precipitation of heavy metal ions in the anodic compartment. In addition, it has been demonstrated that if this precipitation process is carried out excessively, functional bacterial activities are suppressed, which significantly decreases the rates of voltage generation and COD elimination (Zhang *et al.*, 2019).

Activated carbon and carbon fiber are examples of conductive carbon materials which are suitable for use in the MES as they enhance the transfer of electrons in the medium. It also gives microorganisms a suitable space for attachment. Electrogenic bacteria multiply on activated carbon anodes, which is crucial for maintaining the high performance of the MES. Compared to valuable metals (like platinum), activated carbon is a cheaper catalyst with good catalytic activity. Because of its porous structure, better area of surface, graphitization content, and functional groups activated carbon is considered for its use in MES. Deployment of conductive carbon fibers in soil is the latest concept in the field of MES. In addition to successfully forming these fibers they are also easy to obtain from the repaired soil and reuse, and they also support viable microbial communities in the METs. The development of a silica colloid network was made possible by the addition of silica gel to the soil, which decreased the soil's resistivity. The application of conductive materials such as magnetic micro porous surfaces to enhance electron transport in MES has received more attention (Zhang *et al.*, 2019).

9.5.7 Effects of external resistance

External resistance affects MES performance because it inhibits electrons from moving freely from the anode to the cathode. All the three electrical parameters voltage, current, and power are interdependent (Ohm's law $V = IR$) and are equally impacted by external resistance. Increased external resistance in MES reduces both the power density and the effectiveness of the treatment.

The external resistance controls the voltage of the anode, which affect the anode's ability as an electron acceptor. As a result, when various external resistances are applied to the circuits, modifications are made to the growth competition between the electrogenic and non-electrogenic microbial communities (Potrykus *et al.*, 2021). The microbial community patterns that evolve under different external resistance will ultimately effectively impact the anode biofilm, which utilizes organic substrates and generates electrons and protons in the process. The microbial community in the anode biofilm is especially susceptible to the modification of external resistance when the anode voltage is low. Based on reports, the bulk of known *Geobacter* strains are significant at anode potentials below 1.5 V.

A change in external resistance either low or high will affect power generation. The anode's low redox potential and low external resistance (about 10 kΩ) probably made it challenging for microbes to capture electrons. There have been reports of improved generation of power, decreased production of methane, and enhanced columbic efficiency in MES running at optimal external resistance. However, because the electrons in a MES are pushed by the external energy supplied to the system, low resistance would not stop the movement of electrons in the system (Li & Chen, 2018).

9.5.8 Effect of electrode type

The main criteria of good electrodes are that it should serve as both conductors and homes for electrogenic microbes. To increase MES performance while lowering reactor costs, a variety of electrode materials have been researched, and to effectively transmit electrons between microorganisms and electrodes, ideal electrode materials should exhibit a few unique properties of surface, such as a larger surface area, significant surface roughness, and good biocompatibility.

Because of important features, carbonaceous substances are frequently employed as anode materials. Carbon paper, carbon cloth, and graphite brush are the three most commonly employed carbonaceous electrode materials. Carbon paper has a surface that is generally smooth, and it is thin, firm, yet slightly fragile. Carbon cloth, on the other contrary, is more pliable and porous. In comparison with the previous two materials, graphite brush, a fiber fabric, has a significantly thicker structure yet offers a larger surface area for microbial attachment and growth. The statistical findings show that carbon cloth electrodes are the best for MES power generation (945–1550 mW/m^2) and MES hydrogen production (2–6 m^3 $H_2/m^3/d$). Carbon cloth and carbon paper are thought to have greater electricity generation and COD removal capabilities because of their high surface areas. Reactors utilizing graphite brushes perform marginally worse than those using carbon paper as electrodes. The high cost of these carbonaceous materials, particularly carbon fabric, makes them unaffordable for use in practical applications. The higher electrical resistance of these carbonaceous materials, results in electrode ohmic losses in large-scale systems, is a further cause for worry (Li & Chen, 2018).

9.5.9 Effect of reactor configuration

Numerous configurations have been created and developed in order to improve the performance of MES. Single-chamber and two-chamber designs are most popular. In contrast to single-chamber systems, which merely have oxygen in the air as an electron acceptor (an air cathode), two-chamber systems require the proton to pass through a membrane before it reaches the cathode chamber. The findings indicate that while one-chamber METs may produce more biohydrogen (2–6 m^3 $H_2/m^3/d$), two-chamber MES often yield higher power (692–1400 mW/m^2). The high electrical resistance of the two-chamber systems than the single-chamber designs makes them a little problematic. In reality, due to their simplicity and stability, two-chamber systems were utilized more frequently in earlier investigations. Two-chamber systems have been used to examine substrates and operating circumstances, and frequently the substrates (such as acetate) and conditions are ideal (Li & Chen, 2018).

9.5.10 Effects of membrane

In a dual-chamber MES, an area known as a membrane separates the anode and cathode liquids. Furthermore, it permits the passage of hydrogen, oxygen, and protons to the cathodic chamber from the anodic chamber. Common metrics used to evaluate membrane performance include proton conductivity, water permeability, ion transport number, biofouling, internal resistance, mechanical strength, chemical resistance, and oxygen diffusion. It is also reported that the power density increases with the membrane's surface area. For proton-exchange membrane (PEM) areas of 3.5, 6.2, and 30.6 cm^2, they reported values of 45.0, 68.0, and 190.0 mW/m^2, respectively. Therefore, a large surface area is an important characteristic of a PEM. Several membrane materials under the brand names Zirfon, Ultrex, and Nafion are reported to be used in the reactor set-up which serve as separators in systems have been recently reported. Due to its availability and high performance, Nafion has become the most widely used membrane. The sulfonic acid groups in the perfluorinated Nafion membrane promote proton conductivity, transmit protons across the chambers, and cations have a high permeability to water, and inhibit the passage of electrons (Borja-Maldonado & López Zavala, 2022).

9.6 ANALYSIS TECHNIQUES AND DOWNSTREAM PROCESS

The structure and composition of the raw materials employed have an important effect on the rate and yield of biogas production in every system. Therefore, a wide range of analytical techniques must be used to optimize the process design and yield. Various methods such as cyclic voltammetry, electrochemical impedance spectrometry, chemical oxygen demand, and scanning electron microscopy are used.

9.6.1 Cyclic voltammetry

The possible difference between various electron acceptors and donors determines how far an electron may travel from each other. This causes redox reactions to occur inside the reactor. To understand the mechanism underlying the transfer of electrons and redox reactions taking place throughout the process of producing electricity, all types of MET sets are thus frequently studied using cyclic voltammetry (CV). Different levels of this analysis are possible, like on the electrode (biofilm mediated) or in the media (reaction via biological shuttles and mediators). This different analysis can be planned in accordance with the vision of the experiment to check the redox activity in the MET setup. Three electrode assemblies can be used to analyze the electrochemical characteristics of the anode biofilm and the analyte in the second and third phases of the study of the systems without interference or the effects of other parameters (Table 9.4). The voltagram, a type of graphic data produced by the CV analysis, is used for further investigation (Arkatkar *et al.*, 2019).

Table 9.4 Different analytical techniques used for optimization of MET system.

Method	Application	Requirements	Working	Reference
Cyclic voltammetry	To understand the redox reactions occurring in the MET system.	Three electrode system. Working, counter, and reference electrode.	Wires are connected to respective electrodes. The change in response is measured in the form of current value. CV experiment reveals the response of a system at a different voltage, while an EIS experiment explains the response over multiple amplitude.	Mayall and Birss (2017), Arkatkar *et al.* (2019), Elgrishi *et al.* (2018)
Electrochemical impedance spectroscopy	To determine internal resistance of the MET system.			
Chemical oxygen demand	Used to measure how much oxygen is needed to oxidize the organic material present in wastewater.	The sample with reagents and digestor depending upon the method (closed flux or open flux as per ASTM standards	The COD testing method is based on the idea that practically any organic component will oxidize to carbon dioxide under acidic conditions when exposed to a powerful oxidizing agent	Arkatkar *et al.* (2019)
Scanning electron microscopy	Used to show the adherence of microbial community on the anode electrode.	Biofilm on the electrode	The back scattered or secondary electron is captured after electron bombardment on the electrode surface and a two-dimensional mode is produced, which gives visualization of the biofilm surface.	Kannan (2018)

9.6.2 Electrochemical impedance spectroscopy

Without endangering the setup, electrochemical impedance spectroscopy (EIS) is employed to ascertain the chemical systems' electrical response. We can learn more about the temporal response of chemical processes by applying the current at various frequencies. A known voltage is delivered from the working electrode to the counter electrode in the three-electrode arrangement used for this analysis. An electrolytic solution must be present in the system for the working, reference, and counter electrodes to be submerged. Through EIS, a quantitative change in the chemical reaction occurring at the electrode–electrolyte interface may be quantified. EIS is useful in determining a wide range of dielectric and electrical properties of components in areas of research including batteries, corrosion, and so on (Lazanas & Prodromidis, 2022). The interpretation of graphical representation as Niquits or Bode plot and further fitting of electrically simulated circuit will give the data regarding the internal resistances faced by the electrons while passing through different interfaces.

9.6.3 Substrate utilization

In accordance to the substrate added in the anode chamber the method can be chosen for the analysis of degradation of the substrate. The addition of simple sugar like glucose leads to the choice of glucose estimation using chemical methods like 3,5-dinitrosalicylic acid method (DNSA method) (Arkatkar *et al.*, 2020). The capacity of water to absorb oxygen during the breakdown of organic compounds in it, is known as COD (Table 9.4). Indirect information about pollutants (organics) in a water sample is provided by the analysis. When it comes to measuring COD, there are several various approaches that may be taken following the American Society for Testing and Materials (ASTM) Standards. Online testing and offline laboratory techniques utilizing environmental analyzers are some of them. Environmental analyzers are very accurate scientific tools that are used in modern COD testing procedures (Li *et al.*, 2018). With regard to MET, the COD data give an idea regarding the possibility of electron generating capacity of the wastewater and the steps it may follow during its utilization for electricity generation.

9.6.4 Microscopy

The visualization which may be paired with the fluorescent dye for the understanding of active biofilm components can be very effective. The scanning electron microscope (SEM) (Table 9.4) and epifluorescent microscope can be used for this purpose. Most frequently, a section of the electrode surface is selected for visualization and produces the image. The employment of backscattered and secondary electrons in sample imaging is a common practice. Secondary electrons are best used to show a sample's morphology and topography, while backscattered electrons are best used to show contrasts in composition in multiphase samples (Kannan, 2018).

9.7 MES AS SUPPORTIVE TECHNOLOGY

The MET set-up can be employed along with other treatment technologies at a secondary treatment or product utilization step. The effluent or the by-product of the first or initial treatment rector can be utilized by MET.

9.7.1 Bio-refractory pollutant removal

Refractory compound removal has been a top priority for wastewater treatment plants across the world. Techniques used for enhanced wastewater treatment, such as electrochemical oxidation and advanced oxidation, are expensive. Adsorption strategies like utilizing activated charcoal can be used to remove refractory organics from different industrial wastewaters, which can be counted upon as comparatively cheaper alternative. However, this raises the issue of how to dispose of the contaminated adsorbents, additionally the cost of the system also increases. Utilizing MET to remove the organics is a sustainable idea because it allows for simultaneous wastewater treatment and direct one-step bioelectricity production (Das *et al.*, 2019).

9.7.2 Usage of source-separated urine

A consistent and sustainable supply of fertilizers is necessary to ensure the food chain of a growing universal community. It is difficult to reuse the fertilizer, especially in the case of fertilizers providing nitrogen (N), phosphorous (P), and potassium (K). A chart change toward improvement and rework is necessary due to their existing usage and growing price. Effective recovery in the case of origin-separated excreta might provide 20% of the macronutrients now consumed and eliminate 50–80% of the nutrients found in wastewater. A technologically sound, and economically sustainable option is required for this purpose. The recovery of N from urine using METs has proven to be both high tech and efficiently beneficial, paving the way for new dispersed structure centered on nutrient improvement and rework (Ledezma *et al.*, 2015).

9.7.3 Microbial fuel cell with photobioreactor

Wastewater inorganics cannot be removed by microbial fuel cells (MFCs), despite their successful usage in energy assembly and decay water analysis. The removal of organic and inorganic materials, CO_2 fixation, power output, and biodiesel can be achieved simultaneously in a microalgae–microbial fuel cell (m-MFC) that combines MFC with microbic culture. An anion transfer membrane and a cation transfer membrane were used as separators to build a duplex-membrane cylinder photo-microbial fuel cell (DCP-MFC). The DCP-MFC shows two major advantages: (1) the microalgae produce oxygen via the process of photosynthesis, this oxygen can be used as an electron acceptor by the MFC in the process of electricity generation; and (2) it synchronously transforms organic and inorganic materials into useful forms of electricity and microalgae-based bioenergy (Li *et al.*, 2021).

Nitrogen and phosphorus can be removed from wastewater by microalgae, and membrane-less MFCs (ML-MFCs) operate in steady up-flow fashion. Researchers developed the up-flow ML-MFC and photobioreactor-coupled system and studied its performance in producing biomass, power, and treating wastewater. To discard the remaining phosphate and nitrogen and synchronously produce biomass, the contaminated water was first treated with an up-flow ML-MFC to produce electricity while also removing COD, phosphorus, and nitrogen (Jiang *et al.*, 2013).

9.7.4 Methane-driven microbial fuel cell

Anaerobic membrane bioreactors (AnMBR) are a potential low-power wastewater treatment technique that associates membrane isolation and anaerobic biological treatment in a particular unit process. AnMBRs do, however, discharge wastewater that is supersaturated or saturated with dissolved methane, which raises GHG emissions. Other methods for managing dissolved methane through biological treatment are anaerobic oxidation of methane (AOM), aerobic methanotrophy (AM) and anaerobic denitrifying methane oxidation (DAMO).

The AnMBRs and another common anaerobic reaction do not directly remove nutrients. An AM system is a numerous-step metabolic process that uses soluble or particulate monooxygenase to first oxidize methane to methanol. In BES, microorganisms precisely drop electrons onto an anode after the erosion of organic and inorganic substances. However, the remaining COD from the MFC does not offer enough organic material for the anaerobic analysis organization to create vaporous methane and regain energy. MFCs have been approved with anaerobic post-treatment as a linked technique for household wastewater treatment. To regain energy and stop exhalation of GHG from dissolved methane, a methane-guided MFC could be deployed.

Additionally, it might be used to turn gaseous methane into electricity or power subaquatic sensors. The possibility of using air-cathode MFCs as a post-treatment technique analysis for energy regain and GHG exhalation reduction from anaerobic drain water was examined. The presence of DAMO-archaea and *Geobacter* on the anode suggests that they may cooperate to drop electrons in installments on the anode. The anode and cathode biofilms' microbial population structure and movement were assessed using great-throughput sequencing and reverse transcription-quantitative polymerase chain reaction (RT-qPCR) (Chen & Smith, 2018).

9.7.5 Microbial fuel cell with other existing technologies

There are reports and research in the area which tries the combination of different technologies with the microbial fuel technology. This combination of different technology gives a direct effect on the substrate utilization/degradation and the production of different valuable products.

9.7.5.1 Microbial fuel cell-capacitive deionization system

The treatment of sewage and deionization through energy production were tested in conjunction with each other using the combined microbial fuel cell-capacitive deionization (MFC-CDI) system. To enhance the number of freshwater sources, salt can be eliminated from an aqueous solution using CDI, a potential deionization substitute. The electrostatic separation of the ions from the water and the adsorption of the ions at the electrode-solution contact are essential to this process. To enhance the CDI's ability to eliminate electrolytes, continuous-flow MFCs were used. In a different investigation, a matching assembly of dual MFCs was used to provide an ultimate productivity voltage of 0.63 V, which was capable of eliminating 60% of NaCl from wastewater (Patwardhan *et al.*, 2021).

9.7.5.2 Membrane bioreactor-MFCs

To enlarge the quality of wastewater, several studies have been conducted to integrate MFCs with various kinds of bioreactors. Examples include the conventional activated sediment process connected along an MFC, a sequencing batch reactor (SBR), and a system that combines an MFC, an MFC-based biological aerated filter, and an up-flow anaerobic sediment blanket reactor (UASB-MFC-BAF). The integration of the membrane bioreactor (MBR) and MFC can be carried out to lower energy costs and improve effluent quality. A system called the MBR combines the benefits of an MFC and an MBR, to achieve 4.35 W/m^3 maximum power density.

While addressing a chemical contaminant, methyl-red removal from polluted water, a biocathode-equipped MFC and a tubular membrane were combined to create an MFC–tubular MBR system. In the system of algal pollution, electrolysis pretreatment and MFC were combined. Energy was generated, and sewage water was treated using hydrogen bioreactors. Value-added biochemicals such as methane and hydrogen were produced using solid–liquid partition. The organic content of wastewater was reduced, and electricity was produced using the supernatant. By utilizing an ARS/PPy-mutated cathode membrane, the MFC-MBR-integrated setup enhanced wastewater treatment and decreased membrane fouling. Sludge reformation and a fluid-bed membrane were employed to lessen membrane fouling, improving effluent quality while consuming little energy (Patwardhan *et al.*, 2021).

9.7.5.3 Forward osmosis-MFC integration

The alliance of MFC with the forward osmosis (FO) sheath is called, osmotic MFCs (OsMFCs). The membrane is set up with both internal and external placements; the external placement is employed to turn organic components in influent-corrupt water into alcohol and short-chain fatty acids. Such a system produced 43 W/m^3 of electricity, which was more than both anion-exchange membrane (AEM) and cation-exchange membrane (CEM). However, the maximum power density only reached 4.38 W/m^3 when a lower concentration of polluted water was used. Organic matter and phosphorus had a 97% decontamination capacity; nitrogen was less effective since the FO membrane eliminated nitrogen at a lower rate. The integration of MFC with anaerobic FO membrane bioreactors (OMBR) gave methane output that was increased by 1.6 times, while the concentration of methane rose from 55% to 90%. An OMBR with an electrocatalytic-assisted MEC is a great way to generate biodiesel on an industrial scale. An experiment revealed that the power consistency of the OsMFC was 18.7% greater than that of a standard MFC (Patwardhan *et al.*, 2021).

9.7.5.4 Integration of MFCs with dark fermentation

A combination of cloudy agitation, MFCs, and microbial electrolysis cells (MECs) was reported, that achieved bioenergy production by utilizing crude glycerol (which is a disuse by-product obtained during the manufacture of biodiesel). When cheap glycerol with an initial carbon oxygen appeal concentration of 7610 mg/L was utilized in DF, the highest H_2 generation of 332 mL/L was attained. The junk was degraded in MFCs to achieve a power output of 92 mW/m^2 and a carbon oxygen demand elimination of 49% after 50% dilution. To simultaneously generate power and biohydrogen from wastewater treatment, a single-stage DF technique was created. A proton-exchange membrane fuel cell (PEMFC) was used to turn the biohydrogen gas into energy.

Investigations were made on how hydraulic retention time (HRT) affected the production of power and biohydrogen. The highest volumetric biohydrogen production rate (VHPR) for an 8-day HRT was 0.44 L H_2/Ld (0.66 L H_2/g COD eliminated), while the output of electricity was 530 mV (100 mW/m^2). The combined system's biohydrogen production allowed the PEMFC to achieve a greater voltage of 459 mV (367 mW) with an ultimate cell effectiveness of 44%. For the purpose of producing H_2 and energy simultaneously from *Saccharina japonica* in a single reactor, a hybrid technique called sDFMFC that combines DF and MFC was investigated. The coproduction of H_2 and power in the sDFMFC was validated by a timeline of CA concentration, with an H_2 yield of 110 mL/g-VS and a maximum powder consistency of 1.82 W/m^2. The results indicate that the sDFMFC may be a potential single-reactor technique for generating energy and H_2 from various biomasses (Patwardhan *et al.*, 2021).

9.7.5.5 Sediment microbial fuel cells

Different technologies, such as sediment microbial fuel cells (SMFCs), can incorporate MFCs. Reimers *et al.* used an MFC to extract energy from an oceanic sediment and test the viability of extracellular electron transmission (Reimers *et al.*, 2001). The energy production from circling a cathode electrode to increase oxygen generation in river sludge was 49 mW/m^2, with a mean energy production of 0.016 W. The other forms of sludge MFCs are, benthic MFCs (BMFCs), floating macrophyte-dependent MFCs (FMFCs), and soil-based MFCs (SL-MFCs). The BMFCs are more sophisticated form of SMFCs in an oceanic setting, particularly for the bioremediation of pollutants.

Sediment MFCs and plant MFCs have an impact on the system design of BMFCs, which can be used as an inspection system in the sea. The outflow of an H_2 bioreactor was treated with FMFCs to eliminate VFA and leftover organic debris. Organic pesticides were eliminated from the environment by Huang *et al.* using SL-MFCs, with an elimination rate of 71.15% and an energy yield of 77.5 mW/m^2 (Huang *et al.*, 2011). By switching the anodes, Quaglio *et al.* successfully improved power extraction with SMFCs. With this setup, energy harvesting was increased, and an average power consistency of 23.5 mW/m^2 was attained (Quaglio *et al.*, 2021).

Another study used a multianode SMFC to examine the power reproduction potential of mixed-culture algal biomass. The highest power density ever recorded in SMFC research was produced by the SMFC, which generated a high-power consistency of 2965 mW/m^2. These results imply that the use of algal biomass as a viable feedstock in SMFCs could greatly increase electricity generation.

9.7.5.6 Integration of microbial fuel cell with microbial desalination cell

Desalination is a method which can be used for both purification of potable water and for the analysis of contaminated water. Due to its expensive upkeep and high energy requirements, it is not appropriate in all situations. The possibility that MFCs can be implemented for synthesis of sustainable energy, such as CH_4, H_2, and electric power, this set-up has the potential to be integrated with the desalination cells. The microbial desalination cell (MDC), which is based on the migration of ions from water in accordance with the electrons given by bacteria, was described by Cao and his coworkers in 2009 (Cao *et al.*, 2009). For the analysis of synthetic wastewater and saline sewage, Zhang and He developed an osmotic MFC with a FO membrane combined with a MDC (Zhang & He, 2013) set-up. The study obtained an energy generation of 0.160 kW h/m^3 and a 95.9% decrease in electrical conductivity.

9.7.5.7 Integrated constructed wetland MFC

Due to the low treatment efficacy of constructed wetland (CW) treatment, it is an underutilized passive sewage control technique. Yadav presented a hybrid approach that combines an MFC and a CW to increase analysis efficacy while lowering the necessary landmass (Yadav *et al.*, 2012). However, power generation has also been identified as another CW resource. The main goal of incorporating the MFC into the CW is to increase the CW's therapeutic potential. The goal of MET lands, which aims to incorporate conductive material into wetlands, is to incorporate CW-MFCs, which are still under development.

Mittal *et al.* created a novel two-chambered earthen-sheath-based wetland cum microbial fuel cell (CW-MFC), that was able to remove azo dye with an output of 544.6 mA/m^3 and 148.29 mW/m^3 (Mittal *et al.*, 2022). A wetland–MFC system with an up-flow construction was created to remove Cr(VI) and generate energy. Utilizing fillers like bio-ceramic (CW-MFC1), zeolite (CW-MFC2), calcite (CW-MFC3), and volcanic rock (CW-MFC4) allowed for the long-term absorption of pollutants. The rate of Cr(VI) removal was as follows, and all systems removed more than 93% COD: CW-MFC4 (99.0%) is followed by CW-MFC2 (95.5%), CW-MFC3, and CW-MFC1 (72.2%).

9.7.5.8 Integration of MFCs with microalgae

Microalgae MFC is an innovative technology that uses the metabolic activities of photosynthetic microorganisms to transform solar energy into electrical energy. MFCs are incorporated into algal bioreactors, which release oxygen at the cathode while bacteria oxidize organics at the anodic compartment. The integrated system can be utilized to create oxygen, remove nitrogen, and sequester carbon dioxide from contaminated water. In a single-chambered photosynthetic MFC comprising *Synechococcus* sp., *Dunaliella tertiolecta*, and *Synechocystis* sp., Wu *et al.* investigated the energy generation (Wu *et al.*, 2013). An MFC harboring *Synechococcus* sp. had a maximum energy output of 10.3 mW/m^2 when exposed to 10 W/m^2 of white light. Using dissolved CO_2 as an electron donor and an active photobiocathode, an anoxic MFC was produced. Ammonia reduction and biogas upgrading were accomplished simultaneously by an integrated BES set-up. This shows that combining the production of hydrogen, biogas upgrading, and ammonia mitigation with BES appears to be a successful strategy.

9.7.5.9 Anaerobic–anoxic–oxic integrated with MFC

A crossbreed system with a total volume of 1 m^3 for the analysis of home wastewater and the production of electricity was created by integrating an MFC into an anaerobic–anoxic–oxic (AO/O) system. A built wetland and an MFC were used by Tang *et al.* to reduce COD by 92% and recover 0.448 W/m^3 of energy (Tang *et al.*, 2019). The utility and long-term viability of a connected septic tank–MFC–disinfection structure for residential wastewater treatment were established by Valladares Linares *et al.* Gravity-fed flow of raw influent from a five-person house led to a 1300 L septic tank, a 700 L Aquox® MFC, and ultimately a sodium hypochlorite disinfection system (Linares *et al.*, 2019). The system was made practical and sustainable by a power storage (management) device composed of capacitors and microcontrollers that received and stored energy from the MFC.

9.8 CHALLENGES AND FUTURE PERSPECTIVE

MESs are a collection of machinery that have the exclusive capability of modified chemical energy into electric energy using microorganisms. This characteristic has the potential to provide an alternative to the manufacture of chemical goods using hazardous substances and fossil fuels to generate power (Hernandez & Osma, 2020).

9.8.1 Separations and MET

Emerging green technologies called METs harness renewable resources to produce valuable goods without harming the environment or handling garbage. The separator, a crucial component of METs that has a big impact on how well they work, commonly used PEM is Nafion. However, a number of issues with the Nafion PEM have been identified, including its high cost, powerful oxygen and substrate crossovers, the migration of cations other than protons, and biofouling. Alternative separators have been proposed using a range of materials, including porous materials, salt bridges, glass fibers, composite membranes, and ion-exchange membranes. Cheap porous materials are said to perform better than PEM. These include ceramics that allow for non-ion selective charge transfer, like J-cloth, nylon filters, glass fiber mats, and non-woven fabrics. This chapter offers a current review of porous separators and lays out possible research paths (Daud *et al.*, 2015).

9.8.2 Remediation of pollutants

The hazard of persistent organic pollutants (polychlorodiphenyls, hexachlorobenzene, brominated compounds, dioxins, furans, aldrin, endrin, etc.) polluting the environment has grown significantly. Researchers have been more interested in bioremediation over the past several decades as a means of removing these dangerous contaminants from the environment. Conventional methods like phytoremediation, composting (biopiles and windrows), landfarming, and so on are used for the removal of these pollutants. Over the past few decades, scientific empathy of the microbial reactions to specific pollutants has contributed to the reduction of environmental pollution. For the greatest outcomes, it is crucial to find innovative bioremediation procedures with biotechnological inputs because historic bioremediation approaches have limitations for their applications. The advancement of several approaches is anticipated to boost the effectiveness of bioremediation methods and provide environmentally friendly tactics. This study examines the profile of microorganisms found in contaminated locations using a variety of methods, including omics- and culture-based approaches. Additionally, it offers current scientific literature on METs, which are currently regarded as the most effective method for pollutant remediation (Mishra *et al.*, 2020).

9.8.3 Scaling-up issues and challenges

The upgradation of the system faces practical issues, like the high internal resistances and internal electron consumption by the microbes. The scale-up process in real field condition may face challenges such as, maintenance of pure bacterial culture, to maintain the quantity of exoelectrogens in MET system, the frequent change in the substrate concentration, maintenance of membrane, and so on.

To fulfil the rising need for energy in our century, a search for unconventional and alternative renewable energy sources is continuing. A novel method for recovering energy from trash, bioelectrochemical systems may be used to remediate wastewater and recover precious resources. The majority of MET types require an external power source, which raises the overall cost of operation (Logan & Rabaey, 2012).

9.8.4 The abatement of organics in water

In the past ten years, new water regulations and enthusiastic goals, such as the sustainable development goals (SDGs) of the United Nations, have evolved to address water scarcity and pollution. Though many methods for cleaning up water pollution have been developed, it is still necessary to increase their efficiency counter to toxic and biorefractory organic molecules. The conduct of electrocatalytic anodes for direct electrochemical oxidation, the oxidation mediated by electro-generated effective chlorine, the electrocatalytic reduction, as well as coupled access for synchronous anodic and cathodic procedure combined with homogeneous and heterogeneous catalysis can be studied and applied for cleaning up the contaminated water sites (Martínez-Huitle *et al.*, 2023).

9.9 CONCLUSION

The two major applications of MET are the treatment and generation of value-added products, in which electricity and valuable gases can be considered as major shareholders. The utilization of MET for mass-scale production of valuable gases faces challenges like the microbes may utilize the valuable gases within the reactor, the collection system, that should be developed for the collection of valuable gas, production of valuable gases on a large scale, and consistency or continuity in the production magnitude.

The application of MET in waste management sector is well recognized but, a stand-alone large-scale application of MET system is difficult. The system can work efficiently as a hybrid set-up, with other established technologies. Instead of going for large-scale application, the future may demand a decentralized as well as bench-top application of MET for valuable gas production. In this scenario the future may see small MET systems being incorporated at domestic or industrial level that will produce valuable gases and utilize it at the source.

REFERENCES

Arkatkar A., Mungray A. K. and Sharma P. (2019). Effect of treatment on electron transfer mechanism in microbial fuel cell. *Energy Sources, Part A: Recovery, Utilization and Environmental Effects*, **45**, 3843–3858, https://doi.org/10.1080/15567036.2019.1668878

Arkatkar A., Mungray A. K. and Sharma P. (2020). Bioelectrochemical behaviour of a sequentially added biocatalytic coculture in a microbial fuel cell. *Journal of Basic Microbiology*, **60**(7), 562–573, https://doi.org/10.1002/jobm.202000042

Bolognesi S., Cecconet D. and Capodaglio A. G. (2020). Agro-industrial wastewater treatment in microbial fuel cells. In: Integrated Microbial Fuel Cells for Wastewater Treatment, R. Abbassi, A. K. Yadav, F. Khan, and V. Garaniya (eds.), Butterworth-Heinemann, Oxford, United Kingdom, pp. 93–133, https://doi.org/10.1016/B978-0-12-817493-7.00005-9

Borja-Maldonado F. and López Zavala M. Á. (2022). Contribution of configurations, electrode and membrane materials, electron transfer mechanisms, and cost of components on the current and future development of microbial fuel cells. Heliyon **8**, Elsevier Ltd, e09849, https://doi.org/10.1016/j.heliyon.2022.e09849

Cao X., Huang X., Liang P., Xiao K., Zhou Y., Zhang X. and Logan B. E. (2009). A new method for water desalination using microbial desalination cells. *Environmental Science & Technology*, **43**(18), 7148–7152, https://doi.org/10.1021/es901950j

Chen S. and Smith A. L. (2018). Methane-driven microbial fuel cells recover energy and mitigate dissolved methane emissions from anaerobic effluents. *Environmental Science: Water Research and Technology*, **4**(1), 67–79, https://doi.org/10.1039/c7ew00293a

Das I., Das S., Chakraborty I. and Ghangrekar M. (2019). Bio-refractory pollutant removal using microbial electrochemical technologies: a short review. *Journal of the Indian Chemical Society*, **96**, 493–497, https://doi.org/www.indianchemicalsociety.com/portal/uploads/journal/2019_04_13_Extended_1556602782.pdf

Daud S. M., Kim B. H., Ghasemi M. and Daud W. R. W. (2015). Separators used in microbial electrochemical technologies: current status and future prospects. *Bioresource Technology*, **195**, 170–179, https://doi.org/10.1016/j.biortech.2015.06.105

Daut I., Razliana A. R. N., Irwan Y. M. and Farhana Z. (2012). A study on the wind as renewable energy in Perlis, northern Malaysia. *Energy Procedia*, **18**, 1428–1433, https://doi.org/10.1016/j.egypro.2012.05.159

Elgrishi N., Rountree K. J., McCarthy B. D., Rountree E. S., Eisenhart T. T. and Dempsey J. L. (2018). A practical beginner's guide to cyclic voltammetry. *Journal of Chemical Education*, **95**(2), 197–206, https://doi.org/10.1021/acs.jchemed.7b00361

Gadkari S., Fontmorin J. M., Yu E. and Sadhukhan J. (2020). Influence of temperature and other system parameters on microbial fuel cell performance: numerical and experimental investigation. *Chemical Engineering Journal*, **388**, 124176 https://doi.org/10.1016/j.cej.2020.124176

Geary P. M. (1998). Domestic Wastewater-Treatment and Re-use On-site Treatment of Domestic Wastewater View project. https://www.researchgate.net/publication/327435111

Gupta V. K., Potumarthi R., O'Donovan A., Kubicek C. P., Sharma G. D. and Tuohy M. G. (2014). Chapter 2 – bioenergy research: an overview on technological developments and bioresources. In: Bioenergy Research: Advances and Applications, V. K. Gupta, M. G. Tuohy, C. P. Kubicek, J. Saddler and F. Xu (eds.), Elsevier, Amsterdam, pp. 23–47, https://doi.org/10.1016/B978-0-444-59561-4.00002-4

Harnisch F., Aulenta F. and Schröder U. (2011). Microbial fuel cells and bioelectrochemical systems: industrial and environmental biotechnologies based on extracellular electron transfer. In: Comprehensive Biotechnology, 2nd edn (Vol. **6**). Elsevier B.V., Oxford, UK, pp. 643–659, https://doi.org/10.1016/B978-0-08-088504-9.00462-1

Hernandez C. A. and Osma J. F. (2020). Microbial electrochemical systems: deriving future trends from historical perspectives and characterization strategies. In: Frontiers in Environmental Science (Vol. **8**). Frontiers Media S.A, https://doi.org/10.3389/fenvs.2020.00044

Huang D.-Y., Zhou S.-G., Chen Q., Zhao B., Yuan Y. and Zhuang L. (2011). Enhanced anaerobic degradation of organic pollutants in a soil microbial fuel cell. *Chemical Engineering Journal*, **172**(2), 647–653, https://doi.org/10.1016/j.cej.2011.06.024

Islam M. T., Shahir S. A., Uddin T. M. I. and Saifullah A. Z. A. (2014). Current energy scenario and future prospect of renewable energy in Bangladesh. In: Renewable and Sustainable Energy Reviews (Vol. **39**). Elsevier Ltd, pp. 1074–1088, https://doi.org/10.1016/j.rser.2014.07.149

Jack J. (2016). Understanding Factors that Affect Microbial Fuel Cell Performance: Inoculum Characteristics and Methanogenesis. Environmental & Water Resources Engineering Masters Projects. 78, https://doi.org/10.7275/6qbn-eg79

Jiang H. M., Luo S. J., Shi X. S., Dai M. and Guo R. B. (2013). A system combining microbial fuel cell with photobioreactor for continuous domestic wastewater treatment and bioelectricity generation. *Journal of Central South University*, **20**(2), 488–494, https://doi.org/10.1007/s11771-013-1510-2

Kalak T. (2023). Potential use of industrial biomass waste as a sustainable energy source in the future. In: Energies (Vol. **16**, Issue 4). MDPI, p. 1783, https://doi.org/10.3390/en16041783

Kannan M. (2018). Scanning Electron Microscopy: Principle, Components and Applications (pp. 81–92).

Kaunda C. S., Kimambo C. Z. and Nielsen T. K. (2012). Hydropower in the context of sustainable energy supply: a review of technologies and challenges. *ISRN Renewable Energy*, **2012**, 1–15, https://doi.org/10.5402/2012/730631

Khare V. and Bhuiyan M. A. (2022). Tidal energy-path towards sustainable energy: a technical review. *Cleaner Energy Systems*, **3**, 100041, https://doi.org/10.1016/j.cles.2022.100041

Kong F., Ren H. Y., Pavlostathis S. G., Nan J., Ren N. Q. and Wang A. (2020). Overview of value-added products bioelectrosynthesized from waste materials in microbial electrosynthesis systems. In: Renewable and Sustainable Energy Reviews (Vol. **125**). Elsevier Ltd, p. 109816, https://doi.org/10.1016/j.rser.2020.109816

Lazanas A. C. and Prodromidis M. I. (2022). Electrochemical impedance spectroscopy – a tutorial. *ACS Measurement Science Au*, **3**, 162–193, https://doi.org/10.1021/acsmeasuresciau.2c00070

Ledezma P., Kuntke P., Buisman C. J. N., Keller J. and Freguia S. (2015). Source-separated urine opens golden opportunities for microbial electrochemical technologies. In: Trends in Biotechnology (Vol. **33**, Issue 4). Elsevier Ltd, pp. 214–220, https://doi.org/10.1016/j.tibtech.2015.01.007

Li S. and Chen G. (2018). Factors affecting the effectiveness of bioelectrochemical system applications: data synthesis and meta-analysis. *Batteries*, **4**(3), 34, https://doi.org/10.3390/batteries4030034

Li J., Luo G., He L. J., Xu J. and Lyu J. (2018). Analytical approaches for determining chemical oxygen demand in water bodies: a review. *Critical Reviews in Analytical Chemistry*, **48**, 47–65, https://doi.org/10.1080/10408347.2017.1370670

Li G., Chang H., Zou Y., Yue J., Zheng Y., Feng H., Wu H., Wu W., Lou J., Qi X. and Zhang X. (2021). Conversion of organics and minerals into electricity and microalgae using a dual-membrane cylinder photo-microbial fuel cell. *Environmental Chemistry Letters*, **19**(5), 3531–3537, https://doi.org/10.1007/s10311-021-01248-1

Linares R. V., Domínguez-Maldonado J., Rodríguez-Leal E., Patrón G., Castillo-Hernández A., Miranda A., Romero D. D., Moreno-Cervera R., Camara-chale G., Borroto C. G. and Alzate-Gaviria L. (2019). Scale up of microbial fuel cell stack system for residential wastewater treatment in continuous mode operation. *Water (Switzerland)*, **11**(2), 217, https://doi.org/10.3390/w11020217

Liu Z. (2015). Chapter 1 – Global energy development: the reality and challenges. In: Global Energy Interconnection, Z. Liu (ed.), Academic Press, Boston, pp. 1–64, https://doi.org/10.1016/B978-0-12-804405-6.00001-4

Logan B. E. and Rabaey K. (2012). Conversion of wastes into bioelectricity and chemicals by using microbial electrochemical technologies. *Science*, **337**(6095), 686–690, https://doi.org/10.1126/science.1217412

Louro R. O., Costa N. L., Fernandes A. P., Silva A. V., Trindade I. B., Fonseca B. M. and Paquete C. M. (2018). Exploring the molecular mechanisms of extracellular electron transfer for harnessing reducing power in METs: methodologies and approaches. In: Biomass, Biofuels, Biochemicals: Microbial Electrochemical Technology: Sustainable Platform for Fuels, Chemicals and Remediation. Elsevier B.V., Amsterdam, Netherlands, pp. 261–293, https://doi.org/10.1016/B978-0-444-64052-9.00010-8

Martínez-Huitle C. A., Rodrigo M. A., Sirés I. and Scialdone O. (2023). A critical review on latest innovations and future challenges of electrochemical technology for the abatement of organics in water. *Applied Catalysis B: Environmental*, **328**, 122430, https://doi.org/10.1016/j.apcatb.2023.122430

Mayall R. M. and Birss V. I. (2017). The Chalkboard: Relating Cyclic Voltammetry and Electrochemical Impedance Spectroscopy in a Mass Transport Controlled System. *Electrochemical Society Interface*, **26**, 53–56, www.electrochem.org

Mishra B., Varjani S., Kumar G., Awasthi M. K., Awasthi S. K., Sindhu R., Binod P., Rene E. R. and Zhang Z. (2020). Microbial approaches for remediation of pollutants: innovations, future outlook, and challenges. *Energy & Environment*, **32**(6), 1029–1058, https://doi.org/10.1177/0958305X19896781

Mittal Y., Dash S., Srivastava P., Mishra P. M., Aminabhavi T. M. and Yadav A. K. (2022). Azo dye containing wastewater treatment in earthen membrane based unplanted two chambered constructed wetlands-microbial fuel cells: a new design for enhanced performance. *Chemical Engineering Journal*, **427**, 131856, https://doi.org/10.1016/j.cej.2021.131856

Patwardhan S. B., Savla N., Pandit S., Gupta P. K., Mathuriya A. S., Lahiri D., Jadhav D. A., Rai A. K., Kanupriya Ray R. R., Singh V., Kumar V. and Prasad R. (2021). Microbial fuel cell united with other existing technologies for enhanced power generation and efficient wastewater treatment. *Applied Sciences*, **11**, 10777, https://doi.org/10.3390/app112210777

Potrykus S., León-Fernández L. F., Nieznański J., Karkosiński D. and Fernandez-Morales F. J. (2021). The influence of external load on the performance of microbial fuel cells. *Energies*, **14**(3), 612, https://doi.org/10.3390/en14030612

Prakash J., Sharma R., Ray S., Koul S. and Kalia V. C. (2018). Wastewater: a potential bioenergy resource. *Indian Journal of Microbiology*, **58**, 127–137, https://doi.org/10.1007/s12088-017-0703-z

Quaglio M., Ahmed D., Massaglia G., Sacco A., Margaria V. and Pirri C. F. (2021). Enhanced power extraction with sediment microbial fuel cells by anode alternation. *Fuels*, **2**(2), 168–178, https://doi.org/10.3390/fuels2020010

Rahimi N. (2022). Microbial Electrochemical Systems for Enhanced Wastewater Treatment and Energy Recovery.

Reimers C. E., Tender L. M., Fertig S. and Wang W. (2001). Harvesting energy from the marine sediment–water interface. *Environmental Science & Technology*, **35**(1), 192–195, https://doi.org/10.1021/es001223s

Rout P. R., Bhunia P., Lee E. and Bae J. (2021). Microbial electrochemical systems (MESs): promising alternatives for energy sustainability. In: BT - Alternative Energy Resources: The Way to a Sustainable Modern Society, P. Pathak, and R.R. Srivastava, (eds.), Springer International Publishing, Cham, Switzerland, pp. 223–251, https://doi.org/10.1007/698_2020_614

Salazar S. S., Muñoz Y. and Ospino A. (2017). Analysis of geothermal energy as an alternative source for electricity in Colombia. *Geothermal Energy*, **5**(1), 1–12, https://doi.org/10.1186/s40517-017-0084-x

Sathya K., Nagarajan K. S. K., Geor G. C., Rajalakshmi S. and Lakshmi P. R. (2022). A comprehensive review on comparison among effluent treatment methods and modern methods of treatment of industrial wastewater effluent from different sources. *Applied Water Science*, **12**, 1–27, https://doi.org/10.1007/s13201-022-01594-7

Shaikh M. R., Shaikh S., Waghmare S., Labade S. and Tekale A. (2017). A review paper on electricity generation from solar energy. *International Journal for Research in Applied Science and Engineering Technology*, **887**, 1884–1889, https://doi.org/10.22214/ijraset.2017.9272

Stefanova A., Angelov A., Bratkova S., Genova P. and Nikolova K. (2018). Influence of Electrical Conductivity and Temperature in a Microbial Fuel Cell for Treatment of Mining Waste Water. University of Mining and Geology. https://www.researchgate.net/publication/329801332

Tang C., Zhao Y., Kang C., Yang Y., Morgan D. and Xu L. (2019). Towards concurrent pollutants removal and high energy harvesting in a pilot-scale CW-MFC: insight into the cathode conditions and electrodes connection. *Chemical Engineering Journal*, **373**, 150–160, https://doi.org/10.1016/j.cej.2019.05.035

Wu X. Y., Song T. S., Zhu X. J., Wei P. and Zhou C. C. (2013). Construction and operation of microbial fuel cell with *Chlorella vulgaris* biocathode for electricity generation. *Applied Biochemistry and Biotechnology*, **171**(8), 2082–2092, https://doi.org/10.1007/s12010-013-0476-8

Yadav A. K., Abbassi R., Kumar N., Satya S., Sreekrishnan T. R. and Mishra B. K. (2012). The removal of heavy metals in wetland microcosms: effects of bed depth, plant species, and metal mobility. *Chemical Engineering Journal*, **211–212**, 501–507, https://doi.org/10.1016/j.cej.2012.09.039

Zhang B. and He Z. (2013). Improving water desalination by hydraulically coupling an osmotic microbial fuel cell with a microbial desalination cell. *Journal of Membrane Science*, **441**, 18–24, https://doi.org/10.1016/j.memsci.2013.04.005

Zhang X., Li X., Zhao X. and Li Y. (2019). Factors affecting the efficiency of a bioelectrochemical system: a review. *RSC Advances*, **9**, 19748–19761, https://doi.org/10.1039/c9ra03605a

Zhou H., Meng A., Long Y., Li Q. and Zhang Y. (2014). Classification and comparison of municipal solid waste based on thermochemical characteristics. *Journal of the Air and Waste Management Association*, **64**(5), 597–616, https://doi.org/10.1080/10962247.2013.873094

doi: 10.2166/9781789063813_0175

Chapter 10

Application of microbial electrolysis cells for the production of biogas and other valuables from industrial wastewater and sludge

Akash Tripathi[1], *Shraddha Yadav*[2], *Santosh Kumar*[3] *and M. M. Ghangrekar*[1,2,3]*

[1]*Department of Civil Engineering, Indian Institute of Technology Kharagpur, Kharagpur 721302, India*
[2]*School of Environmental Science and Engineering, Indian Institute of Technology Kharagpur, Kharagpur 721302, India*
[3]*P. K. Sinha Centre for Bioenergy and Renewables, Indian Institute of Technology Kharagpur, Kharagpur 721302, India*
**Corresponding author: ghangrekar@civil.iitkgp.ac.in*

ABSTRACT

A microbial electrolysis cell (MEC) has been developed as an effective technology for the microbial conversion of organic matter contained in industrial wastes and sludge into valuable products such as biogas or biochemicals. Exploiting industrial wastewater to produce biogas or chemicals will improve circular economy and concurrently attract industries to pre-treat their wastewater before discharging. However, the products of an MEC such as CH_4, H_2, H_2O_2, and other chemicals are not thermodynamically favourable reactions; hence, an external voltage is required to support these reduction reactions. In this regard, optimizing MEC's different operational parameters and reactor configuration is obligatory to reduce the system's external energy demand, and associated energy loss and simultaneously achieving high conversion efficiency. The critical aspects of this chapter are discussing the basic concept of an MEC and its configurations with a detailed description of the thermodynamic aspect of the process. Furthermore, the description of industrial wastewater characteristics provides a better assessment regarding adopting this technology in industries. Overall, this chapter highlights the significance of MECs as a potential solution for transforming industrial wastewater and sludge into valuable resources, while encouraging environmentally friendly and sustainable practices.

Keywords: biochemical, biogas, industrial wastewater, microbial conversion, microbial electrolysis cell.

10.1 INTRODUCTION

The energy requirement of the world is at its peak, and it is estimated to increase exponentially due to the continuous growth of the world's population. Currently, the majority of the energy requirement is being fulfilled from fossil fuels (petroleum and natural gas) (Sánchez *et al.*, 2023). As a result, the depletion of these fossil fuels and their associated impact on the environment poses a threat to future generations. Moreover, many industries and commercial institutions rely heavily on these fossil fuels to meet their energy needs, generating wastewater, sludge, and greenhouse gases as consequential by-products. Waste produced by industries further requires treatment before disposing of it in the environment, involving additional energy use. In this regard, the treatment of wastewater and sludge has been considered the most energy-consuming process; for example, it consumes around 3% of global electricity (Nakkasunchi *et al.*, 2021). Therefore, investigation on minimizing energy consumption and simultaneously harvesting energy from wastewater has led to the development of microbial electrochemical technologies (METs). METs are emerging bioengineering methods that possess the ability to convert organic compounds present in wastewater into electrical energy by utilizing microorganisms as biocatalysts as shown in Figure 10.1 (Jadhav *et al.*, 2017). The mechanism of METs depends on transferring the microbe's metabolic electrons to solid-state electrodes to facilitate various biotechnological applications. The discrepancy in redox potential between anodic oxidation and cathodic reduction is a critical factor in determining the net production of electrical energy in MET configurations, such as microbial fuel cells (MFCs) and microbial carbon-capture cells. Conversely, in cases such as microbial electrolysis cells (MECs), where the desired reactions require external electrical energy input, this disparity is also influential (Kumar *et al.*, 2023).

MFCs use the metabolic activity of exoelectrogens to transform the chemical energy stored in wastewater and sludge in an anodic chamber into electric power. However, in the case of MECs, a biotic anode executes the anodic oxidation of organic contaminants by the action of exoelectrogens to produce bioelectricity as well as an cathodic chamber facilitates different terminal reduction reactions for the production of biochemicals in the presence of an external electric current. Hence, the production of biofuels, chemicals, and electrical energy depends on different cathodic arrangements.

The general configuration of MECs consists of an anodic chamber, a cathodic chamber, and a membrane for separating them. Electrons are transferred from the anode to the cathode via an external circuit; simultaneously, protons travel through the separation membrane from the anodic chamber to the cathodic chamber and participate in the reduction reaction of the terminal electron acceptor at the cathode surface, thereby generating electricity or utilizing electricity for the synthesis of chemicals (Jadhav *et al.*, 2020). The reduction reactions at the cathode differ as per different energy and chemical requirements. For example, the hydrogen evaluation reaction occurs in MECs, in which the protons and electrons generated at the anode get reduced at the cathode and produce hydrogen (H_2) (Chakraborty *et al.*, 2020). The H_2 produced at the cathode could be collected and can be further utilized for energy production as a primary (fuel cell, internal combustion engine) as well as a secondary energy carrier (synthesis of chemicals) for carbon-free energy production (Nikoo *et al.*, 2015).

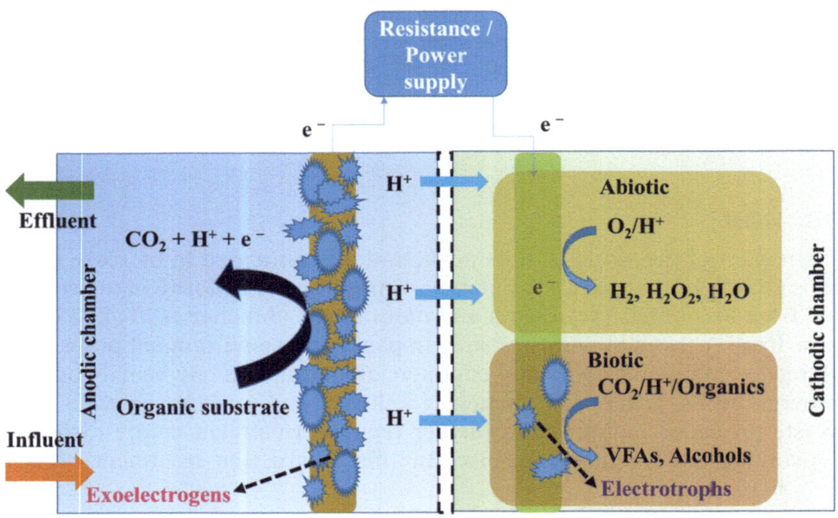

Figure 10.1 Schematic representation of METs.

Besides hydrogen, MECs could also produce other chemicals such as hydrogen peroxide (H_2O_2) and methane (CH_4). However, the production of these through MECs is an endothermic process and requires external voltage application to initiate the chemical reaction. Therefore, recent investigation has focused on producing suitable cathode catalysts, membranes, and hybrid technologies to reduce energy requirement. In spite of the external energy requirement, an MEC has been found to be cost-effective, with a high yield for the production of H_2 and other compounds compared to other commercial technologies. Thus, the use of MECs for industrial wastewater treatment could be economically beneficial due to the less energy-consuming methodology and flexibility to treat different effluents with varying characteristics. Different characteristics of industrial wastewaters with thermodynamic limitations and application of MECs for biogas production are discussed in this chapter. In addition, the current trend in reactor configuration has been explained for a holistic understanding of the process. Finally, future prospects with emphasis on the commercialization of the technology have been discussed.

10.2 CHARACTERIZATION OF INDUSTRIAL WASTEWATER AND SLUDGE

The characteristics of industrial effluents are crucial for assigning suitable treatment technology, and it varies from industry to industry. The manufacturing processes in each industrial category have been identified to emphasize the specific operations that are accountable for wastewater generation. In most industries, wastewater results at different stages: (a) sanitary, (b) cooling, (c) process, and (d) cleaning. Industrial effluent often contains high solids (total, suspended, and volatile) and possesses dark colour, thermal variation, and odour. Chemically, it comprises high organic chemicals (both biodegradable and non-biodegradable), heavy metals, cyanides, ammonia, phenols, toxic compounds, and emerging contaminants (Correia *et al.*, 1994).

10.2.1 High organic strength

In contrast to domestic wastewater (mostly containing chemical oxygen demand (COD) <500 mg/L), industrial operations generate wastewater with high organic strength having COD in the range of 100–60,000 mg/L. Industries such as polyester manufacturing, olive mills, textile production, and beverage production frequently produce highly concentrated wastewater ranging from strong (COD exceeding 1000 mg/L) to extremely strong levels (Sar *et al.*, 2022; Sipma *et al.*, 2010). The definition of high organic strength varies from industry to industry. For instance, the petrochemical effluent with COD ~1000 mg/L is considered as high strength, whereas the food industry effluent with the same COD is considered moderate. This distinction arises from chemical industries containing 'hard' COD with high content of non-biodegradable compounds such as pesticides, synthetic dyes, antibiotics, and surfactants (Mutamim *et al.*, 2013). Due to such extremities, it is advisable to treat the high-strength industrial wastewater before disposing into municipal sewerage systems or surface runoff.

10.2.2 Variable pH and salinity

Industrial effluent can have variable pH levels due to the presence of acids and alkalis during industrial processes. The extreme pH can affect the efficiency of wastewater treatment processes and have profound environmental implications if not correctly disposed of. Industries such as metal plating, mining, and chemical manufacturing generate effluent often containing sulphuric acid, hydrochloric acid, nitric acid, and other organic acids. This ultimately reduces the pH below 7.0, even to 2.0, resulting in acidic discharge from industries (Zouch *et al.*, 2018). However, industries such as paper pulp, food processing, and textiles deploy the use of alkalis, such as sodium hydroxide, potassium hydroxide, and ammonia. Overall, they raise the pH to 11.0–12.0 due to the release of alkaline wastewater. This variation in pH can cause corrosion, scale formation, and clogging of pipes and it is detrimental to aquatic species (Boczkaj & Fernandes, 2017). Also, wastewaters from industries, such as textiles, chemical, and printing, are reported to have high salinity from 0.6 to 6% by wt. (Yin *et al.*, 2022). The presence of salts with organic pollutants in wastewater always serves as a disadvantageous condition due to high operational cost, energy requirements, lower efficiency, and membrane fouling problem during the treatment (Yin *et al.*, 2022).

10.2.3 Thermal transients

The observation of thermal transition in industrial wastewater is related to specific industrial operations or activities. For instance, high-temperate industries, such as power generation, metal smelting, and steam-based manufacturing coolants, can raise the temperature of wastewater up to 60°C; however, due to the large volume of water utilization and sufficient recirculation result in 9.5–10°C elevated wastewater temperature than ambient (Madden *et al.*, 2013). Moreover, specific industrial processes deploy water for cooling purposes that ends up in moderately warm water. Effective monitoring and maintenance of the temperature of industrial effluent is crucial to meet regulatory requirements and to optimize the performance of wastewater treatment scheme.

10.2.4 Toxic and emerging contaminants

Mining and industrial activities are accounted for significant toxic contamination and severely affect ecological health worldwide. Also, the release of chemicals from industries led to the contamination of rivers, lakes, and coastal waters as well. Various environmentally significant compounds, including halogenated pollutants and organic solvents, have been detected in the effluent discharged by the textiles industries. Some of these contaminants are endocrine-disrupting chemicals, such as octyl phenol, nonylphenol, bisphenol A, diethyl phthalate, and nonylphenol ethoxylate. Moreover, the effluents also contained azo and anthraquinone dyes or surfactants involved during production processes. Similar to textiles industries, compounds such as sulphonates (naphthalene and benzene) and linear alkylbenzene sulphonate exist in tannery effluent. Also, degreasing surfactants, leather lubricants, and benzothiazole fungicides are usually found in tannery discharge (Dhruv Patel & Bhatt, 2022).

Discharge from other industries, including petrochemical and similar, contains compounds involved during the distillation of crude oil, for example, naphthalene, alkylated derivatives, quinoline derivatives, polycyclic aromatic hydrocarbons, and indane (Tian *et al.*, 2020). The range of contaminants also encompassed precursors utilized in chemical synthesis, plastic manufacturing, as well as by-products generated during the synthesis procedure adopted in petrochemical plants. In summary, the chemical makeup of effluents from petrochemical plants and refineries exhibited significant heterogeneity (Tian *et al.*, 2020).

Another streamline of industrial contaminants includes heavy metals and cyanides, which can generate through multiple sources such as metal-finishing industries, electroplating industries, iron and steel mills, mining, pharmaceuticals, coal coking, steel tempering, and so on (Dash *et al.*, 2009). Chemicals generally deployed in the rubber industry, for example, benzothiazoles and phenolic and aniline derivatives, which act as catalysts in the vulcanization process were identified in the discharge from the rubber and tyre production plants (Zou *et al.*, 2022). Additionally, phthalic acid esters (plasticizers), di-*tert*-butylmethylphenol (antioxidants), and toluene (solvent) and their structurally related compounds have the potential to be present in wastewater (Raza *et al.*, 2019). Overall, numerous groups of compounds, such as phenol, benzene and naphthalene sulphonates, phthalic acid esters, alkylphenols, chlorinated benzenes, and volatile organic solvents, were found in various industrial effluents. This indicates their non-specific nature and suggests that their presence in the environment cannot be solely attributed to specific emission sources (Dsikowitzky & Schwarzbauer, 2014).

10.3 THERMODYNAMICS OF BIOMETHANE AND BIOHYDROGEN PRODUCTION FROM MECs

MECs facilitate the thermodynamic conversion of organic substrates in wastewater via electrogenic microorganisms into valuable biogas comprising CH_4 and H_2. The nature of biogas generation in MECs highly depends on the reactor configuration, inoculum, and substrate used. There are several distinct advantages of biogas production in MECs over traditional anaerobic digestion (AD) processes (Zhang &

Angelidaki, 2014). Firstly, both organic matter oxidation and CH_4 production are separate processes in MECs, enabling the production of biogas with higher CH_4 content. Secondly, MECs operate efficiently at ambient temperature, eliminating the need for energy-intensive heating and resulting in significant energy savings (Zhang & Angelidaki, 2014). Thirdly, MECs are capable of effectively treating waste streams with low organic content, which would typically be unsuitable for AD processes (Madondo *et al.*, 2023). However, MECs are unlikely to replace conventional AD processes, as AD processes are better suited for the waste stream of high strength. Instead, MECs can serve as a valuable complement to AD, functioning as a downstream process to refine and enhance the quality of biogas production and the quality of treated effluent (Clauwaert & Verstraete, 2009; Zhang & Angelidaki, 2014).

In an MEC, the CH_4 production is expected to occur through two distinct pathways. The first pathway involves AD of sludge and volatile fatty acid (VFA) consumption, where microorganisms break down organic matter in the absence of oxygen, resulting in the production of CH_4. The second pathway involves the reaction of e^- from organics with carbon dioxide (CO_2) to produce CH_4 (Feng *et al.*, 2015). In a CH_4-producing MEC, the reaction is not thermodynamically spontaneous and requires an input of electrical energy to facilitate biochemical reactions (as shown in below equations).

Anodic chamber:

$$CH_3COO^- + 2H_2O \rightarrow 2CO_2 + 7H^+ + 8e^- \quad (E = -0.28\,eV) \tag{10.1}$$

Cathodic chamber:

$$CO_2 + 8H^+ + 8e^- \rightarrow CH_4 + 2H_2O \quad (E = -0.244\,eV) \tag{10.2}$$

The generated e^- then travels to the cathodic chamber through an electronic circuit with the help of a power supply and reacts with free protons to generate H_2. However, to facilitate H_2 production in an MEC, it is necessary to maintain the negative cathode potential of 0.414 eV or more compared to the normal hydrogen electrode (NHE) under standard biological conditions (pH = 7.0 and $T = 25°C$) (Koul *et al.*, 2022):

Cathodic chamber:

$$8H^+ + 8e^- \rightarrow 4H_2 \quad (E = -0.41\,eV) \tag{10.3}$$

Furthermore, the indirect extracellular e^- transfers and intermediate H_2 produced are responsible for the generation of CH_4 on reaction with CO_2:

$$CO_2 + 4H_2 \rightarrow CH_4 + 2H_2O \tag{10.4}$$

The overall reaction for the production of CH_4 from CO_2 and H_2O is

$$CO_2 + 2H_2O \rightarrow CH_4 + 2O_2 \tag{10.5}$$

Thus, the indirect extracellular e^- transfer mechanism for CH_4 production was found to be less energy efficient as compared to a direct extracellular transfer. Under standard biological conditions (pH = 7.0 and $T = 25°C$), the minimum thermodynamic energy input of 32.7 MJ/m^3 of CH_4 is necessary, that is, equivalent to 9.1 kWh/m^3 of CH_4 (Liang *et al.*, 2009).

Recent research studies are more inclined toward the H_2 generation through MECs and suppression of methanogenesis during the process. Similarly, the thermodynamics involved during the production of H_2 in MECs and the half-cell potential can be determined using the Nernst equation as shown below:

Anodic Chamber of MEC for H_2 production:

$$CH_3COO^- + 4H_2O \rightarrow 2HCO_3^- + 9H^+ + 8e^- \tag{10.6}$$

Cathodic Chamber of MEC for H_2 production:

$$8H^+ + 8e^- \rightarrow 4H_2 \tag{10.7}$$

The combined reaction of MEC for H_2 production:

$$CH_3COO^- + 4H_2O \rightarrow 2HCO_3^- + H^+ + 4H_2 \quad (10.8)$$

$$E_{cathode} = E^0_{cathode} - \frac{RT}{2F} \ln \frac{p_{H_2}}{[H^+]^2} = -0.414\,V \quad (10.9)$$

where R is the Rydberg's constant (~8.314 J/K mol); T is the temperature (~298.15 K); F is the Faraday's constant (~96,485 C/mol e⁻); p_{H_2} is the partial pressure of H_2 (1 atm.); and E^0 is the standard reduction potential at the cathode (0.0 V):

$$E_{anode} = E^0_{anode} - \frac{RT}{2F} \ln \frac{[CH_3COO^-]}{[HCO_3^-]^2[H^+]^9} = -0.300\,V \quad (10.10)$$

Thus, by determining the redox potential at the anode and cathode, the necessary cell voltage can be estimated using the below equation:

$$E_{cell} = E_{cathode} - E_{anode} = -0.114\,V \quad (10.11)$$

As the E_{cell} is negative, the Gibbs free energy will be positive, thereby making the reaction non-spontaneous. Hence, an external energy of 0.114 V including the voltage required for acetate oxidation (−0.300 V) will be required to favour H_2 production. However, additional voltage input is required to overcome the losses such as activation loss, ohmic loss, and mass transport loss (Savla *et al.*, 2022).

10.4 REACTOR CONFIGURATION

10.4.1 Dual-chambered MECs

Dual-chambered MEC (DC-MEC) reactors consist of separate anodic and cathodic chambers, each equipped with electrodes and electrolytes, and these chambers are separated by a proton-exchange membrane (PEM) to prevent any substrate crossover effectively (Figure 10.2) (Koul *et al.*, 2022). DC-MECs have been predominantly applied in laboratory-scale investigations owing to

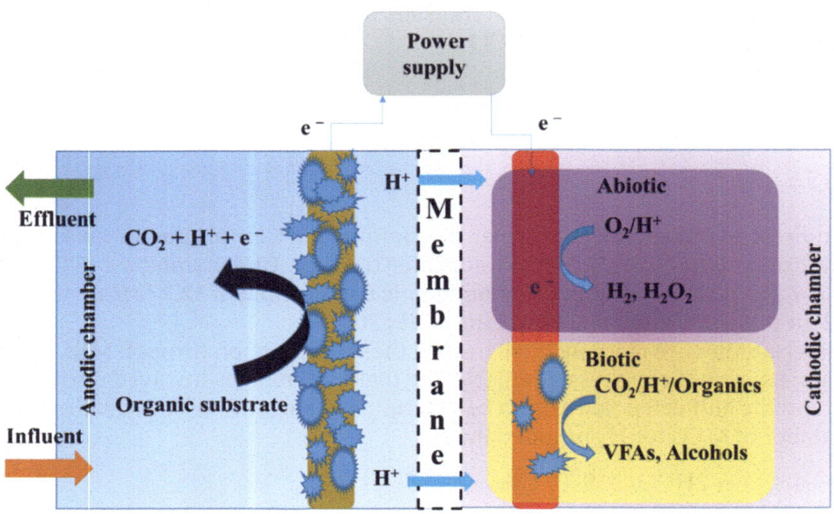

Figure 10.2 Schematic representation of a DC-MEC.

their configuration enabling control over reactions and performance optimization. For example, a DC-MEC was fabricated for the treatment of the sugar industry effluent using Nafion 117 as a PEM and it achieved 56.6% of COD removal and 56% of coulombic efficiency (Jayabalan *et al.*, 2020). In contrast, the implementation of the DC-MEC is impeded by the high installation cost associated with electrodes and PEM, which accounts for about 60% of the overall expenditures. Moreover, the reactor's performance is restricted owing to challenges such as excessive electrode spacing, membrane fouling, and increased ohmic losses (Varanasi *et al.*, 2019).

To overcome the limitations associated with DC-MECs, different configurations such as barrel-shaped, circular-like discs, and concentric tube-shaped, have been investigated. To illustrate the lower mass transfer losses, an experimental set-up consisting of a concentric tube-shaped MEC with a relatively closer electrode distance was employed for the distillery wastewater treatment and a high current density (908 mA/m^2) was harnessed compared to a traditional DC-MEC (811 mA/m^2) (Samsudeen *et al.*, 2020). In addition to configuration modifications, separator membranes of different materials have also been assessed in MECs to reduce the fabrication cost and to evaluate their efficiency. In this context, a cost-effective ceramic membrane emerges as a promising choice and stands out as one of the most extensively researched PEM. Such a ceramic membrane was fabricated by Neethu and co-researcher using activated carbon and clay. This membrane exhibited notable cost-effectiveness and displayed a power density that was twice as high as that of Nafion 117. In addition, much greater proton diffusion coefficient (36×10^{-6} cm^2/s) and coulombic efficiency (13.0%) were reported in comparison to Nafion 117, which displayed proton diffusion coefficient and coulombic efficiency of 4.64×10^{-6} cm^2/s and 9.3%, respectively (Neethu *et al.*, 2019). However, the lower coulombic efficiency is still a limitation and further investigation is required for up-scaling MECs.

10.4.2 Single-chambered MECs

To make MECs ready for a practical implementation, a membrane-less or single-chambered configuration offers the lowest installation and operational costs due to its simplified design and the absence of membrane. A single-chambered MEC becomes more compact and the distance between the electrodes also reduces (Figure 10.3). The lower distance between the electrodes facilitates the easy e^- flow and reduces the ohmic losses (Lee & Rittmann, 2010). Moreover, the membrane-less configuration also minimized the potential loss and manufacturing cost associated with the membrane. To illustrate, Rani *et al.* (2022) operated a single-chambered MEC for the dairy industry effluent for H_2 production (35 mL/L/day) and achieved high COD removal (95%) and power density (152 mW/cm^2).

The most significant issue encountered in single-chambered systems when aiming for H_2 as the desired product is the proliferation of methanogens (Lee & Rittmann, 2010). This is due to the anaerobic conditions within MECs, which provide a favourable environment for the growth of methanogenic microorganisms. Moreover, methanogens and exoelectrogens engaged in competitive interactions for both substrate utilization and product generation. Furthermore, hydrogenotrophic methanogens utilize H_2 to generate methane. Therefore, to avoid the uptake of H_2 by methanogens, an approach is to decrease the retention time of H_2 in the system. Alternatively, suppressing the methanogens through various means, such as acidic pH, intermittent exposure to air, chemical inhibitors, short solid retention time, and high organic loading rate can effectively enhance the purity of H_2 (Jadhav *et al.*, 2019).

These methodologies also pose risks to various outputs of MECs and increase the cost of the operation (Lee *et al.*, 2009). Thus, the most effective approach to mitigate H_2 loss in a single-chambered MEC involves the recapture of H_2 as soon as it is released from the cathode in a manner that outperforms methanogens and other H_2-consuming microorganisms. In this regard, different reactor configurations such as up-flow single-chambered MECs and cathode on top of a single-chambered MEC have been investigated for the quick recovery of H_2 (Call & Logan, 2008). However, the presence of methanogens and oxidation of H_2 by electroactive bacteria are still limiting factors; hence, more investigations are required to focus on eliminating these drawbacks.

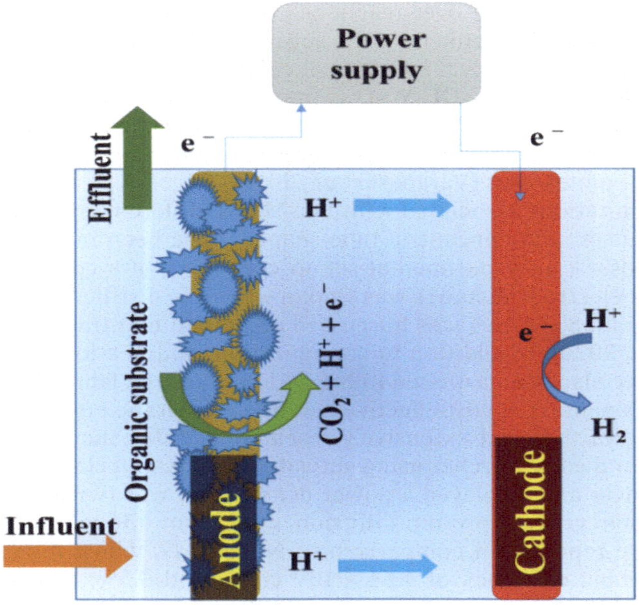

Figure 10.3 Schematic representation of a single-chambered MEC.

10.4.3 Other configurations of MECs

Stacked MECs are composed of several MEC units that are organized in either a series or parallel configuration, and each unit is composed of distinct anodic and cathodic chambers (Guo & Kim, 2019). A stacked configuration leads to improved scalability, enhanced H_2 yield, and coulombic efficiency. Similar to a stacked configuration, a multielectrode configuration also increases the total electrode surface area, thereby enhancing prevailing current and H_2 production (Opoku *et al.*, 2023). Utilization of a multielectrode system is generally employed in continuous flow MECs for the purpose of conducting pilot-scale investigations. For example, a 1000 L pilot-scale MEC for the treatment of winery wastewater was operated with the 144-electrode pair at the applied cell voltage of +0.9 V and the 60% COD removal was achieved under 24 h of retention time. Moreover, the current generation (7.4 A/m^3) and hydrogen recovery (0.19 L/day) were found to be lower owing to high losses in the operations (Cusick *et al.*, 2011).

The more innovative integration of MECs with other technologies has also been investigated to explore the production of CH_4, VFAs, and other commercially viable chemicals. For example, integrating a biocathode with the applied voltage has resulted in microbial electrosynthesis (MES) in biological carbon chain elongation and VFA production. Similarly, the combination of AD and MEC has led to enhanced methanogenesis and increased production of CH_4 as a result of improvements in the microorganism complex such as exoelectrogens and hydrogenotrophic methanogens. The details of these integrations are described later.

10.5 APPLICATION OF MECs

10.5.1 Hydrogen production

Compared to conventional electrolysis techniques, an MEC presents distinct advantages for H_2 production, as it can function at ambient temperatures and pressures, leading to decreased energy demands and the associated costs (Koul *et al.*, 2022). In addition, microorganisms have the ability to oxidize a diverse array of organic substrates, such as agricultural waste, industrial effluents, and wastewater, thereby exhibiting versatility in their potential applications. For example, Kadier *et al.* (2022b) operated an MEC with the palm oil mill wastewater as a substrate and optimized the process at different parameters (temperature: 30.23°C; pH: 6.63; influent dilution: 50.71%) and obtained 1.16 m^3 H_2/m^3/day at the applied cell potential of +1.1 V. A number of investigations have been carried out and thoroughly reviewed regarding the production of bio-H_2 through MECs (Koul *et al.*, 2022).

The analysis of MECs indicates that the applied voltage and choice of cathode material or catalyst are the primary factors that exert a significant influence on H_2 production. In this regard, Jayabalan and co-researcher investigated the sugar industry effluent as a substrate in an MEC using three different cathode materials (stainless-steel mesh, nickel foam, and nickel plate) for H_2 production at distinctive applied cell potentials (0 to +1.2 V). The investigation revealed that the optimal voltage for achieving the highest rate of H_2 production was +1 V; additionally, nickel foam demonstrated a superior efficiency in comparison to stainless-steel mesh and nickel plate, with H_2 production rates of 1.6, 0.8, and 1.3 mmol/L/day, respectively (Jayabalan *et al.*, 2019). To further reduce the cost associated with metal-based electrodes, carbonaceous materials such as carbon cloth, carbon brush, carbon paper, and graphite felt have also been employed as electrodes (Jiwanti *et al.*, 2021). In addition, the utilization of carbon-based electrodes in MECs for hydrogen production is regarded as a viable option due to low cost, ease of fabrication, exceptional conductivity, biocompatibility, high surface area, and corrosion resistance (Jiwanti *et al.*, 2021). However, the electrochemical reduction of H^+ on the surface of a carbon-based cathode still encounters thermodynamic obstacles, necessitating the introduction of catalysts to mitigate the energy barrier.

Electrode catalysts usually consist of metals such as Ni, Fe, Zn, Zr, Co, Mo, and Ti, to illustrate Ni-based catalysts such as Ni nanoparticles, Ni–Fe, and $NiMoO_4$ nanoparticles, have also been employed for H_2 evolution reaction, and the production was enhanced up to 2 L H_2/L/day with a higher coulombic efficiency (58.23%) (Jayabalan *et al.*, 2021; Lu *et al.*, 2016). Furthermore, introduction of non-metal catalysts such as graphene, graphite carbon nitride, metal–organic frameworks (MOFs), and other polymers with metals has also been investigated and has achieved the comparable efficiency (Pan & Wang, 2021). However, the utilization of these catalysts in industrial wastewater-based MECs still needs to be evaluated.

It is crucial to emphasize that MECs can effectively integrate sustainable energy sources, such as wind or solar electricity, without disruption. The integration described herein facilitates the adoption of an environmentally conscientious and carbon-neutral methodology for H_2 production using industrial waste (Kadier *et al.*, 2022a; Mahmoud *et al.*, 2022). Furthermore, investigations need to be directed towards comprehending the microbial consortia involved in the process, examining diverse electrode materials and configurations to enhance the efficiency and output of H_2 production.

10.5.2 Methane production

AD is an age-old and an efficient technology for treating sludge, food waste, and different wastewaters for energy recovery and resource utilization in the form of CH_4 and fertilizer, respectively (Maria *et al.*, 2023). However, AD faces limitations due to VFA accumulation, lower hydrolysis rate, and high dependency on acetoclastic methanogenesis (Haque *et al.*, 2022). Therefore, recent investigations have focused on the hybrid of AD and MEC to overcome these limitations (Figure 10.4). The addition of an MEC with externally applied voltage influences the microbial community, especially enriching the electrophilic hydrogenotrophic methanogens and exoelectrogenic bacteria (Joicy *et al.*, 2022). Hydrogenotrophic methanogens exhibit a relatively low sensitivity to variations in operational parameters, including pH, temperature, and VFAs, when compared to acetoclastic methanogens (Wang *et al.*, 2022). Furthermore, the electrically polarized electrodes in reactors also promote key enzyme production, thus maintaining the robust microbial community for methane production and eliminating the limitation of the traditional AD.

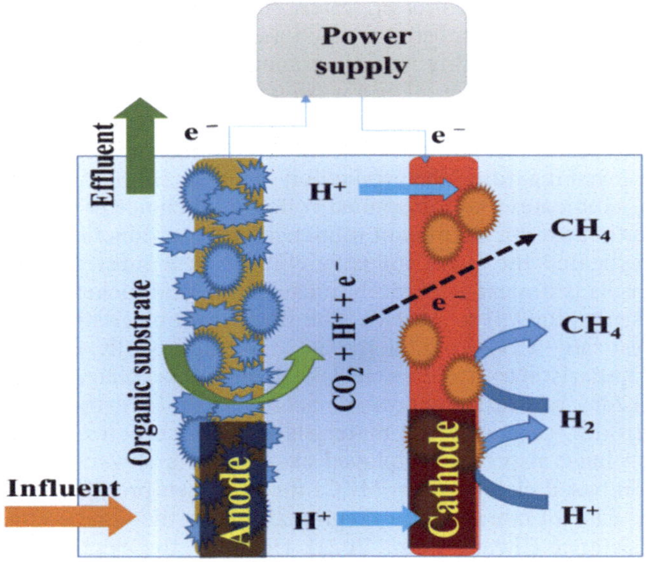

Figure 10.4 Combined system of AD and an MEC.

Investigation of synthetic wastewater showed an increased methane production, for example, in an AD–MEC, while utilizing acetate and glucose as carbon sources, a 2.59 times higher CH_4 production was reported as compared to AD (1.0 L/day) (Zhao *et al.*, 2021a). To further enhance the production of CH_4, considerable effort has been directed towards exploring various aspects, including reactor configuration, cathode catalyst, and applied potential. For example, the synthesis of an MOF-derived catalyst, specifically MOF–Ni/Co–NC, as a cathode catalyst can improve CH_4 production. The utilization of an Ni/Co–NC cathode at a potential of +0.6 V resulted in a notable CH_4 production rate of 0.57 m^3 CH_4/m^3/day with a yield of 0.34 m^3 CH_4/kg COD (Xiaomei *et al.*, 2022). The observed augmentation in CH_4 generation can be ascribed to the advantageous surface charge, elevated conductivity, and effective H_2 evolution facilitated by the cathode catalyst (Xiaomei *et al.*, 2022). In addition, the catalyst's roughness and substantial surface area contribute to the adhesion of a cathodic biofilm through hydrogen bonding and electrostatic interactions, this phenomenon leads to a significant increase in the concentration of *Methanobacterium* (79.6%), particularly on the cathodic biofilm. This suggests a transition in the CH_4 production pathway from acetoclastic to hydrogenotrophic methanogenesis.

The utilization of sludge as a viable substrate has been the subject of investigation in the AD–MEC for the purpose of CH_4 production. For example, Wang *et al.* (2021) treated the raw sewage sludge for the production of CH_4 at a different applied cell voltages (0 to +0.8 V) and maximum CH_4 production of 6.3 mL/g_{VSSin}/day was obtained at +0.8 V. The potential yield of CH_4 can be further increased by implementing pre-treatment methods such as physical, chemical, and biological on sludge. A pre-treatment process facilitates the conversion of the intricate organic compounds present in sludge into simple sugar substrates. In this regard, CH_4 production of alkali pre-treated sludge was compared to that of untreated sludge. It was found that the alkali pre-treated sludge exhibited a 55% increase in methane production (1187 mL) at a cell voltage of +0.8 V (Xu *et al.*, 2020). The higher output of CH_4 in alkali pre-treated sludge was found to be in accordance with the high rate of hydrolysis and VFA production.

Multiple investigations have been conducted to explore the efficacy of different pre-treatment methodologies, including ultrasonication, microwave, and thermal treatment, in enhancing CH_4 production from sludge (Bao *et al.*, 2020; Yin *et al.*, 2019). However, it is essential to consider the effectiveness and durability of extensive AD–MEC utilized for sludge treatment over a prolonged period (Wang *et al.*, 2022). This will aid in recognizing possible obstacles and assessing appropriate solutions for this technology.

10.5.3 Hydrogen peroxide production

Hydrogen peroxide (H_2O_2) production through MECs is a growing field of research that shows great potential for a wide range of applications, and researchers have undertaken an analysis of the viability of employing MECs as a potential method for generating H_2O_2 as an alternative to traditional manufacturing processes (Gupta *et al.*, 2020). The proposed methodology entails the alteration of MEC systems to integrate oxidation reduction reaction (ORR) at the cathode, thereby facilitating the production of H_2O_2 instead of H_2. Furthermore, based on the theoretical principles of Gibbs free energy, it is postulated that the production of H_2O_2 can be achieved by MECs without necessitating any external energy input. In this regard, Rozendal *et al.* (2009) operated a DC-MEC with acetate as a substrate and reported that the cathode achieved the standard potential for H_2O_2 formation (−0.28 V at a pH of 7.0). However, with an external voltage (−0.5 V) supply, H_2O_2 was produced at a rate of 1.9 kg H_2O_2/m^3/day (Rozendal *et al.*, 2009). Similarly, another investigation optimized the H_2O_2 production with synthetic wastewater as a substrate using a DC-MEC and obtained 233 mg/L/day of H_2O_2 at an external voltage of −0.8 V (Gupta *et al.*, 2020).

Earlier investigations also illustrate that H_2O_2 production through MECs demands lower energy than traditional electrochemical methods. Hence, the investigation of employing industrial wastewater as a viable substrate in MECs has also been undertaken for the purpose of H_2O_2 production (Ki *et al.*, 2019). For example, Ling *et al.* (2016) operated an MEC with textile industry wastewater as a substrate and a steady production of H_2O_2 (750 mL/min) was achieved at −0.55 V of external supply. Another prominent industrial wastewater compound, chlorinated phenol (10 mg/L), was utilized as a substrate in an MEC, and 74% Total Organic Carbon (TOC) removal was achieved with a recovery of 253.5 mW/m^2 power density (Miran *et al.*, 2017). Moreover, H_2O_2 production (13.3 g/L/m^2) in the cathodic chamber was optimal at −0.4 V of external voltage supply in 6 h of operation time (Miran *et al.*, 2017). In addition to wastewater, activated sludge also exhibits potential as a substrate and efforts were made to optimize the reactor for sludge, resulting in the attainment of 230 mg/L of H_2O_2 within a 6 h batch operation (Ki *et al.*, 2017). However, H_2O_2 production from MECs is still in much lower quantity than the expected level for practical commercial applications.

Cathode materials and cathode catalysts have been incorporated to improve ORRs and with the optimized operational parameters (pH, temperature, and electrode potential), a higher concentration of H_2O_2 is expected within MECs. For instance, a three-dimensional graphite cathode achieved a 2.12 kg/m^3/day of H_2O_2 production; moreover, with the acid pre-treatment, it can be further enhanced by up to 50%. A comprehensive examination of various metallic and non-metallic catalysts employed in the production of H_2O_2 has been extensively documented in the existing body of literature (Das *et al.*, 2020; Zhao *et al.*, 2021b). Furthermore, life-cycle analysis demonstrates that the production of H_2O_2 in MECs exhibits a higher degree of environmental friendliness when compared to alternative methods (Zhao *et al.*, 2021b). However, additional investigation is required to enhance the product efficacy and expandability of H_2O_2 derived from MECs. Nevertheless, the progress achieved so far in MEC-assisted H_2O_2 production highlights the future application of this technology in the future.

10.5.4 Formic acid

Formic acid (HCOOH) production through MECs is another intriguing avenue in sustainable chemical synthesis and offers a promising approach for converting carbon dioxide (CO_2) or carbon monoxide (CO) into HCOOH (Lu *et al.*, 2017). This process leverages the electrochemical reduction of CO_2 in the cathodic chamber with cathode catalysts, utilizing electricity as an energy source to drive conversion reactions. Zhao *et al.* (2012) provided the proof of concept for MEC-based CO_2 reduction by fulfilling the energy requirements through five MFC units connected in series (2.73 V).

The investigation achieved 0.09 mM/L/h of HCOOH production with 64.8% coulombic efficiency (Zhao *et al.*, 2012). Another experiment focusing on the optimization of the applied voltage in MECs has reported that the maximum production rate (46.84 mg/h) of HCOOH was achieved at an applied cell voltage of 1.8 V (Lu *et al.*, 2017). However, the production rate and concentration of HCOOH remain suboptimal in the present development stage. Therefore, subsequent investigation efforts may focus on the implementation of strategies aimed at improving the solubility of CO_2 and, consequently, facilitating mass transfer by mitigating thermodynamic constraints (Shemfe *et al.*, 2018).

10.5.5 Other chemical synthesis

Electro-fermentation has emerged as a promising methodology that integrates conventional fermentation with an MEC, gaining substantial interest and recognition (Bajracharya *et al.*, 2022). Through the utilization of electro-fermentation, the potential to augment microbial metabolism is explored through electron transfer as an effective means to change the intercellular redox state and $NADH/NAD^+$ ratio, thereby optimizing the product yields and providing a potential solution to overcome the obstacles encountered in traditional fermentation processes (Gong *et al.*, 2020). In addition, this technology presents a viable and effective approach for the synthesis of various chemicals, including ethanol, butanol, acetate, and other VFAs, which hold a significant value as biofuels and chemical precursors (Virdis *et al.*, 2022).

Electro-fermentation processes can be divided into unbalanced/anodic fermentation, MES, or cathodic fermentation. Anodic fermentation achieves efficient chemical synthesis by transferring the excess electrons to the electrode, thus maintaining an intracellular redox balance. Detailed electron transfer mechanisms have been reviewed extensively in the literature (Gong *et al.*, 2020). The application of anodic fermentation has revealed excellent results for synthesizing ethanol, butanol, and acetate by the oxidation of various industrial wastewater compounds such as glycerol, cellobiose, lactose, and glucose (Vassilev *et al.*, 2021). For example, modified strains of *Shewanella oneidensis* and *Escherichia coli* were used for the fermentation of glycerol to produce a more oxidized product at +0.2 V versus NHE potential, and ethanol (production rate: 12.12 mg/h, yield: 35%) and acetate (production rate: 8.94 mg/h, yield: 20%) were recovered (Sturm-Richter *et al.*, 2015). Similar experiments regarding the unbalanced fermentation of glucose were also performed in a DC-MEC by utilizing an engineered strain exhibiting the shuttle-mediated electron transfer at +0.4 V versus Ag/AgCl and D-lactate (0.25 mM) was obtained (Nakagawa *et al.*, 2015).

Further investigations to increase the yield of fermentation were carried out by Speers *et al.* (2014), where 90% of yield was observed due to synergistic metabolic activity of *Clostridium cellobioparum* and *Geobacter sulfurreducens* and it resulted in high ethanol production (10 g/L) and glycerol consumption (50 g/L). Anodic fermentation has the potential to enhance product yield, through improved metabolic pathways; therefore, continued research and development efforts are essential to unravel the underlying mechanisms, optimize reactor configurations, and explore new applications; thus, driving the adoption of anodic-electro-fermentation as a sustainable and efficient platform for biochemical production in the future.

Similar to anodic fermentation, an MES also provides a promising approach for the production of value-added chemicals from the reduction of different carbon sources such as CO_2, acetate, distillery wastewater, and starch-processing wastewater (Kong *et al.*, 2020). An MES exploits a biocathode under strictly anaerobic conditions for the production of these value-added products utilizing the protons and substrate at different applied potentials. Generally, CO_2 is used as a substrate in MECs for the production of acetate; however, different industrial wastewaters have also been investigated.

10.6 FUTURE OUTLOOK

The potential for long-term sustainable uses of MEC-based technologies in chemical synthesis and wastewater treatment has been reported in earlier investigations. Furthermore, it has been asserted that the utilization of MECs in chemical synthesis requires a lower energy input compared to alternative conventional methodologies. On the contrary, up-scaling of this technology is encountering obstacles stemming from a range of factors, including the constrained rates of biogas and chemical production, high costs associated with reactor installation, and fluctuations observed during operations at the pilot scale. For example, pilot-scale reactors exhibited a lower hydrogen production yield and coulombic efficiency compared to lab-scale investigations. Hence, it is imperative to consider further research and development to enhance H_2 production, separation and collection processes while minimizing energy losses for the future advancement of reactors. In this regard, genetically modifying electroactive bacteria can increase the substrate utilization and electron transfer efficiency. Moreover, genetic modifications have also ability to open up new avenues for innovative microbial pathways in MES-based reactors, thereby facilitating more reliable chemical synthesis. Similar results can be obtained in other MEC-based reactors, such as AD–MEC and electro-fermentation.

Another gap in AD–MEC operation pertains to the alteration of the microbial community when different voltage levels are applied. Therefore, it is mandatory to prioritize this aspect for future reference. In addition, implementing reactor modifications in remote villages to enhance the performance of locally operated AD reactors and biogas plants, thereby increasing CH_4 production, may serve as a viable approach to restoring the carbon cycle. H_2O_2 production has been observed to rely on the morphology of the cathode surface to facilitate the two-electron reduction of oxygen. Therefore, further investigations into the accurate mechanism of the electrode surface behaviour for the two-electron pathway are required. Currently, metallic catalysts such as nickel (Ni) and cobalt (Co) are employed in the production of H_2O_2. However, the utilization of metal catalysts incurs significant expenses, and the issue of high leaching necessitates attention. Furthermore, it is imperative to investigate the catalyst behaviours in electro-fermentation processes in order to optimize the production efficiency of single-product synthesis. The membrane also plays a pivotal role in bioelectrochemical technologies due to its significant financial contribution and its frequent encounter with fouling during long-term operations. Therefore, it is imperative to investigate the feasibility of utilizing a cost-effective membrane option to facilitate this technology's commercialization.

10.7 CONCLUSION

Industrial wastewater is a significant contributor to the wastewater matrix, exhibiting compositional variations influenced by a range of factors. Therefore, it is imperative to promote on-site industrial wastewater treatment to alleviate the strain on municipal wastewater treatment facilities. MECs present a lucrative opportunity to the industrial community as they generate additional recourses such as biofuels and chemicals with wastewater treatment. Furthermore, the technology's ability to operate at a low cost, its carbon-negative characteristics, and its energy-efficient attributes have garnered significant interest. The compatibility of MECs with traditional technologies such as AD, dark fermentation, and advanced oxidation also renders them an edge over similar technologies.

However, the bottleneck points of MEC reactors, such as costly membrane and electrode assembly with higher ohmic losses, need to be addressed. Furthermore, the technology has also been less efficient in higher technological levels than in lab-phase investigations. Thus, efforts must be diverted towards the optimal reactor design, substrate selection, and electrode–membrane assembly to improve the performance. With the optimization of these shortcomings, the technology has the capability to pave a new way for industrial wastewater treatment and chemical synthesis at the grass root level with minimal external energy requirement.

REFERENCES

Bajracharya S., Krige A., Matsakas L., Rova U. and Christakopoulos P. (2022). Advances in cathode designs and reactor configurations of microbial electrosynthesis systems to facilitate gas electro-fermentation. *Bioresource Technology*, **354**, 127178, https://doi.org/10.1016/j.biortech.2022.127178

Bao H., Yang H., Zhang H., Liu Y., Su H. and Shen M. (2020). Improving methane productivity of waste activated sludge by ultrasound and alkali pretreatment in microbial electrolysis cell and anaerobic digestion coupled system. *Environmental Research*, **180**, 108863, https://doi.org/10.1016/j.envres.2019.108863

Boczkaj G. and Fernandes A. (2017). Wastewater treatment by means of advanced oxidation processes at basic pH conditions: a review. *Chemical Engineering Journal*, **320**, 608–633, https://doi.org/10.1016/j.cej.2017.03.084

Call D. and Logan B. E. (2008). Hydrogen production in a single chamber microbial electrolysis cell lacking a membrane. *Environmental Science Technology*, **42**, 3401–3406, https://doi.org/10.1021/es8001822

Chakraborty I., Sathe S. M., Khuman C. N. and Ghangrekar M. M. (2020). Bioelectrochemically powered remediation of xenobiotic compounds and heavy metal toxicity using microbial fuel cell and microbial electrolysis cell. *Materials Science for Energy Technologies*, **3**, 104–115, https://doi.org/10.1016/j.mset.2019.09.011

Clauwaert P. and Verstraete W. (2009). Methanogenesis in membraneless microbial electrolysis cells. *Applied Microbiology Biotechnology*, **82**, 829–836, https://doi.org/10.1007/s00253-008-1796-4

Correia V. M., Stephenson T. and Judd S. J. (1994). Characterisation of textile wastewaters-a review. *Environmental Technology*, **15**, 917–929, https://doi.org/10.1080/09593339409385500

Cusick R. D., Bryan B., Parker D. S., Merrill M. D., Mehanna M., Kiely P. D., Liu G. and Logan B. E. (2011). Performance of a pilot-scale continuous flow microbial electrolysis cell fed winery wastewater. *Applied Microbiology Biotechnology*, **89**, 2053–2063, https://doi.org/10.1007/s00253-011-3130-9

Das S., Mishra A. and Ghangrekar M. M. (2020). Production of hydrogen peroxide using various metal-based catalysts in electrochemical and bioelectrochemical systems: mini review. *Journal of Hazardous, Toxic, Radioactive Waste*, **24**, 06020001, https://doi.org/10.1061/(ASCE)HZ.2153-5515.0000498

Dash R. R., Gaur A. and Balomajumder C. (2009). Cyanide in industrial wastewaters and its removal: a review on biotreatment. *Journal of Hazardous Materials*, **163**, 1–11, https //doi.org/10.1016/j.jhazmat.2021.127151

Dhruv Patel D. and Bhatt S. (2022). Environmental pollution, toxicity profile, and physico-chemical and biotechnological approaches for treatment of textile wastewater. *Biotechnology Genetic Engineering Reviews*, **38**, 33–86, https://doi.org/10.1080/02648725.2022.2048434

Dsikowitzky L. and Schwarzbauer J. (2014). Industrial organic contaminants: identification, toxicity and fate in the environment. *Environmental Chemistry Letters*, **12**, 371–386, https://doi.org/10.1007/s10311-014-0467-1

Feng Y., Zhang Y., Chen S. and Quan X. (2015). Enhanced production of methane from waste activated sludge by the combination of high-solid anaerobic digestion and microbial electrolysis cell with iron-graphite electrode. *Chemical Engineering Journal*, **259**, 787–794, https://doi.org/10.1016/j.cej.2014.08.048

Gong Z., Yu H., Zhang J., Li F. and Song H. (2020). Microbial electro-fermentation for synthesis of chemicals and biofuels driven by bi-directional extracellular electron transfer. *Synthetic Systems Biotechnology*, **5**, 304–313, https://doi.org/10.1016/j.synbio.2020.08.004

Guo H. and Kim Y. (2019). Stacked multi-electrode design of microbial electrolysis cells for rapid and low-sludge treatment of municipal wastewater. *Biotechnology for Biofuels*, **12**, 1–10, https://doi.org/10.1186/s13068-019-1368-0

Gupta A., Das S. and Ghangrekar M. M. (2020). Optimal cathodic imposed potential and appropriate catalyst for the synthesis of hydrogen peroxide in microbial electrolysis cell. *Chemical Physics Letters*, **754**, 137690, https://doi.org/10.1016/j.cplett.2020.137690

Haque S., Singh R., Pal D. B., Harakeh S., Alghanmi M., Teklemariam A. D., Abujamel T. S., Srivastava N. and Gupta V. K. (2022). Recent update on anaerobic digestion of paddy straw for biogas production: advancement, limitation and recommendations. *Environmental Research*, **215**, 114292, https://doi.org/10.1016/j.envres.2022.114292

Jadhav D. A., Ray S. G. and Ghangrekar M. M. (2017). Third generation in bio-electrochemical system research – a systematic review on mechanisms for recovery of valuable by-products from wastewater. *Renewable Sustainable Energy Reviews*, **76**, 1022–1031, https://doi.org/10.1016/j.rser.2017.03.096

Jadhav D. A., Chendake A. D., Schievano A. and Pant D. (2019). Suppressing methanogens and enriching electrogens in bioelectrochemical systems. *Bioresource Technology*, **277**, 148–156, https://doi.org/10.1016/j.biortech.2018.12.098

Jadhav D. A., Das I., Ghangrekar M. M. and Pant D. (2020). Moving towards practical applications of microbial fuel cells for sanitation and resource recovery. *Journal of Water Process Engineering*, **38**, 101566, https://doi.org/10.1016/j.jwpe.2020.101566

Jayabalan T., Matheswaran M. and Mohammed S. N. (2019). Biohydrogen production from sugar industry effluents using nickel based electrode materials in microbial electrolysis cell. *International Journal of Hydrogen Energy*, **44**, 17381–17388, https://doi.org/10.1016/j.ijhydene.2018.09.219

Jayabalan T., Matheswaran M., Preethi V. and Mohamed S. N. (2020). Enhancing biohydrogen production from sugar industry wastewater using metal oxide/graphene nanocomposite catalysts in microbial electrolysis cell. *International Journal of Hydrogen Energy*, **45**, 7647–7655, https://doi.org/10.1016/j.ijhydene.2019.09.068

Jayabalan T., Matheswaran M., Radhakrishnan T. K. and Mohamed S. N. (2021). Influence of nickel molybdate nanocatalyst for enhancing biohydrogen production in microbial electrolysis cell utilizing sugar industrial effluent. *Bioresource Technology*, **320**, 124284, https://doi.org/10.1016/j.biortech.2020.124284

Jiwanti P. K., Sultana S., Wicaksono W. P. and Einaga Y. (2021). Metal modified carbon-based electrode for CO_2 electrochemical reduction: a review. *Journal of Electroanalytical Chemistry*, **898**, 115634, https://doi.org/10.1016/j.jelechem.2021.115634

Joicy A., Seo H., Lee M. E., Kim D.H., Cho S. K. and Ahn Y. (2022). Enhanced methane production using pretreated sludge in MEC–AD system: performance, microbial activity, and implications at different applied voltages. *International Journal of Hydrogen Energy*, **47**, 40731–40741, https://doi.org/10.1016/j.ijhydene.2022.07.154

Kadier A., Wang J., Chandrasekhar K., Abdeshahian P., Islam M. A., Ghanbari F., Bajpai M., Katoch S. S., Bhagawati P. B., Li H. and Kalil M. S. (2022a). A novel pico-hydro power (PHP)-microbial electrolysis cell (MEC) coupled system for sustainable hydrogen production during palm oil mill effluent (POME) wastewater treatment. *International Journal of Hydrogen Energy*, **47**(34), pp.15464–15479, https://doi.org/10.1016/j.ijhydene.2022.09.023

Kadier A., Singh R., Song D., Ghanbari F., Zaidi N. S., Aryanti P. T., Jadhav D. A., Islam M. A., Kalil M. S., Nabgan W. and Hamid A. A. (2022b). Performance optimization of microbial electrolysis cell (MEC) for palm oil mill effluent (POME) wastewater treatment and sustainable bio-H_2 production using response surface methodology (RSM). *International Journal of Hydrogen Energy*, **47**, 15464–15479, https://doi.org/10.1016/j.ijhydene.2021.09.259

Ki D., Popat S. C., Rittmann B. E. and Torres C. I. (2017). H_2O_2 production in microbial electrochemical cells fed with primary sludge. *Environmental Science Technology*, **51**, 6139–6145, https://doi.org/10.1021/acs.est.7b00174

Ki D., Kupferer R. and Torres C. I. (2019). High-rate stabilization of primary sludge in a single-chamber microbial hydrogen peroxide producing cell. *Environmental Science: Water Research Technology*, **5**, 1124–1131, https://doi.org/10.1039/C9EW00100J

Kong F., Ren H. Y., Pavlostathis S. G., Nan J., Ren N. Q. and Wang A. (2020). Overview of value-added products bioelectrosynthesized from waste materials in microbial electrosynthesis systems. *Renewable Sustainable Energy Reviews*, **125**, 109816, https://doi.org/10.1016/j.rser.2020.109816

Koul Y., Devda V., Varjani S., Guo W., Ngo H. H., Taherzadeh M. J., Chang J. S., Wong J. W., Bilal M., Kim S. H. and Bui X. T. (2022). Microbial electrolysis: a promising approach for treatment and resource recovery from industrial wastewater. *Bioengineered*, **13**, 8115–8134, https://doi.org/10.1080/21655979.2022.2051842

Kumar S., Priyadarshini M., Ahmad A. and Ghangrekar M. M. (2023). Advanced biological and non-biological technologies for carbon sequestration, wastewater treatment, and concurrent valuable recovery: a review. *Journal of CO2 Utilization*, **68**, 102372, https://doi.org/10.1016/j.jcou.2022.102372

Lee H.-S. and Rittmann B. E. (2010). Characterization of energy losses in an upflow single-chamber microbial electrolysis cell. *International Journal of Hydrogen Energy*, **35**, 920–927, https://doi.org/10.1016/j.ijhydene.2009.11.040

Lee H.-S., Torres C. I., Parameswaran P. and Rittmann B. E. (2009). Fate of H_2 in an upflow single-chamber microbial electrolysis cell using a metal-catalyst-free cathode. *Environmental Science Technology*, **43**, 7971–7976, https://doi.org/10.1021/es900204j

Liang M., Luo B. and Zhi L. (2009). Application of graphene and graphene-based materials in clean energy-related devices. *International Journal of Energy Research*, **33**, 1161–1170, https://doi.org/10.1002/er.1598

Ling T., Huang B., Zhao M., Yan Q. and Shen W. (2016). Repeated oxidative degradation of methyl orange through bio-electro-Fenton in bioelectrochemical system (BES). *Bioresource Technology*, **203**, 89–95, https://doi.org/10.1016/j.biortech.2015.12.031

Lu L., Hou D., Fang Y., Huang Y. and Ren Z. J. (2016). Nickel based catalysts for highly efficient H_2 evolution from wastewater in microbial electrolysis cells. *Electrochimica Acta*, **206**, 381–387, https://doi.org/10.1016/j.electacta.2016.04.167

Lu Y., Luo H., Yang K., Liu G., Zhang R., Li X. and Ye B. (2017). Formic acid production using a microbial electrolysis desalination and chemical-production cell. *Bioresource Technology*, **243**, 118–125, https://doi.org/10.1016/j.biortech.2017.06.059

Madden N., Lewis A. and Davis M. (2013). Thermal effluent from the power sector: an analysis of once-through cooling system impacts on surface water temperature. *Environmental Research Letters*, **8**, 035006, https://doi.org/10.1088/1748-9326/8/3/035006

Madondo N. I., Rathilal S., Bakare B. F. and Tetteh E. K. (2023). Application of bioelectrochemical systems and anaerobic additives in wastewater treatment: a conceptual review. *International Journal of Molecular Sciences*, **24**, 4753, https://doi.org/10.3390%2Fijms24054753

Mahmoud M., El-Kalliny A. S. and Squadrito G. (2022). Stacked titanium dioxide nanotubes photoanode facilitates unbiased hydrogen production in a solar-driven photoelectrochemical cell powered with a microbial fuel cell treating animal manure wastewater. *Energy Conversion Management*, **254**, 115225, https://doi.org/10.1016/j.enconman.2022.115225

Maria M. P., Torres N. H., Nascimento V. R., Chagas T. S., Saratale G. D., Mulla S. I., Bharagava R. N., Cavalcanti E. B. and Ferreira L. F. (2023). Current advances in the brewery wastewater treatment from anaerobic digestion for biogas production: a systematic review. *Environmental Advances*, **13**, 100394, https://doi.org/10.1016/j.envadv.2023.100394

Miran W., Nawaz M., Jang J. and Lee D. S. (2017). Chlorinated phenol treatment and in situ hydrogen peroxide production in a sulfate-reducing bacteria enriched bioelectrochemical system. *Water Research*, **117**, 198–206, https://doi.org/10.1016/j.watres.2017.04.008

Mutamim N. S. A., Noor Z. Z., Hassan M. A., Yuniarto A. and Olsson G. (2013). Membrane bioreactor: applications and limitations in treating high strength industrial wastewater. *Chemical Engineering Journal*, **225**, 109–119, https://doi.org/10.1016/j.cej.2013.02.131

Nakagawa G., Kouzuma A., Hirose A., Kasai T., Yoshida G. and Watanabe K. (2015). Metabolic characteristics of a glucose-utilizing *Shewanella oneidensis* strain grown under electrode-respiring conditions. *PLoS ONE*, **10**, e0138813, https://doi.org/10.1371/journal.pone.0138813

Nakkasunchi S., Hewitt N. J., Zoppi C. and Brandoni C. (2021). A review of energy optimization modelling tools for the decarbonisation of wastewater treatment plants. *Journal of Cleaner Production*, **279**, 123811, https://doi.org/10.1016/j.jclepro.2020.123811

Neethu B., Bhowmick G. D. and Ghangrekar M. M. (2019). A novel proton exchange membrane developed from clay and activated carbon derived from coconut shell for application in microbial fuel cell. *Biochemical Engineering Journal*, **148**, 170–177, https://doi.org/10.1016/j.bej.2019.05.011

Nikoo M.K., Saeidi S. and Lohi A. (2015). A comparative thermodynamic analysis and experimental studies on hydrogen synthesis by supercritical water gasification of glucose. *Clean Technologies and Environmental Policy*, **17**(8), 2267–2288, https://doi.org/10.1007/s10098-015-0965-2

Opoku P. A., Jingyu H., Yi L., Ewusi-Mensah D. and Miwornunyuie N. (2023). Scalability of the multi-anode plug flow microbial fuel cell as a sustainable prospect for large-scale design. *Renewable Energy*, **207**, 693–702, https://doi.org/10.1016/j.renene.2023.03.018

Pan Q. and Wang L. (2021). Recent perspectives on the structure and oxygen evolution activity for non-noble metal-based catalysts. *Journal of Power Sources*, **485**, 229335, https://doi.org/10.1016/j.jpowsour.2020.229335

Rani G., Banu J. R., Kumar G. and Yogalakshmi K. N. (2022). Statistical optimization of operating parameters of microbial electrolysis cell treating dairy industry wastewater using quadratic model to enhance energy generation. *International Journal of Hydrogen Energy*, **47**, 37401–37414, https://doi.org/10.1016/j.ijhydene.2022.03.120

Raza W., Lee J., Raza N., Luo Y., Kim K. H. and Yang J. (2019). Removal of phenolic compounds from industrial waste water based on membrane-based technologies. *Journal of Industrial Engineering Chemistry*, **71**, 1–18, https://doi.org/10.1016/j.jiec.2018.11.024

Rozendal R. A., Leone E., Keller J. and Rabaey K. (2009). Efficient hydrogen peroxide generation from organic matter in a bioelectrochemical system. *Electrochemistry Communications*, **11**, 1752–1755, https://doi.org/10.1016/j.elecom.2009.07.008

Samsudeen N., Spurgeon J., Matheswaran M. and Satyavolu J. (2020). Simultaneous biohydrogen production with distillery wastewater treatment using modified microbial electrolysis cell. *International Journal of Hydrogen Energy*, **45**, 18266–18274, https://doi.org/10.1016/j.ijhydene.2019.06.134

Sánchez A. S., Junior E. P., Gontijo B. M., de Jong P. and dos Reis Nogueira I. B. (2023). Replacing fossil fuels with renewable energy in islands of high ecological value: the cases of Galápagos, Fernando de Noronha, and Príncipe. *Renewable Sustainable Energy Reviews*, **183**, 113527, https://doi.org/10.1016/j.rser.2023.113527

Sar T., Harirchi S., Ramezani M., Bulkan G., Akbas M. Y., Pandey A. and Taherzadeh M. J. (2022). Potential utilization of dairy industries by-products and wastes through microbial processes: a critical review. *Science of the Total Environment*, **810**, 152253, https://doi.org/10.1016/j.scitotenv.2021.152253

Savla N., Guin M., Pandit S., Malik H., Khilari S., Mathuriya A. S. and Gupta P. K , Thapa B. S., Bobba R. and Jung S.P. (2022). Recent advancements in the cathodic catalyst for the hydrogen evolution reaction in microbial electrolytic cells. *International Journal of Hydrogen Energy*, **47**, 15333–15356, https://doi.org/10.1016/j.ijhydene.2022.03.058

Shemfe M., Gadkari S., Yu E., Rasul S., Scott K., Head I. M., Gu S. and Sadhukhan J. (2018). Life cycle, techno-economic and dynamic simulation assessment of bioelectrochemical systems: a case of formic acid synthesis. *Bioresource Technology*, **255**, 39–49, https://doi.org/10.1016/j.biortech.2018.01.071

Sipma J., Osuna M. B., Emanuelsson M. A. and Castro P. M. (2010). Biotreatment of industrial wastewaters under transient-state conditions: process stability with fluctuations of organic load, substrates, toxicants, and environmental parameters. *Critical Reviews in Environmental Science Technology*, **40**, 147–197, https://doi.org/10.1080/10643380802039329

Speers A. M., Young J. M. and Reguera G. (2014). Fermentation of glycerol into ethanol in a microbial electrolysis cell driven by a customized consortium. *Environmental Science Technology*, **48**, 6350–6358, https://doi.org/10.1021/es500690a

Sturm-Richter K., Golitsch F., Sturm G., Kipf E., Dittrich A., Beblawy S., Kerzenmacher S. and Gescher J. (2015). Unbalanced fermentation of glycerol in *Escherichia coli* via heterologous production of an electron transport chain and electrode interaction in microbial electrochemical cells. *Bioresource Technology*, **186**, 89–96, https://doi.org/10.1016/j.biortech.2015.02.116

Tian X., Song Y., Shen Z., Zhou Y., Wang K., Jin X., Han Z. and Liu T. (2020). A comprehensive review on toxic petrochemical wastewater pretreatment and advanced treatment. *Journal of Cleaner Production*, **245**, 118692, https://doi.org/10.1016/j.jclepro.2019.118692

Varanasi J. L., Veerubhotla R., Pandit S. and Das D. (2019). Biohydrogen production using microbial electrolysis cell: recent advances and future prospects. In: Microbial Electrochemical Technology: sustainable platform for fuels, chemicals and remediation, S. V. Mohan, S. Varjani and A. Pandey (eds.), Elsevier, **2018**, 843–869, https://doi.org/10.1016/B978-0-444-64052-9.00035-2

Vassilev I., Averesch N. J., Ledezma P. and Kokko M. (2021). Anodic electro-fermentation: empowering anaerobic production processes via anodic respiration. *Biotechnology Advances*, **48**, 107728, https://doi.org/10.1016/j.biotechadv.2021.107728

Virdis B., Hoelzle R. D., Marchetti A., Boto S. T., Rosenbaum M. A., Blasco-Gómez R., Puig S., Freguia S. and Villano M. (2022). Electro-fermentation: sustainable bioproductions steered by electricity. *Biotechnology Advances*, **59**, 107950, https://doi.org/10.1016/j.biotechadv.2022.107950

Wang X.-T., Zhao L., Chen C., Chen K. Y., Yang H., Xu X. J., Zhou X., Liu W. Z., Xing D. F., Ren N. Q. and Lee D. J. (2021). Microbial electrolysis cells (MEC) accelerated methane production from the enhanced hydrolysis and acidogenesis of raw waste activated sludge. *Chemical Engineering Journal*, **413**, 127472, https://doi.org/10.1016/j.cej.2020.127472

Wang X.-T., Zhang Y. F., Wang B., Wang S., Xing X., Xu X. J., Liu W. Z., Ren N. Q., Lee D. J. and Chen C. (2022). Enhancement of methane production from waste activated sludge using hybrid microbial electrolysis cells-anaerobic digestion (MEC-AD) process – a review. *Bioresource Technology*, **346**, 126641, https://doi.org/10.1016/j.biortech.2021.126641

Xiaomei Z., Rujing L., Jun X., Yingying H., Xinying Z. and Li X. (2022). Enhanced methane production by bimetallic metal–organic frameworks (MOFs) as cathode in an anaerobic digestion microbial electrolysis cell. *Chemical Engineering Journal*, **440**, 135799, https://doi.org/10.1016/j.cej.2022.135799

Xu X.-J., Wang W. Q., Chen C., Xie P., Liu W. Z., Zhou X., Wang X. T., Yuan Y., Wang A. J., Lee D. J. and Yuan Y. X. (2020). Bioelectrochemical system for the enhancement of methane production by anaerobic digestion of alkaline pretreated sludge. *Bioresource Technology*, **304**, 123000, https://doi.org/10.1016/j.biortech.2020.123000

Yin C., Shen Y., Yuan R., Zhu N., Yuan H. and Lou Z. (2019). Sludge-based biochar-assisted thermophilic anaerobic digestion of waste-activated sludge in microbial electrolysis cell for methane production. *Bioresource Technology*, **284**, 315–324, https://doi.org/10.1016/j.biortech.2019.03.146

Yin Y., Zhang Z., Yang K., Gu P., Liu S., Jia Y., Zhang Z., Wang T., Yin J. and Miao H. (2022). Deeper insight into the effect of salinity on the relationship of enzymatic activity, microbial community and key metabolic pathway during the anaerobic digestion of high strength organic wastewater. *Bioresource Technology*, **363**, 127978, https://doi.org/10.1016/j.biortech.2022.127978

Zhang Y. and Angelidaki I. (2014). Microbial electrolysis cells turning to be versatile technology: recent advances and future challenges. *Water Research*, **56**, 11–25, https://doi.org/10.1016/j.watres.2014.02.031

Zhao H.-Z., Zhang Y., Chang Y. Y. and Li Z. S. (2012). Conversion of a substrate carbon source to formic acid for carbon dioxide emission reduction utilizing series-stacked microbial fuel cells. *Journal of Power Sources*, **217**, 59–64, https://doi.org/10.1016/j.jpowsour.2012.06.014

Zhao L., Wang X. T., Chen K. Y., Wang Z. H., Xu X. J., Zhou X., Xing D. F., Ren N. Q., Lee D. J. and Chen C. (2021a). The underlying mechanism of enhanced methane production using microbial electrolysis cell assisted anaerobic digestion (MEC-AD) of proteins. *Water Research*, **201**, 117325, https://doi.org/10.1016/j.watres.2021.117325

Zhao Q., An J., Wang X. and Li N. (2021b). In-situ hydrogen peroxide synthesis with environmental applications in bioelectrochemical systems: a state-of-the-art review. *International Journal of Hydrogen Energy*, **46**, 3204–3219, https://doi.org/10.1016/j.ijhydene.2020.05.227

Zou R., Tang K., Hambly A. C., Wünsch U. J., Andersen H. R., Angelidaki I. and Zhang Y. (2022). When microbial electrochemistry meets UV: the applicability to high-strength real pharmaceutical industry wastewater. *Journal of Hazardous Materials*, **423**, 127151, https://doi.org/10.1016/j.jhazmat.2021.127151

Zouch H., Cabrol L., Chifflet S., Tedetti M., Karray F., Zaghden H., Sayadi S. and Quéméneur M. (2018). Effect of acidic industrial effluent release on microbial diversity and trace metal dynamics during resuspension of coastal sediment. *Frontiers in Microbiology*, **9**, 3103, https://doi.org/10.3389/fmicb.2018.03103

doi: 10.2166/9781789063813_0193

Chapter 11

Amalgamation of constructed wetland and microbial fuel cell for cost-effective and low carbon emitting treatment of industrial wastewater

Anil Dhanda[1], *Akash Tripathi*[1], *Rishabh Raj*[2] *and M. M. Ghangrekar*[1,2]*

[1]*Department of Civil Engineering, Indian Institute of Technology Kharagpur, Kharagpur 721302, West Bengal, India*
[2]*School of Environmental Science and Engineering, Indian Institute of Technology Kharagpur, Kharagpur 721302, West Bengal, India*
**Corresponding author: ghangrekar@civil.iitkgp.ac.in*

ABSTRACT

The increasing demand for potable water and the negative impacts of industrial wastewaters on freshwater sources have necessitated the development of new technologies for holistic wastewater treatment. Constructed wetland–microbial fuel cell (CW–MFC) is an innovative and sustainable technology that combines the benefits of both CW and MFC. The wetland component provides conducive habitat for microbial growth, while the fuel cell generates electricity from microbial activity while treating wastewater. The same approach can potentially remove a wide range of recalcitrant pollutants from industrial effluents in addition to organic matter, nitrogen, and phosphorus. This technology consumes less energy, has low operating/maintenance costs and simultaneously produces renewable energy from waste streams, which makes it superior to traditional wastewater treatment technologies. The current chapter explores different types of CW–MFC systems evolved for wastewater application, emphasizing the fundamental principles, design considerations, and operation mechanisms of these systems. Further, the efficacy of CW–MFC for the treatment of a variety of industrial wastewaters, such as dairy, brewery, and pharmaceutical wastewater, is also elucidated. The chapter also highlights the challenges and limitations of the CW–MFC-based technologies and the measures required to improve their performance and scalability aspects.

Keywords: constructed wetland, industrial wastewater treatment, microbial fuel cell, persistent organic pollutants.

11.1 INTRODUCTION

Discharge of inadequately treated industrial wastewater into natural waterbodies is an imminent environmental concern due to its high ecotoxicological potential that can affect aquatic life and humans. Treating toxic industrial effluents via traditional technologies is costly, energy-intensive, and requires extensive maintenance (Wu *et al.*, 2015). In recent years, constructed wetlands (CW) have emerged as a promising technology for treating industrial wastewater. The CW are engineered systems that mimic natural wetland processes, using plants and microorganisms to remove pollutants from water, though at a more rapid rate (Vymazal, 2011). These systems can be designed to treat a variety of waste streams, including those from agricultural, municipal, industrial, landfill leachate, and mine drainage (Wu *et al.*, 2014). The CW can effectively remove organic and inorganic contaminants, nutrients, and heavy metals from wastewater while providing wildlife habitat (Wu *et al.*, 2014). One of the significant advantages of CW is their low-energy requirement compared to traditional treatment approaches and their aptness to treat wastewater even in cold climates (Arslan *et al.*, 2023). Further, the CW system can also be designed to be self-sustaining, using solar energy and natural processes to power the wastewater treatment operation. Additionally, the presence of plants in CW can improve air quality and reduce greenhouse gas emissions by sequestering CO_2.

For instance, Calherios and co-researchers employed a pilot CW (0.6 m^3 effective volume) with five different vegetation types, including *Typha latifolia, Phragmites australis, Canna indica, Iris, and Stenotaphrum secundatum* (Calheiros *et al.*, 2007). The system investigated two different hydraulic loading rates (HLRs) of 3 and 6 cm/d for treating tannery wastewater. The CW with an HLR of 6 cm/d performed better with 54–73% chemical oxygen demand (COD) removal for all vegetation types than the CW with an HLR of 3 cm/d (41–67%). Additionally, complete removal of phosphorous and chromium was observed for all vegetation types. However, the CW had a very low nutrient removal efficiency. The observed total Kjeldahl nitrogen (TKN) and ammonia nitrogen (NH_4^+-N) removal efficiencies were 18–42% and 11–27% for HLR of 3 cm/day, and 16–30% and 2–16% for HLR of 6 cm/d, respectively. Additionally, the CW have limitations regarding their treatment efficiency for persistent organic pollutants (POPs) commonly generated in different industrial processes (Lei *et al.*, 2022). Moreover, wastewater treatment in CW is a relatively slow process due to the absence of electron acceptors, which necessitates a larger land area. These shortcomings require technical amendments in the CW design to make it a suitable technology for treating industrial effluents with a high organic load.

Microbial fuel cell (MFC) technology is an innovative solution for wastewater treatment (Logan *et al.*, 2006). By utilizing bacterial metabolism, this cutting-edge approach not only facilitates the degradation of pollutants but also generates electrical energy cost-effectively and sustainably. Investigations have demonstrated successful applications of MFCs in treating industrial effluents, particularly those containing POPs that are challenging to degrade through conventional technologies (Selvasembian *et al.*, 2022). Another benefit of MFC is its potential for the recovery and reuse of valuable resources present in industrial wastewater, such as nitrogen and phosphorus (Baby & Ahammed 2022; Dhanda *et al.*, 2023). This is particularly important in the current scenario where more impetus is on waste-to-wealth recovery platforms that fit into the circular economy concept. The potential for combining the functions of wastewater treatment and resource recovery makes the MFC a highly promising technology for tackling critical environmental issues faced by industries worldwide (Das *et al.*, 2021). Moreover, the MFCs can be easily combined with other treatment technologies and readily retrofitted in existing infrastructure with nominal adjustments (Das *et al.*, 2022; Ghangrekar *et al.*, 2022). For example, the CW system can be clubbed with MFC in the so-called CW–MFC system, eliminating the drawback of individual technologies and uplifting the overall performance.

In a CW–MFC hybrid system, the MFC is integrated into the wetland by providing an electrode assembly for the migration of electrons (Mosquera-Romero *et al.*, 2023; Raj *et al.*, 2023). The MFC provides an anaerobic environment for microorganisms to break down organic matter and generate electricity while removing pollutants from the wastewater. The hybrid CW–MFC system effectively removes various contaminants, including organic matter, nutrients, and heavy metals. It can also improve treatment efficiency for POP (Yadav *et al.*, 2018). Additionally, the generated electrical energy can be harvested using a power management system to power the treatment process or exported to the grid. This improves the sustainability of the CW–MFC hybrid systems and can also bring down the overall treatment costs. Rathour and co-researchers employed a CW–MFC vegetated with *Fimbristylis dichotoma* to treat textile wastewater. The CW–MFC attained 82% of dye and 70% of COD removal efficiency, which was 9% and 7.4% higher than the sole CW. Additionally, the CW–MFC was able to harvest a maximum power density of 198 mW/m^2, proving the suitability of the system for textile wastewater treatment (Rathour *et al.*, 2019).

Thus, combining the MFC with a CW can help to overcome the challenges posed by the CW and make them a suitable technology for industrial wastewater treatment and simultaneous electricity generation. This book chapter explores the use of CW–MFC hybrid systems for industrial wastewater treatment and the current state of research in this field, highlighting the advantages and limitations of the technology with more stress on factors that affect performance of system. The potential for future research and development of CW–MFC and its broader implications for sustainable wastewater treatment and energy production are also discussed.

11.2 PHYSICOCHEMICAL AND BIOLOGICAL PARAMETERS OF THE INDUSTRIAL WASTEWATER

Industrial wastewater is a complex chemical and biological constituent mixture with characteristics very different from typical municipal sewage. The discharge of untreated industrial effluent can lead to the contamination of water bodies, soil, and air, causing severe environmental and health hazards. Therefore, the treatment of industrial wastewater is essential for preventing environmental deterioration. Industrial wastewater can contain a wide range of contaminants, such as heavy metals, chemicals, organic matter, suspended or dissolved solids, oils, and greases. These contaminants are present in varying concentrations and compositions, depending on the type of industry and its activities. Heavy metals, such as lead, cadmium, and mercury, are toxic even in trace amounts and can accumulate in living organisms, causing severe health disorders. Similarly, organic matter and other pollutants can cause oxygen depletion in the receiving water bodies, which can endanger aquatic life. Chemicals used in industrial processes, such as acids and solvents, can cause skin irritation, respiratory problems, and other health issues. The detailed characteristics of different industrial effluents are presented in Table 11.1.

Given the complexity of wastewater generated by different industries, it is crucial to identify the most appropriate treatment technology for the respective industry. The treatment scheme can involve physical, chemical, and biological processes to remove contaminants from wastewater and make them safe for disposal or reuse. The physicochemical treatments use physical or chemical processes such as coagulation, flocculation, sedimentation, and filtration to remove pollutants from wastewater. On the contrary, biological treatments (e.g., activated sludge process) use microorganisms to degrade organic matter and other contaminants. However, these conventional physicochemical and biological processes are energy intensive and not economically viable. In this regard, the CW–MFC can be a suitable technology to treat industrial wastewater and simultaneously generate electricity.

11.3 STATE OF THE ART OF CW–MFC

The conglomeration of CW and MFC was first investigated by Yadav and co-researchers as a vertical up-flow CW-MFC and demonstrated that the lower portion of the CW and its filler material is prevailed by the anaerobic zone (Yadav *et al.*, 2012). At the same time, the aerobic zone prevails in the vicinity of the water–air interface (Figure 11.1). Thus, the electrode assembly similar to MFC in respective anaerobic and aerobic zones can make the CW function like a single-chamber MFC without any membrane (Kesarwani *et al.*, 2023). The electrodes in respective regions act as an inexhaustible electron acceptor and donor and support the redox potential-dependent microbial process. Easy migration of metabolically generated electrons promotes the growth of electroactive bacteria on the anodic side, which enhances the wastewater treatment and pollutants removal efficiency compared to sole CW or MFC (Zhang *et al.*, 2023). The electron reaching the cathode (high redox potential) is further utilized to reduce the electron acceptor, typically oxygen, which is colloquially known as the oxygen reduction reaction (ORR). Plant roots provide the oxygen required for the ORR in the CW–MFC in the form of radical oxygen loss (ROL) or through natural aeration at the water surface.

Table 11.1 Characteristics of different industrial wastewaters.

Wastewater Source	BOD (mg/L)	COD (mg/L)	pH	TKN (mg/L)	TSS (mg/L)	Other Pollutants (mg/L)	References
Poultry processing	925–5000	2133–12 500	6.0–8.0	163–563	313–8000	CFU (30–4020)	Fatima *et al.* (2021)
Slaughter-house	150–4635	500–15 900	4.9–8.1	50–841	270–6400	Potassium (0.01–100), Turbidity (200–300)	Bustillo-Lecompte and Mehrvar (2015)
Textile	80–6000	150–30 000	9.95–11.8	70–80	15–8000	Oil and grease (5–5.5), Sulphates (600–1000), Na (≈7000)	Yaseen and Scholz (2019)
Dairy	565–5772	785–7619	6.2–11.3	–	326–3560	Alkalinity (225–1550)	Danalewich *et al.* (1998)
Tannery	1500–2000	3000–4000	8–11	–	2000–3000	Sulphide (50–100), Total Cr (60–100)	Zhao and Chen (2019)
Sugarcane industry	3000–5000	5000–9000	6–10	–	6000–9000	–	Yadav *et al.* (2021)
Rice mill	157–1937	1114–4216	4.3–7.6	N (6.7–106)	135–1265	Lignin (60–172), Phenol (5–19)	Kumar and Deswal (2021)
Petroleum refinery	205–448	744–1673	7.5–9.41	Nitrates (82–95)	280–340	Grease (48–97), Sulphates (40–50)	Kondaveeti *et al.* (2023)

BOD – biochemical oxygen demand; COD – chemical oxygen demand; CFU – colony forming units; TKN – total Kjeldahl nitrogen; TSS – total suspended solids

To illustrate the efficiency of CW-MFC, Saz *et al.* (2018) employed different plants such as *Typha latifolia, J. gerardii, C. divisa* and *Typha angustifolia* in CW–MFC and planted CW–MFC achieved higher (around 80%) COD removal compared to unplanted CW–MFC (70 mg/L). Moreover, *Typha angustifolia* was found efficient with a coulombic efficiency of 8.28% and a power density of 7.47 mW/m^2 (Saz *et al.*, 2018). The CW–MFC is commonly operated in up-flow mode to maximize the redox gradient between the anode and cathode (Figure 11.2a). Regrettably, using the inherent redox gradient facilitated by an up-flow regime leads to substantial electrode spacing, which increases the ohmic resistance of the system drastically (Gupta *et al.*, 2021). To minimize the ohmic resistance and improve the performance of the CW–MFC, different CW–MFC design configurations have been investigated and are illustrated in Figure 11.2. In this regard, Villasenor *et al.* (2013) and co-researchers operated the horizontal flow CW–MFC with a bentonite layer separating the cathodic and anodic zones for the treatment of synthetic wastewater in a continuous mode of operation (Figure 11.2b). Using this configuration 90% of COD removal efficiency was achieved with a maximum power density of 43 mW/m^2 and a current density of 1.22 mA/m^2. In another investigation, the down-flow vertical CW–MFC (Figure 11.2c) with macrophyte-based cathode arrangement was found to strategically improve the aerobic–anaerobic environment for total nitrogen (TN) removal (Wang *et al.*, 2017).

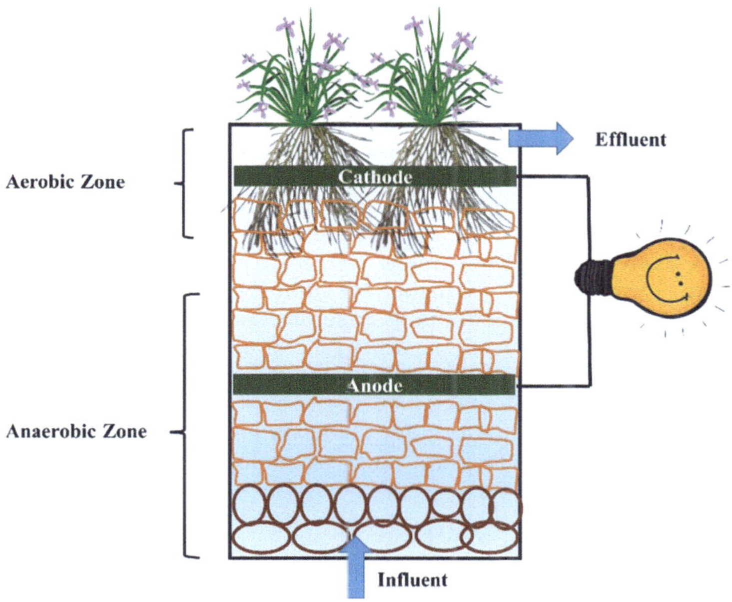

Figure 11.1 Schematic representation of a CW–MFC.

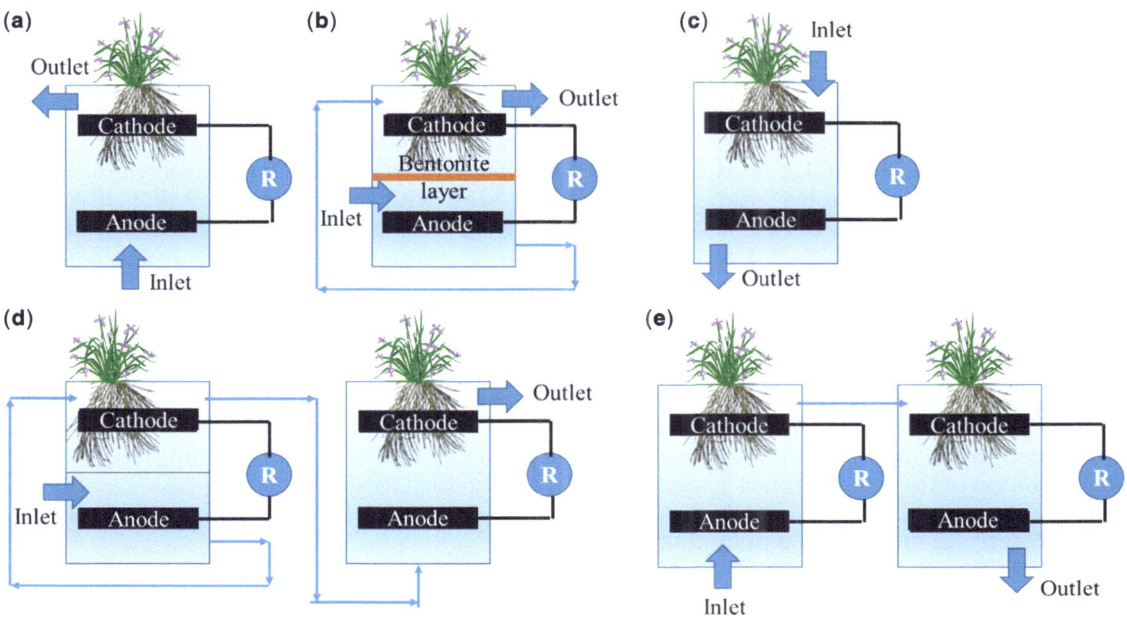

Figure 11.2 Different configurations used for CW–MFC.

In another investigation, Srivastava and co-researchers explored hybrid horizontal and vertical flow-based CW–MFC, which effectively removed total organic carbon (67.8%), COD (99.5%), TN (90%), and NH_4^+-N (94.4%) with a power density of 224 mW/m³ (Srivastava *et al.*, 2020). The excellent performance of the hybrid configuration was attributed to the horizontal bed providing a large surface area for ammonium oxidation, while allowing denitrification in the subsequent up-flow reactor (Figure 11.2d). Similarly, the hybrid vertical up-flow–down-flow reactors (Figure 11.2e) have also been investigated for the efficient removal of pollutants from wastewater. For example, Liu *et al.* (2022a, 2022b) operated three different CW–MFCs to treat mariculture wastewater in a hybrid up-flow/down-flow CW–MFC. The hybrid combination achieved a higher power density (114.56 mW/m²) compared to single up-flow (43.96 mW/m²) and down-flow (79.53 mW/m²) CW–MFC (Liu *et al.*, 2022a). In addition, the hybrid system also achieved a higher TN (almost 95%) and total phosphorous (TP) (92%) removal, which can be attributed to a continuous anaerobic–aerobic–anaerobic environment.

Apart from these prominent modifications, other reactor designs such as stacked modular, multi-anode, and multi-cathode CW–MFC have also been investigated to improve wastewater treatment efficiency with higher electricity production. Nevertheless, the higher internal resistance of a CW–MFC hybrid system is the protuberant reason for the lower energy output and must be addressed to obtain meaningful bioelectricity from this innovative technology.

11.4 FACTORS GOVERNING THE CW–MFC

The performance of CW–MFC is influenced by many factors, including organic loading rate (OLR), electrode materials, electrode spacing, vegetation type, and external resistance. A detailed discussion of how these parameters affect the CW–MFC is elaborated below.

11.4.1 Organic loading rate

The OLR governs the substrate availability to microorganisms, thus affecting the metabolic activity and electron generation rate in the CW–MFC. A low OLR can limit the substrate availability for the microorganisms, resulting in reduced microbial activity, an extended startup period, and, consequently, a lower power output. At the same time, a higher OLR in the CW–MFC can also result in lower power generation due to the migration of the organic substrate towards the cathode and the growth of methanogens in the reactor, thus hindering the electricity production (Gupta *et al.*, 2021). The accumulation of organic matter at the cathode is entailed by biofouling and depletion of dissolved oxygen concentration, which hampers the cathodic reduction reactions (Doherty *et al.*, 2015). Thus, maintaining an optimum OLR is crucial for the efficiency of CW–MFC. Similar observations were made by Xu and co-researcher, who operated the CW–MFC with the varying OLR (9.2, 18.4, 27.6, 55.2, and 92 g COD/m²-d) and the maximum power density (10.77 mW/m²) was observed at an OLR of 18.4 g COD/m²-d (Xu *et al.*, 2017a).

Apart from the cathode-based power inhibiting issues, performance drop can also be caused by a high COD with other pollutants, such as dye, pharmaceutical compounds, and surfactants can also cause a toxic effect on exo-electrogens. For example, Zhang and co-researchers investigated the effect of high concentrations of citric acid (5, 10, 25, and 40 mM) on electricity generation and higher power density (611 mW/m²) was reported at 25 mM of citric acid with 75% of COD removal (Zhang *et al.*, 2021a). Further increasing the citric acid concentration to 40 mM, the performance of CW–MFC started decreasing due to the induced toxicity to the microorganisms. Thus, maintaining an optimized concentration of these pollutants is necessary for scaling up the CW–MFC.

11.4.2 Electrode material and positioning

The electrode materials in the CW–MFC perform three significant roles: (1) support for the electroactive biofilm, (2) electron transfer and (3) reduction of oxygen. Thus, an ideal electrode must possess excellent electro-conductivity and high surface area, and simultaneously, be biocompatible, notably the anode. Metal-based electrodes provide higher electro-conductivity; however, the high cost and toxicity towards the exoelectrogens limit their application in the CW–MFC (Gupta *et al.*, 2021). In this context, carbon-based electrodes (such as carbon cloth, graphite felt, activated carbon and graphite) have gained interest due to optimal electro-activity and biocompatibility. For instance, Wang *et al.* (2016) explored the feasibility of four different electrode materials, that is, carbon fibre felt, foamed nickel, stainless-steel mesh, and graphite rod, for application in CW–MFC. The carbon fibre felt outperformed the other electrodes with a maximum output voltage of 177 mV compared to stainless-steel mesh (41 mV), graphite rod (51 mV) and foamed nickel (148 mV) (Wang *et al.*, 2016). Further, the CW–MFC are usually operated without any catalyst support over the electrode. However, utilizing a cathode/anode catalyst can increase the ORR kinetics, COD removal, and pollutant removal efficiency (Su *et al.*, 2023).

The inter-electrode gap is another crucial parameter for the CW–MFC due to the different operating conditions of both electrodes. The membrane-less operation of the CW–MFC makes it challenging to differentiate between the anaerobic and aerobic zone, which limits a smaller electrode spacing. Moreover, increasing the electrode spacing can significantly increase the internal resistance of the system. Therefore, the inter-electrode distance must be optimized for the efficient operation of CW–MFC, which results in maximum wastewater treatment and electricity output (Doherty *et al.*, 2015). It is observed that the cathode placed on the free-flowing water surface can maintain better aerobic conditions and helps to promote the ORR. For example, Corbella and the co-researcher found that the cathode placed above the support media and near free-flowing water generated 131 mWh/m^2-d power density compared to the cathode placed on the support media (80 mWh/ m^2-d) and the placement was also found effective in coping with the evapotranspiration and seasonal variations (Corbella *et al.*, 2016). Thus, CW–MFCs with appropriate electrode material placed at a suitable distance can aid in enhancing the performance of the reactor and bolster the commercialization prospects of this hybrid technology.

11.4.3 Effect of vegetation

Plants play a crucial role in the CW–MFC by releasing oxygen, exudating through their roots and regulating the biogeochemical cycles in and around the root layer, which is known as the rhizosphere. Moreover, the dense network of roots provides additional surface area for microbial flora and improves the final effluent quality through filtration and bioaccumulation (Gupta *et al.*, 2021). The ROL, or the oxygen produced in the rhizosphere, is exploited to facilitate ORR at the cathode. In addition, the exudates serve as electron donors during the denitrification process, thus positively influencing nitrogen removal. However, the comparison between the ROL and artificial aeration has shown that artificial aeration achieves a higher COD removal (Kesarwani *et al.*, 2023). The high removal can be attributed to the development of aerobic microbes in the system. In contrast, the ROL works on developing the redox gradient and maintaining the anaerobic regions in the CW–MFC.

Furthermore, it has been observed that different plant species exhibit varying effects in CW–MFC due to ROL and phytoremediation processes. For example, the submerged macrophyte (*Elodea nuttallii*) has a greater capacity to release ROL compared to the emergent macrophyte (*Typha latifolia*) (Oon *et al.*, 2017). The plant also helps to reduce internal resistance through the release of ROL. In this regard, Fang and co-researchers operated the CW–MFC with *Ipomoea aquatica* and a lower internal resistance (217.7 Ω) was observed compared to an unplanted control reactor (272.9 Ω) (Fang *et al.*, 2013). Moreover, the effect of the day and night can also be observed in the CW–MFC as the photosynthetic activity increases in the daytime, and a potential drop of 200 mV was observed between them (Villasenor *et al.*, 2013). Overall, further investigations must focus on evaluating the plant species to optimize the performance of CW–MFC.

11.4.4 External resistance

External resistance plays a crucial role in determining the efficacy of CW–MFCs. A low external resistance promotes a higher electron flux from the anode, which enriches the growth of electroactive biofilm (Liu *et al.*, 2016). In contrast, a higher external resistance assists in achieving a higher COD removal, though at the cost of lower power density. By increasing the external resistance, the flow of electrons is impeded, resulting in a higher potential difference across the system. This promotes more efficient electrochemical reactions, facilitating the breakdown of organic compounds and increasing COD removal. Due to the large volume and size of CW–MFC, the internal resistance is high compared to the sole MFC. Therefore, the ohmic loss must be administered through the electrode distance and external resistance (Gupta *et al.*, 2021). In this regard, Yang and co-researchers operated the CW–MFC with varying external resistance from 200 to 2000 Ω, and the power density of 107.54 mW/m^3 was obtained at an optimum resistance of 1000 Ω (Yang *et al.*, 2022).

On the contrary, the investigation by Zhang and co-researchers has shown that external resistance does not affect COD removal, and only a slight variation can be observed (Zhang *et al.*, 2021b). Therefore, the operational controllability of external resistance in CW–MFC is challenging due to the involvement of a number of factors, including large size and volume (Zhang *et al.*, 2023). Nevertheless, it is possible to manage the losses within the system by selecting eco-friendly, highly conductive filling materials and implementing design modifications (Doherty *et al.*, 2015). Hence, future investigations must be conducted to enhance the efficacy of CW–MFC and address the issue of elevated resistance in CW–MFC by modifying its internal components.

11.5 TREATMENT OF THE DIFFERENT INDUSTRIAL WASTEWATER IN CW–MFCs

The CW–MFC has garnered significant attention in sustainable development research due to its intrinsic capacity to treat industrial wastewater. To illustrate, Venkata Mohan and co-researchers employed a CW–MFC with *Eichhornia crassipes*, snails, and microorganisms as the biological components for treating the distillery effluent (Venkata Mohan *et al.*, 2011). The CW–MFC generated a maximum power density of 69.70 mW/m^2 and 86.67% and 72.32% COD and volatile fatty acids (VFA) removal efficiency, respectively. It was also observed that increasing the concentration of organic matter in the system resulted in more COD removal efficiency. The COD removal initially at 71.82% enhanced to 82.20% and further increased to 86.67% with an increase in COD concentrations from 316 to 674 mg/L and 1170 mg/L, respectively. The enhanced COD and VFA removal were attributed to the fibrous root system of *Eichhornia,* which can partially remove organic pollutants in wastewater through physical, chemical, and biological processes, altering their chemical nature. The root system can also foster beneficial insects and micro-biota while developing biofilm, which can help in effective organic matter removal.

Further, the system also performed well in terms of nitrate removal (79.88%), colour removal (83.33%), and turbidity removal (91.95%). The nitrate removal was attributed to simultaneous nitrification by the macrophytes and denitrification by the denitrifying bacteria in the system. At the same time, the colour removal was attributed to the *Eichhornia* turf root system, which releases a significant amount of polymeric root exudates that destabilize colloidal suspensions, resulting in coagulation and sedimentation of suspended particles. Additionally, the observed reduction in turbidity in the system may be attributed to snails that graze on suspended solids while moving along the walls of the floating macrophyte-based ecological microenvironment. Thus, the integration of CW with MFC demonstrated an excellent option for treating distillery wastewater.

In another investigation, Yadav and co-researchers explored the efficiency of CW–MFC planted with *Canna indica* plants for the treatment of synthetic wastewater (COD-1500 mg/L) highly concentrated with methylene blue (MB) dye (500–2000 mg/L) (Yadav *et al.*, 2012). The CW–MFC produced a maximum power density of 15.73 mW/m^2 (MB-1000 mg/L). The system efficiently removed 93.15% of MB with an initial MB concentration of 500 mg/L, whereas the observed removal was nearly 80% for higher concentrations of MB (1000–2000 mg/L) with a hydraulic retention time of 96 h. The MB can act as an electron mediator in MFC, promoting the electron transfer rate to the electrode and thus enhancing performance of the system. Therefore, in this case, the electrons generated from the organic matter degradation can be accepted by the MB and converted to a colourless reduced form. Moreover, MB removal can be further accelerated by adsorption on gravel surfaces, plants, and biomass in CW–MFC. A maximum of 75% COD removal was also observed at an MB concentration of 1500 mg/L. However, the COD removal was less than 60% at an MB concentration of 2000 mg/L, probably due to induced toxicity towards the microbes at higher MB concentrations.

Further, the CW–MFC can be bio-augmented with different strains to improve system's efficiency and remove the targeted pollutants. Recently, Kongthale and co-researchers employed the CW–MFC for phenol removal and winery wastewater treatment (Kongthale *et al.*, 2022). To remove ethanol, the system was modified using ethanol-tolerant yeast isolated from honey mead, pineapple, palm vinegar, mulberry, and traditional beverage starter. The CW–MFC produced a maximum of 139 mA/m^2 current density and 38 mW/m^2 power density (pineapple extracted yeast), simultaneously achieving a 79% of COD removal efficiency while the influent concentration was 2027 mg/L. A 85% phenol removal was also reported with influent concentration of 3639 mg/L), which was highest among the ethanol-tolerant yeast strains. Thus, the CW–MFC demonstrated the ability to efficiently treat the wastewater, remove phenol and produce electricity simultaneously. The performance of other CW–MFCs investigated to treat different industrial wastewaters with different types of pollutants is presented in Table 11.2.

Table 11.2 Industrial wastewater treatment using CW–MFC.

Vegetation Used	Total Volume# (L)	Wastewater Source	COD in mg/L (% removal)	Maximum Power (mW/m^2)	Pollutant Removed (% removal)	References
Epipremnum	1.7	Synthetic winery	2027 (79)	38	Phenol (85)	Kongthale *et al.* (2022)
Typha latifolia	19	Synthetic dye	560 (74)	1.58	Dye (96)	Oon *et al.* (2020)
Iris tectorum	17.8	Ciprofloxacin wastewater	(87.10)	3.55	Ciprofloxacin (91.2), TP (97.1)	Dai *et al.* (2022)
Phragmites australis	8.25	Swine wastewater	1058 (76.48)	9.4	TSS (92.92), TP (65.89), TN (49.73)	Zhao *et al.* (2013)
Canna indica	14.2	SDBS poised synthetic wastewater	–	4	SDBS (56.80)	Wang *et al.* (2023)
Acorus calamus	3.5	Cr (VI)-poised synthetic wastewater	(86.51)	15.84	Cr (94.10), NH_4^+-N (85.29), TP (54.08)	Liu *et al.* (2022b)
Fimbristylis dichotoma	4.3	Dyestuff wastewater	1580 (70)	198.8	Dye (82.20)	Rathour *et al.* (2019)
Cyperus	22 and 77$	Swine wastewater	1600 (72)	9 mW/m^3	TN (47), TP (85)	Ren *et al.* (2021)
Unplanted	3.2	Synthetic wastewater with Azo dye	550 (94.04)	148.29	Azo dye (94.22)	Mittal *et al.* (2022)
Canna indica	6.5	Textile wastewater	5600 (83)	102.08	Dye (90), TDS (84)	Sonu *et al.* (2021)
Fimbristylis ferruginea	60	Textile wastewater	1058 (74.10)	194.19	Dye (97.32)	Patel *et al.* (2021)

#Value calculated from the data provided in the article; $First stage and second stage volume of the reactor; Cr – chromium; NH_4^+ – N – ammonium nitrogen; SDBS – sodium dodecyl benzenesulphonate; TDS – total dissolved solids; TN – total nitrogen; TP – total phosphorus; TSS – total suspended solids.

11.6 FUTURE PROSPECTS

The CW–MFC presents an excellent industrial wastewater treatment option and concurrently produces electricity (Ahmad *et al.*, 2022). However, the CW–MFCs face the same challenges during their operation as the conventional MFCs. The cathode is considered a significant constraint, encompassing charge transfer losses attributable to sluggish reaction kinetics, ohmic losses arising from elevated internal resistance, and mass transportation losses due to inadequate oxidant availability (i.e., O_2). These constraints led to low power densities generation and coulombic efficiencies of less than 200 mW/m^2 and less than 10%, respectively, which are many folds less than conventional MFCs (Patel *et al.*, 2021). The net energy recovery (NER) of MFCs may be a more suitable parameter for cross-study comparisons as it is based on effluent characteristics and it is less dependent on MFC dimensions. The highest NER obtained by a CW–MFC is only 0.047 kWh/kg of COD, which is two orders of magnitude lower than the theoretical energy required (3.86 kWh/kg of COD) for COD oxidation to CO_2 and water (Doherty *et al.*, 2015).

Different operational and technical factors limit the scalability and commercialization of the CW–MFC system. For instance, a linear increase in internal resistance is observed as the reactor size and distance between the electrodes increase (Doherty *et al.*, 2015). Further, cathode deterioration occurs over time due to excessive heterotrophic bacterial growth, which limits oxygen reduction at the cathode (Xu *et al.*, 2017b). Also, insufficient contact between bacteria and the anode results in overpotential losses (Nitisoravut and Regmi, 2017). The presence of fermentative, methanogenic, or other microbial communities or high-potential compounds such as nitrate can act as alternative electron acceptors and limit the amount of current production (Saket *et al.*, 2022). Finally, high organic matter concentration can increase the acidity of the system, creating an inappropriate environment for the growth of electroactive bacteria at the anode (Mittal *et al.*, 2022).

Notably, the CW–MFC cathode can use nitrate as a final electron acceptor. At the wetland surface, enough oxygen for nitrification and low-strength wastewater as a food source should be supplied to support the development of nitrifiers and electrotrophs. The development of controlled biofilm at the cathode may generate sufficient anoxic pockets where denitrifiers that use the cathode as an electron donor can flourish. This may enhance the electrical performance of CW–MFCs and help surmount the limitation of CW–MFCs in nitrification/denitrification for industrial wastewater treatment (Doherty *et al.*, 2015).

Furthermore, the reactor design, circuits, and placement of electrodes are other operational challenges that obstruct the development of CW–MFC into a full-scale technology (Doherty *et al.*, 2015). These constraints must be considered to achieve the most favourable outcome in wastewater treatment and the concurrent generation of bioelectricity. Various strategies have been employed to address these obstacles, including the implementation of multiple electrodes, an increase in the projected electrode area, the use of series and parallel circuit combinations, stacked CW–MFCs, cathodic aeration, and the incorporation of a power management system to enable uninterrupted power generation (Srivastava *et al.*, 2017; Tamta *et al.*, 2020; Tang *et al.*, 2019). Unfortunately, the power production from a CW–MFC is still meagre and requires technological improvisation from budding researchers. Moreover, only limited pilot-scale studies explicitly targeting complex industrial wastewater exist. More exploration must be taken up to tackle the limitation of CW–MFC diligently.

11.7 CONCLUSION

The CW–MFC has emerged as a promising solution for industrial wastewater treatment. This technology offers a sustainable and cost-effective approach for treating wastewater while generating energy. Researchers have explored the potential of CW–MFC technology to remove a broad range of pollutants, including organic matter, nitrogen, phosphorus, dyes, and heavy metals, from industrial wastewater. One of the main advantages of the CW–MFC technology is its low footprint, as it can operate without external energy sources, making it highly cost-effective. Moreover, the CW–MFC offers a green solution by simultaneously removing pollutants and generating energy. Although the CW–MFC technology shows great potential for industrial wastewater treatment, several challenges must be addressed before contemplating real-life applications. The proportional increase in internal resistance with reactor size, cathode deterioration over time, and insufficient contact between bacteria and the anode limits the scalability and commercialization of the technology.

Further research and development efforts are necessary to optimize this technology and overcome the associated challenges. Additionally, there is limited know-how on the operation of CW–MFCs at a magnified scale, especially for complex effluents like industrial wastewater. In conclusion, the CW–MFC system has tremendous potential to be a long-term solution for handling industrial liquid waste. However, significant technical improvements are necessary in engineering and material science aspects to make this novel hybrid technology more robust and reliable.

REFERENCES

Ahmad A., Priyadarshani M., Das S. and Ghangrekar M. M. (2022). Role of bioelectrochemical systems for the remediation of emerging contaminants from wastewater: a review. *Journal of Basic Microbiology*, **62**(3–4), 201–222, https://doi.org/10.1002/jobm.202100368

Arslan M., Wilkinson S., Naeth M. A., Gamal El-Din M., Khokhar Z., Walker C. and Lucke T. (2023). Performance of constructed floating wetlands in a cold climate waste stabilization pond. *Science of the Total Environment*, **880**, 163115, https://doi.org/10.1016/j.scitotenv.2023.163115

Baby M. G. and Ahammed M. M. (2022). Nutrient removal and recovery from wastewater by microbial fuel cell-based systems – a review. *Water Science & Technology*, **86**(1), 29–55, https://doi.org/10.2166/wst.2022.196

Bustillo-Lecompte C. F. and Mehrvar M. (2015). Slaughterhouse wastewater characteristics, treatment, and management in the meat processing industry: a review on trends and advances. *Journal of Environmental Management*, **161**, 287–302, https://doi.org/10.1016/j.jenvman.2015.07.008

Calheiros C. S., Rangel A. O. and Castro P. M. (2007). Constructed wetland systems vegetated with different plants applied to the treatment of tannery wastewater. *Water Research*, **41**(8), 1790–1798, https://doi.org/10.1016/j.watres.2007.01.012

Corbella C., Garfi M. and Puigagut J. (2016). Long-term assessment of best cathode position to maximise microbial fuel cell performance in horizontal subsurface flow constructed wetlands. *Science of the Total Environment*, **563–564**, 448–455, https://doi.org/10.1016/j.scitotenv.2016.03.170

Dai M., Wu Y., Wang J., Lv Z., Li F., Zhang Y. and Kong Q. (2022). Constructed wetland-microbial fuel cells enhanced with iron carbon fillers for ciprofloxacin wastewater treatment and power generation. *Chemosphere*, **305**, 135377, https://doi.org/10.1016/j.chemosphere.2022.135377

Danalewich J. R., Papagiannis T. G., Belyea R. L., Tumbleson M. E. and Raskin L. (1998). Characterization of dairy waste streams, current treatment practices, and potential for biological nutrient removal. *Water Research*, **32**(12), 3555–3568, https://doi.org/10.1016/S0043-1354(98)00160-2

Das S., Raj R., Das S. and Ghangrekar M. M. (2021). A sustainable approach for the production of green energy with the holistic treatment of wastewater through microbial electrochemical technologies: a review. *Frontiers in Sustainability*, **2**, 792028, https://doi.org/10.3389/frsus.2021.792028

Das S., Raj R., Das S. and Ghangrekar M. M. (2022). Evaluating application of photosynthetic microbial fuel cell to exhibit efficient carbon sequestration with concomitant value-added product recovery from wastewater: a review. *Environmental Science and Pollution Research International*, **30**(44), 98995–99012.

Dhanda A., Raj R., Sathe S., Dubey B. and Ghangrekar M. (2023). Graphene and biochar-based cathode catalysts for microbial fuel cell: performance evaluation, economic comparison, environmental and future perspectives. *Environmental Research*, **231**, 116143, https://doi.org/10.1016/j.envres.2023.116143

Doherty L., Zhao Y. Q., Zhao X. H., Hu Y. S., Hao X. D., Xu L. and Liu R. B. (2015). A review of a recently emerged technology: constructed wetland – microbial fuel cells. *Water Research*, **85**, 38–45, https://doi.org/10.1016/j.watres.2015.08.016

Fang Z., Song H. L., Cang N. and Li X. N. (2013). Performance of microbial fuel cell coupled constructed wetland system for decolorization of azo dye and bioelectricity generation. *Bioresource Technology*, **144**, 165–171, https://doi.org/10.1016/j.biortech.2013.06.073

Fatima F., Du H. B. and Kommalapati R. R. (2021). Treatment of poultry slaughterhouse wastewater with membrane technologies: a review. *Water*, **13**(14), 1905, https://doi.org/10.3390/w13141905

Ghangrekar M. M., Dhanda A., Sathe S. M. and Chakraborty I. (2022). Enhancing microbial fuel cell performance by carbon nitride-based nanocomposites. In: Nanostructured Carbon Nitrides for Sustainable Energy and Environmental Applications, S. Chowdhury and Mu. Naushad (eds) Elsevier, AE Amsterdam, Netherlands, pp. 63–79.

Gupta S., Srivastava P., Patil S. A. and Yadav A. K. (2021). A comprehensive review on emerging constructed wetland coupled microbial fuel cell technology: potential applications and challenges. *Bioresource Technology*, **320**(Pt B), 124376, https://doi.org/10.1016/j.biortech.2020.124376

Kesarwani S., Panwar D., Mal J., Pradhan N. and Rani R. (2023). Constructed wetland coupled microbial fuel cell: a clean technology for sustainable treatment of wastewater and bioelectricity generation. *Fermentation Basel*, **9**(1), 6, https://doi.org/10.3390/fermentation9010006

Kondaveeti S., Govindarajan D., Mohanakrishna G., Thatikayala D., Abu-Reesh I. M., Min B. and … Aminabhavi T. M. (2023). Sustainable bioelectrochemical systems for bioenergy generation via waste treatment from petroleum industries. *Fuel*, **331**, 125632, https://doi.org/10.1016/j.fuel.2022.125632

Kongthale G., Sotha S., Michu P., Madloh A., Wetchapan P. and Chaijak P. (2022). Electricity production and phenol removal of winery wastewater by constructed wetland – microbial fuel cell integrated with ethanol tolerant yeast. *Biointerface Research in Applied Chemistry*, **13**(2), p. 157.

Kumar S. and Deswal S. (2021). A review on current techniques used in India for rice mill wastewater treatment and emerging techniques with valuable by-products. *Environmental Science and Pollution Research*, **28**(7), 7652–7668, https://doi.org/10.1007/s11356-020-11898-3

Lei Y., Rijnaarts H. and Langenhoff A. (2022). Mesocosm constructed wetlands to remove micropollutants from wastewater treatment plant effluent: effect of matrices and pre-treatments. *Chemosphere*, **305**, 135306, https://doi.org/10.1016/j.chemosphere.2022.135306

Liu T., Yu Y. Y., Li D. Z., Song H., Yan X. L. and Chen W. N. (2016). The effect of external resistance on biofilm formation and internal resistance in *Shewanella* inoculated microbial fuel cells. *RSC Advances*, **6**(24), 20317–20323, https://doi.org/10.1039/C5RA26125B

Liu F., Lu T. and Zhang Y.-X. (2022a). Performance assessment of constructed wetland-microbial fuel cell for treatment of mariculture wastewater containing heavy metals. *Process Safety and Environmental Protection*, **168**, 633–641, https://doi.org/10.1016/j.psep.2022.10.026

Liu S. T., Qiu D. F., Lu F. F., Wang Y., Wang Z., Feng X. J. and Pyo S. H. (2022b). *Acorus calamus* L. Constructed wetland-microbial fuel cell for Cr (VI)-containing wastewater treatment and bioelectricity production. *Journal of Environmental Chemical Engineering*, **10**(3), 129324.

Logan B. E., Hamelers B., Rozendal R., Schroder U., Keller J., Freguia S. and … Rabaey K. (2006). Microbial fuel cells: methodology and technology. *Environmental Science & Technology*, **40**(17), 5181–5192, https://doi.org/10.1021/es0605016

Mittal Y., Dash S., Srivastava P., Mishra P. M., Aminabhavi T. M. and Yadav A. K. (2022). Azo dye containing wastewater treatment in earthen membrane based unplanted two chambered constructed wetlands-microbial fuel cells: a new design for enhanced performance. *Chemical Engineering Journal*, **427**, 131856, https://doi.org/10.1016/j.cej.2021.131856

Mosquera-Romero S., Ntagia E., Rousseau D. P. L., Esteve-Nunez A. and Prevoteau A. (2023). Water treatment and reclamation by implementing electrochemical systems with constructed wetlands. *Environmental Science and Ecotechnology*, **16**, 100265, https://doi.org/10.1016/j.ese.2023.100265

Nitisoravut R. and Regmi R. (2017). Plant microbial fuel cells: a promising biosystems engineering. *Renewable & Sustainable Energy Reviews*, **76**, 81–89, https://doi.org/10.1016/j.rser.2017.03.064

Oon Y. L., Ong S. A., Ho L. N., Wong Y. S., Dahalan F. A., Oon Y. S. and … Nordin N. (2017). Role of macrophyte and effect of supplementary aeration in up-flow constructed wetland-microbial fuel cell for simultaneous wastewater treatment and energy recovery. *Bioresource Technology*, **224**, 265–275, https://doi.org/10.1016/j.biortech.2016.10.079

Oon Y. L., Ong S. A., Ho L. N., Wong Y. S., Dahalan F. A., Oon Y. S. and … Thung W. E. (2020). Constructed wetland-microbial fuel cell for azo dyes degradation and energy recovery: influence of molecular structure, kinetics, mechanisms and degradation pathways. *Science of the Total Environment*, **720**, 137370, https://doi.org/10.1016/j.scitotenv.2020.137370

Patel D., Bapodra S. L., Madamwar D. and Desai C. (2021). Electroactive bacterial community augmentation enhances the performance of a pilot scale constructed wetland microbial fuel cell for treatment of textile dye wastewater. *Bioresource Technology*, **332**, 125088, https://doi.org/10.1016/j.biortech.2021.125088

Raj R., Das S., Behera M. and Ghangrekar M. M. (2023). Bio-electrochemical systems for micropollutant removal. In: Microconstituents in the Environment: Occurrence, Fate, Removal Management, R. Surampalli, T. Zhang, C. M. Kao, M. Ghangrekar, P. Bhunia, M. Behera and P. Rout (eds.) John Wiley & Sons, Inc., Hoboken, NJ, USA, pp. 455–490.

Rathour R., Patel D., Shaikh S. and Desai C. (2019). Eco-electrogenic treatment of dyestuff wastewater using constructed wetland-microbial fuel cell system with an evaluation of electrode-enriched microbial community structures. *Bioresource Technology*, **285**, 121349, https://doi.org/10.1016/j.biortech.2019.121349

Ren B., Wang T. and Zhao Y. (2021). Two-stage hybrid constructed wetland-microbial fuel cells for swine wastewater treatment and bioenergy generation. *Chemosphere*, **268**, 128803, https://doi.org/10.1016/j.chemosphere.2020.128803

Saket P., Mittal Y., Bala K., Joshi A. and Kumar Yadav A. (2022). Innovative constructed wetland coupled with microbial fuel cell for enhancing diazo dye degradation with simultaneous electricity generation. *Bioresource Technology*, **345**, 126490, https://doi.org/10.1016/j.biortech.2021.126490

Saz C., Ture C., Turker O. C. and Yakar A. (2018). Effect of vegetation type on treatment performance and bioelectric production of constructed wetland modules combined with microbial fuel cell (CW-MFC) treating synthetic wastewater. *Environmental Science and Pollution Research*, **25**(9), 8777–8792, https://doi.org/10.1007/s11356-018-1208-y

Selvasembian R., Mal J., Rani R., Sinha R., Agrahari R., Joshua I. and … Pradhan N. (2022). Recent progress in microbial fuel cells for industrial effluent treatment and energy generation: fundamentals to scale-up application and challenges. *Bioresource Technology*, **346**, 126462, https://doi.org/10.1016/j.biortech.2021.126462

Sonu K., Sogani M. and Syed Z. (2021). Integrated constructed wetland-microbial fuel cell using biochar as wetland matrix: influence on power generation and textile wastewater treatment. *Chemistryselect*, **6**(32), 8323–8328, https://doi.org/10.1002/slct.202102033

Srivastava P., Dwivedi S., Kumar N., Abbassi R., Garaniya V. and Yadav A. K. (2017). Performance assessment of aeration and radial oxygen loss assisted cathode based integrated constructed wetland-microbial fuel cell systems. *Bioresource Technology*, **244**(Pt 1), 1178–1182, https://doi.org/10.1016/j.biortech.2017.08.026

Srivastava P., Yadav A. K., Garaniya V., Lewis T., Abbassi R. and Khan S. J. (2020). Electrode dependent anaerobic ammonium oxidation in microbial fuel cell integrated hybrid constructed wetlands: a new process. *Science of the Total Environment*, **698**, 134248, https://doi.org/10.1016/j.scitotenv.2019.134248

Su C., Xian Y., Qin R., Zhou Y., Lu M., Wan X. and … Chen M. (2023). Fe(III) enhances Cr(VI) bioreduction in a MFC-granular sludge coupling system: experimental evidence and metagenomics analysis. *Water Research*, **235**, 119863, https://doi.org/10.1016/j.watres.2023.119863

Tamta P., Rani N. and Yadav A. K. (2020). Enhanced wastewater treatment and electricity generation using stacked constructed wetland-microbial fuel cells. *Environmental Chemistry Letters*, **18**(3), 871–879, https://doi.org/10.1007/s10311-020-00966-2

Tang C., Zhao Y. Q., Kang C., Yang Y., Morgan D. and Xu L. (2019). Towards concurrent pollutants removal and high energy harvesting in a pilot-scale CW-MFC: insight into the cathode conditions and electrodes connection. *Chemical Engineering Journal*, **373**, 150–160, https://doi.org/10.1016/j.cej.2019.05.035

Venkata Mohan S., Mohanakrishna G. and Chiranjeevi P. (2011). Sustainable power generation from floating macrophytes based ecological microenvironment through embedded fuel cells along with simultaneous wastewater treatment. *Bioresource Technology*, **102**(14), 7036–7042, https://doi.org/10.1016/j.biortech.2011.04.033

Villasenor J., Capilla P., Rodrigo M. A., Canizares P. and Fernandez F. J. (2013). Operation of a horizontal subsurface flow constructed wetland – microbial fuel cell treating wastewater under different organic loading rates. *Water Research*, **47**(17), 6731–6738, https://doi.org/10.1016/j.watres.2013.09.005

Vymazal J. (2011). Constructed wetlands for wastewater treatment: five decades of experience. *Environmental Science & Technology*, **45**(1), 61–69, https://doi.org/10.1021/es101403q

Wang J., Song X., Wang Y., Abayneh B., Ding Y., Yan D. and Bai J. (2016). Microbial community structure of different electrode materials in constructed wetland incorporating microbial fuel cell. *Bioresource Technology*, **221**, 697–702, https://doi.org/10.1016/j.biortech.2016.09.116

Wang J. F., Song X. S., Wang Y. H., Zhao Z. M., Wang B. D. and Yan D. H. (2017). Effects of electrode material and substrate concentration on the bioenergy output and wastewater treatment in air-cathode microbial fuel cell integrating with constructed wetland. *Ecological Engineering*, **99**, 191–198, https://doi.org/10.1016/j.ecoleng.2016.11.015

Wang H. X., Wang X. O., Wang M. Y., Zhang C. P., Li J. Y., Xue M. and … Xie H. J. (2023). Degradation and transformation of linear alkyl-benzene sulfonates (LAS) in integrated constructed wetland-microbial fuel cell systems. *Chemosphere*, **321**, 138135, https://doi.org/10.1016/j.chemosphere.2023.138135

Wu S., Kuschk P., Brix H., Vymazal J. and Dong R. (2014). Development of constructed wetlands in performance intensifications for wastewater treatment: a nitrogen and organic matter targeted review. *Water Research*, **57**, 40–55, https://doi.org/10.1016/j.watres.2014.03.020

Wu H., Zhang J., Ngo H. H., Guo W., Hu Z., Liang S. and … Liu H. (2015). A review on the sustainability of constructed wetlands for wastewater treatment: design and operation. *Bioresource Technology*, **175**, 594–601, https://doi.org/10.1016/j.biortech.2014.10.068

Xu D., Xiao E. R., Xu P., Zhou Y., Zhou Q. H., Xu D. and Wu Z. B. (2017a). Effects of influent organic loading rates and electrode locations on the electrogenesis capacity of constructed wetland-microbial fuel cell systems. *Environmental Progress & Sustainable Energy*, **36**(2), 435–441, https://doi.org/10.1002/ep.12481

Xu L., Zhao Y., Wang T., Liu R. and Gao F. (2017b). Energy capture and nutrients removal enhancement through a stacked constructed wetland incorporated with microbial fuel cell. *Water Science & Technology*, **76**(1–2), 28–34, https://doi.org/10.2166/wst.2017.168

Yadav A. K., Dash P., Mohanty A., Abbassi R. and Mishra B. K. (2012). Performance assessment of innovative constructed wetland-microbial fuel cell for electricity production and dye removal. *Ecological Engineering*, **47**, 126–131, https://doi.org/10.1016/j.ecoleng.2012.06.029

Yadav A. K., Srivastava P., Kumar N., Abbassi R. and Mishra B. K. (2018). Constructed wetland-microbial fuel cell: an emerging integrated technology for potential industrial wastewater treatment and bio-electricity generation. In: Constructed Wetlands for Industrial Wastewater Treatment, S. Alexandros (ed.), John Wiley & Sons, Schrobenhausen, Germany, pp. 493–510.

Yadav M., Yadav R. K. and Gole V. L. (2021). Sugar industry wastewater treatment: current practices and advances. In: Microbial Ecology of Wastewater Treatment Plants, M. Shah and S. Rodriguez-Couto (eds), Amsterdam, Netherlands, Elsevier, pp. 151–174.

Yang H., Chen J., Yu L., Li W., Huang X., Qin Q. and Zhu S. (2022). Performance optimization and microbial community evaluation for domestic wastewater treatment in a constructed wetland-microbial fuel cell. *Environmental Research*, **212**(Pt B), 113249, https://doi.org/10.1016/j.envres.2022.113249

Yaseen D. A. and Scholz M. (2019). Textile dye wastewater characteristics and constituents of synthetic effluents: a critical review. *International Journal of Environmental Science and Technology*, **16**(2), 1193–1226, https://doi.org/10.1007/s13762-018-2130-z

Zhang X. L., Liu Y. F., Zheng L. S., Zhang Q. C. and Li C. J. (2021a). Simultaneous degradation of high concentration of citric acid coupled with electricity generation in dual-chamber microbial fuel cell. *Biochemical Engineering Journal*, **173**, 108095.

Zhang K., Wu X., Wang W., Chen J., Chen J. and Luo H. (2021b). Roles of external circuit and rhizosphere location in CH_4 emission control in sequencing batch flow constructed wetland-microbial fuel cell. *Journal of Environmental Chemical Engineering*, **9**(6), 106583, https://doi.org/10.1016/j.jece.2021.106583

Zhang L., Liu Y., Lv S., Wang R., Wang Y., Lin K. and … Liu L. (2023). An overview on constructed wetland-microbial fuel cell: greenhouse gases emissions and extracellular electron transfer. *Journal of Environmental Chemical Engineering*, **11**, 109551, https://doi.org/10.1016/j.jece.2023.109551

Zhao C. Q. and Chen W. Y. (2019). A review for tannery wastewater treatment: some thoughts under stricter discharge requirements. *Environmental Science and Pollution Research*, **26**(25), 26102–26111, https://doi.org/10.1007/s11356-019-05699-6

Zhao Y. Q., Collum S., Phelan M., Goodbody T., Doherty L. and Hu Y. S. (2013). Preliminary investigation of constructed wetland incorporating microbial fuel cell: batch and continuous flow trials. *Chemical Engineering Journal*, **229**, 364–370, https://doi.org/10.1016/j.cej.2013.06.023

doi: 10.2166/9781789063813_0207

Chapter 12

Application of microbial fuel cell towards bioremediation and clean energy production for industrial wastewater

Riti Thapar Kapoor[1]* *and Mohd Rafatullah*[2,3]

[1]*Centre for Plant and Environmental Biotechnology, Amity Institute of Biotechnology, Amity University Uttar Pradesh, Noida 201 313, Uttar Pradesh, India*
[2]*Environmental Technology Division, School of Industrial Technology, Universiti Sains Malaysia, Penang 11800, Malaysia*
[3]*Renewable Biomass Transformation Cluster, School of Industrial Technology, Universiti Sains Malaysia, Penang 11800, Malaysia*
**Corresponding author: rkapoor@amity.edu*

ABSTRACT

A huge amount of wastewater released from industries flow into different water resources such as rivers. Industrial effluents can be regarded as an important resource for water, nutrients and energy. Microbial fuel cell (MFC), a green eco-friendly approach can be applied for the treatment of wastewater with electricity generation concomitantly. It is a novel strategy to generate clean, renewable, safe green energy to maintain a clean environment. MFC technology can be used for effluent treatment, biochemical oxygen demand and chemical oxygen demand elimination, sulphate and removal of toxic metal and denitrification. MFC has an advantage as compared with other wastewater treatment methods because of certain unique properties such as energy and economic benefits, less effect on the environment and high stability. However, the operation of MFCs also has multiple setbacks such as short life span, high cost, membrane fouling and so on. MFC technology shows a pivotal function in solving problems of energy crisis and waste management. This chapter describes current applications of MFC technology for the treatment of industrial effluents with cost-effective energy generation and covers the gap by highlighting key future research areas to improve its performance.

Keywords: bioremediation, bioenergy, microbial fuel cell, wastewater

12.1 INTRODUCTION

Water is a precious source on earth for life and human development. The increase in population, industrialization, anthropogenic activities and urbanization has increased the demand for water. The industrial sector is a major source of global pollution as it releases large quantities of chemicals, dyes and heavy metals in its effluent without any prior treatment and shows adverse effects on the aquatic ecosystem (Mistry *et al.*, 2023). Toxic metals and dyes are the major contaminants of industrial wastewater which reflect severe health hazards on plants, animals and human beings (Wang *et al.*, 2022). Dyes are coloured aromatic compounds, resistant to degradation and highly visible even when present at very low concentrations. During textile processing, 80–85% dye attaches to the fibre, the remaining 15–20% is washed and discarded as effluents (Zafar *et al.*, 2022). Dyes show negative effects on soil microbial communities and prevent seed germination and development of plants (Lellis *et al.*, 2019). In addition, heavy metals are persistent, non-degradable and accumulate in tissues of living organisms (Gu *et al.*, 2022). Dyes and heavy metals gain entry in animals and human beings via the food chain and have chronic toxic effects such as allergy, respiratory disorders, dysfunction of kidneys and the immune system, ailments of the nervous system and cancer (Gazwi *et al.*, 2020). The demand of water and energy is continuously rising for domestic and industrial purposes. Hence, treatment and reutilization of wastewater, recovery of resources and energy have become important to increase water availability (Mohammed & Ismail, 2018). Several physico-chemical procedures such as precipitation, coagulation, advanced oxidation processes, adsorption, reverse osmosis, ozonation, ultra-filtration, and biological methods like bioremediation, phytoremediation and anammox have been developed for treatment of polluted water (Kapoor *et al.*, 2022). The above-mentioned procedures are complex, energy and time consuming, need chemicals and sophisticated equipments and produces sludge, thus not suitable for large-scale applications (Katheresan *et al.*, 2018).

It has been estimated that the demand for electricity will increase up to 90% by 2040 worldwide (IEA, 2019). Major energy requirements have been catered by fossil fuels but combustion of fossil fuel generates carbon dioxide and other greenhouse gases which have reflected adverse repercussions on the ecosystem and disturbed ecological balance. It has been reported that various energy resources were responsible for around 40% of CO_2 emissions in 2019 at the global level (Kurniawan *et al.*, 2022). Therefore, attempts to develop renewable energy resources that are cheap, eco-friendly and which can reduce carbon footprint have become a prime necessity (Gielen *et al.*, 2019). United Nations Sustainable Development Goals have also emphasized affordable clean water and energy for every citizen. Therefore, carbon neutrality for economic growth can be achieved via application of renewable energy resources. Thus, application of renewable energy resources plays a significant role in handling climate change with net-zero emission.

Microbial fuel cell (MFC) is a sustainable technique which transforms substrate energy into electricity via catalytic activity of electroactive microbes. The transformation of wastes by using MFC for energy generation is clean, reliable, efficient, cost-effective, renewable and environmental-benign strategy which does not produce any toxic by-products (Palanisamy *et al.*, 2019; Yaqoob *et al.*, 2020). MFC was first developed by Professor M.C. Potter of University of Durham in 1911 by using microbes for degradation of pollutants and energy generation. MFC is a bioelectrochemical tool that utilizes microorganisms as biocatalysts and converts chemical energy of organic or inorganic complexes into electricity (Kumar *et al.*, 2016). Organic substances present in the anodic chamber get oxidized by microbes in MFC. The electrons and protons generated in an anodic chamber enter the cathodic chamber through circuit and proton-exchange membrane (PEM), respectively, reduces the electron acceptor in the cathode chamber (Penteado *et al.*, 2017). Due to the unique features of MFC, it is more advantageous as compared to other available technologies. MFC degrades contaminants via anaerobic oxidation by using microbes such as bacteria, algae and so on and generate electrons in the process. These electrons passing through the external circuit, allows electricity production, hence MFC reduces contamination of water and generates green energy. MFCs can be fabricated in different designs as they do not require supply of an external electricity and can remediate organic, organometallic, inorganic contaminants with their high removal rate (Lim *et al.*, 2021). MFC can also be used for treatment of polluted soil and toxic gases (Abbas & Rafatullah, 2021). MFC-based hybrid reactors use algae and plants for removal of pollutants and valuable products from wastes can also be obtained by algae-based MFC reactors (Arun *et al.*, 2020). In comparison to aerobic treatment procedure, MFC can work efficiently under varying temperature conditions, pH with different electron acceptors and consumes low energy with reduced sludge production. MFC is better as compared to anaerobic digestion technology due to its operation flexibility at low temperature and less substrate concentration. Hence, the objective of this chapter is to highlight the advancements in a field of MFC technology for wastewater treatment with energy production and need of further investigations for improvement of its performance.

12.2 COMPONENTS OF MFC AND ITS ROLE IN ELECTRICITY GENERATION

MFC contains anode, cathode and ion-exchange membrane (Figure 12.1). The function of MFC depends on MFC design, type of membrane, electrode material, microbes, substrate, type of electrolyte and operational conditions. The electrodes of MFC are made up of graphite rods or plates or granules, carbon cloth and brush and stainless steel. Graphene oxide has been found to be a dynamic material for electrodes due to their large surface area, more conductivity with thermal and mechanical durability (Sun *et al.*, 2019).

The electrode material with high conductivity, mechanical, chemical and thermal solidity, more surface area, less resistance and high biocompatibility can be used in MFC (Liu *et al.*, 2019). Electrodes can be engineered with conductive nanomaterials to increase their surface area to improve MFC performance. The nano-modification enhances formation of biofilm at the anode, decreases start-up time and promotes transfer of electrons, reduces internal resistance and enhances the functioning of MFC. The analyte present in the anode chamber should be chemically inert. The catalyst found on the cathode acts as an electron acceptor. The use of electron acceptors may not only increase power generation but reduce the operating cost and expand the application scope of MFCs. A catalyst (platinum) enhances the oxygen reduction rate and can be utilized as a cathode electrode. Because of more redox potential with wide availability and a cheap source, oxygen can be used as an electron acceptor. Ferricyanide has been applied as an electron acceptor in MFC. However, the application of ferricyanide has some limitations as it can enter the anode chamber via ion-exchange membrane and reduce the function of the MFC. A

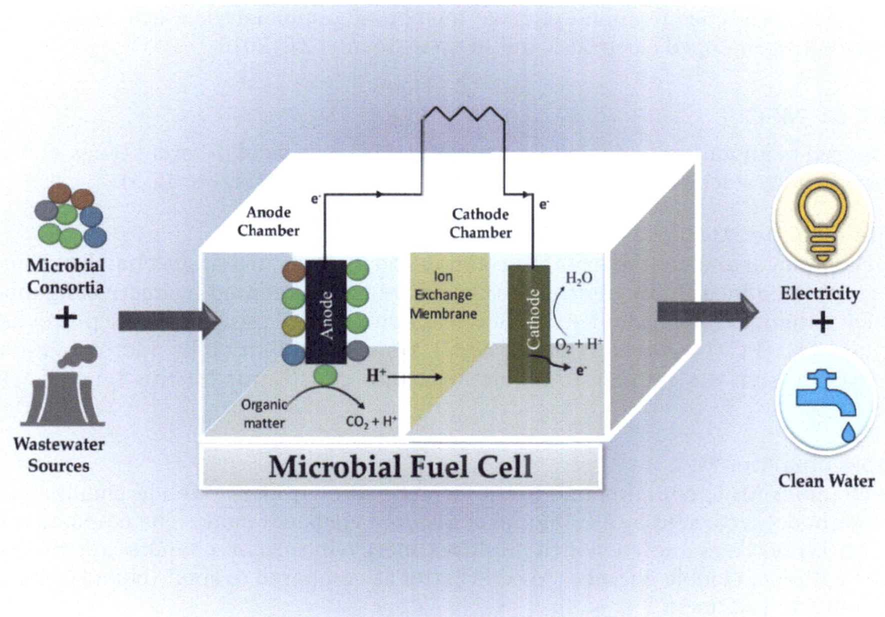

Figure 12.1 Design of MFC (Reprinted from Malik *et al.*, 2023).

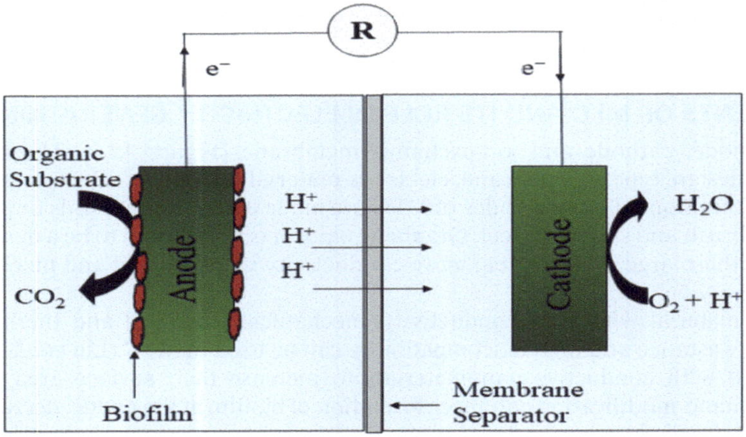

Figure 12.2 Components of MFC (Reprinted from Patwardhan *et al.*, 2021).

PEM is present in between the anode and cathode chambers through which protons migrate from the anode to the cathode in MFC. The Nafion, Ultrex CMI-7000 is mostly used as PEM. Bacteria transfer electron to the anode from where it collects at the cathode chamber by external circuit (Figure 12.2). However, proton moves directly from the anode to the cathode chamber. MFC chambers will be joined through multimeter and resistor box, for measurement of power and voltage. Such substrates which can be oxidized into electrons can provide high power output in MFC. Recently, modifications in MFC have been done to develop novel designs of MFC to decrease system resistance, affordable electrode with large surface area, cheap cation-exchange membrane, nanomaterials use in electrode, and the development of nitrogen-doped electrodes and so on (Kumar *et al.*, 2016).

12.3 TYPES OF MFC

MFC is composed of anode, cathode and electrolytes. MFC may be of different types such as single or dual chamber, upflow, stacked, multi-electrode and flat-plate MFC (Figure 12.3).

12.3.1 Single-chamber MFC

It is simple, economical and contains both the anode and cathode in a single chamber. The anode and cathode are separated by a PEM. Internal resistance can be decreased by decreasing interelectrode spacing, which enhances the power density. Single-chamber MFC produces more power as compared to double-chamber MFC (Prasad & Tripathi, 2022). However, pollution by microbes and the reverse channel of oxygen from the cathode to anode are major constraints for this type of MFC (Kumar *et al.*, 2017).

12.3.2 Double-chamber MFC

The double-chamber MFC contains one bottle or cube known as the anode chamber whereas the other one as cathode, separated by a PEM and conducted via batch mode. The medium present in the anode chamber is known as anolyte which produces energy. In the two-chamber air cathode MFC, air is given in the cathode. Double-chamber MFC is better as compared to single chamber for wastewater or industrial effluent treatment.

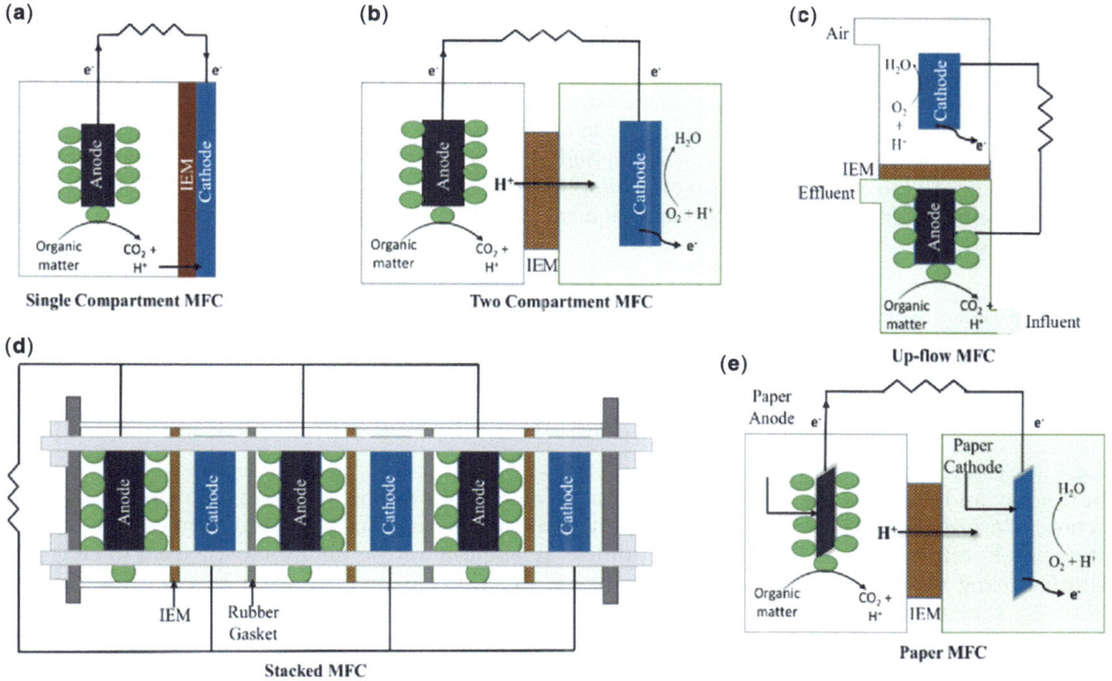

Figure 12.3 Types of MFCs (Reprinted from Malik *et al.*, 2023).

12.3.3 Upflow MFC

Upflow MFC is cylinder shaped in which the anode chamber is kept at the bottom and the cathode chamber at the top and these are separated by glass bead layers. The substrate moves from the anode to cathode and a gradient is generated between the electrodes which helps in MFC function. The proton transmission-related problems have not been observed as there is no separate anolyte and catholyte in this MFC. Upflow mode MFC is used for wastewater treatment instead of power generation and can be enhanced as compared to other designs of MFCs. The main drawback of this process is high energy cost to pump substrate as compared to power generation (Zhou *et al.*, 2013).

12.3.4 Stacked MFC

When MFCs are joined in series or parallel to increase energy production is known as stacked MFC. However, there will be voltage loss after connection of cells either in series or parallel and voltage will not be the same as sum of voltage of each cell. More current production has been reported in parallel connected stacked MFCs in comparison to fuel cells that are assembled in a series connection. The parallel stacked MFC showed high rate of bioelectrochemical reaction rate and significant COD removal with increased wastewater treatment efficiency. Major problem of stacked MFC is to obtain high-voltage output as due to voltage reversal there may be substrate reduction in the cell, which reduces bacterial capacity for generation of high voltage.

12.3.5 Other designs of MFC

Multi-electrode MFC is composed of four anode and cathode, respectively, which are connected in parallel mode. The removal efficiency of organic matter and nitrogen compounds in MFC can be enhanced by increasing interface between microbes and electrodes (surface area, volume and number of electrodes). Kim *et al.* (2020) reported multiple electrode MFC remove the pollutants from wastewater with higher efficiency as compared to an MFC with a single inserted electrode. The flat-plate MFC is designed to decrease ohmic resistance which may be because of spacing between the electrodes. In this MFC, the anode and cathode are composed of flat plates and Nafion membrane, which is kept between the two plates. It is applied in chemical fuel cells which generates significant energy as compared to other designs. Arun *et al.* (2020) reported that microalgal-based photosynthetic MFCs can be used for generation of biofuel, sequestration of carbon dioxide, and recovery of valuable end-products.

12.4 FACTORS AFFECTING PERFORMANCE OF MFC

Function of MFC depends on different variables like design and configuration, type of wastewater, material and surface of electrode, anodic potential, operational parameters such as temperature, pH, time and external resistance, substrate concentration, microbial diversity and so on.

12.4.1 Materials for electrodes

The anode and cathode should be conductive, non-corrosive, cost-effective and non-fouling for bacteria. Carbon paper or cloth or mesh, graphite plate, granular graphite or activated carbon, carbon felt or brush, reticulated vitrified carbon, stainless-steel mesh and so on are mostly used in electrode manufacturing (Yaqoob *et al.*, 2020). Performance of MFC can be increased by choosing suitable electrode. Electrode modification with nanomaterial or catalyst in the anode support formation of biofilm, thus boost electron transfer mechanism for increased power output. The platinum-coated cathode can produce high power output in comparison to simple cathode but because of more cost it is not feasible for application at industrial scale.

12.4.2 Effect of substrate and inoculum

The substrate influences the activity of microbes present in the biofilm of anode and affects performance of MFC and electricity production (Malik *et al.*, 2023). The most frequently used substrates are glucose, acetate, butyrate, lignocellulosic biomass, landfill leachates, starch processing wastewater, inorganic substrate and dye wastewater (Obileke *et al.*, 2021). The substrate provides nutrients and energy to microbes. The exoelectrogenic bacteria such as *Shewanella putrefaciens, Clostridium butyricum, Rhodoferax ferrireducens, Geobacter metallireducens, Pseudomonas aeruginosa, Escherichia coli, Desulfovibrio desulfuricans, Shewanella oneidensis* are capable of transferring electrons to the anode using mediators, nanowires or direct contact with the electrodes (Chaudhuri & Lovley, 2003). Non-exoelectrogenic bacteria use mediators produced via exoelectrogenic bacteria and transfer electrons to the electrode.

The inoculum contributes significant role in the performance of MFC. The presence of mixed microbial communities in MFC have gained considerable attention due to their adaptability and stability. The mixed electrogenic strain inocula showed high current yield along with concurrent substrate removal efficiency (Mathuriya, 2013).

12.4.3 Proton exchange membrane

PEM is present in between the anode and cathode chamber. The proton membrane contains charged side walls with pores, which assists in proton movement from the anode to cathode. Due to the diffusion of anolyte to catholyte via the membrane, there might be fouling of the membrane, which restricts protons movement to cathode and reduces power output in MFC. Raghavulu *et al.* (2013) observed application of Nafion membrane as PEM in MFC because this membrane is selectively permeable and regulates charges between anolyte and catholyte. MFCs with PEM reduced internal resistance and generates more energy. However high cost of PEM is one of the major drawbacks, hence in future membrane-less MFCs can be fabricated for large-scale application.

12.4.4 Impact of pH

During long duration of MFC operation, proton concentration enhances in the anolyte because of slow flow of protons via PEM which reduces anode chamber pH and restricts growth of microbes. However, rise in pH in the cathode chamber may decrease power output. Less pH is beneficial for reduction of oxygen but high power output can be produced from MFC (Kumar *et al.*, 2016). Earlier studies have shown that pH 6–9 is suitable for microbial growth and for obtaining high current output (Obileke *et al.*, 2021).

12.4.5 Effect of temperature

Temperature plays a pivotal role in power production and treatment of wastewater in MFC (Tang *et al.*, 2015). The optimum temperature for MFCs is 25–30°C (Malekmohammadi & Mirbagheri, 2021). Conductivity of anolyte and catholyte and power density can be enhanced with the rise in temperature but it decreases ohmic resistance of MFC. High temperature reduces start-up time of MFC and helps in the formation of stable biofilm. Yong *et al.* (2014) found that temperature range between 30°C and 45°C was useful for achieving high electricity production by MFC due to high catalytic activity of bacterial biofilm.

12.4.6 Feed rate and shear stress

MFC can be conducted in batch and continuous mode. Substrate is given with the initiation of cycle whereas substrate is given with regular interval in continuous mode. MFC performance depends on the concentration of feed solution. The concentration of carbon source and rate at which microbes use the substrates dictate power generation. Biofilm formation occurs on the surface of the anode which enhances current production. The reduced feed rate will decline power generation in MFC (Choudhury *et al.*, 2020). Reports revealed that less shear rate led to thick biofilm formation and dense biofilm reflected more stable attachment of bacteria at the anode (Logan *et al.*, 2005). Raghavulu *et al.* (2013) observed that high shear rate declines diversity of microbes in MFC and showed formation of homogeneous biofilm.

12.4.7 Flow rate and hydraulic retention time

Flow rate and hydraulic retention time play a pivotal role for achieving more power generation from MFC. The studies suggested that high flow rate reduced hydraulic retention time, power output, elimination of chemical oxygen demand and coulombic efficiencies (Kumar *et al.*, 2017). Removal efficiency of pollutants such as nitrogen, nitrate, ammonium and pharmaceutical products was enhanced with time (Chang *et al.*, 2018). High HRT shows more diversity of microbes (HaiLiang *et al.*, 2018). With increase in flow rate, time reduces which may lead to rise in dilution rate and reduction in electricity production (You *et al.*, 2018). Hydraulic retention time is associated with contaminants elimination capacity but inversely with power output.

12.4.8 External and internal resistance

The voltage reduces by reducing external resistance, but power output enhances. Different operational parameters such as pH, temperature and properties of wastewater can modify internal resistance. For commercialization of MFC, external resistance is a significant parameter. The coulombic efficiency and current output can be enhanced if external resistance is optimum. Maximum power density can be achieved if internal and external resistances are equivalent. By reducing external resistance, contaminants and COD elimination efficiency and electricity production were enhanced in MFC (Buitron *et al.*, 2017). Rapid microbial growth has been observed at high resistance; thus, it is recommended that MFC should start with more resistance after this low resistance can be maintained to obtain more power output (Suzuki *et al.*, 2018).

The distance between electrodes affects electricity generation process. During scaling up, if electrodes size is enhanced but distance between electrodes is less, then MFC becomes heavy and power output gets reduced (Madondo *et al.*, 2023). Further investigations are needed to investigate interactive impacts of various variables for design of experiments. Different fractional and full factorial design, response surface methodology can be utilized to reduce number of experiments. Boudaghpour and Malekmohammadi (2020) stated that ANOVA and neural networks can be used for analysis of results. The analyses of the interactive effects of each parameter will assist in the development of MFC design for experiments with more accuracy.

12.5 APPLICATIONS OF MFC

MFCs have shown potential for industrial and domestic wastewater treatment, electricity generation, production of biosensors and hydrogen. Application of wastewater as a substrate has many advantages like it generates cost-effective electricity and also leads to treatment of wastewater (Figure 12.4). MFC can remove 98% chemical oxygen demand from effluents. Reports revealed that MFC can significantly remove heavy metals, organic pollutants, dye, sulphides and can be applied for heavy metals and nutrients recovery from industrial effluent (Singh & Kaushik, 2021). However, highly toxic wastewater cannot be remediated completely in MFC, but they can decrease COD of wastewater to meet the regulatory guidelines prior its release into the surroundings.

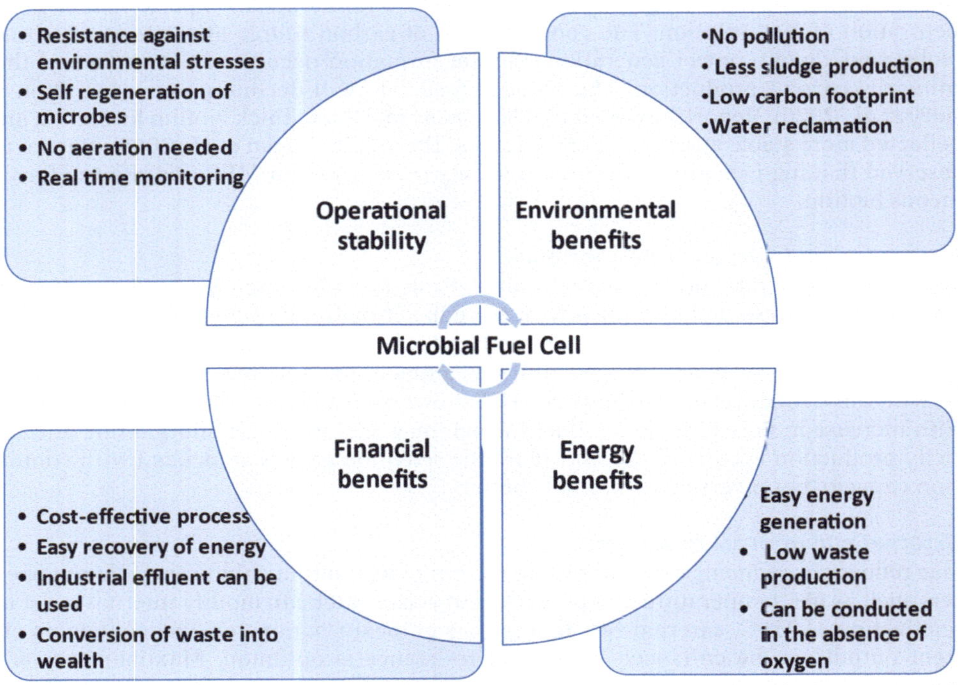

Figure 12.4 Functions of MFC.

Table 12.1 Removal of heavy metals with energy output by MFC.

Metal Pollutants	Type of MFC	Microbes/ Inoculum	Initial Concentration	Operating Time (h)	Temperature (°C)	pH	Removal Efficiency (%)	Power Density (mW/m^2)	References
Ag	DMFC	Sludge mixture	50–200 mg/L	8	25	7	99	4250	Choi and Cui (2012)
Cr (VI)	DMFC	Anaerobic sludge	100 mg/L	240	30	6	75	970 ± 61	Zhang *et al.* (2012)
Au (III)	DMFC	*Tetrachloroaurate* wastewater	2000 mg/L	12	25	2.8	99	6580	Choi and Hu (2013)
Ag+	Tubular MFC	Anaerobic sludge	1000 mg/L	21	26	9.2	99	300	Nancharaiah *et al.* (2015)
Se	SMFC	Anaerobic sludge	75 mg/L	48	25	7	99	2900	Nancharaiah *et al.* (2015)
Cu (II)	Tubular MFC	Anaerobic sludge	200 mg/L	264	25	4.7	96	399	Nancharaiah *et al.* (2015)
Cd (II)	DMFC	Contaminated soil	100 mg/L	3432	25	6.8	31	7500	Habibul *et al.* (2016)
Pb (II)	DMFC	Contaminated soil	900 mg/L	2592	25	6.9	44	3600	Habibul *et al.* (2016)
Cr (VI) and Cu (II)	Sedimental MFC	Sediment sample	250 mg/L	2160	37	2	96	400–450	Abbas *et al.* (2016)
Cr (VI) and Cu (II)	Sedimental MFC	Sediment sample	250 mg/L	2160	37	2	96	400–450	Abbas *et al.* (2016)
V (V)	DMFC	*Klebsiella* and *Dysgonomonas*	200 mg/L	168	22	–	61	529±12	Qiu *et al.* (2017)
Cr (VI)	DMFC	Primary clarifier effluent	20 mg/L	2	22	7	76	970	Kumar *et al.* (2018)
Cu^{2+}	SMFC	Anaerobic sludge bed	13 mg/L	5	35	6	98	200	Wu *et al.* (2018)
Cr (VI)	DMFC	Anaerobic pure culture	385 μM	24	30	6	73	14	Huang *et al.* (2018)
Cd (II)	DMFC	Anaerobic pure culture	179 μM	24	30	6	61	14	Huang *et al.* (2018)
Cd (II)	DMFC	Mixed microbial culture	50 mg/mL	168	25	7.1	60	700–750	Gai *et al.* (2018)
Hg (II)	DMFC	Mixed microbial culture	25 mg/mL	360	25	6.8	55	800	Gai *et al.* (2018)
Cr (VI)	DMFC	*Klebsiella pneumonia*	10 mg/mL	3.5	30	2	99	52	Li *et al.* (2019)
Pt	DMFC	Anaerobic sludge bed	17 mg/mL	24	25	7	90	844	Liu *et al.* (2019)

MFC = microbial fuel cell, SMFC = single-chamber microbial fuel cell, DMFC = double-chamber microbial fuel cell.

12.5.1 Removal of metals through MFC

Toxic metals are released into the environment from industrial and domestic wastewater which show adverse effect on soil and aquatic ecosystem due to their non-biodegradable nature (Yaqoob & Ibrahim, 2019). Reduction of soluble heavy metals from its toxic state to non-hazardous insoluble form is carried out by MFCs with the help of electroactive bacteria (Aarthy *et al.*, 2020). Table 12.1 reflects removal of heavy metals with energy generation via MFC. Utilization of MFC for sodium selenite wastewater treatment contained sodium acetate or glucose as carbon resource (Catal *et al.*, 2009). Mixed bacterial culture was used in MFC and 2900 mW/m^2 power density was observed. The increase in selenite concentration (50 mg/L) showed 13–17% reduction in voltage; however, with high selenite concentration (75 mg/L), power and voltage were decreased to 2200 mW/m^2 and 0.41 V, respectively. Cassava wastewater rich in organic content (16,000 mg/L) used for generation of electricity in MFC (Kaewkannetra *et al.*, 2009). Results reflected 88% COD elimination efficiency at 120 h with 1771 mW/m^2 current output.

Bakhshian *et al.* (2011) used molasses as substrate for reactive blue 221 dye (113 mg/L) decolorization by laccase in double-chamber MFC and they reported 28 mW/m^2 power density. The sulphide (60 mg/L) and nitrate (11 mg/L) were removed by using activated sludge in double-chamber MFC. The MFC current output was 138 mA/m^2 and nitrogen and sulphate were obtained at the end (Cai & Zheng, 2013).

Chaturvedi and Verma (2014) used chicken feathers for power generation by *Pseudomonas aeruginosa* in MFC. Highest voltage 141 mV was reported after 14 days of incubation period with current density of 8.6 mA/m^2 with power output of 1206.78 mW/m^2. Xafenias *et al.* (2013) reported 67% chromium (200 mg/L initial concentration) elimination efficiency and 32.5 mA/m^2 current output by substrate lactate and *Shewanella oneidensis* after 192 h. For removal of copper metal from waste effluent, graphite material was used in the anode and cathode (Nancharaiah *et al.*, 2015). They observed 99% elimination capacity of copper (1 mg/L) at 144 h under low pH. With increase in initial concentration of copper was enhanced to 200–600 mg/L, copper removal efficiency was reduced and power density was increased at low initial concentration.

Decrease in hexavalent chromium concentration (100 mg/L) was observed in anaerobic sludge within 48 h of incubation period and recorded 767 mW/m^2 power density (Xafenias *et al.*, 2015). Cucu *et al.* (2016) used anode prepared from cow manure and fruit waste whereas cathode was made up of soil and cow manure for denitrification in MFC. Maximum current and power density were 190 $\pm$ 9.1 mA/m^2 and 31.92 $\pm$ 4 mW/m^2, respectively.

Qiu *et al.* (2017) completely removed vanadium (200 mg/L concentration) from polluted water within 7 days and 529 $\pm$ 12 mW/m^2 power density was achieved. The carbon fibre felt was used as a biocathode for removal of vanadium with power generation in MFC. Huang *et al.* (2018) examined reduction of cadmium and chromium by applying graphite felt as cathode and anode carbon rod in dual-chamber MFC. Chromium and cadmium showed 73 and 61% elimination efficiency, respectively, under anaerobic culture and energy production was 14 mW/m^2 at acidic pH after one day in MFC.

Wang *et al.* (2018) studied the removal efficiency of thallium by MFC. They reported that 67% thallium elimination ability with its initial concentration (100 µg/L) after three days, reflected approximately 458 $\pm$ 15 mW/m^2 energy output. Liu *et al.* (2019) applied graphite as electrode for effluent treatment containing nickel, cadmium and mercury. After treatment of nickel for 30 days 150–200 mW/m^2 energy was obtained whereas cadmium and mercury treated for 7 and 15 days observed 60% and 55% removal with 700–750 and 800 mW/m^2 power output, respectively. Li *et al.* (2019) found chromium (100 mg/L) elimination from wastewater via MFC and converted hexavalent to trivalent chromium at pH 2 and elimination of hexavalent chromium was enhanced up to 99% if potassium dihydrogen phosphate was supplemented and generated current of 52 mW/cm^2. Liu *et al.* (2019) reported 90% removal of platinum (17 mg/L concentration) with MFC with power output of 844 mW/m^2.

Generation of energy and recovery of metal depend on microbial growth and electron production. However, it can be improved by incorporating electrodes with large surface area for better bacterial growth and electrons production. Electrodes generated from carbon are mostly used in MFC but they have shown some disadvantages like less surface area and electrical conductivity for colonization of microbes. Recently, graphene has been widely used due to its large surface area for microbial growth, better conductivity, mechanical and thermal durability in comparison to another material.

12.5.2 Removal of dyes through MFC

Dye containing wastewater is released from different textile, paper and pulp and pharmaceutical industries are major threats for the entire ecosystem. Due to the recalcitrant nature, dyes remain fixed during washing, and showed defiance to microbial deterioration. Yadav *et al.* (2012) designed wetland MFC and reported 76%, 81%, 69% and 93% removal of methylene blue dye after 96 h with methylene blue concentrations like 2000, 1500, 1000, 500 mg/L. Wetland MFC showed 75% COD removal capacity with 1500 mg/L dye and it reflected the highest energy output of 16 mW/m^2 and current density of 70 mA/m^2. Sun *et al.* (2013) explored degradation of Congo Red dye by applying single-chamber MFC as they used glucose and Congo Red (300 mg/L concentration) mixture as fuel in single-chamber MFC. *Methylobacterium*, *Azospirillum*, *Rhodobacter*, *Trichococcus*, *Desulfovibrio* and *Bacteroides* were applied for Congo Red dye removal. They used anaerobic sludge with graphite felt and carbon paper as anode and cathode and observed 72 mW/m^2 power output. Graphene anode in MFC for electricity generation and degradation of methyl orange dye was used by Guo *et al.* (2014). Anaerobic sludge was taken as inoculum with methyl orange dye and they reported 368 mW/m^2 energy production. *Proteus hauseri* were grown on carbon cloth surface electrode of MFC and they reflected 50% removal efficiency of thionine-based textile dye from wastewater and 83 mW/m^2 energy output was observed (Chen *et al.*, 2016). Oon *et al.* (2017) stated when mono and diazo dyes were utilized as electron acceptor, efficiency of COD elimination and degradation of azo dye were enhanced to 73% and 95%, respectively. It was observed that monoazo-based dyes decolourization rate was 50% high as compared to diazo-based dye with a high of 21 mW/m^2 power density with 120 mA/m^2 current density. Air-free single-chamber MFC by applying algal cells as biocathode and carbon fibre as anode was fabricated by Logrono *et al.* (2017). It showed power density of 123 ± 28 mW/m^3 and 42% dye removal efficiency after one month of treatment. Miran *et al.* (2018) stated 90% degradation of textile diazo dye through double-chamber MFC after 24 hours operation in the presence of *Deltaproteobacteria*, *Proteobacteria*, *Desulfovibrio* and energy outcome was 258 ± 10 mW/m^2. Single-chamber MFC application showed 98% degradation of Congo Red dye (300 mg/L) after 36 h of operation time with 103 mW/m^2 electricity production (Kumar *et al.*, 2019). Sulphide-mediated azo dye degradation was observed by Dai *et al.* (2020) with single-chamber MFC. Results reflected Congo Red dye removal and sulphide was 88% and 98% at pH 7 with highest energy output of 24 mW/m^2. Sonu *et al.* (2020) utilized stacked MFC for real textile wastewater treatment. They observed high 82% decolourization rate of textile dye and power generation efficiency was 39 mW/m^2 in series mode whereas in parallel stack arrangement reflected 75% dye decolourization efficiency with 0.47 mW/m^2 power density.

12.5.3 Removal of COD and nitrogen by MFC

The rise in chemical oxygen demand promotes power generation (He *et al.*, 2016). Al-Mamun *et al.* (2017) found that MFC can eliminate COD and nitrate concomitantly. Nguyen *et al.* (2016) stated that MFC can reduce nitrate during denitrification under anaerobic conditions of the anode chamber. Faraghi and Ebrahimi (2012) reported that nitrate is formed from nitrite at the anode chamber and liberates electrons prior to denitrification. Wang *et al.* (2020) found that less nitrite concentration can generate electricity but rise in its concentration checks the activities of bacteria.

12.5.4 Biosensor

The application of MFC technology as a biosensor for detection of pollutants in water resources can be done with treatment of effluent and power production (Zhou *et al.*, 2013). Linear association between MFC coulombic yield and strength of effluent promotes the use of MFC as a BOD sensor. MFC-based biosensor is not expensive and used for long duration without any maintenance. They do not need any transducer which is used in traditional biosensor. Hence, biosensors formed by MFC have been found reliable and stable.

12.5.5 Biohydrogen

To produce hydrogen, microbial electrolysis cell can be assembled with MFC (Kumar *et al.*, 2015). Microbial electrolysis cell (MEC) contains the anode and cathode chambers which are separated by ion-exchange membrane (Zhou *et al.*, 2013). An exoelectrogens metabolize substrate and generate electrons and protons in the anode chamber and protons transfer to the cathode. The electricity is generated at the cathode by MFC and hydrogen is generated from MEC, which can be reserved and utilized for power generation.

12.5.6 Energy generation

MFC is a self-sustainable and promising technique for wastewater treatment and electricity production, thus, can combat the problem of energy shortage. MFC is a bio-electrochemical device that converts chemical energy contained in organic substrates into electrical energy by the activities of microbes. The use of organic material such as wastewater in MFC makes it an eco-friendly device that offers a dual benefit of bioelectricity generation and waste management. The electrons released due to metabolic activities of microbes are captured to maintain a constant power density, without carbon dioxide emission in the environment. MFCs are eco-friendly because they show less CO_2 emission and can continuously generate electricity for long duration, if the fuel and oxidant are provided to the cell. Bacteria acts as biocatalyst for conversion of electrochemical energy and offers practical solution to dealing with energy shortage, resource depletion and environmental pollution, thus contributing to zero-waste paradigm with carbon neutrality. MFC offers a window to promote the deployment of decentralized and sustainable energy solution which can address problem of energy scarcity specifically for rural areas and small-scale industries and helps in achieving decarbonization goals.

12.6 BENEFITS AND CHALLENGES IN COMMERCIALIZATION OF MFC

MFC is emerging as a potential sustainable technology for simultaneous removal of contaminants and electricity production. Application of any technique depends on its commercialization if it is produced at large scale and used by people. In MFC, electricity is generated by application of industrial wastewater, hence its commercialization may offer many benefits like low-cost electricity generation from waste materials as they are easily available at zero cost round the year. This technology can be useful for developing and least developed nations like Africa where the basic infrastructure for production of energy is not available, thus by MFC, people can generate electricity in their homes. Despite several advantages, there are some drawbacks like high cost of the electrodes and membrane, inadequate power output which makes its application limited at large scale. The output of MFC depends on various factors such as configuration, substrate and its concentration, microbes used, material of electrode and membranes and so on.

Nylon is mostly used in MFC membrane which is expensive. The high power output of MFC has been reported with pure substrate whereas it was decreased with wastewater as microbes cannot metabolize contaminants in a proper manner. Limited surface area on the electrodes in MFC gives less space to microbes for attachment which is another disadvantage. The low power output of MFC can be increased by isolation of potential microbes or by genetically modified strains which can significantly transfer electrons to the anode. Many reports have confirmed that consortium of different bacteria have shown better performance as compared to pure culture. The absence of PEM can make the MFC more economical in future. Recently, many MFC hybrid systems have been developed which show effective degradation or mineralization of contaminants along with energy generation as compared to pristine MFC.

12.7 CONCLUSION

Application of MFC for wastewater treatment with electricity production is an important breakthrough without showing any adverse impact on the environment. MFC technology promotes the development of sustainable and decentralized energy solution which can solve the problem of shortage of electricity. MFC can fulfil increasing demand of clean water and electricity together. However, most of the experiments on MFC have been carried out at laboratory scale which lack studies on long-term stability. Hence, further extensive research studies are needed on MFC-based technique for real effluent treatment to achieve significant removal of contaminants with more power generation to reach its application at commercialization stage.

ACKNOWLEDGEMENTS

The authors extend their appreciation to Amity University Uttar Pradesh, Noida, India and Universiti Sains Malaysia, Malaysia for providing all necessary facilities and valuable support.

REFERENCES

Aarthy M., Rajesh T. and Thirunavoukkarasu M. (2020). Critical review on microbial fuel cells for concomitant reduction of hexavalent chromium and bioelectricity generation. *Journal of Chemical Technology and Biotechnology*, **95**, 1298–1307, https://doi.org/10.1002/jctb.6257

Abbas S. Z. and Rafatullah M. (2021). Recent advances in soil microbial fuel cells for soil contaminants remediation. *Chemosphere*, **272**, 129691, https://doi.org/10.1016/j.chemosphere.2021.129691

Abbas S. Z., Rafatullah M. and Ismail N. (2016). Removal of metals (chromium and copper) and power generation through sediment microbial fuel cell. *International Journal of Environmental and Technological Sciences*, **2**, 56–60.

Al-Mamun A., Baawain M. S., Egger F., Al-Muhtaseb A. H. and Ng H. Y. (2017). Optimization of a baffled-reactor microbial fuel cell using autotrophic denitrifying bio-cathode for removing nitrogen and recovering electrical energy. *Biochemical Engineering Journal*, **120**, 93–102, https://doi.org/10.1016/j.bej.2016.12.015

Arun S., Sinharoy A., Pakshirajan K. and Lens P. N. L. (2020). Algae based microbial fuel cells for wastewater treatment and recovery of value added products. *Renewable and Sustainable Energy Reviews*, **132**, 110041 https://doi.org/10.1016/j.rser.2020.110041

Bakhshian S., Kariminia H. R. and Roshandel R. (2011). Bioelectricity generation enhancement in a dual chamber microbial fuel cell under cathodic enzyme catalyzed dye decolorization. *Bioresource Technology*, **102**, 6761–6765, https://doi.org/10.1016/j.biortech.2011.03.060

Boudaghpour S. and Malekmohammadi S. (2020). Modeling prediction of dispersal of heavy metals in plain using neural network. *Journal of Applied Water Engineering and Research*, 8, 28–43, https://doi.org/10.1080/23249676.2020.1719219

Buitron G., Lopez-Prieto I., Zuniga I. T. and Vargas A. (2017). Reduction of start-up time in a microbial fuel cell through the variation of external resistance. *Energy Procedia*, **142**, 694–699, https://doi.org/10.1016/j.egypro.2017.12.114

Cai J. and Zheng P. (2013). Simultaneous anaerobic sulfide and nitrate removal in microbial fuel cell. *Bioresource Technology*, **128**, 760–764, https://doi.org/10.1016/j.biortech.2012.08.046

Catal T., Bermek H. and Liu H. (2009). Removal of selenite from wastewater using microbial fuel cells. *Biotechnology Letters*, **31**, 1211–1216, https://doi.org/10.1007/s10529-009-9990-8

Chang T. J., Chang Y. H., Chao W. L., Jane W. N. and Chang Y. T. (2018). Effect of hydraulic retention time on electricity generation using a solid plain-graphite plate microbial fuel cell anoxic/oxic process for treating pharmaceutical sewage. *Journal of Environmental Science and Health – Part A Toxic/Hazardous Substances and Environmental Engineering*, **53**, 1185–1197.

Chaturvedi V. and Verma P. (2014). Metabolism of chicken feathers and concomitant electricity generation by *Pseudomonas aeruginosa* by employing microbial fuel cell. *Journal of Waste Management*, 928618, https://doi.org/10.1155/2014/928618

Chaudhuri S. K. and Lovley D. R. (2003). Electricity generation by direct oxidation of glucose in mediatorless microbial fuel cells. *Nature Biotechnology*, **21**, 1229–1232, https://doi.org/10.1038/nbt867

Chen B. Y., Ma C. M., Han K., Yueh P. L., Qin L. J. and Hsueh C. C. (2016). Influence of textile dye and decolorized metabolites on microbial fuel cell-assisted bioremediation. *Bioresource Technology*, **200**, 1033–1038, https://doi.org/10.1016/j.biortech.2015.10.011

Choi C. and Cui Y. (2012). Recovery of silver from wastewater coupled with power generation using a microbial fuel cell. *Bioresource Technology*, **107**, 522–525, https://doi.org/10.1016/j.biortech.2011.12.058

Choi C. and Hu N. (2013). The modeling of gold recovery from tetrachloroaurate wastewater using a microbial fuel cell. *Bioresource Technology*, **133**, 589–598, https://doi.org/10.1016/j.biortech.2013.01.143

Choudhury P., Ray R. N., Bandyopadhyay T. K. and Bhunia B. (2020). Fed batch approach for stable generation of power from dairy wastewater using microbial fuel cell and its kinetic study. *Fuel*, **266**, 117073, https://doi.org/10.1016/j.fuel.2020.117073

Cucu A., Tiliakos A., Tanase I., Serban C. E., Stamatin I., Ciocanea A. and Nichita C. (2016). Microbial fuel cell for nitrate reduction. *Energy Procedia*, **85**, 156–161, https://doi.org/10.1016/j.egypro.2015.12.286

Dai Q., Zhang S., Liu H., Huang J. and Li L. (2020). Sulfide-mediated azo dye degradation and microbial community analysis in a single-chamber air cathode microbial fuel cell. *Bioelectrochemistry*, **131**, 107349, https://doi.org/10.1016/j.bioelechem.2019.107349

Faraghi N. and Ebrahimi S. (2012). Nitrite as a candidate substrate in microbial fuel cells. *Biotechnology Letters*, **34**, 1483–1486, https://doi.org/10.1007/s10529-012-0939-y

Gai R., Liu Y., Liu J., Yan C., Jiao Y., Cai L. and Zhang L. (2018). Behavior of copper, nickel, cadmium and mercury ions in anode chamber of microbial fuel cells. *International Journal of Electrochemical Science*, **13**, 3050–3062, https://doi.org/10.20964/2018.03.69

Gielen D., Boshell F., Saygin D., Bazilian M. D., Wagner N. and Gorini R. (2019). The role of renewable energy in the global energy transformation. *Energy Strategy Reviews*, **24**, 38–50, https://doi.org/10.1016/j.esr.2019.01.006

Gazwi H. S. S., Yassien E. E. and Hassan H. M. (2020). Mitigation of lead neurotoxicity by the ethanolic extract of *Laurus* leaf in rats. *Ecotoxicology and Environmental Safety*, **192**, 110297, https://doi.org/10.1016/j.ecoenv.2020.110297

Gu X., Lin C., Wang B., Wang J. and Ouyang W. (2022). A comprehensive assessment of anthropogenic impacts, contamination, and ecological risks of toxic elements in sediments of urban rivers: a case study in Qingdao, East China. *Environmental Advances*, **7**, 100143, https://doi.org/10.1016/j.envadv.2021.100143

Guo W., Cui Y., Song H. and Sun J. (2014). Layer-by-layer construction of graphene-based microbial fuel cell for improved power generation and methyl orange removal. *Bioprocess Biosystems Engineering*, **37**, 1749–1758, https://doi.org/10.1007/s00449-014-1148-y

Habibul N., Hu Y. and Sheng G. P. (2016). Microbial fuel cell driving electrokinetic remediation of toxic metal contaminated soils. *Journal of Hazardous Materials*, **318**, 9–14, https://doi.org/10.1016/j.jhazmat.2016.06.041

HaiLiang S., Hua L., Shuai Z., YuLi Y., LiMin Z., Han X. and XiaoLi Y. (2018). Fate of sulfadiazine and its corresponding resistance genes in up-flow microbial fuel cell coupled constructed wetlands: effects of circuit operation mode and hydraulic retention time. *Chemical Engineering Journal*, **350**, 920–929, https://doi.org/10.1016/j.cej.2018.06.035

He W., Wallack M. J., Kim K. Y., Zhang X., Yang W., Zhu X., Feng Y. and Logan B. E. (2016). The effect of flow modes and electrode combinations on the performance of a multiple module microbial fuel cell installed at wastewater treatment plant. *Water Research*, **105**, 351–360, https://doi.org/10.1016/j.watres.2016.09.008

Huang L. Z., Hou P., Quan X. and Logan B. E. (2018). Removal of binary Cr (VI) and Cd (II) from the catholyte of MFCs and determining their fate in EAB using fluorescence probes. *Bioelectrochemistry*, **122**, 61–68, https://doi.org/10.1016/j.bioelechem.2018.02.010

IEA. (2019). World Energy Outlook 2019. IEA, Paris, France, p. 2019. Available online: https://www.iea.org/reports/world-energy-outlook-2019 (accessed on 16 November 2019).

Kaewkannetra P., Imai T., Garcia-Garcia F. J. and Chiu T. Y. (2009). Cyanide removal from cassava mill wastewater using *Azotobactor vinelandii* TISTR 1094 with mixed microorganisms in activated sludge treatment system. *Journal of Hazardous Materials*, **172**, 224–228, https://doi.org/10.1016/j.jhazmat.2009.06.162

Kapoor R. T., Rafatullah M., Aljuwayid A. M., Habila M. A., Wabaidur S. M. and Alam M. (2022). Removal of patent blue dye using *Ananas comosus*-derived biochar: equilibrium, kinetics and phytotoxicity studies. *Separations*, **9**, 426, https://doi.org/10.3390/separations9120426

Katheresan V., Kansedo J. and Lau S. Y. (2018). Efficiency of various recent wastewater dye removal methods: a review. *Journal of Environmental Chemical Engineering*, **6**, 4676–4697, https://doi.org/10.1016/j.jece.2018.06.060

Kim T., An J., Jang J. K. and Chang I. S. (2020). Determination of optimum electrical connection mode for multi-electrode-embedded microbial fuel cells coupled with anaerobic digester for enhancement of swine wastewater treatment efficiency and energy recovery. *Bioresource Technology*, **297**, 122464, https://doi.org/10.1016/j.biortech.2019.122464

Kumar R., Singh L., Wahid Z. A. and Fadhil M. (2015). Exoelectrogens in microbial fuel cells towards bioelectricity generation: a review. *International Journal of Energy Research*, **39**, 1048–1067, https://doi.org/10.1002/er.3305

Kumar R., Singh L. and Wahid Z. A. (2016). Exoelectrogens: recent advances in molecular drivers involved in extracellular electron transfer and strategies used to improve it for microbial fuel cell applications. *Renewable and Sustainable Energy Reviews*, **56**, 1322–1336, https://doi.org/10.1016/j.rser.2015.12.029

Kumar R., Singh L. and Zularisam A. W. (2017). Microbial fuel cells: types and applications. In: Waste biomass management – a holistic approach, L. Singh and V. C. Kalia (eds.), ISBN: 978-3-319-49595-8, Springer, Cham. pp. 367–384, https://doi.org/10.1007/978-3-319-49595-8_16

Kumar R., Singh L., Zularisam A. and Hai F. I. (2018). Microbial fuel cell is emerging as a versatile technology: a review on its possible applications, challenges and strategies to improve the performances. *International Journal of Energy Research*, **42**, 369–394, https://doi.org/10.1002/er.3780

Kumar S. S., Kumar V., Malyan S. K., Sharma J., Mathimani T., Maskarenj M. S., Ghosh P. C. and Pugazhendhi A. (2019). Microbial fuel cells for bioelectrochemical treatment of different wastewater streams. *Fuel*, **254**, 115526, https://doi.org/10.1016/j.fuel.2019.05.109

Kurniawan T. A., Othman M. H. D., Liang X., Ayub M., Goh H. H., Kusworo T. D., Mohyuddin A. and Chew K. W. (2022). Microbial fuel cells: a potential game-changer in renewable energy development. *Sustainability*, **14**, 16847, https://doi.org/10.3390/su142416847

Lellis B., Fávaro-Polonio C. Z., Pamphile J. A. and Polonio J. C. (2019). Effects of textile dyes on health and the environment and bioremediation potential of living organisms. *Biotechnology Research and Innovation*, **3**(2), 275–290, https://doi.org/10.1016/j.biori.2019.09.001

Li F., An X., Wu D., Xu J., Chen Y., Li W., Cao Y., Guo X., Lin X., Li C., Liu S. and Song H. (2019). Engineering microbial consortia for high-performance cellulosic hydrolyzates-fed microbial fuel cells. *Frontiers in Microbiology*, **10**, 409, https://doi.org/10.3389/fmicb.2019.00409

Lim S. S., Fontmorin J. M., Pham H. T., Milner E., Abdul P. M., Scott K., Head I. and Yu E. H. (2021). Zinc removal and recovery from industrial wastewater with a microbial fuel cell: experimental investigation and theoretical prediction. *Science of the Total Environment*, **776**, 145934, https://doi.org/10.1016/j.scitotenv.2021.145934

Liu X., Wang S., Xu A., Zhang L., Liu H. and Ma L. Z. (2019). Biological synthesis of high-conductive pili in aerobic bacterium *Pseudomonas aeruginosa*. *Applied Microbiology and Biotechnology*, **103**, 1535–1544, https://doi.org/10.1007/s00253-018-9484-5

Logan B. E., Murano C., Scott K., Gray N. D. and Head I. M. (2005). Electricity generation from cysteine in a microbial fuel cell. *Water Research*, **39**, 942–952, https://doi.org/10.1016/j.watres.2004.11.019

Logrono W., Perez M., Urquizo G., Kadier A., Echeverría M., Recalde C. and Rakhely G. (2017). Single chamber microbial fuel cell with a cathodic microalgal biofilm: a preliminary assessment of the generation of bioelectricity and biodegradation of real dye textile wastewater. *Chemosphere*, **176**, 378–388, https://doi.org/10.1016/j.chemosphere.2017.02.099

Madondo N. I., Rathilal S., Bakare B. F. and Tetteh E. K. (2023). Effect of electrode spacing on the performance of a membrane-less microbial fuel cell with magnetite as an additive. *Molecules*, **28**, 2853, https://doi.org/10.3390/molecules28062853

Malekmohammadi S. and Mirbagheri S. A. (2021). A review of the operating parameters on the microbial fuel cell for wastewater treatment and electricity generation. *Water Science & Technology*, **84**(6), 1309, https://doi.org/10.2166/wst.2021.333

Malik S., Kishore S., Dhasmana A., Kumari P., Mitra T., Chaudhary V., Kumari R., Bora J., Ranjan A., Minkina T. and Rajput V. D. (2023). A perspective review on microbial fuel cells in treatment and product recovery from wastewater. *Water*, **15**, 316, https://doi.org/10.3390/w15020316

Mathuriya A. S. (2013). Inoculum selection to enhance performance of a microbial fuel cell for electricity generation during wastewater treatment. *Environmental Technology*, **34**(13–14), 1957–1964, https://doi.org/10.1080/09593330.2013.808674

Miran W., Jang J., Nawaz M., Shahzad A. and Lee D. S. (2018). Sulfate-reducing mixed communities with the ability to generate bioelectricity and degrade textile diazo dye in microbial fuel cells. *Journal of Hazardous Materials*, **352**, 70–79, https://doi.org/10.1016/j.jhazmat.2018.03.027

Mistry G., Popat K., Patel J., Panchal K., Ngo H. H., Bilal M. and Varjani S. (2023). New outlook on hazardous pollutants in the wastewater environment: occurrence, risk assessment and elimination by electrodeionization technologies. *Environmental Research*, **219**, 115112, https://doi.org/10.1016/j.envres.2022.115112

Mohammed A. J. and Ismail Z. Z. (2018). Slaughterhouse wastewater biotreatment associated with bioelectricity generation and nitrogen recovery in hybrid system of microbial fuel cell with aerobic and anoxic bioreactors. *Ecological Engineering*, **125**, 119–130, https://doi.org/10.1016/j.ecoleng.2018.10.010

Nancharaiah Y., Mohan S. V. and Lens P. (2015). Metals removal and recovery in bioelectrochemical systems: a review. *Bioresource Technology*, **195**, 102–114, https://doi.org/10.1016/j.biortech.2015.06.058

Nguyen V. K., Park Y., Yang H., Yu J. and Lee T. (2016). Effect of the cathode potential and sulfate ions on nitrate reduction in a microbial electrochemical denitrification system. *Journal of Industrial Microbiology and Biotechnology*, **43**, 783–793, https://doi.org/10.1007/s10295-016-1762-6

Obileke K. C., Onyeaka H., Meyer E. L. and Nwokolo N. (2021). Microbial fuel cells, a renewable energy technology for bio-electricity generation: a mini-review. *Electrochemistry Communications*, **125**, 107003, https://doi.org/10.1016/j.elecom.2021.107003

Oon Y. S., Ong S. A., Ho L. N., Wong Y. S., Oon Y. L., Lehl H. K., Thung W. E. and Nordin N. (2017). Microbial fuel cell operation using monoazo and diazo dyes as terminal electron acceptor for simultaneous decolourisation and bioelectricity generation. *Journal of Hazardous Materials*, **325**, 170–177, https://doi.org/10.1016/j.jhazmat.2016.11.074

Palanisamy G., Jung H. Y., Sadhasivam T., Kurkuri M. D., Kim S. C. and Roh S. H. A. (2019). Comprehensive review on microbial fuel cell technologies: processes, utilization, and advanced developments in electrodes and membranes. *Journal of Cleaner Production*, **221**, 598–621, https://doi.org/10.1016/j.jclepro.2019.02.172

Patwardhan S. B., Savla N., Pandit S., Gupta P. K., Mathuriya A. S., Lahiri D., Jadhav D. A., Rai A. K., KanuPriya Ray R. R., Singh V., Kumar V. and Prasad R. (2021). Microbial fuel cell united with other existing technologies for enhanced power generation and efficient wastewater treatment. *Applied Sciences*, **11**(22), 10777, https://doi.org/10.3390/app112210777

Penteado E. D., Fernandez-Marchante C. M., Zaiat M., Gonzalez E. R. and Rodrigo M. A. (2017). Influence of carbon electrode material on energy recovery from winery wastewater using a dual-chamber microbial fuel cell. *Environmental Technology*, **38**, 1333–1341, https://doi.org/10.1080/09593330.2016.1226961

Prasad J. and Tripathi R. K. (2022). Review on improving microbial fuel cell power management systems for consumer applications. *Energy Reports*, **8**, 10418–10433, https://doi.org/10.1016/j.egyr.2022.08.192

Qiu R., Zhang B., Li J., Lv Q., Wang S. and Gu Q. (2017). Enhanced vanadium reduction and bioelectricity generation in microbial fuel cells with biocathode. *Journal of Power Sources*, **359**, 379–383, https://doi.org/10.1016/j.jpowsour.2017.05.099

Raghavulu S. V., Annie Modestra J., Amulya K., Reddy C. N. and Venkata Mohan S. (2013). Relative effect of bioaugmentation with electrochemically active and nonactive bacteria on bioelectrogenesis in microbial fuel cell. *Bioresource Technology*, **146**, 696–703, https://doi.org/10.1016/j.biortech.2013.07.097

Singh A. and Kaushik A. (2021). Removal of Cd and Ni with enhanced energy generation using biocathode microbial fuel cell: insights from molecular characterization of biofilm communities. *Journal of Cleaner Production*, **315**, 127940, https://doi.org/10.1016/j.jclepro.2021.127940

Sonu K., Zainab S. and Monika S. (2020). Up-scaling microbial fuel cell systems for the treatment of real textile dye wastewater and bioelectricity recovery. *International Journal of Environmental Studies*, **77**(4), 1–11, https://doi.org/10.1080/00207233.2020.1736438

Sun J., Li Y., Hu Y., Hou B., Zhang Y. and Li S. (2013). Understanding the degradation of Congo red and bacterial diversity in an air-cathode microbial fuel cell being evaluated for simultaneous azo dye removal from wastewater and bioelectricity generation. *Applied Microbiology and Biotechnology*, **97**, 3711–3719, https://doi.org/10.1007/s00253-012-4180-3

Sun J., Xu W., Yang P., Li N., Yuan Y., Zhang H., Ning X., Zhang Y., Chang K., Peng Y. and Chen K. (2019). Enhancing the performance of photo-bioelectrochemical fuel cell using graphene oxide/cobalt/polypyrrole composite modified photo-biocathode in the presence of antibiotic. *International Journal of Hydrogen Energy*, **44**(3), 1919–1929, https://doi.org/10.1016/j.ijhydene.2018.11.111

Suzuki K., Kato Y., Yui A., Yamamoto S., Ando S., Rubaba O., Tashiro Y. and Futamata H. (2018). Bacterial communities adapted to higher external resistance can reduce the onset potential of anode in microbial fuel cells. *Journal of Bioscience and Bioengineering*, **125**, 565–571, https://doi.org/10.1016/j.jbiosc.2017.12.018

Tang J., Chen S., Yuan Y., Cai X. and Zhou S. (2015). In situ formation of graphene layers on graphite surfaces for efficient anodes of microbial fuel cells. *Biosensors and Bioelectronics*, **71**, 387–395, https://doi.org/10.1016/j.bios.2015.04.074

Wang Z., Zhang B., Jiang Y., Li Y. and He C. (2018). Spontaneous thallium (I) oxidation with electricity generation in single-chamber microbial fuel cells. *Applied Energy*, **209**, 33–42, https://doi.org/10.1016/j.apenergy.2017.10.075

Wang J., Ren K., Zhu Y., Huang J. and Liu S. (2022). A review of recent advances in microbial fuel cells: preparation, operation, and application. *BioTech*, **11**(4), 44, https://doi.org/10.3390/biotech11040044

Wang R., Wang X., Zhou X. and Yao J. (2020). Effect of anolytic nitrite concentration on electricity generation and electron transfer in a dual-chamber microbial fuel cell. *Environmental Science Pollution Research*, **27**(9), 9910–9918, https://doi.org/ 10.1007/s11356-019-07323-z

Wu Y., Zhao X., Jin M., Li Y., Li S. and Kong F. (2018). Copper removal and microbial community analysis in single-chamber microbial fuel cell. *Bioresource Technology*, **253**, 372–377, https://doi.org/10.1016/j.biortech.2018.01.046

Xafenias N., Zhang Y. and Banks C. J. (2013). Enhanced performance of hexavalent chromium reducing cathodes in the presence of *Shewanella oneidensis* MR-1 and lactate. *Environmental Science and Technology*, **47**, 4512–4520, https://doi.org/10.1021/es304606u

Xafenias N., Zhang Y. and Banks C. J. (2015). Evaluating hexavalent chromium reduction and electricity production in microbial fuel cells with alkaline cathodes. *International Journal of Science and Technology*, **12**, 2435–2446.

Yadav A. K., Dash P., Mohanty A., Abbassi R. and Mishra B. K. (2012). Performance assessment of innovative constructed wetland-microbial fuel cell for electricity production and dye removal. *Ecological Engineering*, **47**, 126–131, https://doi.org/10.1016/j.ecoleng.2012.06.029

Yaqoob A. A. and Ibrahim M. N. M. (2019). A review article of nanoparticles; synthetic approaches and wastewater treatment methods. *International Research Journal of Engineering and Technology*, **6**, 1–7.

Yaqoob A. A., Khatoon A., Mohd Setapar S. H., Umar K., Parveen T., Mohamad Ibrahim M. N., Ahmad A. and Rafatullah M. (2020). Outlook on the role of microbial fuel cells in remediation of environmental pollutants with electricity generation. *Catalysts*, **10**, 819, https://doi.org/10.3390/catal10080819

Yong X. Y., Feng J., Chen Y. L., Shi D. Y., Xu X. Y., Zhou J., Wang S. Y., Xu S., Yong Y. C., Sun Y. M., Shi C. L., OuYang P. K. and Zheng T. (2014). Enhancement of bioelectricity generation by cofactor manipulation in microbial fuel cell. *Biosensors and Bioelectronics*, **56**, 19–25, https://doi.org/10.1016/j.bios.2013.12.058

You J., Greenman J. and Ieropoulos I. (2018). Novel analytical microbial fuel cell design for rapid in situ optimisation of dilution rate and substrate supply rate, by flow, volume control and anode placement. *Energies*, **11**(9), 2377, https://doi.org/10.3390/en11092377

Zafar S., Bukhari D. A. and Rehman A. (2022). Azo dyes degradation by microorganisms – an efficient and sustainable approach. *Saudi Journal of Biological Sciences*, **29**, 103437, https://doi.org/10.1016/j.sjbs.2022.103437

Zhang W., Li X., Liu T. and Li F. (2012). Enhanced nitrate reduction and current generation by *Bacillus* sp. in the presence of iron oxides. *Journal of Soils and Sediments*, **12**, 354–365, https://doi.org/10.1007/s11368-011-0460-2

Zhou M., Wang H., Hassett J. D. and Gu T. (2013). Recent advances in microbial fuel cells and microbial electrolysis cells for wastewater treatment, bioenergy and bioproducts. *Journal of Chemical Technology and Biotechnology*, **88**, 508–518, https://doi.org/10.1002/jctb.4004

doi: 10.2166/9781789063813_0225

Chapter 13

Use of industrial waste to synthesize novel electrode and membrane materials for microbial electrochemical technologies

Himanshi Sen[1], Monika Sogani[1], Jayana Rajvanshi[1], Nishan Sen Gupta[1], Zainab Syed[1], Sovik Das[2] and Kumar Sonu[3]*

[1]*Department of Biosciences, Manipal University Jaipur, Jaipur, Rajasthan, India*
[2]*Department of Civil Engineering, Indian Institute of Technology, Delhi, India*
[3]*Department of Mechanical Engineering, Kashi Institute of Technology, Varanasi, India*
**Corresponding author: monika.sogani@jaipur.manipal.edu*

ABSTRACT

The escalating environmental degradation resulting from an exponential growth in waste generation and improper disposal practices necessitates immediate consideration and innovative solutions. The concept of a circular economy, which emphasizes the potential of solid wastes to be transformed into valuable raw materials, offers a promising approach to address this pressing issue. In this chapter, we explore the transformation of industrial waste into electrode and membrane materials for microbial electrochemical technologies (METs). Various synthetic techniques have been identified, enabling the conversion of industrial waste into viable materials for METs. The proposed framework encompasses critical aspects, including waste pre-treatment, raw material extraction, fabrication, and characterization processes, as well as the performance evaluation of electrodes and membranes derived from waste sources. We thoroughly examine the advantages and limitations associated with waste-derived electrodes and membranes, providing valuable insights into their potential applications and challenges. By harnessing waste resources for electrode and membrane production, we not only contribute to environmental sustainability but also advance toward greener practices and a more sustainable future. The favorable material properties inherent in waste resources and the availability of suitable processing facilities render this approach particularly promising. This chapter elucidates the vast potential of using industrial waste to synthesize novel electrode and membrane materials for METs. By adopting a circular economy mindset and promoting waste recycling and reutilization, we can achieve significant benefits in terms of environmental preservation and resource optimization. The integration of waste-derived materials into METs paves the way for a more sustainable and efficient approach to address environmental challenges.

Keywords: electrode fabrication, industrial waste, membrane fabrication, microbial electrochemical technologies.

13.1 INTRODUCTION

The branch of research that brings together the disciplines of microbiology and electrochemistry to develop novel technologies for the generation of energy, the treatment of waste, and the recovery of resources is known as microbial electrochemical technology (MET) (Yaqoob *et al.*, 2020). The technique is based on the capability of some microorganisms that oxidize organic substrates through their metabolic processes and release electrons as by-products. Some microorganisms, known as exoelectrogens can transfer these electrons to anodes through extracellular electron transfer pathways. This transfer can occur via a direct contact between the microbial cells and the electrode surface or through the secretion of redox-active compounds such as mediators or conductive pili. The electrons flow from the microbial cell to the anode, generating an electric current. However, in the cathode a reduction reaction occurs. This reaction can involve the transfer of electrons from the cathode to the microorganisms or a terminal electron acceptor, such as oxygen (O_2) or other oxidized species. This completes the electron flow circuit initiated at the anode (Hernandez & Osma, 2020). METs have the potential to be used for resource recovery, bioremediation, and the production of sustainable energy (Yaqoob *et al.*, 2020). However, the use of conventional electrodes in METs comes with certain limitations including biofouling and surface degradation of electrodes which lowers proton transfer across the membrane and also depletes dissolved oxygen in the cathode reaction leading to reduced performance over time. The accumulation of biomass and extracellular polymeric substances can interfere with electron transfer and limit the system's longevity. Additionally, the use of conventional electrodes may result in electrode degradation due to corrosion or chemical reactions in the operating environment (Kalathil *et al.*, 2018). Thus, to create novel electrode and membrane materials for METs, there is significant interest in industrial waste as an alternative source (Muhammad *et al.*, 2022).

Industrial waste is a beneficial resource for METs because it is economical, abundant, and sustainable. By the process of pyrolysis or carbonization, waste can be converted into carbon-based materials, and certain surface modifications are also performed by incorporating a coating with conductive materials such as metals, carbon-based materials, conductive polymers, metal oxides, and composite materials such as carbon nanofiber (CNF)/Nafion and activated carbon nanofiber (ACNF)/Nafion nanocomposite membranes. Conductive materials and composite materials have various advantages such as enhanced conductivity, increased surface area, enhanced stability and durability, compatibility, and adhesion (Pandit *et al.*, 2021). This approach can be cost-effective enabling the use of a wide range of waste materials as a sustainable resource for energy generation, bioremediation, and resource recovery (Muhammad *et al.*, 2022). Researchers have been evaluating the use of industrial waste as a feedstock for the synthesis of innovative electrode and membrane materials to resolve this problem (Pandit *et al.*, 2021). One illustration is the creation of electrodes for METs using carbonaceous materials obtained from waste biomass, such as rice husk, sawdust, and sewage sludge (SS). These substances have demonstrated good performance in terms of power output and long-term stability, and they can be used as affordable and environmentally friendly alternatives for conventional graphite or platinum electrodes (Behera *et al.*, 2010). In addition to biomass waste, agricultural waste and plastic waste are also under investigation for their potential in synthesizing novel electrode and membrane materials for METs. Activated carbons are derived from bio-based polymers synthesized from agricultural and forestry wastes, as well as metal and metal–oxide nanoparticles produced from industrial effluent. These alternative waste materials demonstrate promise for advancing METs, offering sustainable and cost-effective solutions for electrode and membrane fabrication (Muhammad *et al.*, 2022).

Overall, the usage of industrial waste as a feedstock for the synthesis of novel electrode and membrane materials has the potential to enhance the efficiency and sustainability of METs as well as contribute to the circular economy by reducing waste and generating value-added products from waste streams (Ramirez *et al.*, 2021).

13.2 CONVENTIONAL MATERIALS

13.2.1 Electrode materials

13.2.1.1 Anode materials used in METs

Numerous anode materials exhibit suitability for implementation in METs. Ideal anode materials should demonstrate both conductivity and porosity, thereby facilitating ample surface area for bacterial adhesion and efficient electron transfer (Butti *et al.*, 2016). Some commonly used anode materials for METs are discussed below.

13.2.1.1.1 Carbonaceous anodes

Carbon-based materials are commonly utilized for the construction of anode electrodes due to their favorable characteristics. These include a large surface area, high conductivity coupled with low porosity, excellent biocompatibility, remarkable chemical and thermal stability, easy availability, and efficient electron transfer kinetics. These properties make carbon-based materials highly desirable for the fabrication of anode electrodes in various applications. A wide range of materials can be employed as anode electrodes, including activated carbon cloth, carbon paper, carbon rods, carbon meshes, carbon brushes, reticulated vitreous carbon, and various graphitic forms such as felt, granule cloth, sheet, paper, fiber, and graphite oxide. These diverse options offer flexibility in choosing the most suitable material based on specific requirements and applications. Each of these materials possesses unique characteristics that contribute to their effectiveness as anode electrode materials in various electrochemical systems (Kalathil *et al.*, 2018). In METs, graphite is frequently utilized as the anode electrode material. Graphite possesses desirable attributes such as mechanical strength, biocompatibility, and a relatively large surface area, which make it an attractive option for constructing anode electrodes in METs. However, certain limitations, including its high cost and suboptimal conductivity, pose challenges in meeting the growing energy demands in commercial applications (Santoro *et al.*, 2017). The most common type of carbon utilized in MET anode electrode fabrication is carbon cloth. Compared to normal carbon paper, this type of carbon has a larger surface area, but because it has so many wide void areas, it has a comparatively high porosity. Overall, carbon cloth outperforms conventional carbon paper in terms of flexibility, electrical conductivity, and mechanical stability. The drawbacks of this material are its chemical instability resulting in fouling and lowering the anode electrode's long-term stability (Ul *et al.*, 2023). Carbon paper is a prevalent electrode material utilized in numerous electrochemical applications, notably in METs. This material, composed of carbon fibers acting as a catalyst substrate, is renowned for its conductivity. Due to its cost-effectiveness and extensive surface area, carbon paper electrodes are commonly selected as anodes in METs. To increase the electrocatalytic activity of carbon paper electrodes, a layer of catalyst materials, such as platinum or carbon black, is frequently applied. By serving as an interface between the bacteria and the electrode surface, the catalyst layer encourages electron transport and increases power output (Kalathil *et al.*, 2018). Carbon mesh is commercially accessible at an affordable cost, but it has weak mechanical stability and limited electric conductivity, which makes it less reliable. To achieve good MET performance, it is necessary to subject carbon mesh to pre-treatment methods such as heat treatment, acetone treatment, or an ammonia-gas process. These pre-treatment techniques are employed to enhance the properties of carbon mesh and optimize its performance as an anode electrode in METs (Santoro *et al.*, 2017). To increase the effectiveness of energy production in METs, a carbon veil, a low-cost carbon material with reasonably strong electric conductivity and a highly porous structure, is employed as an anode electrode. Although a single-layer carbon veil is quite fragile, it is versatile and may be folded to create a three-dimensional (3D) anode electrode (Mustakeem, 2015). One of the prevalent materials for anode electrodes in METs is carbon felt. It is characterized by high porosity and good electrical conductivity. Depending on the thickness of the material, it has a high degree of mechanical stability and a relatively low cost (Kalathil *et al.*, 2018).

13.2.1.1.2 Metal/metal–oxide anodes

Electrodes fabricated from silver, molybdenum, iron, nickel, gold, aluminum, copper, stainless steel, and titanium performed exceptionally well due to their high electric conductivity. Even though corrosion renders them less friendly with microorganisms, metals and metal oxides are often utilized as anode electrodes in METs. This, however, limits their applicability and utilization. However, the biocompatibility and durability of these materials are not adequate to prevent bacterial adhesion (Kalathil *et al.*, 2018).

13.2.1.2 Cathode materials used in METs

In METs, the same materials can often be used as both anode and cathode materials. For example, carbon-based materials (e.g. graphite, carbon cloth), metal-based catalysts (e.g. platinum, gold), and metal oxides (e.g. manganese oxide), can serve as both anode and cathode materials in MET systems. These materials play crucial roles in facilitating electron transfer and catalytic reactions at both electrodes. This dual applicability of materials offers flexibility and simplifies the selection process for electrode materials in METs (Mustakeem, 2015).

13.2.2 Membrane and separator materials

A membrane or separator is considered efficient and economical when it consists of a minimal crossover of oxygen and substrate and along with that transfer of protons and ions could be observed efficiently. Separators possess different performance specificities depending on certain characteristics such as configuration, thickness, surface condition, and a few of the operating conditions of whole MET such as electrolyte composition, current discharge, and so on. Cost and ion-mass permeability are also a measure to assess the performance of separators (Fan *et al.*, 2007).

13.2.2.1 Ion-exchange membranes

In ion-exchange membranes (IEMs), the prevalent functional group in anion-exchange membranes (AEMs) is typically positively charged tertiary amines (ANH_3^+), whereas negatively charged sulfonates are commonly found in cation-exchange membranes (CEMs). These functional groups play a crucial role in facilitating the movement of ions with opposite charges across the membrane (Daud *et al.*, 2015).

13.2.2.1.1 Cation-exchange membranes

Nafion membranes: These are utilized in chemical fuel cells and METs as proton-exchange membranes (PEMs). They function very well both as solid electrolytes for the selective transport of cations and as internal separators between the anode and cathode chambers (Ke *et al.*, 2021). Nafion is a sulfonated tetrafluoroethylene copolymer that has hydrophilic sulfonate groups (SO_3) and a hydrophobic fluorocarbon backbone attached to a membrane backbone. The sulfonate groups that are present in Nafion membranes are responsible for the high amount of cation conductivity that they exhibit. The membranes prepared using Nafion are strong and resistant to various chemical attacks (Daud *et al.*, 2015).

Sulfonated polyether ether ketone (SPEEK): It is an excellent alternative for Nafion in applications involving microbial fuel cells (MFCs) because it contains negatively charged sulfonate ions (SO_3) that only allow cations to pass by while limiting anions. This property makes SPEEK a sulfonated polymer. Nafion, which is structurally comparable to SPEEK, has hydrophobic regions of the polymer backbone that coexist with ionic clusters. SPEEK has a higher tensile strength than Nafion due to the presence of aromatic groups in the backbone of the polymer (Zhang *et al.*, 2010).

Sulfonated polystyrene–ethylene–butylene–polystyrene (SPSEBS): It is yet another type of CEM that is widely available and utilized in METs (Ayyaru *et al.*, 2012). The ionomer is constructed from a hydrocarbon network that is chemically, thermally, and physically stable, and it has an aromatic backbone that is not fluorinated (Mishra *et al.*, 2012). It was observed by Ayyaru *et al.* (2012) that Nafion membrane-based METs have a lower power density (300 mW/m^2) and a higher pH difference (4.4–10.3) between the anode and cathode chambers. SPSEBS membrane-based METs have a smaller pH difference (6.4–7.6) and a higher power density (600 mW/m^2) (Ayyaru *et al.*, 2012).

13.2.2.1.2 Anion-exchange membranes

AEMs are a specific type of polymer membrane that allow anions to flow through but block the passage of cations and other molecules. AEMs are polymer electrolytes that transfer anions such as Cl and OH because they include positively charged cationic groups bound covalently to a polymer backbone. AEMs may perform better in microbial desalination cells (MDCs) compared to CEMs, despite their lower utilization in MET processes. Thus, a study by Kim *et al.* (2012) claimed that the power density of an MDC equipped with an AEM, model number AMI-7001, is 25% greater (610 mW/m^2) than that of a CEM-equipped MDC (Kim *et al.*, 2012).

13.2.2.2 Composite membranes

A type of polymer membrane made up of two or more distinct materials is known as a composite membrane. Composite membranes, which function better than Nafion or individual PEMs, are fabricated by fusing several polymeric membranes with additional organic or inorganic components (Ghasemi *et al.*, 2013). Nafion's separator capabilities could be improved by combining it with other substances as mentioned by Mohan *et al.* (2008) revealing, METs that employ CNF/Nafion and ACNF/Nafion nanocomposite membranes have a power density that is 57.6 mW/m^2 higher than those that use Nafion as the material for their membranes (Mohan *et al.*, 2008).

13.2.2.3 Porous membranes

METs have included the use of permeable separator materials that provide non-ion selective charge transfer. Non-woven fabric, glass fiber, clay pots, pottery, natural rubber, and biodegradable bags are a few examples of these types of materials (Daud *et al.*, 2015). Porous separators are classified into two types based on their pore sizes. Textiles, glass fiber, nylon mesh, and cellulose filters are examples of coarse pore filter materials, whereas ultrafiltration membranes and microfiltration membranes are examples of microporous filtration membranes (Sun *et al.*, 2009).

13.3 WASTE SOURCES FOR ELECTRODE AND MEMBRANE MATERIALS AS ALTERNATIVES

13.3.1 Electrode fabrication

Fabricating electrodes from industrial waste typically involves several steps. A specific process may vary depending on the type of waste material being used and the desired electrode composition. However, the general overview of the electrode fabrication steps is mentioned below (Yaqoob *et al.*, 2020):

(1) *Waste collection and preparation*: Collect the waste material that contains carbonaceous components suitable for electrode fabrication. Remove any impurities, contaminants, or non-carbon elements from the waste material.

(2) *Waste material characterization*: Analyze the waste material to determine its carbon content, chemical composition, physical properties, and potential hazards. This characterization helps in selecting appropriate processing parameters and understanding the material's suitability for electrode fabrication.

(3) *Waste treatment and conditioning*: Process the waste material to prepare it for carbonization or pyrolysis. This step may involve shredding, grinding, or other mechanical treatments to increase the material's surface area and facilitate subsequent thermal processes.

(4) *Carbonization*: Subject the waste material to controlled heating in an oxygen-limited environment. Carbonization involves heating the material to high temperatures (typically between 400 and 800°C) to drive off volatile components and transform the carbonaceous material into a more stable form of carbon known as char. This step helps to increase the carbon content and reduce impurities in the material.

(5) *Pyrolysis*: In some cases, pyrolysis may be employed as an alternative or additional step to carbonization. Pyrolysis involves subjecting the waste material to higher temperatures (typically above 800°C) in an inert or reducing atmosphere. This process breaks down the complex organic compounds in the waste material into simpler carbon-rich molecules, gases, and liquids, leaving behind a solid carbonaceous residue that can be used as an electrode precursor.

(6) *Grinding and sizing*: Once the carbonized or pyrolyzed material has been obtained, it is typically ground or milled to achieve a fine powder. Particle-size reduction helps to improve the material's reactivity and homogeneity, making it suitable for electrode fabrication.

(7) *Formulation and mixing*: Combine the carbonized or pyrolyzed material with binders, additives, and conductive agents to create a homogeneous electrode mixture. The formulation aims to optimize the material's electrochemical performance, stability, and conductivity.
(8) *Electrode shaping*: Shape the electrode mixture into the desired form, such as sheets, foils, or 3D structures. This can be achieved through techniques such as calendaring, extrusion, or pressing.
(9) *Electrode drying and curing*: Remove any remaining moisture or solvents from the shaped electrodes through a controlled drying process. Curing may also be applied to enhance the electrode's mechanical and electrochemical properties.
(10) *Electrode testing and quality control*: Perform various tests by electrochemical impedance spectroscopy (EIS) and current measurement by quantification of microbial biomass and metabolic activity assay on the fabricated electrodes to assess their quality, conductivity, electrochemical performance, and durability. This step ensures that the electrodes meet the desired specifications and identifies any defects or inconsistencies.
(11) *Packaging and storage*: Properly package and store the electrodes to maintain their performance and prevent contamination or degradation. Accurate labeling and documentation help track batch information and manufacturing details.

The specific parameters and techniques employed in carbonization, pyrolysis, and electrode fabrication processes can vary depending on the waste material, desired electrode properties, and industry practices.

13.3.1.1 Plastic waste

The increasing accumulation of plastic waste has emerged as a significant environmental concern, drawing widespread attention. In response, numerous treatment technologies have been developed to address this issue, encompassing degradation, recycling, upcycling, and the conversion of plastic waste into value-added products. Among these approaches, the transformation of plastic waste into carbon-based functional materials holds particular appeal due to the practical applications offered by plastic waste-derived carbon materials (PWCMs) in the realms of green energy and sustainable environmental practices. Extensive research has been conducted to explore the potential of PWCMs and their contributions to fostering a greener future and sustainable environmental practices (Chen *et al.*, 2022a).

13.3.1.1.1 Performance of plastic waste-derived electrodes

The performance of plastic waste-derived electrodes can be better than conventional electrodes in several applications due to their unique properties, including high surface area, porosity, and catalytic activity.

(1) *Low cost*: Plastic waste-derived electrodes can be produced from waste materials, making them an affordable alternative to conventional electrodes. In addition, plastic waste is readily available and can be obtained from a variety of sources, reducing the dependency on expensive raw materials (Hopewell *et al.*, 2009).
(2) *High surface area*: Plastic waste-derived electrodes can be produced with a high surface area, allowing for more microbial attachment and improvised electron transfer rates. This can uplift the efficiency and performance of metabolic electrochemistry systems, such as METs (Singh *et al.*, 2016).
(3) *Improved catalytic activity*: Plastic waste-derived electrodes can be produced with enhanced catalytic activity due to the presence of functional groups on the surface of the material. This can improve the performance of the electrode in electrochemical reactions, leading to improved power output in metabolic electrochemistry systems (Singh *et al.*, 2016).

(4) *Sustainability*: The use of plastic waste-derived electrodes is an economical and sustainable means of diverting plastic waste from landfills and reducing the demand for new raw materials. This can improve the environmental sustainability of metabolic electrochemistry systems (Chen *et al.*, 2022a).

Here are some ways in which plastic waste-derived electrodes can outperform conventional electrodes by forming conductive polymers (CPs). Electronics and biosensors frequently use CP materials because they have good conductivity and polymer-like characteristics. CPs have gained more attention recently for altering the anode of METs to improve performance. Among the investigated CPs, polypyrrole, polyaniline (PANI), and composite materials can significantly boost MET performance. In a study by Kaur *et al.* (2021), it was proposed that linkers would be used in the production of porous nanostructured materials. An iron-based metal–organic framework was fabricated known as a terephthalic acid monomer (t), which was generated from waste plastic. Significant advancements were made in the synthesized iron-terephthalate metal–organic framework (Fe-t-MOF) by the use of the CP PANI. The produced nanocomposite, which comprised of Fe-t-MOF and PANI, was coated on a disk made of stainless steel to act as the collector of current for the electrode component of the MET system. This was performed to improve the performance of the MET system (Vinodh *et al.*, 2022). Anode and cathode electrodes made of Fe-t-MOF/PANI composites were successfully produced and used in MET applications for the first time. The fabricated MET had a high open circuit potential value of 0.67 V, a power density of 680 mW/m^2, and a limiting current density of ~3500 mW/m^2 (Kaur *et al.*, 2021). The electrodes were not only inexpensive but also stable, biocompatible, and electroconductive. In addition, it has been shown that the generated Fe-t-MOF/PANI composite is a good alternative for expensive carbon-based materials in the fabrication of electrodes for any kind of fuel cell. This was discovered after extensive research was conducted. The functioning of nanocomposite materials is improved further by the inclusion of MOFs, which is a component of PANI. The cost of electrode material is gradually brought down by the use of MOFs which has been prepared from recycled plastic. There are a variety of newly discovered options for increasing the efficiency of electrodes for METs that are based on advanced MOF materials (Kaur *et al.*, 2021). Due to their distinct qualities, affordability, and sustainability, plastic waste-derived electrodes may perform better overall than traditional electrodes in some applications. The approach chosen here should aid in the quest for an innovative, practical, cost-effective, and sustainable method for converting plastic waste into electrodes for METs to produce bioenergy while treating wastewater (Hopewell *et al.*, 2009).

13.3.1.2 Agricultural waste

In the energy sector, biochar prepared from agricultural waste is gaining popularity due to its affordability, sustainability, and high super-capacitance performance. It also has a variety of environmental applications, including contaminant immobilization, improving soil fertility, and in situ carbon sequestration for wastewater treatment (Khedulkar *et al.*, 2023).

Recent developments in the production of biochar from agricultural waste for supercapacitor electrodes place a strong emphasis on a green, circular economy that promotes net-zero biomass utilization (Khedulkar *et al.*, 2023). Using pyrolysis or hydrothermal carbonization, agricultural waste such as rice husks, wheat straw, and maize stover can be transformed into carbon-based materials and used as electrodes in METs (Chakraborty *et al.*, 2020).

13.3.1.2.1 Performance of agricultural waste-derived electrodes

In a study by Divyashree *et al.* (2016), it was revealed that to create inexpensive, high-performance supercapacitor electrodes, coconut waste was used. To obtain a spherical-shaped component, a straightforward one-step pyrolysis method is used. This method transforms the lignocellulosic characteristics of carbon into porous carbon nanospheres. Coconut leaves, coconut fiber, and coconut sticks have all been studied for their potential application in supercapacitors. Leaves and sticks had particles between 30 and 60 nm in size, whereas fibers were 20 nm or smaller (Divyashree *et al.*, 2016). Various techniques such as EIS, cyclic voltammetry, and chronopotentiometry were used to investigate the electrochemical characteristics of the porous carbon nanospheres. In the same study, improved electrochemical performance can be observed in the porous carbon nanospheres that were created from all three different types of biowaste samples. Porous carbon nanospheres produced from coconut fiber had the highest specific capacitance of 236 F/g when scanned at a rate of 2 mV/s, followed by coconut sticks with 208 F/g and coconut leaves with 116 F/g. The porous carbon nanospheres that were generated from all three distinct kinds of biowaste showed significantly improved electrochemical performance when compared to the original samples. The nanospheres in this order are designed for application in METs. When scanned at a rate of 2 mV/s, porous carbon nanospheres generated from coconut fiber had the greatest specific capacitance, measuring 236 F/g. This was followed by coconut sticks, which measured 208 F/g, and coconut leaves, which measured 116 F/g (Divyashree *et al.*, 2016).

As a result, this study demonstrated a method that is efficient, low-cost, and environment friendly for the production of polychlorinated naphthalene (PCNs) by employing waste from coconuts as a carbon precursor. This method also has the potential to improve the performance of electrodes. The research that was carried out utilizing X-ray diffraction, Raman, and transmission electron microscopy analyses indicates that the PCNs that were synthesized have a great particle size, confirming both their high quality and crystalline nature. Raw coconut fiber is excellent for the construction of electrodes because of its porous structure and fibrous shape, which make it a viable precursor for PCN synthesis without the use of templates or catalysts (Divyashree *et al.*, 2016). Table 13.1 provides a performance analysis of various waste-derived electrodes and the method of activation involved in their fabrication.

13.3.1.3 Biomass waste

Biomass waste refers to organic materials derived from living organisms or their by-products that are no longer used or needed and are considered waste. These organic materials can come from various sources, including municipal waste, bark, sawdust, SS, wood chips, leftover food, cow dung, pig manure, and algal biomass. The production of layered corrugated carbon anode electrodes through carbonization using cost-effective components represents an intriguing development in the field of anode electrode fabrication. This approach involves utilizing organic substances, including biomass wastes, which offer several advantages such as the utilization of recyclable materials, abundant

Table 13.1 Performance analysis of various waste-derived electrodes and method of activation involved in their fabrication.

Type of Samples	Activation Methods	Electrolyte Used for Conduction	Specific Conductance of Sample (F/g)	References
Coconut fiber	KOH activation	1.0 M KOH	236	Mensah-Darkwa *et al.* (2019)
Natural wood	Carbonized and KOH activation	6 M KOH	200	Mensah-Darkwa *et al.* (2019)
Activated carbon nanotube	–	1 M H_2SO_4	145	Chandrabhan Shende *et al.* (2018)
Rice husk	Chemical $ZnCl_2$	–	927	Mensah-Darkwa *et al.* (2019)
Sugarcane bagasse	Physical steam 900°C	–	320	Mensah-Darkwa *et al.* (2019)
Tea leaves	Chemical KOH	–	2352	Bhoyate *et al.* (2017)

availability, and material stability. The incorporation of multiple layers enhances the effectiveness of the electrode's density by providing bacteria with an expansive surface area upon which they can form biofilms. This innovation holds promise in optimizing the performance of anode electrodes by maximizing surface area for biofilm formation while utilizing sustainable and readily available materials (Zhou *et al.*, 2017). Natural waste-derived materials are widely regarded as highly effective for the fabrication of anode electrodes in METs. This is primarily due to their economic advantages over traditional materials, making them a cost-effective choice. Moreover, these materials possess all the necessary properties required for an ideal anode material, making them highly suitable for MET applications. The utilization of natural waste-derived materials offers a sustainable and environmentally friendly approach to anode electrode construction in METs, combining economic benefits with the desired material characteristics. This is because natural waste-derived materials are formed from decomposing organic matter. Materials that arise from natural waste and have a microporous or mesoporous 3D structure show a high electron transfer rate and an effective electrokinetics mechanism, both of which are necessary for the electrochemical processes that take place in microbial electrochemical snorkels (MESs). Because natural waste materials are both inexpensive and high-performing, the vast majority of spongy 3D anode electrodes developed for MESs make use of these materials. Although increasing the anode electrode's surface area will increase the pace of anode kinetics, the overall performance of MESs will only improve if the reaction is carried out with low internal resistance (Mahmoud *et al.*, 2022). Traditional methods for producing graphene oxide (GO) and reduced graphene oxide (rGO) sheets from high pure graphite (HPG) have relied on Hummers' method (see Figure 13.1). HPG can be costly to produce in large quantities. High-quality GO and rGO sheets can be prepared from a wide range of carbonaceous wastes, including those derived from plants (fruit wastes, leaf, bagasse, and wood), animals (bone, cow dung), industrial sources (soot powders created in diesel vehicle exhaust), SS, and semi-industrial sources (newspaper) (Akhavan *et al.*, 2014).

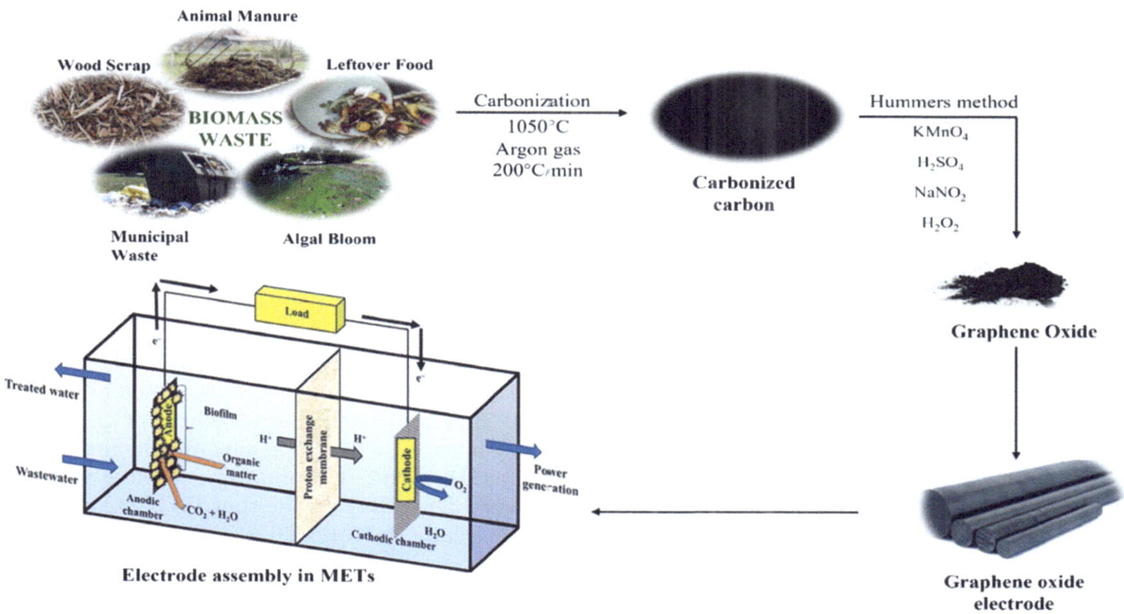

Figure 13.1 Various biomass substrates utilized for the fabrication of electrodes by Hummers' method.

13.3.1.3.1 Performance of biomass waste-derived electrodes

Given its potential applications as a soil amendment and pollutant adsorbent, the process of converting SS into value-added biochar has gained a growing amount of interest in recent years. The use of SS-derived biochar in an MET as an improved bifunctional electrode material (anode and cathode) is a recently suggested application of biochar. In a study by Yuan *et al.* (2015), an SS anode was first pressed into a mold containing varying quantities of coconut shell and then subjected to a heat treatment that resulted in the formation of SS-derived carbon monoliths specific materials (SMs). Powdered conductive materials, referred to as PSMs, have been employed as catalysts in the cathodes of microbial electroremediation cells (MERCs) to reduce oxygen levels. By incorporating an SM anode and a PSM cathode, MERCs achieved a remarkable maximum power density of 96,928 mW/m^2, ~2.4 times higher than an MERC utilizing a graphite anode and a Pt cathode (Yuan *et al.*, 2015). The enhanced electrical conductivity of SMs, achieved through modifications to the coconut shell, facilitated the proliferation of exoelectrogens and reduced electron transfer resistance. These two factors played a crucial role in the exceptional performance exhibited by the prepared electrodes in MERCs. The enhanced electrical conductivity of SMs was brought about by the modification of the coconut shell. The findings of this study point to a potentially useful process for converting SS into bifunctional electrode materials. This opens up a brand-new door for the value-added application of SS-derived biochar (Yuan *et al.*, 2015). Anodes and cathodes are composed primarily of carbonaceous electrodes (carbon and graphite), which have a larger surface area, superior biocompatibility, low price, and superior mechanical strength. A graphene or GO-based nanocomposite can serve as a cost-effective substitute for electrode modifications and an anode and cathode catalyst in MERCs (Akhavan *et al.*, 2014). X-ray photoelectron spectroscopy, atomic force microscopy, and Raman spectroscopy revealed that the single- and multilayer characteristics, chemical state, carbonaceous structure, and electrical properties of graphene sheets synthesized from various waste materials were nearly independent of one another. These findings propose a method for economically manufacturing vast quantities of graphene sheets of high quality from industrial and natural carbonaceous wastes (Akhavan *et al.*, 2014).

13.3.1.4 Metal scrap waste

For use as electrodes, metallic wastes such as used scrap iron, aluminum foil, and scrap mild steel are considered. In general, scrap metal wastes may be utilized to manufacture alternative electrodes that are more environmentally friendly (Deshwal & Panjagari, 2020).

13.3.1.4.1 Performance of metal scrap waste

According to Bani-Melhem *et al.* (2023), evaluation of electrodes constructed out of scrap metal may show drastically lower color, turbidity, chemical oxygen demand (COD), and electric conductivity by around 97.2%, 99%, 88%, and 89%, respectively. For a reaction time of 10 min and electrical current densities ranging from 5 to 20 mA/cm^2, the efficacy of the treatment is evaluated in removing color, electrical conductivity, COD, and turbidity, as well as the consumption of energy and materials, and metal contamination of grey water from electrodes (Bani-Melhem *et al.*, 2023).

More study is needed to determine the optimal way for synthesizing metallic wastes into a form that can be reused in the electrocoagulation (EC) technique and the operating expenses, which are major factors in the EC process due to the quantity of energy that is utilized (Bani-Melhem *et al.*, 2023). These studies demonstrate the potential of industrial waste as a low-cost and sustainable alternative electrode material in metabolic electrochemistry systems. However, it is important to note that the performance of industrial waste-based electrodes may vary depending on the specific waste material used and the conditions of the electrochemical system (Nippatla & Philip, 2020).

13.3.2 Membrane fabrication

Membranes used in METs can be prepared from inexpensive renewable green materials such as biopolymers, clay, ceramics, bio-derived materials, minerals, or naturally occurring soil. Typically, biopolymers with strong thermal, mechanical, and water retention capabilities as well as sustainability are preferred. Additionally, the functional qualities of MET power density are improved by altering or adding different functional groups to biopolymers.

Commercial membranes with high performance in METs include Nafion. However, the costs of these membranes substantially increase the overall cost of METs. Due to their natural abundance and affordability, separators or membranes prepared from naturally occurring earthen sources and biopolymers have emerged as novel and effective concepts (Olayiwola Sirajudeen *et al.*, 2021).

Steps involved in the fabrication of membrane materials from industrial waste for METs (see Figure 13.2) are mentioned below (Samavati *et al.*, 2023):

(1) *Identification and collection of industrial waste*: The first stage involves the identification and collection of suitable industrial refuse that can be used to manufacture membrane materials. These wastes include discarded grain from distilleries, fruit peels, and lignocellulosic biomass refuse from paper mills.

(2) *Pyrolysis*: The prepared industrial waste is then pyrolyzed at temperatures between 400 and 700°C in the absence of oxygen. The process breaks down organic detritus into gases, liquids, and particulates, leaving behind a carbon-rich substance.

(3) *Extraction of membrane material*: Extraction of membrane material can be extracted from industrial detritus via various techniques, including chemical extraction and enzymatic hydrolysis. Typically, the residual material is broken down to extract cellulose or other polysaccharides that can be used as membrane material.

(4) *Preparation of the polymer*: The polymer material is first washed with a suitable solvent, such as *N*-methyl pyrrolidone or dimethylformamide, to remove impurities.

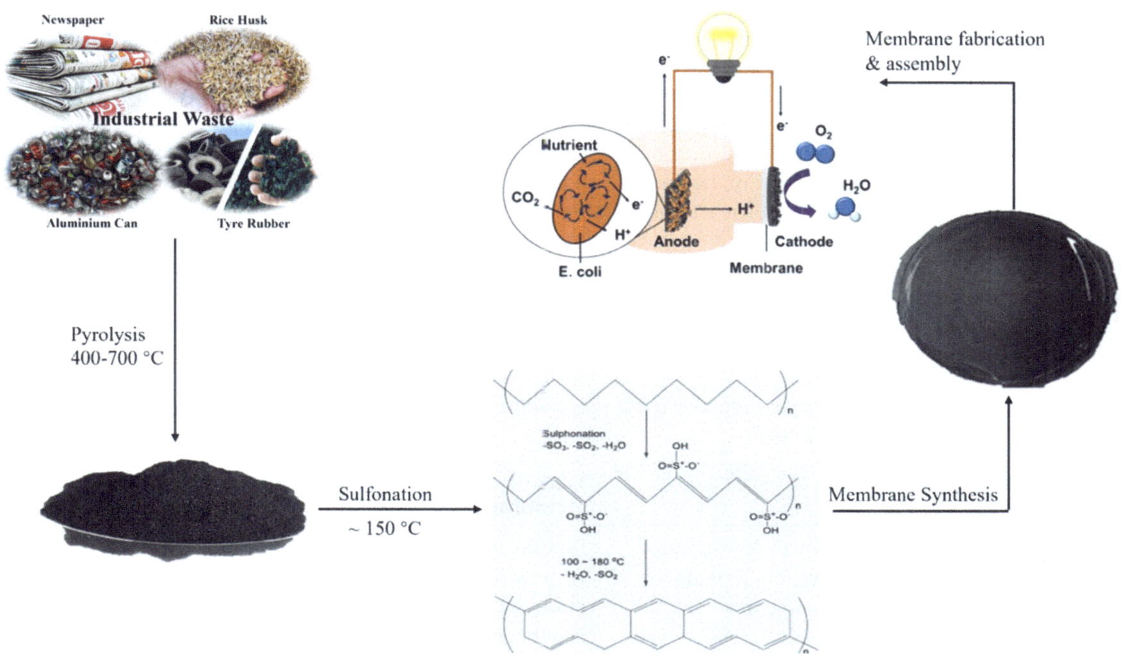

Figure 13.2 Utilization of waste from various industries for fabricating membrane materials.

(5) *Sulfonation*: The prepared polymer is then subjected to a sulfonation reaction in the presence of a sulfonating agent such as sulfuric acid, chlorosulfonic acid, or sulfur trioxide in a solvent such as sulfuric acid at ~150°C. To prevent polymer degradation, the reaction is typically conducted at a high temperature and under controlled conditions. The degree of sulfonation can be controlled by modifying the duration of the reaction, temperature, and concentration of the sulfonating agent.
(6) *Membrane fabrication*: Using techniques such as phase inversion, electrospinning, or casting, the extracted membrane material can be fabricated into membranes. The technique utilized will depend on the properties of the membrane material and the desired characteristics of the final membrane.
(7) *Membrane modification*: It may be necessary to modify the membrane's surface to increase its MET performance. This may involve the addition of surface coatings, such as nanoparticles or polymers, to improve the properties of the membrane, such as its hydrophilicity, selectivity, or antifouling ability.
(8) *Membrane testing*: Evaluation of the membrane's performance characteristics, such as permeability, selectivity, and mechanical strength, requires testing such as thermal stability test, surface morphology analysis performed by scanning electron microscopy, atomic force microscopy, and so on. This will aid in ensuring that the membrane satisfies the requirements for use in METs.
(9) *Assembly into microbial electrochemical systems*: The membranes can be assembled into microbial electrochemical systems, where they can be used to separate anodic and cathodic compartments while permitting the selective transport of ions and other molecules. This enables the production of electrical current or valuable compounds or fuels.

A wide variety of solid wastes, such as newspaper, rice husk ash (RHA), tire rubber, and aluminum cans have been used in the production of waste-derived membranes. During the construction of membranes, these solid wastes might serve as a selective layer, a support layer, pore-forming agents, or binders (Maraveas, 2020). Membranes that are fabricated from recycled materials hold a great deal of potential for use in a wide range of water purification processes. These processes include size-exclusion-based industrial wastewater treatment, desalination, oil–water separation, pollutant adsorption (such as that of heavy metals), virus extraction, gas separation, and catalytic degradation of organic contaminants. Because of increased awareness of the significance of the 'waste-to-resource' approach and the circular economy, the use of membranes that are obtained from waste is becoming an increasingly common method for the purification of water (Samavati *et al.*, 2023).

In general, the fabrication of membrane materials from industrial waste for METs is an environmentally responsible and sustainable method that can convert waste into valuable materials for energy production and effluent treatment. Table 13.2 provides an overview of various waste-derived membranes, the type of membranes that can be fabricated, and their enhanced permeability.

Table 13.2 Overview of membranes prepared using recycled materials such as tire rubber, aluminum cans, wastepaper/newspaper, and rice husk.

Waste Material	Purpose	Type of Membrane	Permeability (L/m²/h)	References
Tire rubber	Dye removal, desalination	Flat sheet	10.64	Lin *et al.* (2020)
Aluminum cans	Testing with pure water	Hollow fiber	129	Aziz *et al.* (2019)
Waste newspaper	Phenol removal	Photocatalytic membrane	–	Rodrigues Filho *et al.* (2008)
Rice husk	Testing with pure water	Hollow fiber	3.00×10^2	Raychaudhuri and Behera (2020)

13.3.2.1 Performance of waste tire rubber

Except for thermoplastic elastomers, all rubber materials are categorized as thermosetting polymers. This classification highlights the significance of the rubber present in waste tires as a valuable resource, particularly considering its potential as a membrane precursor. A study by Samavati *et al.* (2023) proposed that recycled rubber could be employed as a raw material for producing gas-separation membranes. Two types of recovered ground tire rubber (GTR) generated through mechano-chemical and cryo-mechanical processes are used for creating membranes from recycled rubber for gas separation (Samavati *et al.*, 2023).

In the initial stage, toluene was employed to extract the soluble rubber component from the recovered rubber, while centrifugation is used to separate carbon black and other solid impurities. The sol fraction obtained from the recovered GTR is then combined with the curing system in the subsequent stage. Once a solution is obtained, it is spun onto an aluminum oxide substrate to form a protective layer. Subsequently, the solvent is allowed to evaporate, and the membrane composed of recycled rubber undergoes cross-linking (Zhuang *et al.*, 2016). Another study by Lin *et al.* (2020) investigated the performance of the membrane prepared from recycled rubber favoring the successful separation of carbon dioxide from nitrogen and oxygen from nitrogen under high pressure and after prolonged usage. The gas separation experiments indicated that the C7-P2.8-T250 membrane showed the highest selectivity of 4.0 for H_2/CO_2 and an acceptable hydrogen permeance of 1124.61 GPU. These outcomes demonstrate the potential of utilizing reclaimed rubber as a viable, cost-effective, and sustainable precursor for the development of thin-film composite membranes in gas separation applications (Lin *et al.*, 2020). The approach employed in this study enables the identification of suitable reclaimed rubber precursors and the optimization of membrane preparation parameters, thereby enhancing the value of waste tires through recycling efforts.

13.3.2.2 Performance of aluminum cans

The incorporation of membranes derived from recycled aluminum dross residue in purification processes offers benefits such as enhanced mechanical strength and reduced sintering temperatures. A combination of the phase-inversion approach and sintering process is employed to develop a ceramic hollow fiber membrane using aluminum dross, a waste by-product from the aluminum industry. As the sintering temperature increased, the hollow fiber membrane became denser but experienced a decrease in flux (Aziz *et al.*, 2019). The inclusion of spinel in the microstructure of the hollow fiber facilitated achieving lower sintering temperatures. Despite being sintered at lower temperatures (ranging from 1350 to 1400°C) with a ceramic loading of 40%, this alternative ceramic hollow fiber membrane demonstrated mechanical strength comparable to pure alumina membranes. The mechanical strength of the alternative membrane ranged from 78.3 to 155.1 MPa, highlighting its robustness and ability to withstand mechanical stresses. Another method for creating ultrathin membranes from aluminum scrap is the anodization technique. A recent study by Ilango *et al.* (2016) successfully produced ultrathin anodic aluminum oxide (AAO) membranes with uniformly spaced holes using recycled aluminum cans. By adjusting the duration of the second anodization step, the AAO thickness was controlled within a narrow range of 60–650 nm. An anodic voltage of 40 and 25 V resulted in hexagonal AAO membranes with hole densities of 1.121010 and 2.961010 holes/cm^2, respectively. The diameter of the AAO membrane's holes increased from 30 to 95 nm by extending the pore-widening period (Ilango *et al.*, 2016).

13.3.2.3 Performance of waste newspaper

Waste newspaper represents a plentiful resource that is frequently underutilized in terms of its potential for extracting cellulose and its derivatives. In the process of manufacturing cellulose membranes, cellulose can be obtained from recycled newspaper. This extraction can be achieved using a dissolution medium comprised of NaOH/urea. The process involves utilizing H_2SO_4 as a coagulant in the phase-inversion process to form the cellulose membranes for various applications (Rodrigues Filho *et al.*, 2008). A study by Mohamed *et al.* (2015) revealed that the dissolution process of NaOH played a crucial role in facilitating ion-pair interactions. This reduced the strong self-association ability of water and enabled the establishment of hydrogen bonds between urea molecules and cellulose chains. Consequently, cellulose underwent dissolution, and upon contact with an acidic coagulant, its crystalline structure transitioned from cellulose I to cellulose II. The resulting cellulose membrane exhibited a consistent and homogeneous dense structure, characterized by its symmetrical appearance. It had an average pore size of 2.48 nm and a porosity of 41.0% (Mohamed *et al.*, 2015). The observed characteristics of the membrane demonstrated its capacity to effectively control the flow of substances while preserving its structural integrity. The utilization of waste-based adsorbents for wastewater treatment is environmentally significant, but current options often lack efficiency. However, a recent study by Li *et al.* (2023) demonstrated a breakthrough by successfully creating an aminated waste paper (WP-NH_2) membrane. This was accomplished through a process involving the oxidation of diverse waste paper types, followed by grafting and crosslinking with polyethyleneimine. The resulting WP-NH_2 membrane exhibits exceptional adsorption capabilities for various anionic dyes and antibiotics. Remarkably, it achieves high adsorption capacities of 2156.2 mg/g for Methyl Orange and 253.7 mg/g for tetracycline. Furthermore, the WP-NH_2 membrane demonstrates remarkable stability, allowing for multiple regeneration cycles without compromising its adsorption efficiency. This development shows significant promise in addressing wastewater treatment challenges, providing an efficient and reusable solution for removing pollutants (Li *et al.*, 2023).

13.3.2.4 Performance of rice husk

Rice husk, which refers to the outer coating that covers the rice grain, is a significant agricultural solid waste generated during the rice manufacturing process. This waste material presents an opportunity to extract valuable silica either through the pyrolysis of rice husk at high temperatures to produce RHA or by directly extracting sodium silicate from rice husk (Pode, 2016). The complete utilization of RHA is crucial to prevent its negative impact on the environment and human health when disposed of in open areas. Membrane technology, particularly those with a porous structure, has long been recognized as a promising choice for water treatment applications (Samavati *et al.*, 2023). Both amorphous and crystalline silica can be recovered from residual rice husk providing a detailed account of a cost-effective and environmentally friendly ceramic hollow-fiber membrane prepared from waste rice husk for water filtration, utilizing the phase-inversion and sintering processes. A study by Pode (2016) reported inference on ceramic hollow fiber membrane, composed of 37.5 wt% crystalline silica-based RHA content sintered at 1200°C, and also exhibits robust mechanical properties (71.2 MPa) and excellent porosity (50.2%). A study by Raychaudhuri and Behera (2020) revealed ceramic membranes as an economic alternative to CEMs used in MFCs. This study focused on developing low-cost ceramic membranes by blending soil with RHA for their application in MFCs. The ceramic membrane with 10% RHA content exhibited improved proton mass transfer and reduced oxygen diffusion. The presence of silica in RHA enhanced membrane hydration, facilitating efficient proton mobility from the anode to the cathode chamber. To continuously assess their effectiveness in treating rice mill wastewater, modified membranes (MFC_T) with RHA and unmodified membranes (MFC_C) without RHA were compared. The maximum volumetric power densities achieved were 2.14 W/m^3 for MFC_T and 1.33 W/m^3 for MFC_C, with COD removal efficiencies of $70.7 \pm 1.24\%$ and $63.8 \pm 1.08\%$, respectively. EIS revealed the ohmic resistance (R) to be 47.1 Ω for MFC_T and 91.3 Ω for MFC_C. An RHA composite membrane showed less susceptibility to fouling, making it a favorable choice for long-term operation. This study demonstrated that an RHA composite membrane is a promising and cost-effective alternative to expensive polymeric membranes in MFCs, offering higher power performance and minimal oxygen diffusion (Raychaudhuri & Behera, 2020).

13.4 ADVANTAGES AND LIMITATIONS OF WASTE-DERIVED ELECTRODES AND MEMBRANES

Waste-derived electrodes and membranes provide numerous advantages in METs harnessing the potential of waste materials, such as carbon-based materials or waste biomass.

13.4.1 Enhanced catalytic activity

Enhanced catalytic activity significantly improves the efficiency of waste-derived electrodes through multiple mechanisms. It accelerates electrochemical reactions, leading to faster conversion of reactants to products. Catalysis lowers the activation energy, enabling reactions to occur at lower temperatures and milder conditions. Catalytic materials exhibit selectivity, promoting desired reactions while minimizing undesired side reactions. Enhanced catalytic activity also improves the long-term stability and durability of waste-derived electrodes, ensuring consistent performance over extended periods. Additionally, it enables the utilization of low-grade waste materials, expanding resource availability and promoting sustainability. These combined benefits synergistically enhance the overall efficiency and effectiveness of waste-derived electrode systems (Chen *et al.*, 2022b).

13.4.2 Unique surface properties

Bioelectrodes may have high porosity or specific functional groups on their surface, which play a crucial role in promoting biofilm formation and facilitating microbial attachment. The high porosity allows for a larger surface area, providing ample space for the growth and colonization of microorganisms. Additionally, specific functional groups can serve as binding sites for microbial cells, facilitating their attachment and biofilm development on the electrode surface. This enhanced microbial attachment promotes efficient electron transfer between the microorganisms and the electrode, leading to improved bioelectrode performance (Lekshmi *et al.*, 2023).

13.4.3 Tailored selectivity

Waste-derived membranes offer the advantage of tailored selectivity by being customizable to exhibit specific preferences for ions or molecules. Through modifications and engineering, these membranes can achieve enhanced separation efficiency. By controlling their composition, structure, and surface properties, waste-derived membranes selectively allow the passage of desired substances while blocking others. This customization improves processes such as water treatment and resource recovery, making waste-derived membranes valuable in various applications (Lu *et al.*, 2022).

13.4.4 Cost-effective solutions

Waste-derived materials present a cost-effective solution by leveraging waste materials, resulting in substantially reduced production costs compared to conventional materials. By repurposing waste materials, it not only provides a sustainable approach but also offers significant cost savings without compromising performance. This affordability factor renders waste-derived materials highly appealing for the implementation of METs on a larger scale, driving the widespread adoption of these technologies for efficient energy generation and environmentally friendly remediation practices (Xie *et al.*, 2023).

13.4.5 Waste valorization

Waste-derived electrodes and membranes play a vital role in waste valorization by transforming discarded materials into valuable components within METs. Instead of being viewed as mere waste, these materials are repurposed and integrated into the functioning of METs, generating added value. By utilizing waste materials as electrodes and membranes, METs enable the conversion of waste into valuable products, such as energy generation or resource recovery. This approach not only reduces waste accumulation but also promotes sustainability by maximizing the utilization of available resources and represents a significant step toward a circular economy and the efficient management of waste materials (Chandrasekhar *et al.*, 2021).

13.4.5.1 Limitations

Waste-derived electrodes and membranes in METs also face certain limitations such as low columbic efficiency and limited power density, slow kinetic rate of electron transfer from anodic microbes to electrode, limited catalytic activity, biofilm detachment and instability, biofouling and biofilm resistance, high internal resistance, and compatibility issues under harsh operating conditions. These challenges hinder the overall performance and efficiency of MET systems. Overcoming these limitations requires a reduction of biofilm thickness for predominant activities of electrogenic microbes improving power density, optimizing operational conditions for increased biofilm activity for improving electron transfer kinetics, the close proximity of anode to exchange membrane to the cathode for reducing internal resistance, and designing robust membranes for harsh conditions. Thus, these mitigation methods can address these challenges and advance the effectiveness of waste-derived electrodes and membranes in METs.

13.5 CONCLUSION

METs hold immense potential in addressing energy generation, waste treatment, and resource recovery challenges through the synergy of microbiology and electrochemistry. However, conventional electrodes and membranes have limitations, leading researchers to explore innovative solutions using industrial waste materials. By converting waste into novel electrode and membrane materials, METs can become more efficient and eco-friendly, and contribute to a circular economy. The choice of electrode and membrane materials is crucial for METs efficiency, and carbon-based materials, metal/metal oxides, and IEMs are commonly employed. Selecting the appropriate materials allows METs to achieve better performance in various applications, from energy generation to waste treatment and resource recovery. Utilizing waste sources for electrode and membrane materials presents promising alternatives, such as plastic waste-derived electrodes with cost-effectiveness and high catalytic activity, agricultural waste-derived electrodes offering sustainable options for supercapacitor applications, and biomass waste-derived electrodes demonstrating outstanding performance in microbial electrochemical systems. Additionally, graphene sheets from different waste materials serve as cost-effective electrode modifications and catalysts. Embracing waste materials for electrode fabrication promotes a greener, sustainable future by reducing landfill waste and minimizing the demand for expensive raw materials. However, further research and optimization are needed to fully explore the potential of waste-derived materials and enhance their efficiency in various applications. Similarly, utilizing waste materials for membrane fabrication in METs offers a sustainable and environmentally responsible approach. These waste-derived membranes from rice husk, waste tire, newspaper waste, and aluminum cans show enhanced performance in various water purification processes, gas separation, and pollutant removal, contributing to a circular economy and minimizing environmental impact. Overall, waste-derived electrodes and membranes offer numerous advantages, including enhanced catalytic activity, tailored selectivity, cost-effectiveness, and waste valorization. To fully unleash their potential, addressing challenges such as low columbic efficiency and biofilm-related issues requires ongoing research and innovation. In conclusion, waste-derived materials play a crucial role in driving us toward a more sustainable future. By repurposing waste into valuable components within METs, we can address energy and waste challenges while promoting a greener, more efficient, and environmentally conscious society. Through continuous efforts in research and optimization, waste-derived electrodes and membranes have the potential to revolutionize energy production and wastewater treatment, contributing to a cleaner, more sustainable world for generations to come.

REFERENCES

Akhavan O., Bijanzad K. and Mirsepah A. (2014). Synthesis of graphene from natural and industrial carbonaceous wastes. *RSC Advances*, **4**(39), 20441–20448, https://doi.org/10.1039/C4RA01550A

Ayyaru S., Letchoumanane P., Dharmalingam S. and Stanislaus A. R. (2012). Performance of sulfonated polystyrene–ethylene–butylene–polystyrene membrane in microbial fuel cell for bioelectricity production. *Journal of Power Sources*, **217**, 204–208, https://doi.org/10.1016/j.jpowsour.2012.05.053

Aziz M. H. A., Othman M. H. D., Hashim N. A., Rahman M. A., Jaafar J., Hubadillah S. K. and Tai Z. S. (2019). Pretreated aluminium dross waste as a source of inexpensive alumina-spinel composite ceramic hollow fibre membrane for pretreatment of oily saline produced water. *Ceramics International*, **45**(2), 2069–2078, https://doi.org/10.1016/j.ceramint.2018.10.110

Bani-Melhem K., Al-Kilani M. R. and Tawalbeh M. (2023). Evaluation of scrap metallic waste electrode materials for the application in electrocoagulation treatment of wastewater. *Chemosphere*, **310**(October 2022), 136668, https://doi.org/10.1016/j.chemosphere.2022.136668

Behera M., Jana P. S. and Ghangrekar M. M. (2010). Performance evaluation of low cost microbial fuel cell fabricated using earthen pot with biotic and abiotic cathode. *Bioresource Technology*, **101**(4), 1183–1189, https://doi.org/10.1016/j.biortech.2009.07.089

Bhoyate S., Ranaweera C. K., Zhang C., Morey T., Hyatt M., Kahol P. K., Ghimire M., Mishra S. R. and Gupta R. K. (2017). Eco-friendly and high performance supercapacitors for elevated temperature applications using recycled tea leaves. *Global Challenges*, **1**(8), 1700063, https://doi.org/10.1002/GCH2.201700063

Butti S. K., Velvizhi G., Sulonen M. L. K., Haavisto J. M., Oguz Koroglu E., Yusuf Cetinkaya A., Singh S., Arya D., Annie Modestra J., Vamsi Krishna K., Verma A., Ozkaya B., Lakaniemi A. M., Puhakka J. A. and Venkata Mohan S. (2016). Microbial electrochemical technologies with the perspective of harnessing bioenergy: maneuvering towards upscaling. *Renewable and Sustainable Energy Reviews*, **53**, 462–476, https://doi.org/10.1016/j.rser.2015.08.058

Chakraborty I., Sathe S. M., Dubey B. K. and Ghangrekar M. M. (2020). Waste-derived biochar: applications and future perspective in microbial fuel cells. *Bioresource Technology*, **312**(April), 123587, https://doi.org/10.1016/j.biortech.2020.123587

Chandrabhan Shende R., Muruganathan M., Mizuta H., Akabori M. and Sundara R. (2018). Chemical simultaneous synthesis strategy of two nitrogen-rich carbon nanomaterials for all-solid-state symmetric supercapacitor. *ACS Omega*, **3**(12), 17276–17286, https://doi.org/10.1021/acsomega.8b02835

Chandrasekhar K., Kumar A. N., Raj T., Kumar G. and Kim S. H. (2021). Bioelectrochemical system-mediated waste valorization. *Systems Microbiology and Biomanufacturing*, **1**(4), 432–443, https://doi.org/10.1007/S43393-021-00039-7

Chen Z., Wei W., Ni B.-J. and Chen H. (2022a). Plastic wastes derived carbon materials for green energy and sustainable environmental applications. *Environmental Functional Materials*, **1**(1), 34–48, https://doi.org/10.1016/j.efmat.2022.05.005

Chen Z., Yun S., Wu L., Zhang J., Shi X., Wei W., Liu Y., Zheng R., Han N. and Ni B. J. (2022b). Waste-derived catalysts for water electrolysis: circular economy-driven sustainable green hydrogen energy. *Nano–Micro Letters*, **15**(1), 1–37, https://doi.org/10.1007/S40820-022-00974-7

Daud S. M., Kim B. H., Ghasemi M. and Daud W. R. W. (2015). Separators used in microbial electrochemical technologies: current status and future prospects. *Bioresource Technology*, **195**, 170–179, https://doi.org/10.1016/j.biortech.2015.06.105

Deshwal G. K. and Panjagari N. R. (2020). Review on metal packaging: materials, forms, food applications, safety and recyclability. *Journal of Food Science and Technology*, **57**(7), 2377–2392, https://doi.org/10.1007/s13197-019-04172-z

Divyashree A., Manaf S. A. B. A., Yallappa S., Chaitra K., Kathyayini N. and Hegde G. (2016). Low cost, high performance supercapacitor electrode using coconut wastes: eco-friendly approach. *Journal of Energy Chemistry*, **25**(5), 880–887, https://doi.org/10.1016/j.jechem.2016.08.002

Fan Y., Hu H. and Liu H. (2007). Enhanced coulombic efficiency and power density of air-cathode microbial fuel cells with an improved cell configuration. *Journal of Power Sources*, **171**(2), 348–354, https://doi.org/10.1016/j.jpowsour.2007.06.220

Ghasemi M., Daud W. R. W., Ismail A. F., Jafari Y., Ismail M., Mayahi A. and Othman J. (2013). Simultaneous wastewater treatment and electricity generation by microbial fuel cell: performance comparison and cost investigation of using Nafion 117 and SPEEK as separators. *Desalination*, **325**, 1–6, https://doi.org/10.1016/j.desal.2013.06.013

Hernandez C. A. and Osma J. F. (2020). Microbial electrochemical systems: deriving future trends from historical perspectives and characterization strategies. *Frontiers in Environmental Science*, **8**(May), 1–19, https://doi.org/10.3389/fenvs.2020.00044

Hopewell J., Dvorak R. and Kosior E. (2009). Plastics recycling: challenges and opportunities. *Philosophical Transactions of the Royal Society B: Biological Sciences*, **364**(1526), 2115–2126, https://doi.org/10.1098/rstb.2008.0311

Ilango M. S., Mutalikdesai A. and Ramasesha S. K. (2016). Anodization of aluminium using a fast two-step process. *Journal of Chemical Sciences*, **128**(1), 153–158, https://doi.org/10.1007/s12039-015-1006-8

Kalathil S., Patil S. A. and Pant D. (2018). Microbial fuel cells: electrode materials. *Encyclopedia of Interfacial Chemistry*, 309–318, https://doi.org/10.1016/B978-0-12-409547-2.13459-6

Kaur R., Singh S., Chhabra V. A., Marwaha A., Kim K. H. and Tripathi S. K. (2021). A sustainable approach towards utilization of plastic waste for an efficient electrode in microbial fuel cell applications. *Journal of Hazardous Materials*, **417**(November 2020), 125992, https://doi.org/10.1016/j.jhazmat.2021.125992

Ke Y., Yuan W., Zhou F., Guo W., Li J., Zhuang Z., Su X., Lu B., Zhao Y., Tang Y., Chen Y. and Song J. (2021). A critical review on surface-pattern engineering of Nafion membrane for fuel cell applications. *Renewable and Sustainable Energy Reviews*, **145**(March), 110860, https://doi.org/10.1016/j.rser.2021.110860

Khedulkar A. P., Pandit B., Dang V. D. and van Doong R. (2023). Agricultural waste to real worth biochar as a sustainable material for supercapacitor. *Science of the Total Environment*, **869**, 161441, https://doi.org/10.1016/J.SCITOTENV.2023.161441

Kim B., An J., Kim D., Kim T. and Chang I. S. (2012). *Inherent overpotential generated by characteristic of microbial fuel cells with multiple membrane electrode assemblies*. pp. 2–6.

Lekshmi G. S., Bazaka K., Ramakrishna S. and Kumaravel V. (2023). Microbial electrosynthesis: carbonaceous electrode materials for CO_2 conversion. *Materials Horizons*, **10**(2), 292–312, https://doi.org/10.1039/D2MH01178F

Li D., Zhan W., Gao X., Wang Q., Li L. P., Zhang J., Cai G., Zuo W. and Tian Y. (2023). Aminated waste paper membrane for efficient and rapid filtration of anionic dyes and antibiotics from water. *Chemical Engineering Journal*, **455**, 140641, https://doi.org/10.1016/J.CEJ.2022.140641

Lin Y. T., Zhuang G. L., Wey M. Y. and Tseng H. H. (2020). The viable fabrication of gas separation membrane used by reclaimed rubber from waste tires. *Polymers*, **12**(11), 1–15, https://doi.org/10.3390/polym12112540

Lu D., Yao Z., Jiao L., Waheed M., Sun Z. and Zhang L. (2022). Separation mechanism, selectivity enhancement strategies and advanced materials for mono-/multivalent ion-selective nanofiltration membrane. *Advanced Membranes*, **2**, 100032, https://doi.org/10.1016/j.advmem.2022.100032

Mahmoud R. H., Gomaa O. M. and Hassan R. Y. A. (2022). Bio-electrochemical frameworks governing microbial fuel cell performance: technical bottlenecks and proposed solutions. *RSC Advances*, **12**(10), 5749–5764, https://doi.org/10.1039/d1ra08487a

Maraveas C. (2020). Production of sustainable construction materials using agro-wastes. *Materials*, **13**(2), 262, https://doi.org/10.3390/ma13020262

Mensah-Darkwa K., Zequine C., Kahol P. K. and Gupta R. K. (2019). Supercapacitor energy storage device using biowastes: a sustainable approach to green energy. *Sustainability (Switzerland)*, **11**(2), 414, https://doi.org/10.3390/su11020414

Mishra A. K., Bose S., Kuila T., Kim N. H. and Lee J. H. (2012). Silicate-based polymer-nanocomposite membranes for polymer electrolyte membrane fuel cells. *Progress in Polymer Science*, **37**(6), 842–869, https://doi.org/10.1016/j.progpolymsci.2011.11.002

Mohamed M. A., Salleh W. N. W., Jaafar J., Ismail A. F., Abd. Mutalib M. and Jamil S. M. (2015). Feasibility of recycled newspaper as cellulose source for regenerated cellulose membrane fabrication. *Journal of Applied Polymer Science*, **132**(43), 1–10, https://doi.org/10.1002/app.42684

Mohan Y., Manoj Muthu Kumar S. and Das D. (2008). Electricity generation using microbial fuel cells. *International Journal of Hydrogen Energy*, **33**(1), 423–426, https://doi.org/10.1016/j.ijhydene.2007.07.027

Muhammad S., Khalil H. P. S., Abd Hamid S., Albadn Y. M., Suriani A. B., Kamaruzzaman S., Mohamed A., Allaq A. A. and Yahya E. B. (2022). Insights into agricultural-waste-based nano-activated carbon fabrication and modifications for wastewater treatment application. *Agriculture*, **12**(10), 1737, https://doi.org/10.3390/agriculture12101737

Mustakeem. (2015). Electrode materials for microbial fuel cells: nanomaterial approach. *Materials for Renewable and Sustainable Energy*, **4**(4), 1–11, https://doi.org/10.1007/s40243-015-0063-8

Nippatla N. and Philip L. (2020). Electrochemical process employing scrap metal waste as electrodes for dye removal. *Journal of Environmental Management*, **273**(March), 111039, https://doi.org/10.1016/j.jenvman.2020.111039

Olayiwola Sirajudeen A. A., Mohamad Annuar M. S., Ishak K. A., Yusuf H. and Subramaniam R. (2021). Innovative application of biopolymer composite as proton exchange membrane in microbial fuel cell utilizing real wastewater for electricity generation. *Journal of Cleaner Production*, **278**, 123449, https://doi.org/10.1016/J.JCLEPRO.2020.123449

Pandit S., Savla N., Sonawane J. M., Sani, A. M., Muh A., Gupta P. K., Mathuriya, A. S., Rai, A. K., Jadhav, D. A., Jung, S. P., and Prasad R. (2021). Agricultural waste and wastewater as feedstock for bioelectricity generation using microbial fuel cells: recent advances. *Fermentation*, **7**(3), 169, https://doi.org/10.3390/fermentation7030169.

Pode R. (2016). Potential applications of rice husk ash waste from rice husk biomass power plant. *Renewable and Sustainable Energy Reviews*, **53**, 1468–1485, https://doi.org/10.1016/j.rser.2015.09.051

Ramirez J., McCabe B., Jensen P. D., Speight R., Harrison M., Van Den Berg L. and O'Hara I. (2021). Wastes to profit: a circular economy approach to value-addition in livestock industries. *Animal Production Science*, **61**(6), 541–550, https://doi.org/10.1071/AN20400

Raychaudhuri A. and Behera M. (2020). Ceramic membrane modified with rice husk ash for application in microbial fuel cells. *Electrochimica Acta*, **363**, 137261, https://doi.org/10.1016/j.electacta.2020.137261

Rodrigues Filho G., Monteiro D. S., da Meireles C. S., de Assunção R. M. N., Cerqueira D. A., Barud H. S., Ribeiro S. J. L. and Messadeq Y. (2008). Synthesis and characterization of cellulose acetate produced from recycled newspaper. *Carbohydrate Polymers*, **73**(1), 74–82, https://doi.org/10.1016/j.carbpol.2007.11.010

Samavati Z., Goh P. S. and Ismail A. F. (2023). Membrane fabrication by solid waste: opportunities and challenges. *Journal of Resource Recovery*, **1**(1), 1003, https://doi.org/10.52547/jrr.2210.1003

Santoro C., Arbizzani C., Erable B. and Ieropoulos I. (2017). Microbial fuel cells: from fundamentals to applications. A review. *Journal of Power Sources*, **356**, 225–244, https://doi.org/10.1016/j.jpowsour.2017.03.109

Singh J., Kumar M., Sharma A., Pandey G., Chae K. and Lee S. (2016). We are IntechOpen, the world's leading publisher of Open Access books Built by scientists, for scientists TOP 1%. *Intech*, **11**(tourism), 13.

Sun J., Hu Y., Bi Z. and Cao Y. (2009). Improved performance of air-cathode single-chamber microbial fuel cell for wastewater treatment using microfiltration membranes and multiple sludge inoculation. *Journal of Power Sources*, **187**(2), 471–479, https://doi.org/10.1016/j.jpowsour.2008.11.022

Ul Z., Sánchez-Peña P., Baeza M., Sulonen M., Gabriel D., Baeza J. A. and Guisasola A. (2023). Systematic screening of carbon-based anode materials for bioelectrochemical systems. *Journal of Chemical Technology & Biotechnology*, **98**, 1402–1415, https://doi.org/10.1002/jctb.7357

Vinodh R., Babu R. S., Sambasivam S., Muralee Gopi C. V. V., Alzahmi S., Kim H. J., de Barros A. L. F. and Obaidat I. M. (2022). Recent advancements of polyaniline/metal organic framework (PANI/MOF) composite electrodes for supercapacitor applications: a critical review. *Nanomaterials*, **12**(9), 1511, https://doi.org/10.3390/nano12091511

Xie K., Zhang W., Ren K., Zhu E., Lu J., Chen J., Yin P., Yang L., Guan X. and Wang G. (2023). Electrochemical performance of corn waste derived carbon electrodes based on the intrinsic biomass properties. *Materials*, **16**(14), 5022, https://doi.org/10.3390/ma16145022

Yaqoob A. A., Mohamad Ibrahim M. N. and Rafatullah M., Chua, Y. S., Ahmad, A. and Umar, K. (2020). Recent advances in anodes for microbial fuel cells: an overview. *Materials*, **13**(9), 2078, https://doi.org/10.3390/ma13092078

Yuan Y., Liu T., Fu P., Tang J. and Zhou S. (2015). Conversion of sewage sludge into high-performance bifunctional electrode materials for microbial energy harvesting. *Journal of Materials Chemistry A*, **3**(16), 8475–8482, https://doi.org/10.1039/c5ta00458f

Zhang X., Cheng S., Huang X. and Logan B. E. (2010). The use of nylon and glass fiber filter separators with different pore sizes in air-cathode single-chamber microbial fuel cells. *Energy and Environmental Science*, **3**(5), 659–664, https://doi.org/10.1039/b927151a

Zhou Y., Tang L., Liu Z., Hou J., Chen W., Li Y. and Sang L. (2017). A novel anode fabricated by three-dimensional printing for use in urine-powered microbial fuel cell. *Biochemical Engineering Journal*, **124**, 36–43, https://doi.org/10.1016/j.bej.2017.04.012

Zhuang G. L., Wey M. Y. and Tseng H. H. (2016). A novel technique using reclaimed tire rubber for gas separation membranes. *Journal of Membrane Science*, **520**, 314–325, https://doi.org/10.1016/j.memsci.2016.07.044

doi: 10.2166/9781789063813_0245

Chapter 14

Treatment of industrial wastewater with emerging pollutants through microbial fuel cells

Edwin Hualpa-Cutipa[1], Ingrid Maldonado-Jimenez[2], Evelin Yana-Neyra[2], Linda Cristal Sanchez Garcia[3], Nikol Gianella Julca[3], María José Mayhua Mallma[3], Yadira Karolay Ravelo Machari, Susan Raquel Ruiz Alba[4], Lucero Katherine Castro Tena[5] and Luis Manuel Atoche Ordoñez[6]*

[1]*Faculty of Pharmacy and Biochemistry, Biotechnology and Omics in Life Sciences Research Group, Universidad Nacional Mayor de San Marcos, Lima, Peru*
[2]*Universidad Nacional del Altiplano, Puno, Peru*
[3]*Faculty of Pharmacy and Biochemistry, Biotechnology and Omics in Life Sciences Research Group, Universidad Nacional Mayor de San Marcos, Lima, Peru*
[4]*Grupo de investigación Biofilms Bacterianos en el Perú, Facultad de Farmacia y Bioquímica, Universidad Nacional Mayor de San Marcos, Lima, Peru*
[5]*Universidad Nacional José Faustino Sánchez Carrión, Huacho, Peru*
[6]*Universidad Nacional de Piura, Piura, Peru*
**Corresponding author: ehualpa@continental.edu.pe*

ABSTRACT

Currently, environmental pollution is on the rise because of industrial growth to fulfill the demands of a society that is undergoing rapid demographic growth. This population change triggers the use of more resources and the release of pollutants into the environment. From the range of pollutants generated, new hazardous pollutants are being released, these are known as emerging pollutants. Several efforts have been made to address the management, treatment, disposal, and elimination of industrial effluents containing high levels of persistent contaminants, but these efforts have not been effective. A few approaches and/or techniques have been developed to treat these pollutants to mitigate their negative effects. Among these methods, technologies based on microbial fuel cells (MFCs) have proven to be an effective and suitable strategy for treating industrial wastewater containing recalcitrant pollutants (pharmaceuticals, hormones, pesticides, personal care products, etc.). In an MFC system, bacteria have an electroactive role, both in the degradation of these compounds and the donation of electrons for the generation of electricity in the process, this in turn accelerates the removal rate. In addition, emerging contaminants are organic compounds, that is, their composition contains carbon bonds, which can be used as a carbon source by bacteria and can also be degraded by electrochemical redox reaction processes. The aim of this chapter was to discuss the characteristics and properties of MFCs about their application in the treatment of wastewater rich in emerging pollutants that are resistant to spontaneous biodegradation. Finally, the prospects and future challenges of MFCs for applications in large-scale systems will be presented because studies are still being carried out at the laboratory level.

Keywords: industrial wastewater, emerging pollutants, MFC, pollution, ecosystems

14.1 INTRODUCTION

The effective management and remediation of pollutants released through industrial wastewater poses significant environmental concerns (Garg *et al.*, 2022). Traditional pollutants found in wastewater primarily consist of organic matter and heavy metals. However, in recent years, a category of diverse compounds known as 'emerging pollutants' has been recognized (Khan *et al.*, 2022). Emerging contaminants include different chemical compounds of industrial origin, which are of growing concern due to their persistence, toxicity, and potential negative impact on aquatic ecosystems and human health. Among these, pollutants from the pharmaceutical industry such as antibiotics, anti-inflammatories, beta blockers, lipid regulators, X-ray contrast media (concentration range from ng/L to μg/L), among others, which represent great relevance in pollution (Sivaranjanee & Kumar, 2021).

The efficiency of elimination of emerging pollutants is the greatest challenge in wastewater treatment because in addition to emphasizing the recovery of water with ideal levels of quality, it must consider the reduction of pollutants released to the environment. In this sense, several wastewater treatment methods have been developed. Advanced oxidation processes are among the most common methods, which are based on the application of powerful oxidants on the medium, with the aim of degrading, transforming, and reducing the toxic effects of pollutants (Ahmed *et al.*, 2021). Another potential method for wastewater treatment is adsorption with active materials, such as activated carbon or ion-exchange resins, which can selectively capture pollutants in water. Additionally, membrane technology, such as reverse osmosis and nanofiltration, has been shown to be effective in retaining emerging contaminants at the molecular level (Liu *et al.*, 2022). The use of biological treatments, such as activated sludge systems and microbial fuel cells (MFCs), has also gained interest due to their ability to degrade and metabolize various compounds (Oveisi *et al.*, 2021).

There are several alternative methods for the removal and/or biodegradation of emerging contaminants from wastewater. However, it is very important to keep in mind that the selected method must be associated with the nature of the contaminants identified, as well as the physicochemical characteristics of wastewater and the specific treatment requirements. As different methods (chemical and physics) do not achieve the maximum removal efficiency of emerging contaminants in industrial wastewater, alternative methods are being studied, among which the use of MFCs has been highlighted (Jain *et al.*, 2020). MFCs are bioelectrochemical systems that decompose organic matter present in wastewater with the simultaneous generation of electricity using microorganisms as biocatalysts (Sarma *et al.*, 2022). Unlike traditional treatment methods, MFCs offer significant advantages, such as higher energy efficiency, the possibility of recovering valuable products, and the ability to degrade persistent organic compounds (Qiao & Xiong, 2021).

Using MFCs is known to be an innovative and promising strategy in the field of wastewater treatment. The microbial cells that comprise these systems have a metabolic versatility that allows them to take advantage of pollutants and use them as nutrients, and in this process, they can generate electricity or chemical compounds, all through electrochemical reactions. Specifically, microorganisms oxidize the organic compounds present in contaminated effluents, resulting in a transfer of electrons through the electrodes included in the system, which generates usable electric current (Saadi *et al.*, 2020).

An essential component of MFCs is microorganisms, which are called exoelectrogenic or electroactive microbes. The capacity and ability of these microorganisms allows them to directly transfer electrons from organic or pollutant matter onto electrodes, all under oxygen-free (anaerobic) conditions. Within this group of microbes, the following genera have been identified: *Geobacter*, *Shewanella*, and *Anaeromyxobacter*. The mechanism identified for electron transfer is called 'extracellular electron transfer' (Rusyn, 2021).

Various biomolecules (organic acids, sugars, alcohols, among others) present on a contaminated substrate are oxidized by exoelectrogenic microorganisms at the anode of an MFC. During this process, electrons are recovered from the biomolecules and transferred to the electrode surface (anode), ultimately leading to the generation of an electric current (Umar *et al.*, 2020).

This chapter explores the potential of MFCs for the treatment of industrial wastewater with emerging contaminants. Mechanisms involved in the degradation of emerging contaminants, as well as the optimal operating conditions and the associated challenges, are also analyzed. In addition, recent advances in the research and development of this technology, as well as future perspectives in the field of industrial wastewater treatment are discussed. Finally, it is expected to raise awareness about the importance of implementing advanced technologies that minimize environmental impact and promote the preservation of our water resources.

14.2 MFCs: HOW THEY WORK?

MFCs are categorized as emerging technologies with diverse applications because they are involved in the removal of toxic compounds from the aquatic environment and the simultaneous production of electrical energy (Shaikh *et al.*, 2020; Thapa *et al.*, 2022). This bioelectricity system has the ability to produce energy at low cost in comparison to physicochemical methods (Shabani *et al.*, 2020; Tsekouras *et al.*, 2022). The design is basically composed of two zones. The anodic zone is where the oxidation of organic matter, as well as toxic compounds in the medium, is carried out by electroactive bacteria that utilize electrons in the absence of oxygen (Figure 14.1). The cathodic zone is where the catalytic activity will take place, through the reduction of oxygen and carbon dioxide (CO_2) (Huarachi-Olivera *et al.*, 2018). A proton-exchange membrane (PEM) favors the passage of protons from the anode to the cathode. All this integrated to an electrical circuit through which electrons transit (Kong *et al.*, 2021; Thapa *et al.*, 2022). At the cathode, a reduction reaction is required to complete the electrical circuit (Kumar *et al.*, 2017a).

The conversion process carried out in the anodic zone involves various groups of bacteria adapted to achieve a homogeneous development on anodes (Shaikh *et al.*, 2020). In this zone, microorganisms use an anode as an electron acceptor during the oxidation process allowing them to reduce organic and inorganic compounds of toxic nature (Tucci *et al.*, 2021). Bacteria can directly transfer electrons to an extracellular environment, due to the use of membrane-bound proteins located on the surface of the membrane that acts as electron transfer agents; these include cytochromes, iron–sulfur proteins, and conductive pili. Similarly, there are also compounds such as flavins and phenazines or some artificial redox mediators that allow an indirect transport of electrons outside the bacterial structure.

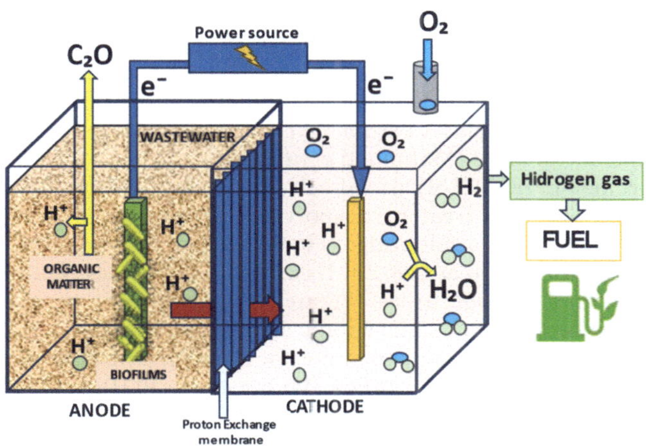

Figure 14.1 Operation of MFCs.

These electron transfer pathways occur in combination with conductive pili produced by bacteria called nanowires (Sharma *et al.*, 2014). After oxidation, electrons flow from the anodic to the cathodic zone through a conductive material (Tucci *et al.*, 2021).

The performance of MFCs is conditioned by different factors, such as bacterial type and concentration, electrode material, nature of the substrate, redox potential, electrolyte chemistry, as well as operationalization conditions (Sharma *et al.*, 2014). Numerous bacterial genera exist with the ability to remove toxic compounds and produce electricity, such as *Escherichia coli*, *Pseudomonas*, *Ochrobactrum*, *Geobacter*, *Shewanella*, *Geopsychrobacter*, *Geothrix*, *Desulfuromonas*, *Desulfobulbus*, *Rhodoferax ferrireducens*, and photosynthetic *Rhodopseudomonas palustris* (Guang *et al.*, 2020). The microorganisms in these systems are called electroactive microorganisms (Rusyn, 2021).

In an aerobic respiration process, the final electron acceptor is oxygen (O_2) (Guang *et al.*, 2020). However, in an anaerobic process, the final electron acceptor can be composed of electrodes that can be developed by four mechanisms: (1) direct connection of bacteria with the electrode (anode) in order to transfer electrons through C-type cytochromes; this process occurs by a direct physical contact of bacteria with the electrode; (2) use of transmembrane proteins as electron transport mediators such as flavins and phenazines; (3) bacteria pili are made of proteins that can conduct electrons; they provide a direct conduit for electrons to travel from bacterial cells to the anode such as *Shewanella* and *Geobacter*; and (4) establishment of electroactive biofilms on the electrode surface (Guang *et al.*, 2020; Mahmoud *et al.*, 2022; Shaikh *et al.*, 2020; Sharma *et al.*, 2014; Tucci *et al.*, 2021; Verma *et al.*, 2021). The electrons released at the anode pass through an electrical circuit to reduce O_2 to H_2O in the cathodic zone. The flow of electrons through the circuit generates electricity, which can be measured and stored (Guang *et al.*, 2020).

14.3 DESIGN AND EFFICIENCY OF MFCs

The material used and configuration of MFC design and operation are factors that affect the efficiency of these systems. The material used is diverse, including plastic-based materials, plexiglass, and so on (Du *et al.*, 2007; Liu & Lee, 2021; Shaikh *et al.*, 2020).

Electrodes are key parts in MFC systems as bacterial communities develop on their surface. Efficiency of electrodes depends on the characteristics of the material used such as surface area, durability, porosity, electrical conductivity, hydrophilicity, biocompatibility, cost, and anticorrosion properties. The conductivity of electrodes increases when the resistance reduces as a result of electron transfer. A suitable material used for anodes should be inexpensive, chemically inert, with a large surface area, low resistance, and biocompatible for optimal bacterial adhesion (Verma *et al.*, 2021). The electrode material is usually fabricated from carbon (Du *et al.*, 2007), graphite rods, granular graphite, carbon paper, cross-linked glassy carbon, carbon mesh, carbon felt, carbon brush, and some metal anode strips on the bottom half (Mahmoud *et al.*, 2022).

An electron transport membrane (ETM) also affects the efficiency of the process because it depends on the main material used, allowing greater transport of protons from an anodic zone to a cathodic zone. Instead of an ETM, sometimes a salt bridge (Figure 14.2) or other materials such as Nafion (Du *et al.*, 2007) and polyvinyl alcohol (Liu & Lee, 2021) can be used. The best-known MFCs are cells with two compartments (anodic and cathodic areas) (Figure 14.2A), which are separated by a PEM (Verma *et al.*, 2021). However, in single-cell MFCs, the anodic area is near the base, and the cathodic area is at the surface part, where it comes in contact with atmospheric oxygen (Kong *et al.*, 2021) (Figure 14.2B). There are also up-flow, stacked, flat, tubular MFCs (Verma *et al.*, 2021).

14.3.1 Design of MFCs

The amount of energy generated through MFCs depends mainly on their design and integrated materials, that is, the distance between the electrodes, the electrode used, the PEM, the mediators,

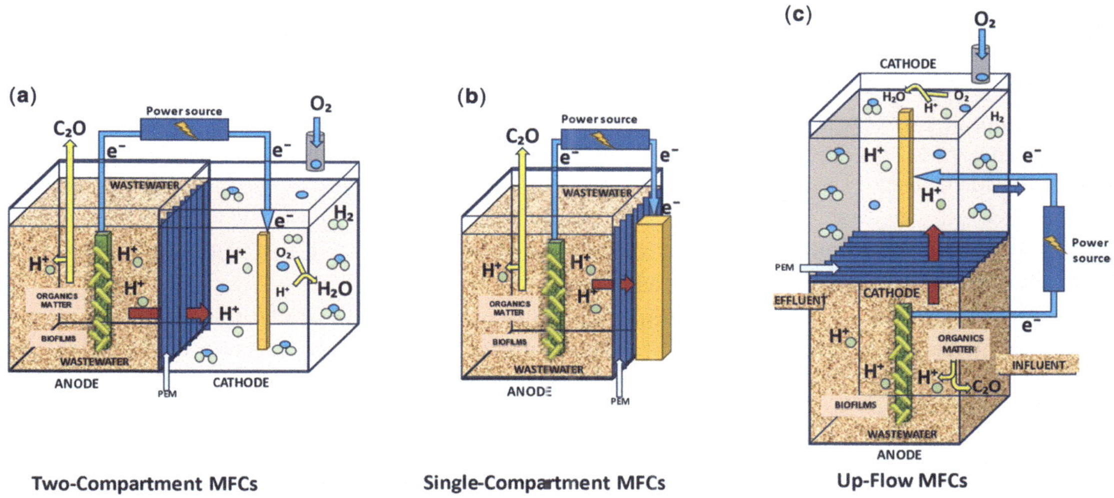

Figure 14.2 Design of MFCs and their types: (A) two-compartment MFCs, (B) single-compartment MFCs, and (C) up-flow MFCs.

the substrate, and the microorganisms used. MFCs are composed of different designs, including stacked, single chamber, double chamber, and so on. PEM is the main component of MFCs, as it affects energy production. It is prepared using nafion, cellophane, agar, and so on. Mediators are the compounds involved in the electron transport and thus induce power density. Examples of mediators: reduced nicotinamide adenine dinucleotide phosphate, oxidized nicotinamide adenine dinucleotide phosphate, cytochromes, and so on (Malik *et al.*, 2023).

14.3.1.1 Two-compartment MFC design

This type of MFC design is composed of two compartments for the anode and cathode (Figure 14.2A), both connected by an ion-exchange membrane (PEM) or a salt bridge (1% salt, either NaCl or KCl, 2.5% agar). The anode compartment is under anaerobic conditions and contains microorganisms, specific substrate (wastewater), and an electrode. The second compartment is under aerobic conditions and contains fresh water or buffers (catholyte), an electrode, and an O_2 supply. The materials for electrodes for both anode and cathode can be stainless-steel mesh, copper, graphite, carbon paper, carbon fiber, and graphite brush (Malik *et al.*, 2023; Thakur *et al.*, 2017).

14.3.1.2 Single-compartment MFC design

Single-compartment MFCs are composed of an anode compartment (under anaerobic conditions) and the cathode is exposed (hence no oxygen supply (O_2) is required), they are simple and cost-effective (Figure 14.2B). Embedded porous carbon can be used as the anode and carbon cloth together with a platinum catalyst works as the cathode. PEM can be prepared using naphlon, and a copper (Cu) wire is used to join electrodes and external circuits (Malik *et al.*, 2023; Thakur *et al.*, 2017).

14.3.1.3 Up-flow MFCs

Up-flow MFCs have a continuous mode design and apply wastewater injection from the bottom with high force upward and the effluent exits from the top of the system (Figure 14.2C). The design can be fabricated from polyacrylic plastic or glass flasks. It features a cation-exchange membrane (CEM). The anode and cathode can be of graphite felt, carbon fiber, carbon brush, and carbon cloth covered with activated carbon. They feature bead and glass wool separators. Aeration is provided by a cathodic layer. The electrodes and the external circuit are connected by a platinum wire. The electrodes can be prepared using a copper wire and/or titanium wire (Malik *et al.*, 2023).

14.3.2 Efficiency of MFCs

The efficiency of MFCs depends on the operating conditions and the materials and microorganisms used in the MFC design (Foudhaili *et al.*, 2019).

14.4 POTENTIAL USE OF WASTEWATER CONTAMINATED WITH ENDOCRINAL DISRUPTORS AS A SUBSTRATE FOR MFCs

Endocrine-disrupting compounds are chemicals, not present in the human body, that alter the hormonal balance of physiological processes even at very low concentrations. And they are being considered as emerging pollutants because they are dispersed in the environment, including in wastewater (Werkneh *et al.*, 2022). Lauretta *et al.* (2019) pointed out that endocrine disruptors are exogenous agents that can interfere in the synthesis, secretion, transport, metabolism, binding, or elimination of hormones present in organisms. These exogenous substances can be of natural or synthetic origin, being acquired through environmental exposures, affecting hormonal systems, and preventing adequate communication between them. Some of these exogenous agents are polychlorinated biphenyls, polybrominated biphenyls, dioxins, bisphenol A, pesticides, and pharmaceutical agents. According to Gonsioroski *et al.* (2020), exposure to these contaminants through drinking water can have detrimental effects on human reproduction, such as low sperm counts or adverse effects during pregnancy.

Tetrabromobisphenol A (TBBPA) is a possible substrate for an MFC, being better utilized in a closed bioanode, having shorter degradation periods with respect to the open circuit (Lin *et al.*, 2021). The generation of bisphenol A during the TBBPA degradation process is reduced. Similarly, TBBPA degradation pathways follow debromination, hydroxylation, aromatic ring opening, and *O*-methylation reactions. Another important disruptor that can be degraded with the use of MFCs is dimethyl phthalate, which is used in the manufacture of repellents and plastics. Sarmin *et al.* (2021) used acrylic acid as a co-substrate and *Pseudomonas aeruginosa* communities as biocatalysts in the degradation process of dimethyl phthalate. The electrical energy levels increased upon implementation of the photocatalytic treatment, favoring the degradation of dimethyl phthalate. Thus, the implementation of wetlands integrating MFC technology will allow better removal of disruptive compounds such as bisphenol A, which are found in large proportions in polluted water bodies (Li *et al.*, 2019).

14.5 POTENTIAL USE OF FOSSIL-FUEL POWER STATION WASTEWATER AS A SUBSTRATE FOR MFCs

Fossil fuels (natural gas, oil, and coal) are derived from waste organic matter that has been compacted and decomposed anaerobically over long periods of time (Thulasinathan *et al.*, 2022). Power plants using fossil fuels are sources of energy; however, they generate large amounts of CO_2 emissions and oil spills generating polluted waters, all of which negatively impact the environment and its associated flora and fauna (Cebrucean *et al.*, 2014). In addition, some fossil-fuel power plants, specifically coal-fired power plants are a major source of industrial wastewater. According to EPA (2020), wastewater comes from activities such as flue gas desulfurization in water treatment plants, inadequate ash handling and air pollution control systems, coal piles, miscellaneous waste drainage. According to Patel *et al.* (2023), some of wastewaters is organic, produced from gasification processes, coal water sludge, and entrained pulverized dry coal sewage. Wastewater from fossil-fuel power plants also contains heavy metals, acidic compounds, coal ash, oil and grease, cooling water additives, heat, chlorine, and suspended solids (EPA, 2020).

Currently, a search for new sustainable energy sources that do not cause more damage to the environment is underway. Therefore, the use of MFCs as bioelectrochemical systems is proposed as an alternative (Sarma *et al.*, 2022). The advantage of these systems is the generation of bioelectricity as well as the treatment of wastewater and the elimination of its pollutants. In this proposed system, microorganisms are the key components, which over the years have been used in the production of biofuels using organic substrates. These electricity-producing microorganisms are called electrogenic, which could convert organic matter into electricity through the inclusion of materials (electrodes) in MFC systems (Thulasinathan *et al.*, 2022).

The application and use of wastewater as a substrate for MFCs has been studied by several authors. Jatoi *et al.* (2021) observed how substrates present in wastewater act in terms of treatment efficiency as well as energy generation. Mahajan and Panwar (2022) demonstrated that there is sufficient information to consider MFCs as sustainable energy sources by using different nanomaterials to increase the performance of these bioelectric systems. All of the above-mentioned benefits and applications of MFCs will contribute to energy efficiency, electricity production, and pollutant removal, grouped together as a set of environmentally friendly techniques (Gude, 2016; Obileke *et al.*, 2021; Panwar & Mahajan, 2022).

14.6 POTENTIAL USE OF FOOD AND AGRICULTURAL WASTEWATER AS A SUBSTRATE FOR MFCs

The growing global demand for energy and the need to reduce dependence on fossil fuels highlight the importance of MFC-based approaches (Pandey *et al.*, 2016; Pandit *et al.*, 2021). In relation to the food and agricultural industry, they are known to generate large amounts of organic waste, which represents a major environmental and public health challenge, due to the potential presence of hazardous pollutants that can contribute to water and soil contamination (Chowdhary *et al.*, 2019; Okereafor *et al.*, 2020; Zhang *et al.*, 2016).

In recent years, the potential use of MFC systems in the treatment of wastewater from food and agricultural industries, which contain large amounts of organic matter, is being investigated these effluents being a very good source of electricity (Pandey *et al.*, 2016; Pandit *et al.*, 2021; Srivastava, 2019). The ability of MFCs to produce electricity from food and agricultural wastewater depends on the nature and concentration of organic compounds present in these waters (Kumar *et al.*, 2017b; Pandit *et al.*, 2021). In general, wastewaters with higher organic matter content are more suitable for energy production in MFCs (Figure 14.3). For example, wastewaters generated in the food industry, such as fruit and vegetable washing and processing waters, often contain high concentrations of carbohydrates and organic acids, which are readily oxidizable by microorganisms in CBMs (Asgharnejad *et al.*, 2021; Pandit *et al.*, 2021; Pervez *et al.*, 2021).

14.6.1 Factors altering the performance of MFCs

Several factors mentioned below can alter the performance of MFCs:

- *Type of microorganisms*: the choice of microorganisms used in MFC systems can significantly affect performance. Some microorganisms are more efficient at producing electricity than others, so proper selection of microorganisms is a critical factor in maximizing the efficiency of MFCs (Choudhury *et al.*, 2017).
- *Substrate type*: MFCs can use different types of substrates, such as wastewater, organic waste, and food, among others (Jung & Pandit, 2019).
- *Substrate concentration*: the concentration of organic compounds in the substrate can also influence the performance of MFCs. Too much substrate can inhibit microbial activity, whereas too little concentration can reduce electricity production (Jung & Pandit, 2019).
- *pH*: the pH of the medium can also influence microbial activity and, therefore the performance of CBMs. The optimal pH range for proper functioning of CBM systems is from 6.5 to 8.0 (Jung & Pandit, 2019; Tremouli *et al.*, 2016).

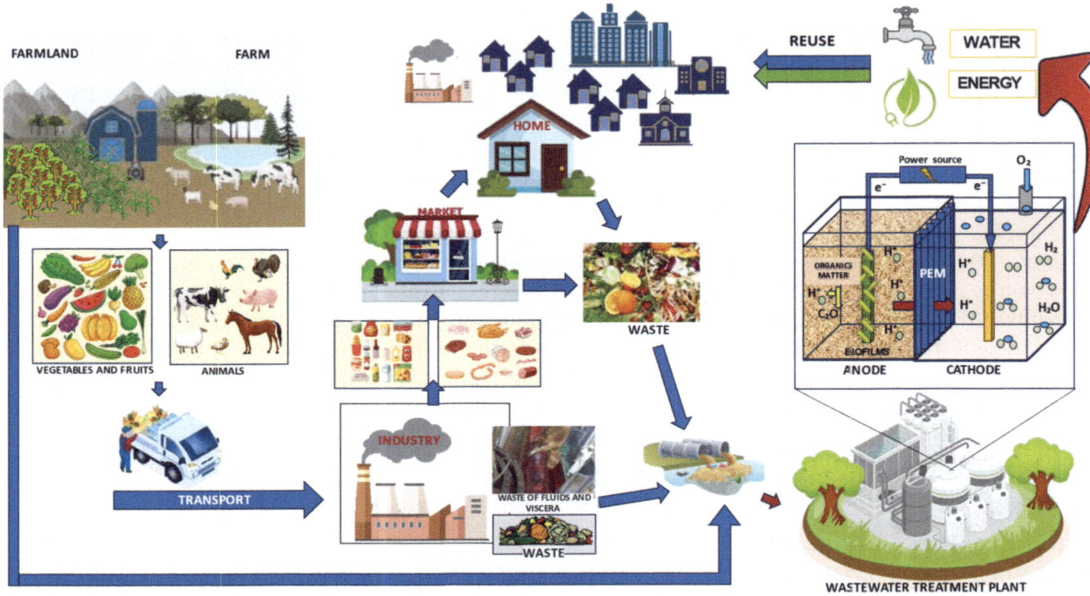

Figure 14.3 Use and reuse of wastewater from the food and agricultural industry integrated with MFCs.

- *Temperature*: it is also an important factor that can affect the microbial activity and performance of CBMs. The optimal temperature for CBMs depends on the type of microorganisms used, but they are typically in the range of 20–40°C (Jung & Pandit, 2019; Tremouli *et al.*, 2016).
- *Salinity*: high-salt concentrations can inhibit microbial activity, whereas low concentrations may not provide sufficient electrical conductivity (Guo *et al.*, 2020; Tremouli *et al.*, 2016).
- *Cell design*: the design of a fuel cell, including the electrode size, the distance between the electrodes, the shape of the electrodes, and the type of membrane used, can affect the efficiency of electricity production (Jung & Pandit, 2019).

The use of food and agricultural wastewater as a substrate in CBMs is an innovative and promising approach to treat organic waste and produce renewable energy in a sustainable manner. Its importance lies in contributing to the transition to a cleaner energy system and in reducing the environmental impact of the food and agricultural industry.

14.7 POTENTIAL USE OF PHARMACEUTICAL WASTEWATER AS A SUBSTRATE FOR MFCs

In recent years, the potential treatment of pharmaceutical wastewater containing a variety of organic and inorganic compounds with the use of MFC systems is being investigated (Li *et al.*, 2020). However, some challenges still need to be evaluated, such as (1) high presence of toxic compounds in the wastewater of the pharmaceutical industry (Li *et al.*, 2020; Thapa *et al.*, 2022), the presence of these compounds inhibit the population increase of microorganisms in MFCs, which will also have an impact on the generation of electricity (Thapa *et al.*, 2022); and (2) the heterogeneous composition (antibiotics, analgesics, anti-inflammatory compounds, etc.) of drug-contaminated effluents can have negative effects on the efficiency and stability of MFCs (Kesarwani *et al.*, 2023). However, certain advantages are observed in the use of drug-contaminated wastewater (Figure 14.4): (1) abundant organic matter and (2) optional effluents for electricity production (Kesarwani *et al.*, 2023; Thapa *et al.*, 2022).

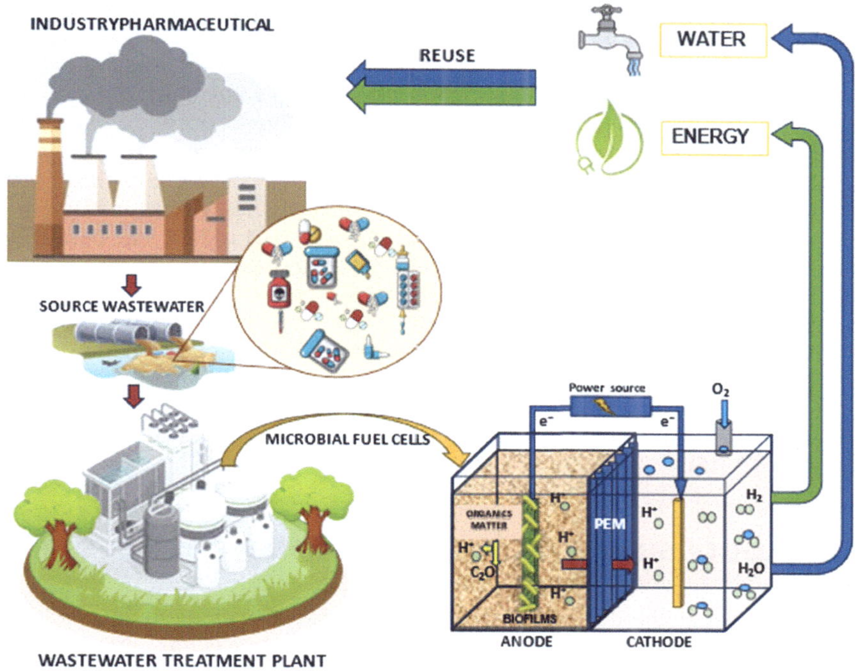

Figure 14.4 Schematic representation of pharmaceutical wastewater treatment by an MFC.

14.7.1 Production of energy from pharmaceutical wastewater

Energy production from pharmaceutical wastewater using MFCs is possible and has high potential as a sustainable alternative for wastewater treatment and renewable energy generation. However, more research is required to determine the feasibility and effectiveness of this technology in different contexts and for different types of organic compounds present in pharmaceutical wastewater (Thapa *et al.*, 2022).

Some of the studies with the use of MFCs for treatment of pharmaceutical products are illustrated below:

(a) *Ibuprofen and naproxen*: A dual-chambered MFC system was operated with a granular activated carbon cathode using pharmaceutical wastewater as a substrate and compared with high-density polyethylene as an alternative. Chemical oxygen demand (COD) removal of 83% was achieved in both cases, demonstrating that adsorption of organics did not affect the efficiency of the MFC (Sonawane *et al.*, 2022).

(b) *Sulfamethoxazole*: Approximately 50,000 tons of antibiotics are discharged into aquatic habitats annually. Sulfamethoxazole is a potent broad-spectrum antibiotic and is released in high concentrations in hospital wastewater. If this antibiotic is released on a large scale into the environment, it can create antibiotic-resistant genes. Transfer of these genes between non-pathogenic and pathogenic bacteria is risky, and these genes could contaminate aquatic life. Therefore, sulfamethoxazole must be removed from the environment. The removal efficiencies of tetracycline and sulfamethoxazole showed that the configuration of an MFC integrated with a constructed wetland could efficiently reduce the concentration of these antibiotics and the accumulation of antibiotic-resistant genes in the cathode. Another study revealed that a concentration of 200 ppm sulfamethoxazole could be treated in a CBM system. Microbial analysis of biofilms revealed the presence of sulfamethoxazole scavengers (Sonawane *et al.*, 2022).

(c) *Gelatin wastewater*: Gelatin is a heterogeneous mixture of proteins obtained from animal bones and collagens and is used in the pharmaceutical, cosmetic, food, enological, and photographic industries. The wastewater obtained during gelatin production consists of calcium, nitrogen, and phosphorus. Conventional methods cannot treat wastewater containing gelatin. Therefore, gelatin can be effectively degraded using MFC systems. Bioelectricity production using gelatin wastewater as a substrate was evaluated using a single-chambered MFC system. Different microbial concentrations (0%, 1.25%, 2.5%, and 5%) were used, and the results show that a concentration of 1.25% produced the highest performance with a maximum biochemical oxygen demand removal of 20.91 ± 0.95 mg L1; 81.64 ± 0.01 mg L1. The performance of CBM fluctuated due to competitive interactions between the different effective microorganisms (*E. coli*, *Aspergillus niger*, *Saccharomyces cerevisiae*, and *Lactobacillus bulgaricus*) (Sonawane *et al.*, 2022).

(d) *Synthetic penicillin wastewater*: Treatment of wastewater with synthetic penicillin with an air-cathode single-chambered MFC system showed that six times more energy was produced as the sole fuel. These results indicated that toxic, biorefractory organic matter containing waste, such as antibiotic-containing wastewater, could also be a good resource for MFC technology (Pandey *et al.*, 2016).

14.8 POTENTIAL USE OF SHALE GAS DRILLING WASTEWATER AS A SUBSTRATE FOR MFCs

Currently, the problem of global depletion of conventional fossil fuels, such as coal, oil, and natural gas, has motivated the search for alternative energy sources. Shale gas has emerged as an important energy source in recent years, with significant growth in gas production from the exploitation of shale reservoirs. However, this extraction process is not without environmental challenges, and one of the main issues to consider is the proper management of wastewater generated during the drilling of shale gas wells (Sun *et al.*, 2019). This drilling wastewater contains a variety of contaminants, including recalcitrant organic compounds and high concentrations of salts. Rather than considering this wastewater as a problematic waste, there is growing interest in exploring its potential use as a substrate for MFCs (Feng *et al.*, 2020).

MFCs offer a promising technology that combines the degradation of organic compounds by microorganisms with electricity generation. The potential use of shale gas drilling wastewater as a substrate for MFCs is reviewed below. Challenges associated with the composition and characteristics of this wastewater are examined, as well as recent advances in the research and development of MFCs for its treatment. In addition, the advantages and limitations of this application are discussed, along with potential environmental implications and technical feasibility considerations.

14.8.1 Characterization of drilling wastewater

Wastewater generated during shale gas drilling has distinctive physical, chemical, and biological characteristics due to the nature of the extraction process (Ji *et al.*, 2023). Some of these characteristics are described below:

14.8.1.1 Physical characteristics

- *Volume*: Shale gas drilling wastewaters are typically high in volume, as large quantities of water are generated during the hydraulic fracturing process.
- *Temperature*: Wastewater can be at elevated temperatures due to the heat generated during shale gas drilling and fracturing.
- *Turbidity*: Due to the presence of suspended solids, such as fine-clay particles and other sediments, wastewater can be turbid and opaque in appearance.

14.8.1.2 Chemical characteristics

- *Organic contaminants*: Shale gas drilling wastewater may contain recalcitrant organic compounds, such as polynuclear aromatic hydrocarbons, phenolic compounds, and chemicals used in the hydraulic fracturing process, such as additives and surfactants.
- *Salts and minerals*: Wastewater may contain high concentrations of dissolved salts, such as chlorides, sulfates, and bicarbonates, which originate from the geological formation of shale gas.
- *Heavy metals*: Lead, cadmium, and mercury in high concentrations are found in the effluents generated by drilling that occurs when there is contact with rocks.

14.8.1.3 Biological characteristics

Microorganisms: A wide variety of microbial populations are found in wastewater (bacteria, fungi, viruses, archaea, protozoa, etc.) which are found or incorporated in the drilling process. This microbiota has great potential in bioremediation processes, as well as in the approach of a treatment based on the use of MFCs. It should be considered that the native microorganisms present in extreme and undisturbed regions have a high potential in the treatment of this type of contaminated effluents (Gude, 2016).

14.8.2 Challenges encountered

Complexity of the wastewater matrix: Shale wastewater is highly complex and contains a wide range of contaminants, including recalcitrant organic compounds, salts, and heavy metals. This presents challenges in adapting microorganisms and optimizing microbial cell conditions (Acharya *et al.*, 2020). Another limitation is the toxicity because some of the contaminants present in shale wastewater can be toxic to microorganisms and affect their performance in microbial cells. The presence of toxic compounds can decrease microbial activity and treatment efficiency (Sun *et al.*, 2019). Finally, variability and heterogeneity of shale wastewater can vary widely depending on geographic location and specific reservoir conditions. This makes it difficult to standardize treatment processes and generalize the results (Hazra *et al.*, 2022).

14.8.3 Identified opportunities

- *Integrated treatment*: Treatment of shale wastewater with microbial cells can be complementary to other existing treatment technologies, such as advanced oxidation or filtration. The combination of different technologies can enable more efficient removal of contaminants present in wastewater (Zhang *et al.*, 2021).
- *By-product valorization*: In addition to electricity generation, microbial cells offer opportunities to produce valuable chemicals, which can be used as alternative energy sources (Liu *et al.*, 2021).
- *Ongoing research*: Research in this field is constantly evolving, which opens up new opportunities to improve the performance of microbial cells in shale wastewater treatment. Optimization of microorganisms, operating conditions, and understanding of degradation mechanisms can further improve the efficiency and feasibility of this technology (Zhou *et al.*, 2022).

14.9 LIMITATIONS AND FUTURE PERSPECTIVES

The efficiency of MFCs is affected by factors such as chamber design and operating conditions such as the type of biocatalyst to be inoculated, the temperature sensitivity of the microbial metabolism, among others. Although MFCs in wastewater treatment show a potential use, large-scale studies in wastewater treatment have not yet been fully carried out (Mahajan, 2022).

Therefore, it is necessary to investigate the limitations of wastewater treatment and to conduct long-term studies evaluating the operating conditions to significantly reduce the environmental impact and, at the same time, obtain a relevant source of green energy.

REFERENCES

Acharya S. M., Chakraborty R. and Tringe S. G. (2020). Emerging trends in biological treatment of wastewater from unconventional oil and gas extraction. *Frontiers in Microbiology*, **11**, 569019, https://doi.org/10.3389/fmicb.2020.569019

Ahmed S. F., Mofijur M., Nuzhat S., Chowdhury A. T., Rafa N., Uddin M. A., … and Show P. L. (2021). Recent developments in physical, biological, chemical, and hybrid treatment techniques for removing emerging contaminants from wastewater. *Journal of Hazardous Materials*, **416**, 125912, https://doi.org/10.1016/j.jhazmat.2021.125912

Asgharnejad H., Khorshidi Nazloo E., Madani Larijani M., Hajinajaf N. and Rashidi H. (2021). Comprehensive review of water management and wastewater treatment in food processing industries in the framework of water-food-environment nexus. *Comprehensive Reviews in Food Science and Food Safety* **20**, 4779–4815, https://doi.org/10.1111/1541-4337.12782

Cebrucean D., Cebrucean V. and Ionel I. (2014). CO_2 capture and storage from fossil fuel power plants. *Energy Procedia*, 12th International Conference on Greenhouse Gas Control Technologies, GHGT-12 63. 18–26, https://doi.org/10.1016/j.egypro.2014.11.003

Choudhury P., Uday U. S. P., Mahata N., Nath Tiwari O., Narayan Ray R., Kanti Bandyopadhyay T. and Bhunia B. (2017). Performance improvement of microbial fuel cells for waste water treatment along with value addition: A review on past achievements and recent perspectives. *Renewable and Sustainable Energy Reviews* **79**, 372389. https://doi.org/10.1016/j.rser.2017.05.098

Chowdhary P., Bharagava R., Mishra S. and Khan N. (2019). Role of industries in water scarcity and its adverse effects on environment and human health. In: Environmental Concerns and Sustainable Development. pp. 235–256, Springer, Singapore.

Du Z., Li H. and Gu T. (2007). A state-of-the-art review on microbial fuel cells: a promising technology for wastewater treatment and bioenergy. *Biotechnology Advances*, **25**(5), 464–482, https://doi.org/10.1016/j.biotechadv.2007.05.004

EPA. (2020). Urban Waste Water Treatment in 2020. https://www.epa.ie/publications/

Feng Q., Xu L., Liu C., Wang H., Jiang Z., Xie Z., … and Qin Y. (2020). Treatment of shale gas fracturing wastewater using microbial fuel cells: mixture of aging landfill leachate and traditional aerobic sludge as catholyte. *Journal of Cleaner Production*, **269**, 121776, https://doi.org/10.1016/j.jclepro.2020.121776

Foudhaili T., Rakotonimaro T. V., Neculita C. M., Coudert L. and Lefebvre O. (2019). Comparative efficiency of microbial fuel cells and electrocoagulation for the treatment of iron-rich acid mine drainage. *Journal of Environmental Chemical Engineering*, **7**, 103149, https://doi.org/10.1016/j.jece.2019.103149

Garg S., Chowdhury Z. Z., Faisal A. N. M., Rumjit N. P. and Thomas P. (2022). Impact of Industrial Wastewater on Environment and Human Health, In: Advanced Industrial Wastewater Treatment and Reclamation of Water: Comparative Study of Water Pollution Index during Pre-Industrial, Industrial Period and Prospect of Wastewater Treatment for Water Resource Conservation, S. Roy, A. Garg, S. Garg and T. A. Tran (eds), Springer International Publishing, Cham, pp. 197209, https://doi.org/10.1007/978-3-030-83811-9_10

Gonsioroski A., Mourikes V. E. and Flaws J. A. (2020). Endocrine disruptors in water and their effects on the reproductive system. *International Journal of Molecular Sciences*, **21**, 1929, https://doi.org/10.3390/ijms21061929

Guang L., Koomson D. A., Jingyu H., Ewusi-Mensah D. and Miwornunyuie N. (2020). Performance of exoelectrogenic bacteria used in microbial desalination cell technology. *International Journal of Environmental Research and Public Health*, **17**(3), 10–12, https://doi.org/10.3390/ijerph17031121

Gude V. G. (2016). Wastewater treatment in microbial fuel cells – an overview. *Journal of Cleaner Production*, **122**, 287–307, https://doi.org/10.1016/j.jclepro.2016.02.022

Guo F., Luo H., Shi Z., Wu Y. and Liu H. (2021). Substrate salinity: A critical factor regulating the performance of microbial fuel cells, a review. *Science of The Total Environment*, **763**, 143021. https://doi.org/10.1016/j.scitotenv.2020.143021

Hazra B., Vishal V., Sethi C. and Chandra D. (2022). Impact of supercritical CO_2 on shale reservoirs and its implication for CO_2 sequestration. *Energy & Fuels*, **36**(17), 9882–9903, https://doi.org/10.1021/acs.energyfuels.2c01894

Huarachi-Olivera R., Dueñas-Gonza A., Yapo-Pari U., Vega P., Romero-Ugarte M., Tapia J., … and Esparza M. (2018). Bioelectrogenesis with microbial fuel cells (MFCs) using the microalga *Chlorella vulgaris* and bacterial communities. *Electronic Journal of Biotechnology*, **31**, 34–43, https://doi.org/10.1016/j.ejbt.2017.10.013

Jain K., Johnson J., Devpura N., Rathour R., Desai C., Tiwari O. and Madamwar D. (2020). Chapter 5 – Emerging bioremediation technologies for the treatment of wastewater containing synthetic organic compounds, In: Emerging Technologies in Environmental Bioremediation. M.P. Shah, S. Rodriguez-Couto and S. S. Şengör, (eds), Elsevier, pp. 131–150, https://doi.org/10.1016/B978-0-12-819860-5.00005-5

Jatoi A. S., Akhter F., Mazari S. A., Sabzoi N., Aziz S., Soomro S. A., Mubarak N. M., Baloch H., Memon A. Q. and Ahmed S. (2021). Advanced microbial fuel cell for waste water treatment–a review. *Environmental Science and Pollution Research* , **28**, 5005–5019, https://doi.org/10.1007/s11356-020-11691-2

Ji X., Tiraferri A., Zhang X., Liu P., Gan Z., Crittenden J.C., Ma J. and Liu B. (2023). Dissolved organic matter in complex shale gas wastewater analyzed with ESI FT-ICR MS: Typical characteristics and potential of biological treatment. *Journal of Hazardous Materials*, **447**, 130823. https://doi.org/10.1016/j.jhazmat.2023.130823

Jung S.P. and Pandit S. (2019). Chapter 3.1 – Important Factors Influencing Microbial Fuel Cell Performance, In: Microbial Electrochemical Technology, Biomass, Biofuels and Biochemicals, S. V. Mohan, S. Varjani and A. Pandey (eds.), Elsevier, pp. 377–406, https://doi.org/10.1016/B978-0-444-64052-9.00015-7

Kesarwani S., Panwar D., Mal J., Pradhan N. and Rani R. (2023). Constructed Wetland Coupled Microbial Fuel Cell: A Clean Technology for Sustainable Treatment of Wastewater and Bioelectricity Generation. *Fermentation* **9**, 6, https://doi.org/10.3390/fermentation9010006

Khan S., Naushad M., Govarthanan M., Iqbal J. and Alfadul S. M. (2022). Emerging contaminants of high concern for the environment: current trends and future research. *Environmental Research*, **207**, 112609, https://doi.org/10.1016/j.envres.2021.112609

Kong Q., Guo W., Sun R., Qin M., Zhao Z., Du Y., Zhang H., Zhao C., Wang X., Zhang R. and Zhang X. (2021). Enhancement of chromium removal and energy production simultaneously using iron scrap as anodic filling material with pyrite-based constructed wetland-microbial fuel cell. *Journal of Environmental Chemical Engineering* **9**, 106–630, https://doi.org/10.1016/j.jece.2021.106630

Kumar A., Hsu L. H. H., Kavanagh P., Barrière F., Lens P. N., Lapinsonnière L., … and Leech D. (2017a). The ins and outs of microorganism–electrode electron transfer reactions. *Nature Reviews Chemistry*, **1**(3), 0024, https://doi.org/10.1038/s41570-017-0024

Kumar R., Singh L., Zularisam A.W. and Hai F.I. (2018). Microbial fuel cell is emerging as a versatile technology: a review on its possible applications, challenges and strategies to improve the performances. *International Journal of Energy Research* **42**, 369–394, https://doi.org/10.1002/er.3780

Lauretta R., Sansone A., Sansone M., Romanelli F. and Appetecchia M. (2019). Endocrine Disrupting Chemicals: Effects on Endocrine Glands. *Front. Endocrinol.* **10**, https://doi.org/10.3389/fendo.2019.00178

Li H., Cai Y., Gu Z., Yang Y.-L., Zhang S., Yang X.-L. and Song H.-L. (2020). Accumulation of sulfonamide resistance genes and bacterial community function prediction in microbial fuel cell-constructed wetland treating pharmaceutical wastewater. *Chemosphere* **248**, 126014, https://doi.org/10.1016/j.chemosphere.2020.126014

Li H., Zhang S., Yang X.-L., Yang Y.-L., Xu H., Li X.-N. and Song H.-L. 2019. Enhanced degradation of bisphenol A and ibuprofen by an up-flow microbial fuel cell-coupled constructed wetland and analysis of bacterial community structure. *Chemosphere*, **217**, 599–608, https://doi.org/10.1016/j.chemosphere.2018.11.022

Lin X.-Q., Li Z.-L., Nan J., Su J.-H., Liang B., Li C.-J. and Wang A.-J. (2021). Biodegradation and metabolism of tetrabromobisphenol A in microbial fuel cell: behaviors, dynamic pathway and the molecular ecological mechanism. *Journal of Hazardous Materials*, **417**, 126104, https://doi.org/10.1016/j.jhazmat.2021.126104

Liu S.-H. and Lee K.-Y., (2021). Optimization preparation of composite membranes as proton exchange membrane for gaseous acetone fed microbial fuel cells. *Journal of Power Sources* **509**, 230368, https://doi.org/10.1016/j.jpowsour.2021.230368

Liu Z., Lin Z. and Nielsen J. (2021). Expression of fungal biosynthetic gene clusters in S. cerevisiae for natural product discovery. *Synthetic and Systems Biotechnology* **6**, 20–22, https://doi.org/10.1016/j.synbio.2021.01.003

Liu C. J., Murray C. C., Marshall R. E., Strathmann T. J. and Bellona C. (2022). Removal of per- and polyfluoroalkyl substances from contaminated groundwater by granular activated carbon and anion exchange resins: a pilot-scale comparative assessment. *Environmental Science: Water Research & Technology*, **8**(10), 2245–2253, https://doi.org/10.1039/D2EW00080F

Mahajan P. (2022). A critical review on sustainable way of power generation from microbial fuel cell technology. *Materials Today: Proceedings*, **71**, 150–154, https://doi.org/10.1016/J.MATPR.2022.08.243

Mahajan P. and Panwar P. (2022). Perspective and future scope of nanotechnology in modification of microbial fuel cell. Materials Today: Proceedings, *World Engineering Conference on Contemporary Technologies TM*, **71**, 179–185, https://doi.org/10.1016/j.matpr.2022.08.369

Mahmoud R. H., Gomaa O. M. and Hassan R. Y. A. (2022). Bio-electrochemical frameworks governing microbial fuel cell performance: technical bottlenecks and proposed solutions. *RSC Advances*, **12**(10), 5749–5764, https://doi.org/10.1039/d1ra08487a

Malik S., Kishore S., Dhasmana A., Kumari P., Mitra T., Chaudhary V., Kumari R., Bora J., Ranjan A., Minkina T. and Rajput V. D. (2023). A perspective review on microbial fuel cells in treatment and product recovery from wastewater. *Water*, **15**, 316, https://doi.org/10.3390/w15020316

Obileke K., Onyeaka H., Meyer E. L. and Nwokolo N. (2021). Microbial fuel cells, a renewable energy technology for bio-electricity generation: a mini-review. *Electrochemistry Communications*, **125**, 107003, https://doi.org/10.1016/j.elecom.2021.107003

Okereafor U., Makhatha M., Mekuto L., Uche-Okereafor N., Sebola T. and Mavumengwana V. (2020). Toxic metal implications on agricultural soils, plants, animals, aquatic life and human health. *International Journal of Environmental Research and Public Health*, **17**, 2204, https://doi org/10.3390/ijerph17072204

Oveisi F., Fallah N. and Nasernejad B. (2021). Biodegradation of synthetic wastewater containing styrene in microbial fuel cell: effect of adaptation of microbial community. *Fuel*, **305**, 121382, https://doi.org/10.1016/j.fuel.2021.121382

Pandey P., Shinde V.N., Deopurkar R.L., Kale S.P., Patil S.A. and Pant D. (2016). Recent advances in the use of different substrates in microbial fuel cells toward wastewater treatment and simultaneous energy recovery. *Applied Energy* **168**, 706–723, https://doi.org/10.1016/j.apenergy.2016.01.056

Pandit S., Savla N., Sonawane J.M., Sani A.M., Gupta P.K., Mathuriya A.S., Rai A.K., Jadhav D.A., Jung S.P. and Prasad R. (2021). Agricultural waste and wastewater as feedstock for bioelectricity generation using microbial fuel cells. *Recent Advances Fermentation*, **7**, 169, https://doi.org/10.3390/fermentation7030169

Patel J., Kumar R. N. and Kumar J. I. N. (2023). A comprehensive review on waste water treatment technologies with special emphasize on biological treatments: Life sciences-microbiology. *International Journal of Life Science and Pharma Research*, L112–L124, https://doi.org/10.22376/ijlpr.2023.13.2.L112-L124

Pervez M.N., Mishu M.R., Stylios G.K., Hasan S.W., Zhao Y., Cai Y., Zarra T., Belgiorno V. and Naddeo V. (2021). Sustainable Treatment of Food Industry Wastewater Using Membrane Technology: A Short Review. *Water* **13**, 3450, https://doi.org/10.3390/w13233450

Pugazhendi A., Jamal M.T., Al-Mur B.A. and Jeyakumar R.B. (2022). Bioaugmentation of electrogenic halophiles in the treatment of pharmaceutical industrial wastewater and energy production in microbial fuel cell under saline condition. *Chemosphere* **288**, 132515, https://doi.org/10.1016/j.chemosphere.2021.132515

Qiao J. and Xiong Y. (2021). Electrochemical oxidation technology: a review of its application in high-efficiency treatment of wastewater containing persistent organic pollutants. *Journal of Water Process Engineering*, **44**, 102308, https://doi.org/10.1016/j.jwpe.2021.102308

Rusyn I. (2021). Role of microbial community and plant species in performance of plant microbial fuel cells. *Renewable and Sustainable Energy Reviews*, **152**, 111697, https://doi.org/10.1016/j.rser.2021.111697

Saadi M., Pézard J., Haddour N., Erouel M., Vogel T. M. and Khirouni K. (2020). Stainless steel coated with carbon nanofiber/PDMS composite as anodes in microbial fuel cells. *Materials Research Express*, **7**(2), 025504, https://doi.org/10.1088/2053-1591/ab6c99

Saeed T., Majed N., Kumar Yadav A., Hasan A. and Jihad Miah M. (2022). Constructed wetlands for drained wastewater treatment and sludge stabilization: role of plants, microbial fuel cell and earthworm assistance. *Chemical Engineering Journal*, **430**, https://doi.org/10.1016/j.cej.2021.132907

Sarma R., Tamuly A. and Kakati B. K. (2022). Recent developments in electricity generation by microbial fuel cell using different substrates. Materials Today: Proceedings, International Conference on Advancement in Materials, Manufacturing and Energy Engineering (ICAMME-2021). **49**, pp. 457–463, https://doi.org/10.1016/j.matpr.2021.02.522

Sarmin S., Tarek M., Cheng C. K., Roopan S. M. and Khan Md. M. R. (2021). Augmentation of microbial fuel cell and photocatalytic polishing technique for the treatment of hazardous dimethyl phthalate containing wastewater. *Journal of Hazardous Materials*, **415**, 125587, https://doi.org/10.1016/j.jhazmat.2021.125587

Shabani M., Younesi H., Pontié M., Rahimpour A., Rahimnejad M. and Zinatizadeh A.A. (2020). A critical review on recent proton exchange membranes applied in microbial fuel cells for renewable energy recovery. *Journal of Cleaner Production* **264**, 121446, https://doi.org/10.1016/j.jclepro.2020.121446

Shaikh R., Rizvi A., Quraishi M., Pandit S., Mathuriya A.S., Gupta P.K., Singh J. and Prasad R. (2021). Bioelectricity production using plant-microbial fuel cell: Present state of art. *South African Journal of Botany* **140**, 393–408, https://doi.org/10.1016/j.sajb.2020.09.025

Sharma M., Bajracharya S., Gildemyn S., Patil S. A., Alvarez-Gallego Y., Pant D., Rabaey K. and Dominguez-Benetton X. (2014). A critical revisit of the key parameters used to describe microbial electrochemical systems. *Electrochimica Acta*, **140**, 191–208, https://doi.org/10.1016/j.electacta.2014.02.111

Sivaranjanee R. and Kumar P. S. (2021). A review on remedial measures for effective separation of emerging contaminants from wastewater. *Environmental Technology & Innovation*, **23**, 101741, https://doi.org/10.1016/j.eti.2021.101741

Sonawane J. M., Mahadevan R., Pandey A. and Greener J. (2022). Recent progress in microbial fuel cells using substrates from diverse sources. *Heliyon*, **8**, https://doi.org/10.1016/j.heliyon.2022.e12353

Srivastava R. K. (2019). Materials science for energy technologies bio-energy production by contribution of effective and suitable microbial system. *Materials Science for Energy Technologies*, **2**(2), 308–318, https://doi.org/10.1016/j.mset.2018.12.007

Sun Y., Wang D., Tsang D. C., Wang L., Ok Y. S. and Feng Y. (2019). A critical review of risks, characteristics, and treatment strategies for potentially toxic elements in wastewater from shale gas extraction. *Environment International*, **125**, 452–469, https://doi.org/10.1016/j.envint.2019.02.019

Thakur S. H., Patil P. and Sangam S. (2017). The design of microbial fuel cell (MFC). *International Journal of Engineering Research and Technology*, **10**(1), 240–246.

Thapa B., Pandit S., Patwardhan S. B., Tripathi S., Mathuriya A. S., Gupta P. K., Lal R. B. and Tusher T. R. (2022). Application of microbial fuel cell (MFC) for pharmaceutical wastewater treatment: an overview and future perspectives. *Sustainability*, **14**, 8379, https://doi.org/10.3390/su14148379

Thulasinathan B., Jayabalan T., Arumugam N., Rasu Kulanthaisamy M., Kim W., Kumar P., Govarthanan M. and Alagarsamy A. (2022). Wastewater substrates in microbial fuel cell systems for carbon-neutral bioelectricity generation: an overview. *Fuel*, **317**, 123369, https //doi.org/10.1016/j.fuel.2022.123369

Tremouli A., Martinos M. and Lyberatos G. (2017). The Effects of Salinity, pH and Temperature on the Performance of a Microbial Fuel Cell. Waste Biomass Valor, **8**, 2037–2043, https://doi.org/10.1007/s12649-016-9712-0

Tsekouras G.J., Deligianni P.M., Kanellos F.D., Kontargyri V.T., Kontaxis P.A., Manousakis N.M. and Elias C.N. (2022). Microbial fuel cell for wastewater treatment as power plant in smart grids: utopia or reality? *Front. Energy Res.* **10**, https://doi.org/10.3389/fenrg.2022.843768

Tucci M., Cruz Viggi C., Esteve Núñez A., Schievano A., Rabaey K. and Aulenta F. (2021). Empowering electroactive microorganisms for soil remediation: challenges in the bioelectrochemical removal of petroleum hydrocarbons. *Chemical Engineering Journal*, **419**, 130008, https://doi.org/10.1016/j.cej.2021.130008

Umar M. F., Abbas S. Z., Mohamad Ibrahim M. N., Ismail N. and Rafatullah M. (2020). Insights into advancements and electrons transfer mechanisms of electrogens in benthic microbial fuel cells. *Membranes*, **10**(9), 205, https://doi.org/10.3390/membranes10090205

Verma P., Daverey A., Kumar A. and Arunachalam K. (2021). Microbial fuel cell – a sustainable approach for simultaneous wastewater treatment and energy recovery. *Journal of Water Process Engineering*, **40**, 101768, https://doi.org/10.1016/j.jwpe.2020.101768

Werkneh A. A., Gebru S. B., Redae G. H. and Tsige A. G. (2022). Removal of endocrine disrupters from the contaminated environment: public health concerns, treatment strategies and future perspectives – a review. *Heliyon*, **8**, e09206, https://doi.org/10.1016/j.heliyon.2022.e09206

Zhang M., Xie L., Yin Z., Khanal S.K. and Zhou Q. (2016). Biorefinery approach for cassava-based industrial wastes: Current status and opportunities. Bioresource Technology, *Waste Biorefinery - Advocating Circular Economy*, **215**, 50–62, https://doi.org/10.1016/j.biortech.2016.04.026

Zhang Y., Zhao E., Cui X., Zhu W., Han X., Yu G. and Wang Y. (2021). Removal of organic compounds from shale gas fracturing flowback water by an integrated electrocoagulation and electro-peroxone process. *Separation and Purification Technology*, **265**, 118496, https://doi.org/10.1016/j.seppur.2021.118496

Zhou X. L., Ji Z. Y., Wang J., Zhang X., Guo X. F., Liu J., … and Yuan J. S. (2022). Highly efficient recovery of bromine from shale gas wastewater by selective electrochemical oxidation. *Journal of Environmental Chemical Engineering*, **10**(3), 107946, https://doi.org/10.1016/j.jece.2022.107946

doi: 10.2166/9781789063813_0261

Chapter 15

Recent developments in the production of proton transfer membranes employed in microbial fuel cells

Nyiko Maurice Chauke and Nomvano Mketo*

Department of Chemistry, College of Science, Engineering and Technology (CSET), Florida Campus, University of South Africa, 1709, Johannesburg, South Africa
**Corresponding author: nyikomau@gmail.com*

ABSTRACT

This chapter discusses the utilization of proton transfer membranes (PTMs) in microbial fuel cells (MFCs) for wastewater treatment and hydrogen generation. Conventional wastewater treatment methods often consume significant energy and generate harmful by-products, but integrated MFCs offer a promising alternative. In MFCs, microorganisms facilitate oxidation and reduction reactions, hydrogen generation, and converting the chemical energy in wastewater into direct electric current. This allows wastewater to be used as a valuable resource for bioelectricity, biofuels, hydrogen, and other valuable by-products. The selection of appropriate microorganisms, electrode materials, designs, PTMs, and process parameters significantly influences the performance of MFCs. Optimal pH and temperature conditions are crucial for promoting bacterial growth and overall efficiency. However, an important challenge in MFCs is the effective separation process. Various types of PTMs, including ion exchange, composite, porous, and ceramic membranes, have been explored. Despite advancements, issues such as high membrane costs, insufficient ion or proton separation, molecule diffusion, substrate crossover, and biofouling still pose challenges. Nevertheless, MFCs offer a promising solution for wastewater treatment by combining efficient treatment processes with the generation of valuable resources. To successfully implement and commercialize MFCs, it is crucial to address challenges related to membrane separation and optimize process parameters. Continued research and development efforts in this field have the potential to revolutionize wastewater treatment practices, enhance resource recovery, and contribute to a more sustainable future.

Keywords: wastewater treatment, microbial electrochemical technologies, microbial fuel cells, bioenergy, membrane technology, proton transfer membrane

15.1 INTRODUCTION

Microbial fuel cells (MFCs) offer a promising avenue for sustainable energy production and wastewater treatment. Central to their success are proton transfer membranes (PTMs), which serve as both proton conductors and barriers against electrical leakage and reactive species (Ramírez-Vargas *et al.*, 2018). However, persistent biofouling on PTMs remains a substantial challenge, undermining performance and necessitating costly replacements (Ramírez-Vargas *et al.*, 2018; Tiquia-Arashiro & Pant, 2020). This chapter focuses on strategies to modify membranes, mitigate biofouling issues and optimize MFC scalability.

The MFCs, as exemplars of microbial electrochemical technologies (METs), harness microbe–electrode interactions, holding great potential for energy conversion and wastewater remediation (Dizge *et al.*, 2019; Ramírez-Vargas *et al.*, 2018). The MFCs may simultaneously generate power and purify wastewater by utilizing the metabolic processes of microorganisms, turning organic matter into energy while removing pollutants (Ramírez-Vargas *et al.*, 2018; Tiquia-Arashiro & Pant, 2020). Microorganisms in the anode oxidize organic compounds, producing electrons that traverse an electrical circuit to the cathode. This process not only generates energy but also enables the production of valuable products like hydrogen (Dizge *et al.*, 2019; Ramírez-Vargas *et al.*, 2018; Tiquia-Arashiro & Pant, 2020).

Crucially, MFCs are eco-friendly waste-processing methods, converting biodegradable substances into electricity and by-products through microbial oxidation and reduction processes (Conway, 2009). Amidst global freshwater scarcity, MFCs emerge as a solution for water recovery and reusing wastewater, harnessing the energy potential of organic matter (Ahmad *et al.*, 2023; Scanlon *et al.*, 2023). Researchers are now focusing on leveraging membrane technology for sustainable MFC development. However, membrane separation and biofouling are significant hurdles (Dizge *et al.*, 2019; Wang *et al.*, 2023). Various membrane types, including polymer electrolyte, composite, ceramic, and ion exchange, are used to facilitate proton transfer between the anodic and cathodic compartments (Wang *et al.*, 2023).

PTMs play a pivotal role in MFCs, conducting protons while insulating against reactive gases. Biofouling, primarily caused by microbial growth on membranes, obstructs proton transfer and hampers efficiency (Rahimnejad *et al.*, 2020; Thapa *et al.*, 2022). Addressing this, researchers have proposed techniques like incorporating antifouling agents and silver nanoparticles. These approaches impede biofilm formation, enhancing long-term performance and reducing replacement costs. The widely used Nafion 117 membrane, known for its use in proton-exchange membrane fuel cells (PEMFCs), has found application in METs due to functional parallels (Virdis *et al.*, 2022).

Effective biofouling mitigation also involves antimicrobial membranes, achieved through antifouling agent integration and nanoparticle modification. These improved membranes improve the performance and longevity of MFCs by preventing the accumulation of biofilm (Nasruddin & Abu Bakar, 2021). To overcome scalability issues, modified membranes must be developed, exceeding the constraints of conventional MFC systems. The review emphasizes the progress in modified membrane technologies, demonstrating their potential to enable large-scale METs application in energy and wastewater domains.

Despite progress, membrane costs remain a challenge; Nafion membranes, for instance, contribute significantly to overall MFC expenses. Exploring cost-effective alternatives with optimal electrochemical properties is pivotal for scaling up MFCs. The review chapter calls for intensified research efforts to produce affordable, high-performing membranes, acknowledging the direct impact of membrane choice on MFC efficiency and cost. Future investigations will greatly benefit from this review's comprehensive overview of state-of-the-art membrane technologies for MFCs.

15.2 MICROBIAL ELECTROCHEMICAL TECHNOLOGIES

15.2.1 Microbial electrochemical fuel cell design

The METs are a promising approach for resource recovery and in situ energy harvesting for treating wastewater (Wang *et al.*, 2023). This method of producing hydrogen *via* cathodic reduction and anodic bio-catalytic oxidation processes is a potential bioelectrochemical method. However, scaling up such systems from laboratory to real-world applications is their main barrier (Virdis *et al.*, 2022). The majority of research to date (about 90%) has been limited to benchtop models only due to poor

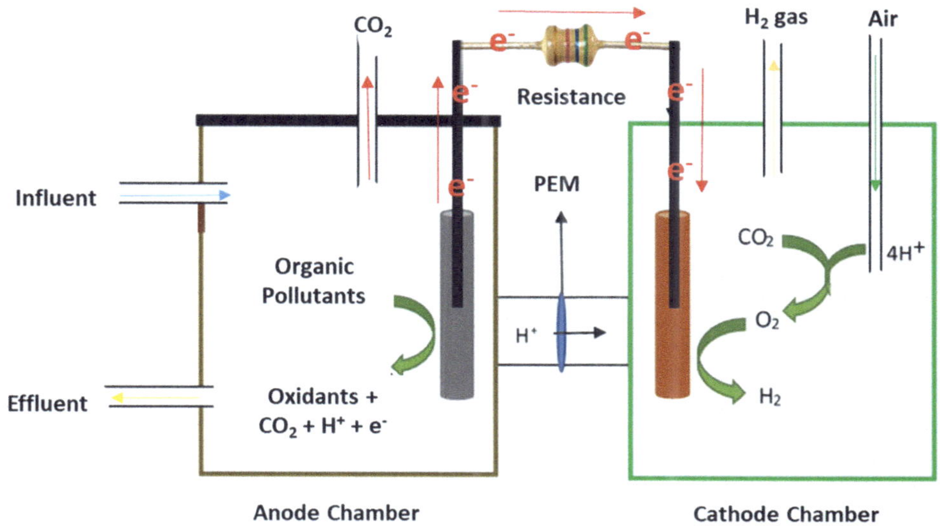

Figure 15.1 Schematic design of a dual-chamber MFCs used in laboratory-scale for METs evaluation.

conversion of chemical energy into electrical energy while employing microbes as catalysts, and budgetary, as well as typical engineering design restrictions. Currently, only a 1.5 m^3 bioelectric toilet and 1000 L microbial electrolysis cell as industrial applications of METs have undergone field tests and their results are promising indications of their readiness for usage in the real world (Yang *et al.*, 2019). Figures 15.1 shows the design of dual-chamber cells typically used on a laboratory scale for METs evaluation to test the individual anode and cathode components.

Here, the surface area-to-volume (*S*/*V*) ratio was drastically reduced when a dual-chamber cell's volume was increased, hence the electrodes must be specially crafted to get around this restriction. Shimoyama *et al.* (2008) built the first cassette-electrode with the cathode chamber sandwiched between two anodes. This group placed several electrodes in a huge chamber, in which they were able to increase their *S*/*V* ratio (Shimoyama *et al.*, 2008). Important factors to take into account while selecting membranes for METs for scaling-up applications are membrane cost, long-term stability, and sustainability. Recently, attempts have been made to improve reactor architectural designs and examine the use of polymer and ceramic membrane materials in order to achieve techno-economical sustainability and efficiency (Jadhav *et al.*, 2022). Recently, modified membranes have been utilized as low-cost materials, electrodes, and separators in some METs applications such as MFCs. When cation-exchange minerals are introduced to membranes, high proton transfer is encouraged along with better membrane characteristics (Jadhav *et al.*, 2022).

15.2.2 Microbial electrochemical fuel cell operation

A physical–chemical oxidation process and a reduction process make up the basic functioning of every electrochemical system (Mohan *et al.*, 2014). A physicochemical process known as an oxidation–reduction reaction involves an electron flow from an anodic to a cathodic electrode and the oxidation of one element or compound followed by the reduction of another element or molecule in the cathodic chamber (Xiao *et al.*, 2020). In contrast to oxidation–reduction processes that naturally and spontaneously occur within the same system, the design of electrochemical systems requires that oxidation and reduction reactions be separated by a membrane or separator (Dessì *et al.*, 2021). The main objective of this membrane separator is to prevent simultaneous oxidation and reduction

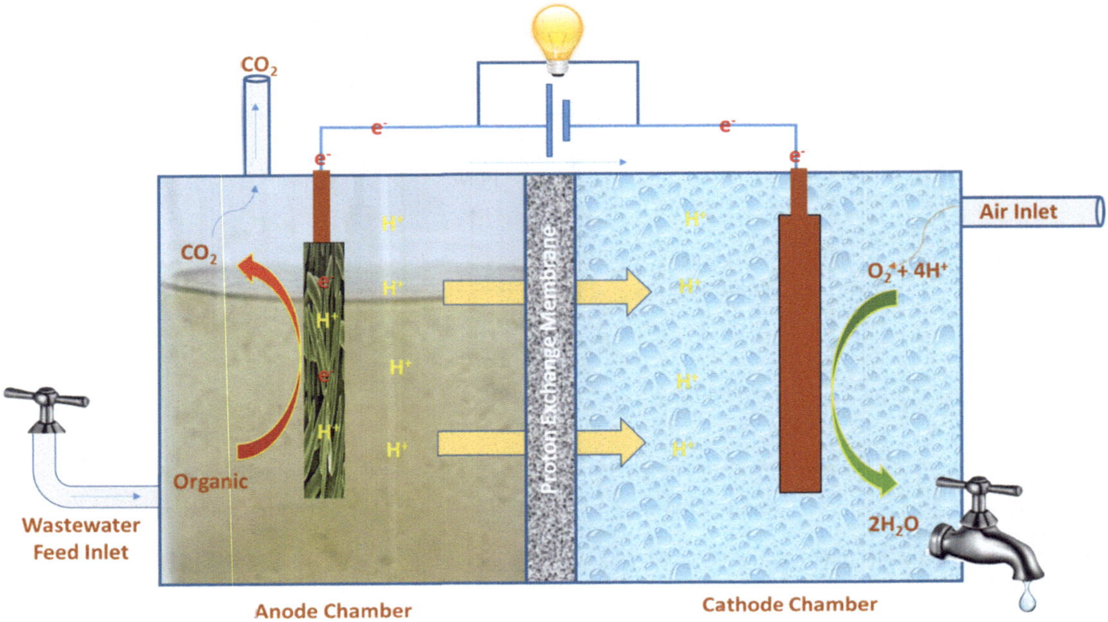

Figure 15.2 Working principle of MET: a schematic diagram of a typical MFCs operation.

processes. The PTMs can therefore be utilized in MFCs between the two chambers, as shown in Figure 15.2. In here, the H^+ (proton) ions produced in the anodic chamber will be able to pass through to the cathodic chamber internally (El Khaloufi & Elasli, 2019).

The generation of electricity, as depicted in Figure 15.2, is a result of the orchestrated flow of electrons across the external circuit. This dynamic process involves two distinct reactions occurring at the anode and cathode interfaces, each exhibiting its unique reaction rate and kinetics. At the anode, organic material oxidation initiates electron release, while at the cathode, reduction reactions facilitate electron acceptance. These parallel yet intricately interconnected reactions underscore the multifaceted nature of microbial electrochemical systems. The anodic and cathodic reactions are integral to the overall electron transfer process, with their individual kinetics playing a pivotal role in determining the overall efficiency and performance of the MFC. A deeper understanding of these reaction dynamics enhances our ability to fine-tune system parameters, optimize electron flux, and ultimately maximize the electricity generation potential of MFCs (Li *et al.*, 2010). Furthermore, as shown by the schematic illustration of Figure 15.2, the wastewater is injected into the anode chamber, where the active microorganisms form a biofilm on the anode surface and oxidize organic material to liberate electrons that are then compelled to flow through an external circuit to the cathode (Wang *et al.*, 2013). When the process is complete, treated water is created when the H^+ flows internally through the proton exchange/transfer membrane and into the cathode chamber. Additionally, the membrane only allows the H^+ ion to pass from the anode to the cathode while resisting all other anions and negatively charged particles (Wee & Lee, 2006). This simultaneously prevents oxygen from crossing across into the anodic chamber from the cathodic chamber. Ionomers, the building blocks of proton exchange membranes, are permeable and solely conduct protons while blocking electrons. Additionally, the membrane stops any gaseous products from flowing, including carbon dioxide (CO_2) in the anodic chamber and oxygen (O_2) in the cathodic chamber (Wee & Lee, 2006).

15.2.3 Microbial electrochemical synthesis

The term microbial electrochemical synthesis (MES) refers to the process of creating valuable compounds from carbon dioxide and other organic substrates through the action of microorganisms that can grow on cathode surfaces (Lekshmi *et al.*, 2023) as illustrated in Figure 15.3. The solid-state electrodes released reducing equivalents serve as a source of energy for the microorganisms, which also serve as catalysts. Methane, ethanol, hydrogen, sodium hydroxide, and hydrogen peroxide can all be produced using microorganisms (Li *et al.*, 2019). This process is used in the production of energy, treated water, and complex organic compounds and can be categorized as the main or secondary METs process depending on the operational circumstances (Morrison *et al.*, 2018). Figure 15.3 shows a schematic diagram of the MES platform for methane, biochemical and biofuels production.

As illustrated in Figure 15.3, the possibilities for using CO_2 to produce a variety of goods with added value can be observed. The process entails combining MES with other bioconversion systems, such as chain elongation mechanisms, as part of the METs process (Jiang *et al.*, 2019). In the context of the circular economy, MES as a production platform becomes desirable when manufacturing can be established at a rapid speed and the end product may be altered to include unique and economically valuable commodities (Bian *et al.*, 2020). While in primary MFCs, the microbial electrochemical processes are carried out by extracellular electron transfer (EET) mechanisms directly from the cell to the acceptor or through the use of electron shuttles (exclusively the Faraday processes) (Kumar *et al.*, 2016; Patil *et al.*, 2012). Similar to how soil remediation or microbial electrolysis of hydrogen is controlled, the secondary MFCs modifies the microbial ambient conditions to regulate the bioelectrochemical processes such as pH, oxygen pressure, metabolite concentrations, and so on (Yasri *et al.*, 2019). Meanwhile, MFCs can be broadly categorized into three groups: the power producers, in which electrons from oxidized organic matter are conducted to the cathode via an external circuit

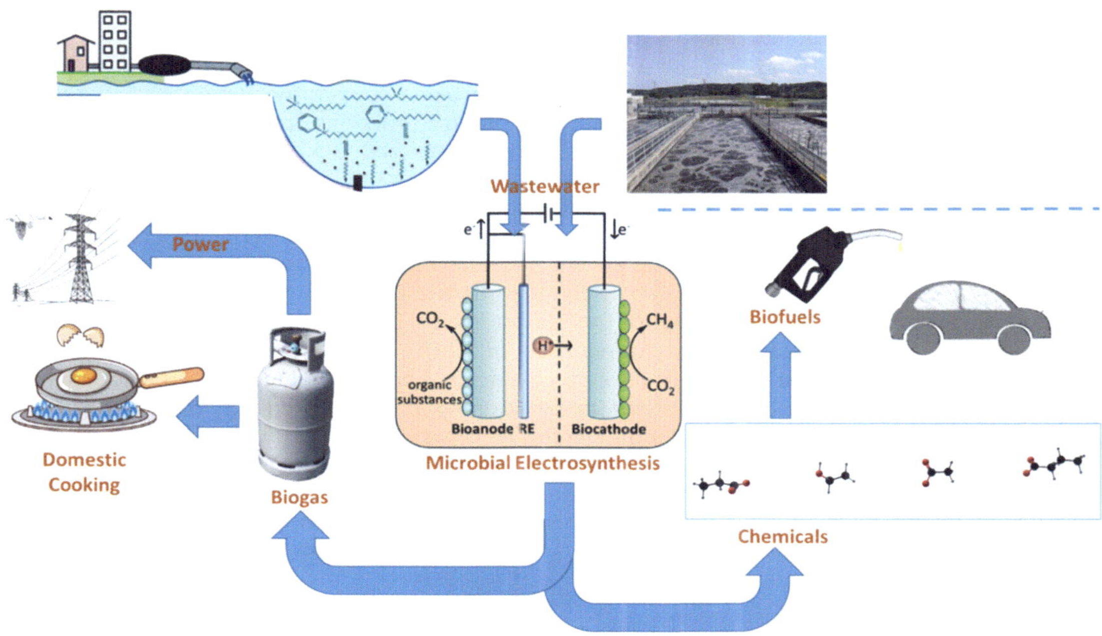

Figure 15.3 Schematic diagram of the MES platform for upgrading methane and biogas as well as production of biochemicals and biofuels. **NB**: RE = reference electrode.

(Lu & Chen, 2020; Rahimnejad *et al.*, 2020), and intermediate systems, which neither produce nor consume power but require stable external power (Ramírez-Vargas *et al.*, 2018). Depending on their intended application, MFCs are referred to using the following processes: MFCs, which produce hydrogen gas or other chemicals by applying external power to lower the cathode potential, microbial desalination cells (MDCs), which desalinate water, and microbial remediation cells (MRCs), which reduce oxidized contaminants through cathodic reduction (Radhika *et al.*, 2022). Given the potential that MFCs offer simultaneous wastewater treatment and energy production, various research studies have looked into the technical viability and benefits of combining this technology with built wetlands since 2012. Constructed wetland is a biologically engineered wastewater treatment system that relies on the presence of plants and microorganisms, as well as how physical, chemical, and biological processes interact with one another and different removal approaches (Lucia *et al.*, 2022).

15.3 MICROBIAL FUEL CELL PTMs

One of the most crucial parts of MFCs is a membrane that allows protons to go from the anode to the cathode in order to sustain an electrical current. This membrane is known as a PTM. The evolution of PTMs in MFCs as emerging wastewater treatment and energy conversion materials for stationary, transportation, and portable electronic applications is desirable as a convenient water and energy source (Pandey *et al.*, 2016). This owes to the remarkably efficient and clean green technology in the production of clean water and hydrogen. The PTMs are smaller and operate at relatively lower temperatures making them easier to contain and reducing thermal losses (Pandey *et al.*, 2016). These typical membranes are perfect for portable and automotive applications. However, the wide development of PTMs is limited by their stability and high cost (Rashid *et al.*). Moreover, the use of precious metals such as Ru, Ir and Pt and exposure to materials in acid media has been costly (Lei *et al.*, 2020). As such, improvement in performance, preparation and design can enhance its cost competitiveness with other available conventional energy conversion and power generating systems on the markets (Kumar *et al.*, 2023).

A unique composite proton transfer or exchange membrane with long-range proton transfer channels was depicted by Huang *et al.* (2018) and they created cooperative action of acid and base functionalized graphene oxide. In their work, two separate functionalized graphene oxides (GO), sulfonic acid functionalized GO (SGO) and amino-functionalized GO (NGO), were produced and applied to the sulfonated poly(arylene ether nitrile) by employing single doping and codoping (SPEN) (Huang *et al.*, 2018). As a result, this group found that the codoping of SGO and NGO enhanced the proton exchange membrane's functioning. In addition, they stated that these two functionalized GO were evenly distributed as fillers in the SPEN matrix and that their interaction created long-range proton transfer channels at the interfaces between the fillers and the SPEN matrix.

Therefore, to create a PTM, it is essential to choose materials with the highest possible transfer efficiency at the lowest possible price (Baroutaji *et al.*, 2015). In this regard, a diverse range of industrial waste materials can be used to develop new desired polymer materials for proton transfers and exchange as well as precisely understand the functioning of these typical fuel cells. The following subsection comprehensively outline different membrane separators (PTMs) using polymeric material modified with different materials such as activated carbon, zeolite incorporation, iron oxide, and titanium oxide. Therefore, the significance of nanocomposite membrane advancement for facilitating proton transfer in fuel cells is addressed. These proton transfer materials (PTMs) encompass various types, including polymer electrolyte membranes, composite membranes, ceramic membranes, and ion-exchange membranes, among others.

15.3.1 Polymer electrolyte membrane in MFCs

The MFCs are exceptional in that they can effectively perform wastewater purification while also producing energy from home and industrial wastewater. As a result, they may be thought of as dual-utility devices. Polymer electrolyte membranes are frequently used as the electrolyte in MFCs. The performance of the cell as a whole depends on the integrity of the polymer electrolyte membrane. These are polymeric materials having functionalized hydrophilic groups that can donate ions when water is added to their hydrophobic backbones (Wang *et al.*, 2013). They have recently received a lot of attention for the separation of macromolecules from water and wastewater treatment processes as PTMs in MFCs due to their remarkable advantages, including their low energy consumption, minimal phase change, lack of additive requirements, and other qualities. However, the main issue preventing their widespread adoption is polymeric membrane biofouling, which is caused by reversible and irreversible solute adsorption on the surface of the membranes or in the membrane pores (Rezakazemi *et al.*, 2018).

The proton-exchange membrane-based water–electrolyte in MFCs has several validated benefits over other water electrolysis techniques since the polymer membrane electrolyzers are smaller in size (Kumar & Himabindu, 2019). To begin with, membrane–electrode assemblies are thin, flexible, and controllable, and anode–cathode interpolar lengths are typically in the 100–200 nm range. In contrast, the oxide ions conducting ceramics utilized in solid oxide technology are not subject to this limitation (Selamet *et al.*, 2011). The electrolyte is commonly a polymer electrolyte membrane constructed of the perfluoro sulfonated acid kind. However, various acid membrane-equipped electrolyzers have also been produced (Le Bideau *et al.*, 2011). The anode produces positively charged hydrogen ions (protons) and oxygen from the water molecules (Le Bideau *et al.*, 2011). As the electrons travel through an external circuit, the hydrogen ions (H^+) move to the cathode through the membrane. There, they combine with the electrons to form hydrogen gas (H_2).

The proton-exchange membrane electrolyzers have a faster dynamic response and a larger load range when compared to alkaline electrolyzers. The alkaline electrolyzer consists of two electrodes working in a liquid alkaline electrolyte solution of sodium hydroxide or potassium hydroxide. However, because they need more expensive catalysts, their technology and capital expenditures remain too high for MFCs applications (Luo *et al.*, 2014). As such, polymer-based membranes in MFCs exhibit less mechanical and chemical stability, as well as proton conductivity, at lower relative humidity levels and temperatures above 80°C. As such, their operating temperature must be increased to improve mass transit, avoid electrode flooding and catalyst poisoning, increase efficiency, reduce system expenses, and encourage system simplicity (Cheng *et al.*, 2007).

15.3.2 Polymer composite membranes in MFCs

Investigations are currently carried out into the viability of employing composite materials made of metal–carbon, metal–polymer, polymer–carbon, polymer–polymer, and carbon–carbon in MFCs. These substances have undergone testing as cathode catalysts, membranes, and anode catalyst supports for MFCs application. Polymer composite materials have better qualities than the individual components alone since they have the traits of each component or even combine them in a synergistic way. The medium-chain poly-3-hydroxyalkanoates (PHA) and multi-walled carbon nanotubes (MCs) functionalized with carboxyl groups were used to make polymer composite membranes (PCMs) for a double-chambered MFC. For the first time, Yusuf *et al.* (2019) employed the PCMs composed of carbon nanotubes and biodegradable medium-chain-length polyhydroxyalkanoates (PHA) as potential PTMs in an MFC (Yusuf *et al.*, 2019). The PCMs as PTMs are favorable due to their lower chemical oxygen demand (COD) removal, enhanced maximum power density of up to 361 mW/m^2, and low internal resistance in comparison to the commonly used commercial Nafion 117 (372 mW/m^2) as the PTMs (Sirajudeen *et al.*, 2021).

The incorporation of inorganic fillers in polymer-based membranes enhances water retention due to the attained hydrophilicity upon modification, which is one of the strategies for maintaining proton transfer performance in an MFC (Nagarale *et al.*, 2006). Recent studies have shown that adding silicon dioxide (SiO_2) to sulfonated graphene oxide (SGO) mixed with poly(vinylidene fluoride) grafted sodium styrene sulfonate (PVDF-g-PSSA) increased the proton conductivity and anti-fouling of PTMs in the MFCs (Daud *et al.*, 2011). Nevertheless, there is more work to be done to gain a fundamental atomistic knowledge of the hydronium diffusion mechanism when using composite membranes in the PTMs environment. Instead of hydrated H_3O^+ ions moving via water channels that resemble liquids as they do in polymeric materials, free protons (H^+) can be conducted through acid sites on the polymer backbone (Zelovich *et al.*, 2021). Therefore, the PCMs has garnered a lot of interest as a potential substitute PTMs for proton transfer in the fuel cell due to the possibility of creating a completely new composition with distinctive features in addition to combining the appealing properties from both the organic and inorganic domains.

15.3.3 Ceramic membranes in MFCs

Ceramic membranes for MFCs promote catholyte formation while providing a less expensive alternative to the more expensive ion or PTMs. Yet, to boost the power output and catholyte quality from MFCs, their physicochemical parameters must be optimized. As such, several low-cost hybrid clay-based ceramic membranes have been developed for use in plant-based microbial fuel cells (PMFCs) (Sarma & Mohanty, 2023). Locally accessible clay can be combined with variable amounts of sodium carbonate, sodium metasilicate, boric acid, bentonite, and fly ash to create these ceramic membranes (Sarma & Mohanty, 2023). Although it is already known that natural clay has ion-exchange properties, it is too fragile to be used as a ceramic membrane in MFCs over the long run. Natural clay membranes have been the subject of numerous research, and it has been determined that they outperform pricey Nafion membranes (Ghadge *et al.*, 2015). The effectiveness of oxygen and sodium hypochlorite (NaOCl) as cathodic electron acceptors on a clayware microbial fuel cell (CMFCs) was assessed by Ghadge *et al.* (2015). In here, NaOCl was employed as the catholyte to achieve the highest power density more than that of oxygen when used as an electron acceptor. They further reported that the hypochlorite catholyte resulted in a power output of 8.7 W/m^3, which was two times more than when using oxygen (4.2 W/m^3) as the electron acceptor (Ghadge *et al.*, 2015).

15.3.4 IEM in MFCs

The performance and commercialization of MFCs still need to be improved, and there are still many obstacles to be overcome. The IEMs in MFCs are one of the primary elements that may have a big impact on the price and effectiveness of MFCs, as is commonly recognized. As such, IEMs are frequently employed as separators in MFCs (Mohan *et al.*, 2018). Because the reaction at the anode typically releases protons, which were once believed to be the species exchanged through the membrane, they are frequently referred to as PTMs (Nagarale *et al.*, 2006). The ion-exchange membrane (IEM) serves as the selective barrier dividing the anode and cathode compartments. The membrane's job is to allow just certain cations, ideally, protons, to pass through as they move from the anode to the cathode. However, this significantly affects how well METs fuel cells function (Nagarale *et al.*, 2006). More specifically, protons (or positive charges) must be able to pass through IEMs due to their high conductivity and low internal resistance, at the same time, the IEMs block substrate or oxygen from moving from the anode to the cathode compartments (Nasef & Hegazy, 2004; Varcoe *et al.*, 2014). This makes them highly selective toward proton transfer. The Nafion membrane, the most used non-porous membrane in METs fuel cell, has demonstrated a high oxygen permeability, which reduces the effectiveness of the resulting cell (Li, 2014). There are no known membranes that can entirely limit oxygen diffusion, hence the only way to counteract the negative effects of oxygen diffusion is to add a chemical oxygen scavenger, like cystein, to the anodic chamber (Li, 2014).

15.3.5 Other types of membranes in MFCs

The main prerequisites for the best electrochemical and microbial performance in the MFCs process are membrane separators for proton transfer. Cation exchange, anion exchange, microfiltration, ultrafiltration, and nanofiltration membranes, as well as porous materials made of textiles, glass fibres, and polymers, are examples of other common membrane separator materials that have been employed as proton transfer in MFCs (Scott, 2016). However, the practical application of membrane material qualities still faces obstacles, particularly in reaching a cheap manufacturing cost, despite advances in our understanding of their properties (Mestre *et al.*, 2019). In the hunt for innovative membranes or separators that may produce/deliver a similar performance to Nafion while being more reasonably priced, several alternatives have been investigated (Ramirez-Nava *et al.*, 2021).

Researchers have evaluated a variety of membranes and separators for use in MFCs, including the aforementioned cation-exchange membranes, anion-exchange membranes, ultrafiltration (UF) and microfiltration (MF) membranes (Ramirez-Nava *et al.*, 2021), bipolar membranes, forward osmosis (FO) membranes, cloth (J-cloth) separators (Dhar & Lee, 2013), glass fiber separators, and cation-exchange layers made of sulfonated poly(ether ether ketone membranes SPEEKs) (Chakrabarty *et al.*, 2010), Selemion HSF polytetrafluoroethylene membranes, mesopore membrane filters, biomax UF discs, glass wool, nylon membranes, polycarbonate membranes, neutral polymers, cation-exchange, anion-exchange, and neutral polymers, as well as porous fabrics and coarse-pore filter material, polytetrafluoroethylene membranes, and cellulose nitrate (Koók *et al.*, 2020). Moreover, the commonly used PTMs, such as Nafion, which is regarded as the gold standard for benchmarking show greater affinity for other cations than for protons (Ke *et al.*, 2021). This results in the cations contending with H^+ for control of the membrane's ion-exchanging functional group. As a result, MFCs may be extremely susceptible to ion transport losses and pH-splitting (a difference in pH between the electrodes and their electrolytes).

To develop the synthesis, production, and engineering of membranes for MFCs applications, researchers should put forth a broad range of workable solutions. This is because MFCs are alternative green methods for processing wastes without adding to environmental degradation while producing usable goods and energy from renewable sources. MFCs use microorganisms with oxidation and reduction processes to transfer chemical energy stored in biodegradable substrates into direct electric currents and by-products. One of the most crucial challenges with MFCs is the separation. However, membranes are often employed, and MFCs have provided a range of membranes that can be used as listed above.

15.4 MEMBRANE PERFORMANCE IN MICROBIAL TECHNOLOGY AND FUEL CELL SYSTEM

Membrane performance plays a crucial role in MET fuel cell systems. METs, also known as MFCs, are bioelectrochemical devices that convert organic matter into electrical energy using microorganisms as catalysts. The membrane in a METs fuel cell system acts as a separator between the anode and cathode compartments while allowing the transport of ions and preventing the mixing of the microbial community with the electrode materials. Following are some key aspects related to membrane performance in METs fuel cell systems:

(1) Selectivity: The membrane should have good ion selectivity, allowing the transport of specific ions (e.g., protons) while blocking the transfer of other species (e.g., bacteria, larger organic molecules). This selectivity is crucial to maintain the integrity of the microbial community in the anode compartment and prevent unwanted reactions at the cathode.
(2) Proton conductivity: The membrane needs to have high proton conductivity to facilitate the efficient transfer of protons from the anode to the cathode. Proton exchange membranes, such as Nafion, are commonly used due to their excellent proton conductivity. However, alternative membranes with improved performance and lower cost are being explored.
(3) Chemical stability: The membrane should be chemically stable under the operating conditions of the MFCs, including the presence of microorganisms, organic matter, and electrochemical reactions. It should resist degradation, fouling, and biofilm formation, which could reduce its performance over time.

(4) Mechanical properties: The membrane should have adequate mechanical strength to withstand the pressure and potential swelling caused by the liquid flow in the system. It should also be flexible enough to conform to the electrode surfaces, ensuring good contact and minimizing electrical losses.
(5) Mass transfer: The membrane should allow efficient mass transfer of substrate molecules, such as organic matter, from the anode to the microbial biofilm. Additionally, it should facilitate the transport of oxygen or other electron acceptors to the cathode, promoting efficient reactions and avoiding limitations.
(6) Membrane fouling: Membrane fouling can occur due to the accumulation of microbial biomass, organic matter, or mineral precipitates on the membrane surface. Fouling reduces the membrane's performance by impeding ion transport and increasing electrical resistance. Strategies to mitigate fouling include surface modification, periodic cleaning, and the use of anti-fouling coatings.
(7) Longevity and cost: The membrane's durability and cost-effectiveness are important considerations for practical implementation. Longevity refers to the membrane's ability to maintain its performance over an extended period without degradation. Lower-cost alternatives to commercially available membranes, such as innovative polymer composites or ionomer blends, are being explored to reduce the overall system cost.

Advancements in membrane technology for METs fuel cell systems are actively pursued to enhance system performance, scalability, and cost-effectiveness. Researchers are working on developing novel materials and engineering approaches to overcome the limitations associated with current membranes, with the aim of improving power output, stability, and longevity of METs fuel cell systems. Microorganisms are used in a technique that converts chemical energy from bioorganic materials into electrical energy (Hassan *et al.*, 2021). Due to their adaptability and the wide variety of materials that can be used as sources, METs devices show promise in several fields, including energy, environment, and sensing. The membrane as a separator in the system is a crucial component of METs. The most significant material that has an impact on the effectiveness of the separator is typically comprised of Nafion proton exchange or transfer membrane (Ramírez-Vargas *et al.*, 2018). Table 15.1 displays different parameters of different proton transfer or IEM material performance in a MFC for METs applications.

Membranes are often utilized, and a range of membranes, including ion-exchange, composite, porous, and ceramic membranes METs as discussed before, have been supplied (Ghangrekar & Nath, 2022). The Nafion PEM, however, has several drawbacks, including its high cost, significant oxygen and substrate crossings, the transport of cations other than protons, and biofouling. Alternative separators have been proposed using a range of materials, including porous materials, salt bridges, glass fiber composite membranes, and IEMs. However, other issues have been noted, which include high membrane costs, ineffective ion separation, undesirable soluble molecule transport, substrate crossings, and biofouling (Lee *et al.*, 2020).

Several membranes were created using various techniques described in the literature, as given in Table 15.1, and their effectiveness as proton exchange or transfer membranes was assessed using various kinds of MFCs. This has indicated that a high-quality fabricated membrane is essential in a membrane-based electrolysis process to guarantee long-lasting operation and adequate output product purity. Perfluorinated sulfonic acid (PFSA) type membranes are now the most widely utilized solid electrolytes for PEMFCs and proton-exchange membrane electrolyzers (PEMEs) (Ahmad Kamaroddin *et al.*, 2021). Theoretically, several restrictions, including one relating to the separator, have prevented the creation of potential METs. Certain qualities such as long-term stability, low cost, less oxygen and fuel crossover, and the ability to transmit more protons than other cations are required of a good separator. Cation-exchange membranes, which are often used in chemical fuel cells, are suitable for METs applications where the aqueous salt solution is used as the electrolyte. Materials with pores

Table 15.1 Parameters of different membrane material performance in MET fuel cell system.

Type of Fuel Cell in MET	Membrane/Separator Composition	Working Temp (°C)	Typical Fuel	Current Density (mA/m^2)	Power Density (mW/m^2)	Comments	References
Double-chamber-MFC	Cs/MMT(2/1)	25	H_2 from municipal wastewater	~9.8	~0.42	Cs/MMT(1/1) showed the highest power density of 85.7 mW/m^2	Terbish *et al.* (2023)
	Cs/MMT(1/1)			~239.5	~85.7		
	Cs/MMT(1/2)			~6.7	~0.12		
	Cs/MMT(1/4)			~11.1	~0.33		
PEMFC	NR211-Hole membrane	50	Dry condition containing H_2 and O_2	~333	~445	The current density was increased by a factor of 1.78 at 0.7 V for knoll morphology as compared to the Hole morphology	Cuynet *et al.* (2017)
		70		~162	~392		
	NR211-Knolls membrane	50		~436	~446		
		70		~477	566		
Double-chamber MFC	Novel clay earthenware (MFC-NCE)	28	H_2 from sludge from anaerobic wastewater	~6000	~2250	MFC–NCEs produced the highest power output compared to the MFC–PEM	Daud *et al.* (2020)
	Nafion 117 (MFC-PEM)			~3000	~1350		
Double-chamber H-cell MFC	AMHPES	37	H_2 from acetate-fed	~428	–	Anionic IEMs outperformed cationic ones at low acetate concentration	Szakács *et al.* (2022)
	AMHPP			~388	–		
	CMHPES			~376	–		
	CMHPP			~365	–		
H-Type MFC	CMI-7000	25	H_2 from synthetic ($C_2H_4NaO_2$, PBS)	~49.30	~12.58	The best results were achieved by the commercial membranes	Di Palma *et al.* (2018)
	PES0			~1.2	~0.08		
	PES5			~6.65	~1.66		
	PES20			~38.38	~9.59		
Dual chamber H-Type MFC	$S_{80}P_{20}$ + 2.5% SiO_2	25	H_2 from raw sewage	–	~1500	The increasing amount of SiO_2 in the polymeric membrane decreased the voltage generation	Nayak *et al.* (2023)
	$S_{80}P_{20}$ + 5% SiO_2			–	~1100		
	$S_{80}P_{20}$ + 7.5% SiO_2			–	~1100		
	$S_{80}P_{20}$ + 10% SiO_2			–	~900		

(Continued)

Table 15.1 Parameters of different membrane material performance in MET fuel cell system (*Continued*).

Type of Fuel Cell in MET	Membrane/Separator Composition	Working Temp (°C)	Typical Fuel	Current Density (mA/m²)	Power Density (mW/m²)	Comments	References
Dual-chamber MFC	BNC	Room temperature	H_2 from glucose, yeast extracts, tryptone and sodium chloride	~	~6.8	The best performance of MFCs was recorded when a BNC/LIG was used	Souza *et al.* (2023)
	BNC/LIG			~121	~18.5		
Single-chamber MFC	CF/$PVA_{0.2}$/bentonite	25	H_2 from yeast extract, glucose, peptone, baker's yeast	–	25.7	The increase in PVA concentration affects the growth in bentonite concentration in the CF, therefore reducing the power density	Christwardana *et al.* (2023)
	CF/PVA_1/bentonite			–	20.35		
	CF/PVA_2/bentonite			–	13.66		
Single-chamber MFC	PANI/CC	Room temperature	H_2 from pre-domesticated anaerobic sludge and artificial wastewater	725	222	The results showed that HPCN-PANI materials are more favorable as compared to HPCN-PP/CC for MFCs, while PANI/CC attained the highest power density	Xu *et al.* (2023)
	HPCN/CC			612	172		
	HPCN-PP/CC			608	169		
Two-chamber MFC	sPSf	Room temperature	H_2 from nutrients and wastewater substrate	~201.33	29.8	In contrast to the 29.8 and 50.8 mW m² obtained for sPSf and sPSf/MILFe membranes the prepared (sPSf/sulfonated MIL100(Fe)) membrane revealed improved PC of power energy of 64.2 mW m² in mixed liquor suspended solid (MLSS) concentration of 4000 mg	Roshanravan *et al.* (2022)
	sPSf/MILFe			~22.61	~50.8		
	sPSf/sMILFe			~253.33	~253.33		
MFC	NiCoAl-LDH	Room temperature	–	–		COF-300@NiAl-LDH/GO had excellent redox activity and catalytic ability, which in turn promoted the output performance of MFC	Chen *et al.* (2022)
	COF-300@NiAl-LDH/GO			–	~481.69		
	Ti_3AlC_2/NiCoAl-LDH			–	~393.82		
	Ti_3AlC_2-MFC			–	~217.73		
	Ti_3AlC_2, NiAl-LDH/GO			–	~181.23		

Note: Cs = chitosan; MMT = montmorillonite; PEMFC = proton-exchange membrane fuel cells; NR211 = Nafion® 211; MFC = microbial fuel cell; NCE = novel clay earthenware; PEM = proton exchange membrane; AMH = anion exchange membranes; IEMs = ion-exchange membranes; CMH = cation-exchange membranes; PES = polyethersulfone; PP = polypropylene; SiO_2 = silica dioxide; sPSf = sulfonated polysulfone; MIL = Mil-100(Fe); PVA = polyvinylalcohol; BNC = bacterial nanocellulose; LIG = lignin; CF = carbon felt; CC = carbon cloth; PANI = polyaniline; HPCN = hierarchical porous carbon nanospheres.

work better as METs separators. Promising materials for the purpose include earthenware and ceramic. They have problems with increased porosity, proton conductivity, and brittleness when used as METs separators.

15.5 FUTURE PROSPECT OF MEMBRANE MATERIALS IN METs AND FUEL SYSTEMS

METs and fuel systems offer effective solutions for treating hazardous wastewater, including industrial effluents containing toxic chemicals and heavy metals. Incorporating membrane technologies into wastewater treatment processes presents an opportunity to collect and recycle valuable components from waste streams, capitalizing on their compactness, versatility, and adaptability for diverse capacity operations. One notable MET, the MFC, employs electrochemically active bacteria to convert organic matter into hydrogen or various by-products with minimal environmental impact, showcasing high efficiency in waste-to-product conversion (Koul *et al.*, 2022). While power generation in METs is limited, their prowess in wastewater treatment remains pioneering.

Optimizing METs, especially MFCs, requires considering factors like temperature, pH, ionic strength, and salinity to achieve the desired electricity generation and wastewater treatment levels. Metabolic engineering strategies can accelerate bacterial metabolism, enhancing cell potential when seeking power generation (Aghababaie *et al.*, 2015). Addressing scalability challenges, such as MFC component design, manufacturing processes, and performance factors, can facilitate large-scale wastewater treatment using these MFCs (Choudhury *et al.*, 2017).

Although METs have made significant strides, challenges related to cost and durability persist. MFCs contend with issues like low power density and mechanical robustness. While the Nafion membrane offers optimal performance, its high cost remains a limitation. Efforts to reduce Nafion and platinum content have often led to compromised performance and increased costs. As such, membrane materials play a pivotal role in MFCs, sustainable and eco-friendly technologies utilizing microorganisms for electrical energy conversion or valuable chemicals. These technologies hold transformative potential across diverse fields, including wastewater treatment, energy generation, and bioremediation. The future of membrane materials in MFCs is promising, driven by ongoing research to enhance performance and expand applications.

(1) Enhanced selectivity: Advancements in membrane materials should aim to improve selectivity, by enabling precise ion or molecule transport while hindering others. Tailoring pore sizes and surface chemistries can enhance the separation of target chemicals, enabling efficient recovery.
(2) Improved durability: Future membrane developments should also focus on enhancing long-term stability by resisting fouling, scaling, and degradation caused by microbial activity or harsh conditions. Novel materials, surface modifications, and advanced cleaning techniques can be explored to prolong membrane lifespan.
(3) Advanced energy generation: Research can aim to optimize power output and energy conversion efficiency in MFCs by developing membranes with enhanced proton conductivity and reduced internal resistance. This can lead to higher power densities and improved energy recovery from organic waste streams.
(4) Integrated system design: The integration of various membrane technologies, like ion exchange, reverse osmosis, and gas separation membranes, also holds potential to maximize resource recovery and process efficiency. Combining MFCs with forward osmosis membranes can enhance water recovery and electricity generation.
(5) Sustainable and low-cost materials: As focus on sustainable, cost-effective, and eco-friendly membrane materials continues, including biomimetic membranes derived from natural sources. These materials offer alternatives to traditional components, reducing environmental impact.
(6) Advanced sensor integration: Integrating membrane materials with advanced sensors enables real-time monitoring and control of MFCs. Sensors for pH, temperature, conductivity, and specific analytes can optimize system performance and facilitate process control.

In general, the future of PTM materials in MFC technologies is promising, with ongoing research poised to design novel PTMs with enhanced selectivity, durability, and advanced functionalities. Developing alternatives to platinum and Nafion membranes is a key focus, given the significance of MFCs in sustainable energy, chemical, and hydrogen production. These advancements will accelerate the adoption of MFCs, addressing environmental challenges and offering sustainable solutions for energy generation and resource recovery.

15.6 CONCLUSION

This chapter underscores the significance of MFCs in enhancing resource recovery from industrial wastewater and mitigating pollutants. MFCs serve as advanced devices for simultaneous power generation and organic material oxidation. Crucial factors influencing MFC performance in METs encompass substrate composition, microorganisms, electron transfer mechanisms, electrode characteristics, membrane types, operating conditions, and geometric design. The chapter also delves into PTMs in MFCs, exploring production methods and manufacturing challenges, suggesting membrane modifications for improvement. Despite PTMs advancements, cost and durability concerns persist, particularly exemplified by Nafion membranes. Challenges include low power density and mechanical resilience. Research should target alternative materials to fully replace Nafion. This study advances MFCs by illuminating their resource recovery potential from industrial wastewater and underscores the need to comprehend performance influencers and PTM limitations. MFCs can become more effective, affordable, and durable by addressing these problems and looking into alternative materials, enabling their wider integration in resource recovery and wastewater treatment applications.

REFERENCES

Aghababaie M., Farhadian M., Jeihanipour A. and Biria D. (2015). Effective factors on the performance of microbial fuel cells in wastewater treatment – a review. *Environmental Technology Reviews*, **4**(1), 71–89, https://doi.org/10.1080/09593330.2015.1077896

Ahmad A., Alshammari M. B. and Ibrahim M. N. M. (2023). Impact of self-fabricated graphene–metal oxide composite anodes on metal degradation and energy generation via a microbial fuel cell. *Processes*, **11**(1), 163, https://doi.org/10.3390/pr11010163

Ahmad Kamaroddin M. F., Sabli N., Abdullah T. A. T., Siajam S. I., Abdullah L. C., Jalil A. A. and Ahmad A. (2021). Membrane-based electrolysis for hydrogen production: a review. *Membranes*, **11**(11), 810, https://doi.org/10.3390/membranes11110810

Baroutaji A., Carton J. G., Sajjia M. and Olabi A. G. (2015). Materials in PEM fuel cells.

Bian B., Bajracharya S., Xu J., Pant D. and Saikaly P. E. (2020). Microbial electrosynthesis from CO_2: challenges, opportunities and perspectives in the context of circular bioeconomy. *Bioresource Technology*, **302**, 122863, https://doi.org/10.1016/j.biortech.2020.122863

Chakrabarty T., Kumar M., Rajesh K. P., Shahi V. K. and Natarajan T. S. (2010). Nano-fibrous sulfonated poly (ether ether ketone) membrane for selective electro-transport of ions. *Separation and Purification Technology*, **75**(2), 174–182, https://doi.org/10.1016/j.seppur.2010.07.019

Chen J., Yang J., Wang X., Yang D., Yang M., Tian J., Tao R., Wang Y., Wei Q., Wang R., Yang Y. and Liu Y. (2022). Cathode catalyst selection for enhancing oxygen reduction reactions of microbial fuel cells: COF-300@NiAl-LDH/GO and Ti3AlC2/NiCoAl-LDH. *International Journal of Hydrogen Energy*, **47**(36), 16179–16188, https://doi.org/10.1016/j.ijhydene.2022.03.107

Cheng X., Shi Z., Glass N., Zhang L., Zhang J., Song D., Liu Z.-S., Wang H. and Shen J. (2007). A review of PEM hydrogen fuel cell contamination: impacts, mechanisms, and mitigation. *Journal of Power Sources*, **165**(2), 739–756, https://doi.org/10.1016/j.jpowsour.2006.12.012

Choudhury P., Prasad Uday U. S., Bandyopadhyay T. K., Ray R. N. and Bhunia B. (2017). Performance improvement of microbial fuel cell (MFC) using suitable electrode and bioengineered organisms: a review. *Bioengineered*, **8**(5), 471–487, https://doi.org/10.1080/21655979.2016.1267883

Christwardana M., Timuda G. E., Darsono N., Widodo H., Kurniawan K. and Khaerudini D. S. (2023). Fabrication of a polyvinyl alcohol-bentonite composite coated on a carbon felt anode for improving yeast microbial fuel cell performance. *Journal of Power Sources*, **555**, 232366, https://doi.org/10.1016/j.jpowsour.2022.232366

Conway G. (2009). The science of climate change in Africa: impacts and adaptation. *Grantham Institute for Climate Change Discussion Paper*, **1**, 24.

Cuynet S., Caillard A., Bigarre J. and Buvat P. (2017). Impact of the patterned membrane morphology on PEMFC performances of ultra-low platinum loaded MEAs. *International Journal of Hydrogen Energy*, **42**(12), 7974–7985, https://doi.org/10.1016/j.ijhydene.2017.02.152

Daud W. R. W., Ghasemi M., Chong P. S., Jahim J. M., Lim S. S. and Ismail M. (2011). *SPEEK/PES Composite Membranes as an Alternative for Proton Exchange Membrane in Microbial Fuel Cell (MFC)*. Springer Nature, IEEE, pp. 400–403.

Daud S. M., Daud W. R. W., Bakar M. H. A., Kim B. H., Somalu M. R., Muchtar A., Md Jahim J. and Muhammed Ali S. A. (2020). Low-cost novel clay earthenware as separator in microbial electrochemical technology for power output improvement. *Bioprocess and Biosystems Engineering*, **43**, 1369–1379, https://doi.org/10.1007/s00449-020-02331-7

Dessì P., Rovira-Alsina L., Sánchez C., Kumaravel Dinesh G., Tong W., Chatterjee P., Tedesco M., Farràs P., Hamelers H. M. V. and Puig S. (2021). Microbial electrosynthesis: towards sustainable biorefineries for production of green chemicals from CO_2 emissions. *Biotechnology Advances*, **46**, 107675, https://doi.org/10.1016/j.biotechadv.2020.107675

Dhar B. R. and Lee H.-S. (2013). Membranes for bioelectrochemical systems: challenges and research advances. *Environmental Technology*, **34**(13–14), 1751–1764, https://doi.org/10.1080/09593330.2013.822007

Di Palma L., Bavasso I., Sarasini F., Tirillò J., Puglia D., Dominici F. and Torre L. (2018). Synthesis, characterization and performance evaluation of Fe_3O_4/PES nano composite membranes for microbial fuel cell. *European Polymer Journal*, **99**, 222–229, https://doi.org/10.1016/j.eurpolymj.2017.12.037

Dizge N., Unal B. O., Arikan E. B., Karagunduz A. and Keskinler B. (2019). Recent progress and developments in membrane materials for microbial electrochemistry technologies: a review. *Bioresource Technology Reports*, **8**, 100308, https://doi.org/10.1016/j.biteb.2019.100308

El Khaloufi Y. and Elasli A. (2019). Microbial fuel cells for electricity generation. *no. April*,

Ghadge A. N., Jadhav D. A., Pradhan H. and Ghangrekar M. M. (2015). Enhancing waste activated sludge digestion and power production using hypochlorite as catholyte in clayware microbial fuel cell. *Bioresource Technology*, **182**, 225–231, https://doi.org/10.1016/j.biortech.2015.02.004

Ghangrekar M. M. and Nath D. (2022). Microbial electrochemical technologies for wastewater treatment: insight into theory and reality. In: *Clean Energy and Resource Recovery*, Elsevier, pp. 179–200, https://doi.org/10.1016/B978-0-323-90178-9.00004-4

Hassan R. Y. A., Febbraio F. and Andreescu S. (2021). Microbial electrochemical systems: principles, construction and biosensing applications. *Sensors*, **21**(4), 1279, https://doi.org/10.3390/s21041279

Huang Y., Cheng T., Zhang X., Zhang W. and Liu X. (2018). Novel composite proton exchange membrane with long-range proton transfer channels constructed by synergistic effect between acid and base functionalized graphene oxide. *Polymer*, **149**, 305–315, https://doi.org/10.1016/j.polymer.2018.07.009

Jadhav D. A., Park S.-G., Eisa T., Mungray A. K., Madenli E. C., Olabi A.-G., Abdelkareem M. A. and Chae K. J. (2022). Current outlook towards feasibility and sustainability of ceramic membranes for practical scalable applications of microbial fuel cells. *Renewable and Sustainable Energy Reviews*, **167**, 112769, https://doi.org/10.1016/j.rser.2022.112769

Jiang Y., May H. D., Lu L., Liang P., Huang X. and Ren Z. J. (2019). Carbon dioxide and organic waste valorization by microbial electrosynthesis and electro-fermentation. *Water Research*, **149**, 42–55, https://doi.org/10.1016/j.watres.2018.10.092

Ke Y., Yuan W., Zhou F., Guo W., Li J., Zhuang Z., Su X., Lu B., Zhao Y., Tang Y., Chen Y., Song J. (2021). A critical review on surface-pattern engineering of Nafion membrane for fuel cell applications. *Renewable and Sustainable Energy Reviews*, **145**, 110860, https://doi.org/10.1016/j.rser.2021.110860

Koók L., Dörgo G., Bakonyi P., Rózsenberszki T., Nemestóthy N., Bélafi-Bakó K. and Abonyi J. (2020). Directions of membrane separator development for microbial fuel cells: a retrospective analysis using frequent itemset mining and descriptive statistical approach. *Journal of Power Sources*, **478**, 229014, https://doi.org/10.1016/j.jpowsour.2020.229014

Koul Y., Devda V., Varjani S., Guo W., Ngo H. H., Taherzadeh M. J., Chang J.-S., Wong J. W. C., Bilal M., Kim S. H., Bui X. T. and Parra-Saldivar R. (2022). Microbial electrolysis: a promising approach for treatment and resource recovery from industrial wastewater. *Bioengineered*, **13**(4), 8115–8134, https://doi.org/10.1080/21655979.2022.2051842

Kumar S. S. and Himabindu V. (2019). Hydrogen production by PEM water electrolysis – a review. *Materials Science for Energy Technologies*, **2**(3), 442–454, https://doi.org/10.1016/j.mset.2019.03.002

Kumar R., Singh L. and Zularisam A. W. (2016). Exoelectrogens: recent advances in molecular drivers involved in extracellular electron transfer and strategies used to improve it for microbial fuel cell applications. *Renewable and Sustainable Energy Reviews*, **56**, 1322–1336, https://doi.org/10.1016/j.rser.2015.12.029

Kumar C. M. S., Singh S., Gupta M. K, Nimdeo Y. M, Raushan R., Deorankar A. V, Ananda Kumar T. M, Rout P. K, Chanotiya C. S, Pakhale V. D and Deepak A. (2023). Solar energy: a promising renewable source for meeting energy demand in Indian agriculture applications. *Sustainable Energy Technologies and Assessments*, **55**, 102905, https://doi.org/10.1016/j.seta.2022.102905

Le Bideau J., Viau L. and Vioux A. (2011). Ionogels, ionic liquid based hybrid materials. *Chemical Society Reviews*, **40**(2), 907–925, https://doi.org/10.1039/C0CS00059K

Lee W. J., Ng Z. C., Hubadillah S. K., Goh P. S., Lau W. J., Othman M. H. D., Ismail A. F. and Hilal N. (2020). Fouling mitigation in forward osmosis and membrane distillation for desalination. *Desalination*, **480**, 114338, https://doi.org/10.1016/j.desal.2020.114338

Lei Z., Wang T., Zhao B., Cai W., Liu Y., Jiao S., Li Q., Cao R. and Liu M. (2020). Recent progress in electrocatalysts for acidic water oxidation. *Advanced Energy Materials*, **10**(23), 2000478, https://doi.org/10.1002/aenm.202000478

Lekshmi G. S., Bazaka K., Ramakrishna S. and Kumaravel V. (2023). Microbial electrosynthesis: carbonaceous electrode materials for CO_2 conversion. *Materials Horizons*, **10**(2), 292–312.

Li J. (2014). Preparation, characterization and application of functional surfaces based on porous polymethacrylates.

Li X., Hu B., Suib S., Lei Y. and Li B. (2010). Manganese dioxide as a new cathode catalyst in microbial fuel cells. *Journal of Power Sources*, **195**(9), 2586–2591, https://doi.org/10.1016/j.jpowsour.2009.10.084

Li Y., Chen Y. and Wu J. (2019). Enhancement of methane production in anaerobic digestion process: a review. *Applied Energy*, **240**, 120–137, https://doi.org/10.1016/j.apenergy.2019.01.243

Lu Y. and Chen J. (2020). Prospects of organic electrode materials for practical lithium batteries. *Nature Reviews Chemistry*, **4**(3), 127–142, https://doi.org/10.1038/s41570-020-0160-9

Lucia C., Laudicina V. A, Badalucco L., Galati A., Palazzolo E., Torregrossa M., Viviani G. and Corsino S. F (2022). Challenges and opportunities for citrus wastewater management and valorisation: a review. *Journal of Environmental Management*, **321**, 115924, https://doi.org/10.1016/j.jenvman.2022.115924

Luo H., Liu G., Zhang R., Bai Y., Fu S. and Hou Y. (2014). Heavy metal recovery combined with H2 production from artificial acid mine drainage using the microbial electrolysis cell. *Journal of Hazardous Materials*, **270**, 153–159, https://doi.org/10.1016/j.jhazmat.2014.01.050

Mestre S., Gozalbo A., Lorente-Ayza M. M. and Sánchez E. (2019). Low-cost ceramic membranes: a research opportunity for industrial application. *Journal of the European Ceramic Society*, **39**(12), 3392–3407, https://doi.org/10.1016/j.jeurceramsoc.2019.03.054

Mohan S. V., Velvizhi G., Modestra J. A. and Srikanth S. (2014). Microbial fuel cell: critical factors regulating bio-catalyzed electrochemical process and recent advancements. *Renewable and Sustainable Energy Reviews*, **40**, 779–797, https://doi.org/10.1016/j.rser.2014.07.109

Mohan S. V., Pandey A. and Varjani S. (2018). *Biomass, Biofuels, Biochemicals: Microbial Electrochemical Technology: Sustainable Platform for Fuels, Chemicals and Remediation*. Elsevier, Amsterdam.

Morrison C., Heitmann E., Armiger W., Dodds D. and Koffas M. (2018). Electrochemical bioreactor technology for biocatalysis and microbial electrosynthesis. In: *Advances in Applied Microbiology*, vol. **105**, Elsevier, pp. 51–86.

Nagarale R. K., Gohil G. S. and Shahi V. K. (2006). Recent developments on ion-exchange membranes and electro-membrane processes. *Advances in Colloid and Interface Science*, **119**(2–3), 97–130, https://doi.org/10.1016/j.cis.2005.09.005

Nasef M. M. and Hegazy E.-S. A. (2004). Preparation and applications of ion exchange membranes by radiation-induced graft copolymerization of polar monomers onto non-polar films. *Progress in Polymer Science*, **29**(6), 499–561, https://doi.org/10.1016/j.progpolymsci.2004.01.003

Nasruddin N. I. S. M. and Abu Bakar M. H. (2021). Mitigating membrane biofouling in biofuel cell system – a review. *Open Chemistry*, **19**(1), 1193–1206, https://doi.org/10.1515/chem-2021-0111

Nayak J. K., Shankar U. and Samal K. (2023). Fabrication and development of SPEEK/PVdF-HFP/SiO_2 proton exchange membrane for microbial fuel cell application. *Chemical Engineering Journal Advances*, **14**, 100459, https://doi.org/10.1016/j.ceja.2023.100459

Pandey P., Shinde V. N., Deopurkar R. L., Kale S. P., Patil S. A. and Pant D. (2016). Recent advances in the use of different substrates in microbial fuel cells toward wastewater treatment and simultaneous energy recovery. *Applied Energy*, **168**, 706–723, https://doi.org/10.1016/j.apenergy.2016.01.056

Patil S. A., Hägerhäll C. and Gorton L. (2012). Electron transfer mechanisms between microorganisms and electrodes in bioelectrochemical systems. *Bioanalytical Reviews*, **4**, 159–192, https://doi.org/10.1007/s12566-012-0033-x

Radhika D., Shivakumar A., Kasai D. R., Koutavarapu R. and Peera S. G. (2022). Microbial electrolysis cell as a diverse technology: overview of prospective applications, advancements, and challenges. *Energies*, **15**(7), 2611, https://doi.org/10.3390/en15072611

Rahimnejad M., Asghary M. and Fallah M. (2020). Microbial fuel cell (MFC): an innovative technology for wastewater treatment and power generation. In: *Bioremediation of Industrial Waste for Environmental Safety: Volume II: Biological Agents and Methods for Industrial Waste Management*, Springer Nature, Singapore, pp. 215–235.

Ramirez-Nava J., Martínez-Castrejón M., García-Mesino R. L., López-Díaz J. A., Talavera-Mendoza O., Sarmiento-Villagrana A., Rojano F. and Hernández-Flores G. (2021). The implications of membranes used as separators in microbial fuel cells. *Membranes*, **11**(10), 738, https://doi.org/10.3390/membranes11100738

Ramírez-Vargas C. A., Prado A., Arias C. A., Carvalho P. N., Esteve-Núñez A. and Brix H. (2018). Microbial electrochemical technologies for wastewater treatment: principles and evolution from microbial fuel cells to bioelectrochemical-based constructed wetlands. *Water*, **10**(9), 1128, https://doi.org/10.3390/w10091128

Rashid T. U., Salem K. S, Islam M. M, Zaman A., Nuruzzaman Khan M., Luna I. Z, Haque P. and Rahman M. M Recent developments of microbial fuel cell as sustainable bio-energy sources. *Microbial Catalysts*, p. 57.

Rezakazemi M., Dashti A., Riasat Harami H. and Hajifari N. (2018). Fouling-resistant membranes for water reuse. *Environmental Chemistry Letters*, **16**, 715–763, https://doi.org/10.1007/s10311-018-0717-8

Roshanravan B., Younesi H., Abdollahi M., Rahimnejad M. and Pyo S.-H. (2022). Incorporating sulfonated MIL-100 (Fe) in sulfonated polysulfone for enhancing microbial fuel cell performance. *Fuel*, **312**, 122962, https://doi.org/10.1016/j.fuel.2021.122962

Sarma P. J. and Mohanty K. (2023). Development and comprehensive characterization of low-cost hybrid clay based ceramic membrane for power enhancement in plant based microbial fuel cells (PMFCs). *Materials Chemistry and Physics*, **296**, 127337, https://doi org/10.1016/j.matchemphys.2023.127337

Scanlon B. R., Fakhreddine S., Rateb A., de Graaf I., Famiglietti J., Gleeson T., Grafton R. Q., Jobbagy E., Kebede S., Kolusu S. R., Konikow L. F., Long D., Mekonnen M., Schmied H. M., Mukherjee A., MacDonald A., Reedy R. C., Shamsudduha M., Simmons C. T., Sun A., Taylor R. G., Villholth K. G., Vörösmarty C. J. and Zheng C. (2023). Global water resources and the role of groundwater in a resilient water future. *Nature Reviews Earth & Environment*, **4**, 1–15.

Scott K. (2016). Membranes and separators for microbial fuel cells. In: *Microbial Electrochemical and Fuel Cells*, Elsevier, pp. 153–178.

Selamet ÖF., Becerikli F., Mat M. D. and Kaplan Y. (2011). Development and testing of a highly efficient proton exchange membrane (PEM) electrolyzer stack. *International Journal of Hydrogen Energy*, **36**(17), 11480–11487, https://doi.org/10.1016/j.ijhydene.2011.01.129

Shimoyama T., Komukai S., Yamazawa A., Ueno Y., Logan B. E. and Watanabe K. (2008). Electricity generation from model organic wastewater in a cassette-electrode microbial fuel cell. *Applied Microbiology and Biotechnology*, **80**, 325–330, https://doi.org/10.1007/s00253-008-1516-0

Sirajudeen A. A. O., Annuar M. S. M., Ishak K. A., Yusuf H. and Subramaniam R. (2021). Innovative application of biopolymer composite as proton exchange membrane in microbial fuel cell utilizing real wastewater for electricity generation. *Journal of Cleaner Production*, **278**, 123449, https://doi.org/10.1016/j.jclepro.2020.123449

Souza L., Antônio R., Hotza D., Carminatti C., Pineda-Vásquez T., Watzko E., Pezzin A. P., Duarte D. and Recouvreux D. (2023). Lignin-incorporated bacterial nanocellulose for proton exchange membranes in microbial fuel cells. *Materials Chemistry and Physics*, **293**, 126963, https://doi.org/10.1016/j.matchemphys.2022.126963

Szakács S., Koók L., Nemestóthy N., Bélafi-Bakó K. and Bakonyi P. (2022). Studying microbial fuel cells equipped with heterogeneous ion exchange membranes: electrochemical performance and microbial community assessment of anodic and membrane-surface biofilms. *Bioresource Technology*, **360**, 127628, https://doi.org/10.1016/j.biortech.2022.127628

Terbish N., Popuri S. R. and Lee C.-H. (2023). Improved performance of organic–inorganic nanocomposite membrane for bioelectricity generation and wastewater treatment in microbial fuel cells. *Fuel*, **332**, 126167, https://doi.org/10.1016/j.fuel.2022.126167

Thapa B. S., Pandit S., Patwardhan S. B., Tripathi S., Mathuriya A. S., Gupta P. K., Lal R. B. and Tusher T. R. (2022). Application of microbial fuel cell (MFC) for pharmaceutical wastewater treatment: an overview and future perspectives. *Sustainability*, **14**(14), 8379, https://doi.org/10.3390/su14148379

Tiquia-Arashiro S. M. and Pant D. (2020). *Microbial Electrochemical Technologies*. CRC Press, Routledge.

Varcoe J. R., Atanassov P., Dekel D. R., Herring A. M., Hickner M. A., Kohl P. A., Kucernak A. R., Mustain W. E., Nijmeijer K., Scott K., Xu T. and Zhuang L. (2014). Anion-exchange membranes in electrochemical energy systems. *Energy & Environmental Science*, **7**(10), 3135–3191, https://doi.org/10.1039/C4EE01303D

Virdis B., Hoelzle R. D., Marchetti A., Boto S. T., Rosenbaum M. A., Blasco-Gómez R., Puig S., Freguia S. and Villano M. (2022). Electro-fermentation: sustainable bioproductions steered by electricity. *Biotechnology Advances*, **59**, 107950, https://doi.org/10.1016/j.biotechadv.2022.107950

Wang Y.-J., Qiao J., Baker R. and Zhang J. (2013). Alkaline polymer electrolyte membranes for fuel cell applications. *Chemical Society Reviews*, **42**(13), 5768–5787, https://doi.org/10.1039/c3cs60053j

Wang Z., Li D., Shi Y., Sun Y., Okeke S. I., Yang L., Zhang W., Zhang Z., Shi Y. and Xiao L. (2023). Recent implementations of hydrogel-based microbial electrochemical technologies (METs) in sensing applications. *Sensors*, **23**(2), 641, https://doi.org/10.3390/s23020641

Wee J.-H. and Lee K.-Y. (2006). Overview of the development of CO-tolerant anode electrocatalysts for proton-exchange membrane fuel cells. *Journal of Power Sources*, **157**(1), 128–135, https://doi.org/10.1016/j.jpowsour.2005.08.010

Xiao F., Bedane A. H., Mallula S., Sasi P. C., Alinezhad A., Soli D., Hagen Z. M. and Mann M. D. (2020). Production of granular activated carbon by thermal air oxidation of biomass charcoal/biochar for water treatment in rural communities: a mechanistic investigation. *Chemical Engineering Journal Advances*, **4**, 100035, https://doi.org/10.1016/j.ceja.2020.100035

Xu Y., Zhu K., Yang X., Zhu Y., Jiang K., Liu L., Song H., Chen Y., Wang S. and Wang Z. (2023). Efficient application of new porous carbon nanoparticle composite polyaniline material in microbial fuel cells. *Industrial Crops and Products*, **192**, 116130, https://doi.org/10.1016/j.indcrop.2022.116130

Yang E., Chae K.-J., Choi M.-J., He Z. and Kim I. S. (2019). Critical review of bioelectrochemical systems integrated with membrane-based technologies for desalination, energy self-sufficiency, and high-efficiency water and wastewater treatment. *Desalination*, **452**, 40–67, https://doi.org/10.1016/j.desal.2018.11.007

Yasri N., Roberts E. P. L. and Gunasekaran S. (2019). The electrochemical perspective of bioelectrocatalytic activities in microbial electrolysis and microbial fuel cells. *Energy Reports*, **5**, 1116–1136, https://doi.org/10.1016/j.egyr.2019.08.007

Yusuf H., Annuar M. S. M., Syed Mohamed S. M. D. and Subramaniam R. (2019). Medium-chain-length poly-3-hydroxyalkanoates-carbon nanotubes composite as proton exchange membrane in microbial fuel cell. *Chemical Engineering Communications*, **206**(6), 731–745, https://doi.org/10.1080/00986445.2018.1521392

Zelovich T., Winey K. I. and Tuckerman M. E. (2021). Hydronium ion diffusion in model proton exchange membranes at low hydration: insights from ab initio molecular dynamics. *Journal of Materials Chemistry A*, **9**(4), 2448–2458, https://doi.org/10.1039/D0TA10565A

IWA Publishing's authorised EU representative for General Product Safety Regulations is Diane D'Arras, 15 rue Duret, 75116 Paris, France, e-mail: safety@iwap.co.uk.

Printed and bound by CPI Group (UK) Ltd, Croydon, CR0 4YY
12/05/2026
02108869-0012